60th International Scientific Conference on Experimental Stress Analysis (EAN2022)

Olsanka, Czech Republic
6-9 June 2022

ISBN: 979-8-3313-1510-8

Printed from e-media with permission by:

Curran Associates, Inc.
57 Morehouse Lane
Red Hook, NY 12571

Some format issues inherent in the e-media version may also appear in this print version.

Printed with permission by Curran Associates, Inc. (2025)

For permission requests, please contact Czech Society for Mechanics (CSM)
at the address below.

Czech Society for Mechanics (CSM)
Dolejskova 5
Praha 8
182 00
Czech Republic

Phone: +420 266 053 045
Fax: +420 286 587 784

csm@csm.cz

Additional copies of this publication are available from:

Curran Associates, Inc.
57 Morehouse Lane
Red Hook, NY 12571 USA
Phone: 845-758-0400
Fax: 845-758-2633
Email: curran@proceedings.com
Web: www.proceedings.com

EAN 2022

60th annual conference on experimental stress analysis

6 – 9 June, 2022, Prague, Czech Republic

Proceedings of the 60th Annual Conference on Experimental Stress Analysis

ČSM

FACULTY OF MECHANICAL ENGINEERING CTU IN PRAGUE

ENGINEERING ACADEMY OF THE CZECH REPUBLIC

60th Annual International Scientific Conference

Experimental Stress Analysis

Hotel Olšanka, Praha, Czech Republic
6.–9. June 2022

EAN 2022
60th annual conference on experimental stress analysis

Proceedings of the 60th Annual Conference on Experimental Stress Analysis

Edited by
Karel Doubrava
Kristýna Kubášová
Zdeněk Padovec
David Blaha
Radek Sedláček
Milan Růžička

Czech Society for Mechanics, Expert Group of Experimental Mechanics
Czech Technical University in Prague, Faculty of Mechanical Engineering

Experimental Stress Analysis 2022 Conference Partners

HBP měřicí technika

DEWESoft® measurement innovation

LABTECH

MICRO-EPSILON

HELAGO® CZ

inova

XSIGHT VIDEO EXTENSOMETERS & DIC SYSTEMS

Zwick / Roell

Program Committee

Chairman

prof. Ing. Milan Růžička, CSc. (CTU in Prague)

Members

- Ing. Karel Doubrava, Ph.D. (CTU in Prague)
- doc. Ing. Peter Frankovský, PhD. (TUKE)
- prof. Ing. Nikolaj Ganev, CSc. (CTU in Prague)
- prof. Ing. Radim Halama, Ph.D. (VSB TUO)
- prof. Dr. Ing. František Holešovský (CTU in Prague)
- prof. Ing. Stanislav Holý, CSc. (CTU in Prague)
- Ing. Lubomír Houfek, Ph.D. (BUT)
- prof. RNDr. Miroslav Hrabovský, DrSc. (IP CAS)
- doc. Ing. Róbert Huňady, PhD. (TUKE)
- prof. Ing. Josef Jíra, CSc. (CTU in Prague)
- Ing. Kamil Kolařík, Ph.D. (CTU in Prague)
- Ing. Aleš Lufinka, Ph.D. (TUL)
- doc. Ing. TomáŇávrat, Ph.D. (BUT)
- doc. Ing. Pavel Padevět, Ph.D. (CTU in Prague)
- doc. Iva Petríková, Ph.D. (TUL)
- Dr. Ing. Roman Růžek (CARC)
- Ing. Jaroslav Václavík (VZÚ)
- doc. Ing. Miloš Vlk, CSc. (BUT)
- doc. Ing. Robert Zemčík, Ph.D. (UWB)

Organizing Committee

Chairman

Ing. Karel Doubrava, Ph.D. (CTU in Prague)

Members

- Ing. Zdeněk Padovec, Ph.D. (CTU in Prague)
- Ing. Radek Sedláček, Ph.D. (CTU in Prague)
- Ing. Kristýna Kubášová (CTU in Prague)

Contents

Statistical Method to Deal with the Variability of Fatigue Damage in Nonstationary In-service Random Loadings

Denis Benasciutti[1] & Julian M. Enzveiler Marques[2]

Abstract: The fatigue damage $D(T)$of a service random record $x(t)$of finite length $T$is a random variable; its expected value $E\left[D(T)\right]$ is in fact unknown since it represents an average over an infinite population of damage values, which obviously is not obtainable in practice. This paper shows how a confidence interval for $E\left[D(T)\right]$ can be constructed from a sample of damage values computed from a stationary or nonstationary in-service random time-history. The approach is validated by numerical simulations and then by records measured on a mountain bike.

Keywords: Fatigue damage; Confidence interval; Nonstationary time-histories

1. Introduction

The safety assessment of structures subjected to in-service random loadings is usually performed on the basis of the fatigue damage calculated on one or more time-histories. For a time-history $x(t)$of length T, the Palmgren-Miner damage is [1]:

$$D\left(T\right)=\sum_{q=1}^{n}\frac{s_q^k}{A} \tag{1}$$

where $n$is the number of rainflow cycles in time interval T, s_q is the amplitude of the q-th stress cycle in $x(t)$, and $s^k N_f = A$ is the S/N curve of the structural detail.

Damage $D\left(T\right)$ is a random variable. It has two sources of variability: material and loading. The first relates to the scatter in fatigue life (e.g. A is log-normally distributed), which can be accounted for by defining a characteristic S/N curve referred to a prescribed failure probability (e.g. 97.7%). The other source of variability comes from the stochastic nature of the loading, and constitutes the focus of this paper.

The random variable $D\left(T\right)$ follows a certain probability density function $f_D(-)$ with expected value $E\left[D(T)\right]$ and variance σ_D^2. The expected value of the damage is obtained by taking the expectation of (1). The expected damage represents an average computed over an infinite ensemble of damage values obtained from infinite time-histories – clearly an abstraction not achievable in practice. Therefore, $E\left[D(T)\right]$ is unknown and can be estimated, for example, by a sample mean damage, say $\bar{D}(T)$, which represents an average computed over a finite-size sample of damage values. Furthermore, as it is often done in statistics, the sample mean can be used to construct a confidence interval $\left[\bar{D}\left(T\right)\pm L\right]$ of width $2L$, which is expected to enclose $E\left[D(T)\right]$ with a given confidence, for example 95%. The following sections will show how to construct such a confidence interval for the expected damage in case only one single time-history is available. Two distinct situations are considered depending on whether the time-history is stationary or nonstationary.

[1] *Prof. Denis Benasciutti, Ph.D.; Dept. of Engineering, University of Ferrara; via Saragat 1, 44122 Ferrara, Italy; denis.benasciutti@unife.it*

[2] *Dr. Julian M. Enzveiler Marques, Ph.D.; Dept. of Mechanics, Biomechanics and Mechatronics, Fac. Of Mech. Engineering, Czech Technical University in Prague; Technická 4, 166 36 Prague, Czech Republic; julianmarcell.enzveilermarques@fs.cvut.cz*

2. Method

2.1. Stationary time-history

The idea is to divide a stationary time-history $x(t)$ of time length T (see Fig. 1) into $N_B \geq 2$ disjoint blocks of length $T_B = \frac{T}{N_B}$. After calculating the damage of each block, a sample of block damages $D_{B,i}(T_B)$, $i = 1, 2, \ldots, N_B$ is obtained and used to construct a $100\,(1-\beta)\,\%$ confidence interval for $E\,[D(T)]$:

$$D\,(T) - t_{\nu,\frac{\beta}{2}} \cdot \hat{\sigma}_D \leq E\,[D\,(T)] \leq D\,(T) + t_{\nu,\frac{\beta}{2}} \cdot \hat{\sigma}_D \tag{2}$$

where $\hat{\sigma}_D^2 \cong N_B \cdot \hat{\sigma}_{D_B}^2$ is an estimate of the variance of damage and $t_{\nu,\frac{\beta}{2}}$ is the quantile of the Student's t-distribution with $\nu = N_B - 1$ degrees of freedom. The previous formula assumes a normally distributed damage [1–3].

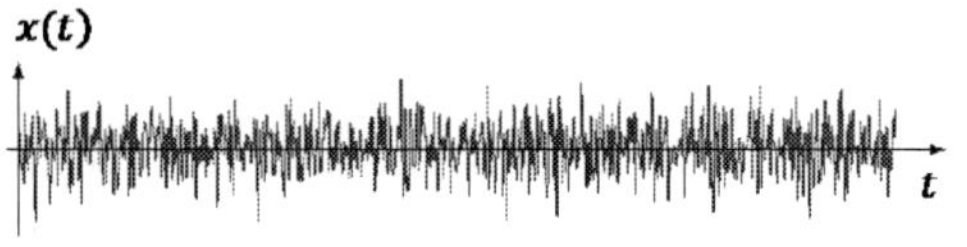

Fig. 1. Example of stationary loading.

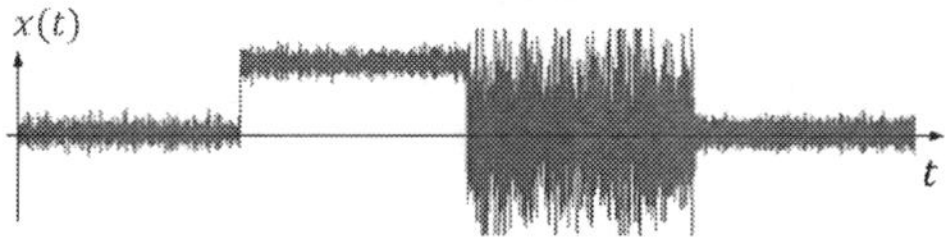

Fig. 2. Example of switching loading.

In Eq.(2), $D\,(T)$ and $\hat{\sigma}_D$ are known: $D\,(T)$ is the fatigue damage of $x(t)$, $\hat{\sigma}_D$ is its standard deviation approximated by the standard deviation of block damage as explained in [3]. The expected damage $E\,[D\,(T)]$ is the unknown quantity to be enclosed by the confidence interval.

2.2. Nonstationary time-history

The result for the stationary case in Eq.(1) can be extended to a nonstationary time-history $x(t)$ formed by a finite number N_S of stationary segments (see Fig. 2), for which the confidence interval is [4]:

$$D\,(T) - t_{\eta,\frac{\beta}{2}} \sum_{j=1}^{N_S} \sqrt{N_B \cdot \hat{\sigma}_{D_{B,j}}^2} \leq E\,[D\,(T)] \leq D\,(T) + t_{\eta,\frac{\beta}{2}} \sum_{j=1}^{N_S} \sqrt{N_B \cdot \hat{\sigma}_{D_{B,j}}^2} \tag{3}$$

where $\hat{\sigma}_{D_{B,j}}^2$ is the estimated variance of the damage for j-th state, and $t_{\eta,\frac{\beta}{2}}$ is a Student quantile referred to a suitable equivalent number of degrees of freedom $\eta \cong (N_B - 1)\,\frac{\left(\sum_{j=1}^{N_S} \hat{\sigma}_{D_{B,j}}^2\right)^2}{\sum_{j=1}^{N_S} \left(\hat{\sigma}_{D_{B,j}}^2\right)^2}$.

In Eq.(3), $D\,(T)$ and $\hat{\sigma}_{D_{B,j}}^2$ are known: $D\,(T)$ is the fatigue damage of the nonstationary time-history $x(t)$, $\hat{\sigma}_{D_{B,j}}^2$ is the sample standard deviation of the block damage for the j-th stationary state. The expected damage $E\,[D\,(T)]$ is the unknown quantity that has to be enclosed by the confidence interval.

3. Results

3.1. Numerical simulations

An enormous set of time-histories is simulated so that their damage values are averaged in order to get an approximate estimate of $E\,[D\,(T)]$ – this estimate is only used for verification purposes. A single time-history is then simulated and used to construct the confidence interval according to the procedure explained so far. Fig. 3 shows an example of 95% confidence intervals for different block subdivisions N_B. All intervals correctly enclose the expected damage value.

The figure suggests that the number of blocks N_B should be the highest possible to make the confidence interval narrower. A similar result, though not shown here, is also obtained for the nonstationary case [4].

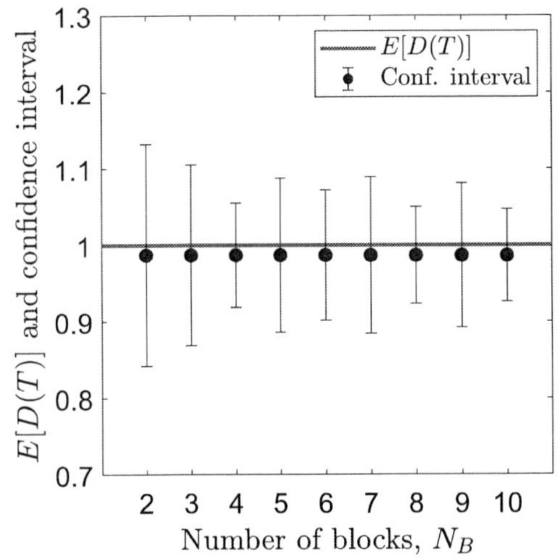

Fig. 3. 95% conf. intervals for $E[D(T)]$. Stationary case, simulations.

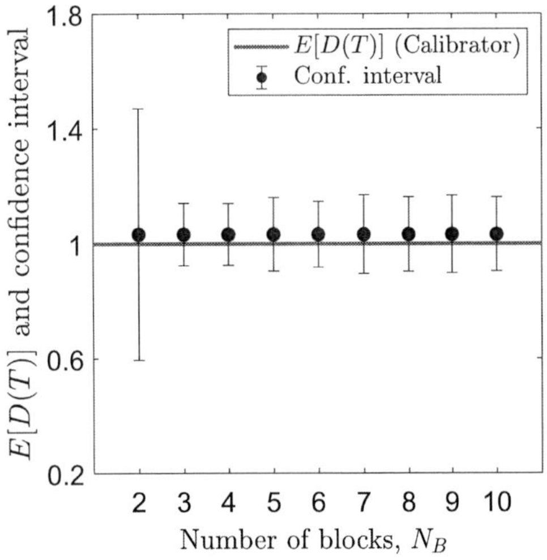

Fig. 4. 95% conf. intervals for $E[D(T)]$. Nonstationary case, experiments.

3.2. Experiments

An instrumented mountain-bike (see Fig. 5) was used to collect measured time-histories. Road surfaces and bicycle speed were suitably selected so as to make the acquired time-histories either stationary or nonstationary (switching) [4]. Besides a single time-history to be analysed, additional 30 time-histories ("calibrator set") were also measured and their damage values used to approximate $E[D(T)]$.

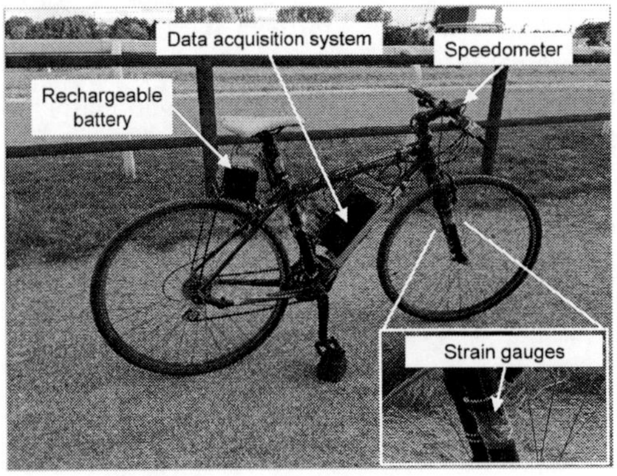

Fig. 5. Instrumented mountain-bike.

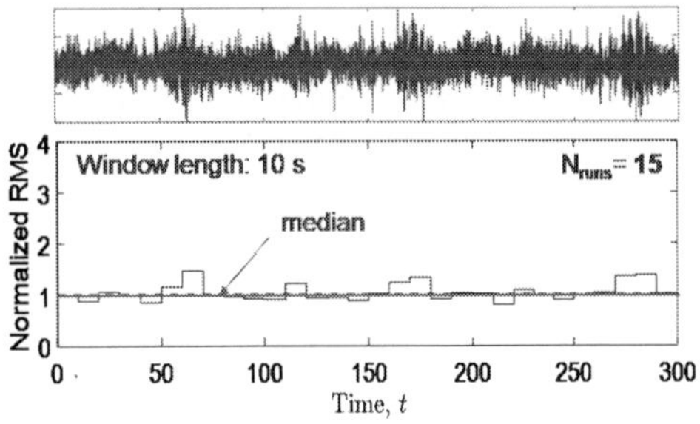

Fig. 6. Outcome of run test of "stationary".

In a preliminary stage, the property of stationarity/nonstationarity of all measured time-histories was verified by two methods: short time Fourier transform, run test. A "run" is defined as a sequence of identical observations followed and preceded by a different observation or no observation at all. The run test can detect whether a random signal has an underlying trend, for example, in its root-mean-square (rms) value. Fig. 6 displays an example of run test, in which the number of runs (15) falls withing the acceptance region (10.7; 21.3) for the run test; the time-history is then classified as stationary.

After the preliminary check, the confidence interval for the damage is computed by the procedure described in section 2. An example for the nonstationary case is shown in Fig. 4. As expected, the confidence intervals enclose $E\left[D\left(T\right)\right]$ for a wide range of block subdivisions N_B.

An application of this confidence interval approach to the strain time-histories measured in a wheel of a telescopic industrial vehicle is described in [5].

4. Conclusions

The paper developed two analytical expressions for the confidence interval of the expected fatigue damage in the case of a single stationary and nonstationary random time-history. The idea was to subdivide a nonstationary time-history into states, and then each state into blocks, and to construct the confidence interval from block damages. Numerical simulations and measurements confirmed the accuracy of the proposed approach.

Acknowledgements

The research activity of one author (J.M.E.M.) was partially funded by CTU Global Postdoc Fellowship Program (research topic #2-11).

References

[1] Low Y.M., "Variance of the fatigue damage due to a Gaussian narrowband process," *Struct. Saf.*, Vol. 34, no. 1, pp. 381–389. ISSN 0167-4730. https://doi.org/10.1016/j.strusafe.2011.09.001.

[2] Crandall S.H., Mark W.D. and Khabbaz G.R., "The variance in Palmgren-Miner damage due to random vibration," in *Proceedings of the 4th US national congress of applied mechanics*, Berkley, USA, 1962 (ASME, New York, 1962), pp. 119–126.

[3] Marques J.M.E., Benasciutti D. and Tovo R., "Variability of the fatigue damage due to the randomness of a stationary vibration load," *Int. J. Fatigue*, Vol. 141, p. 105891 (2020). ISSN 0142-1123. https://doi.org/10.1016/j.ijfatigue.2020.105891.

[4] Benasciutti D. and Marques J.M.E., "Methods for estimating the statistical variability of the fatigue damage computed in one single stationary or non-stationary random time-history record," in *Proceedings of 3rd Int. Conf. on Structural Integrity for Offshore Energy Industry*, Aberdeen, UK, 2021 (ASRANet Ltd., Glasgow, 2021), pp. 28–37. ISBN 978-1-8383226-3-2.

[5] Marques J.M.E., Solazzi L., and Benasciutti D., "Fatigue analysis of nonstationary random loadings measured in an industrial vehicle wheel: Uncertainty of fatigue damage," *Metals*, Vol. 12, p. 616 (2022). ISSN 2075-4701. https://doi.org/10.3390/met12040616.

Strain Measurement Using a Distributed Fiber Optic Method: Influence of Fiber Optic Installation, Strain Measurement Comparison Methods

David Blaha[1], Milan Růžička[2], Milan Dvořák[3],
Nikola Schmidova[4], Petr David[5] & Karel Doubrava[6]

Abstract: This paper deals with the influence of fiber optic installation focusing on two types of adhesives commonly used for the installation of strain gauges. Multiple testing specimens are used to evaluate and compare strain measurement using Distributed Fiber Optic Sensing (DFOS) system, Fiber Bragg Gratings (FBG), Digital Image Correlation (DIC) and Strain Gauges (SG).This article describes the limits and uses of the DFOS method. From the results, the DFOS system is suitable for low-strain monitoring.

Keywords: DFOS; Strain measurement; FBG; DIC; Composite; Carbon fiber; Sensors

1. Introduction

The development of sensors for monitoring is still becoming more precise. Therefore, advanced sensors are used in various applications such as aircraft and bridge monitoring (Fig. 1 and Fig. 2). Monitoring the construction during its lifetime provides real feedback from the critical points of the structure. This information can be used to improve the control and safety of the structure.

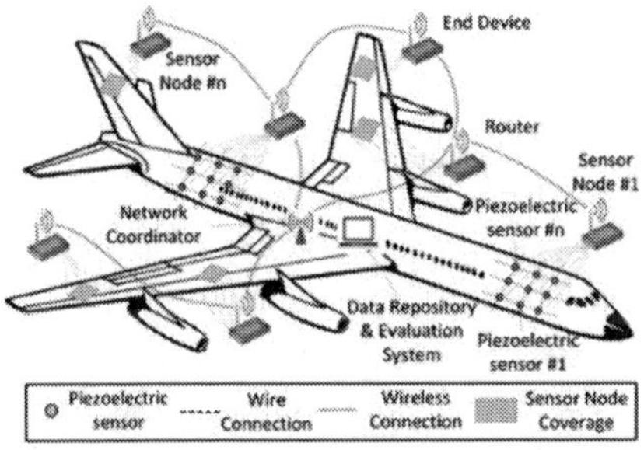

Fig. 1. SHM example [1].

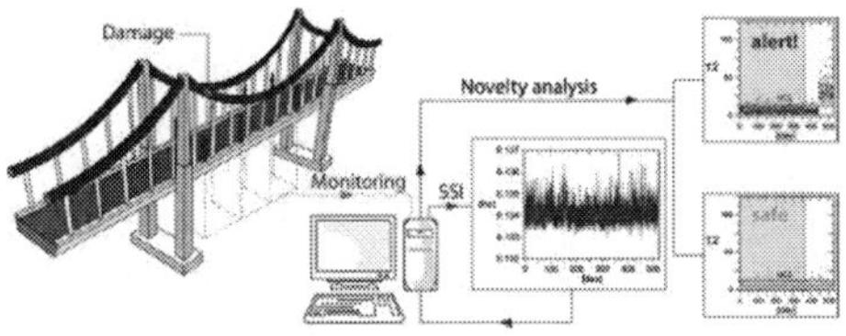

Fig. 2. SHM example [2].

Strain measurement using a Distributed Fiber Optic Sensing (DFOS) method became very attractive in the last years. The basics behind the theory reveal applications not only in strain

[1] *Ing. David Blaha; Division of Strength and Elasticity, Fac. of Mech. Engineering, Czech Technical University in Prague; Technická 4, 166 07 Prague, Czech Republic;* David.Blaha@fs.cvut.cz
[2] *prof. Ing. Milan Růžička; Division of Strength and Elasticity, Fac. of Mech. Engineering, Czech Technical University in Prague; Technická 4, 166 07 Prague, Czech Republic;* milan@ruzickovi.cz
[3] *Ing. Milan Dvořák, Ph.D.; Division of Strength and Elasticity, Fac. of Mech. Engineering, Czech Technical University in Prague; Technická 4, 166 07 Prague, Czech Republic;* Milan.Dvorak@fs.cvut.cz
[4] *Ing. Nikola Schmidova; Division of Strength and Elasticity, Fac. of Mech. Engineering, Czech Technical University in Prague; Technická 4, 166 07 Prague, Czech Republic;* Nikola.Schmidova@fs.cvut.cz
[5] *Ing. Petr David; Division of Strength and Elasticity, Fac. of Mech. Engineering, Czech Technical University in Prague; Technická 4, 166 07 Prague, Czech Republic;* Petr.David@fs.cvut.cz
[6] *Ing. Karel Doubrava, Ph.D.; Division of Strength and Elasticity, Fac. of Mech. Engineering, Czech Technical University in Prague; Technická 4, 166 07 Prague, Czech Republic;* Karel.Doubrava@fs.cvut.cz

or temperature measurements but in acoustic sensing as well. Small changes can be observed along the length of the optical fiber.

This work is focused on strain measurement by optical fibers, especially by the DFOS system. The idea is to validate two types of adhesives for strain measurement, application for different types of materials and different loading. Multiple sensor installation were used in all specimens in order to compare sensor response during loading.

2. Methods

Distributed fiber optic sensing method based on Rayleigh scattering is mainly used to evaluate strain measurement in this article. Random impurities along the glass fiber creates unified key pattern. In the Fig. 3 is presented DFOS schema. Tunable laser as a source of coherent light sweeps through the optical path to the custom made sensor which reflects scattered light back to the detectors. Reflected spectrum is recorded at the non-loaded fiber and during the loading on the detectors. In the Fig. 4 can be seen evaluation of the signal from the detectors. Due to Furier transform, the signal is dependent on position. Then the vector sum of the spectral shift is correlated with a reference signal in all segments to calculate strain change. Froggatt et al showed that spectral resolution of this method is dependent on length of a segment used for correlation. The longer the segment, the better the accuracy of measurement [3].

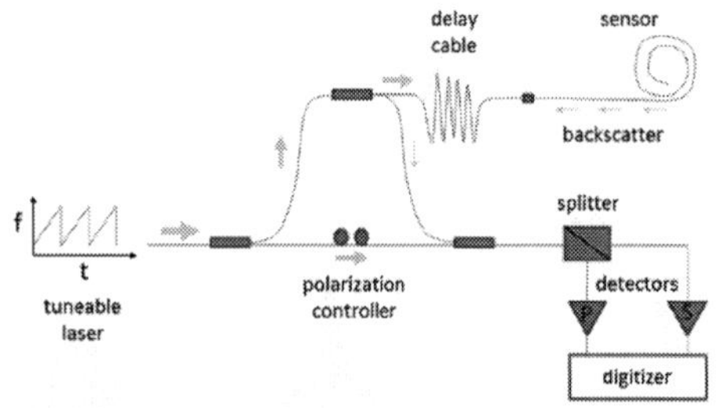

Fig. 3. DFOS schema [4].

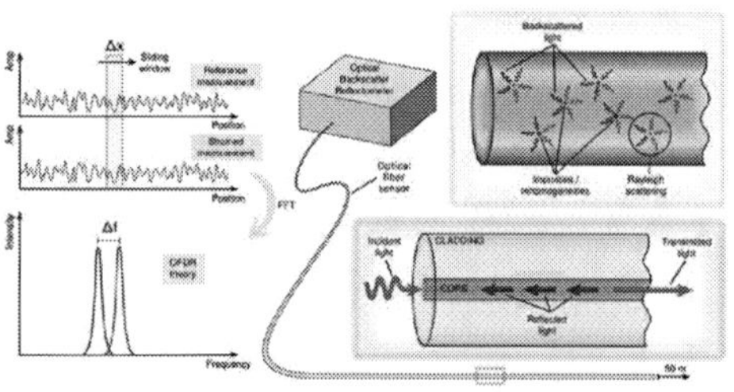

Fig. 4. DFOS principle [5].

Another approach to measure strain with optical fiber is to use Fiber Bragg Gratings (FBG). This method uses gratings that are photo-inscribed into fiber in a defined wavelength. The reflected spectrum is compared with reference stage. Multiple FBG can be inscribed into one fiber which allows to measure more reflections with a single fiber, but number of gratings are limited [6].

Digital Image Correlation (DIC) is used to measure strain and displacement on a surface of a measured specimen. This optical method tracks random pattern of grey scale on a surface of a component which is correlated with a reference image to calculate pattern shift. That shift is transformed to displacement or strains [7]. The main limiting factor of this method is the visual accessibility of the sample.

Carbon Fiber (CF) sensor is used to monitor the change in electrical resistance within the fiber length. The sensor is made of carbon fiber tow and it is coated by nickel electrolyte at the ends of the tow. Change in electrical resistance is measured with Wheatston bridge and resultant strain is integrated along the entire fiber length [8]. Due to conductivity of the fiber, the fiber must be insulated or calibrated for the entire sample. Strain Gauges (SG) are also used to compare with other sensors.

3. Experimental setup and testing

3.1. Sensor installation

In the first place it is necessary to prepare custom made sensors for each specimen individually. A glass fiber with polyimide coating is used for sensor. The fiber must be welded to LC/APC patch cord from one side, and to non-reflective fiber from another side. Before welding process, the fiber must be prepared by fiber optic precise cutter, see Fig. 5. Fitel s179 equipment is used for welding of optical fibers, Fig. 6.

Fig. 5. Fiber optic precise cutter.

Fig. 6. Fitel s179 welding machine.

When the sensor is prepared, surface of component must be thorougly cleaned, similarly as for strain gauges. After surface preparation, the sensor is installed on the surface with recommended adhesive.

After installation, the sensor key is created with an instrument Odisi 6100 series by Luna Innovations. All the DFOS measurement in this article is performed by Odisi 6100 instrument with a frequency 125 Hz and strain pitch 2.6 mm.

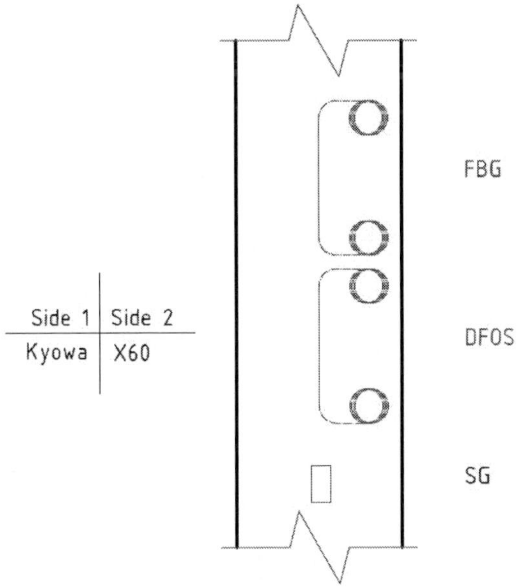

Fig. 7. Metal sheet schema.

3.2. Metal sheet

The first specimen is metal sheet with dimensions 660×69×1 mm. The FBG sensors, DFOS sensors, and strain gauges are applied on the metal sheet. The sensors on the first side of the sheet are installed by Kyowa CC-33A – instantaneous cyanoacrylic adhesive, and the sensors on the second side are installed by HBM X60 – methylmeta-crylate two-component glue, the schema can be seen in the Fig. 7.

Fig. 8. Tira Test 2300 setup.

Fig. 9. A detail of the sensors on the specimen.

The specimen was tested on tension on universal testing machine TIRA TEST 2300. The experiment consists of periodically variying the load between 0.5 kN and 7 kN, after which the specimen is loaded until failure. A setup for testing is shown in the Fig. 8 and Fig. 9.

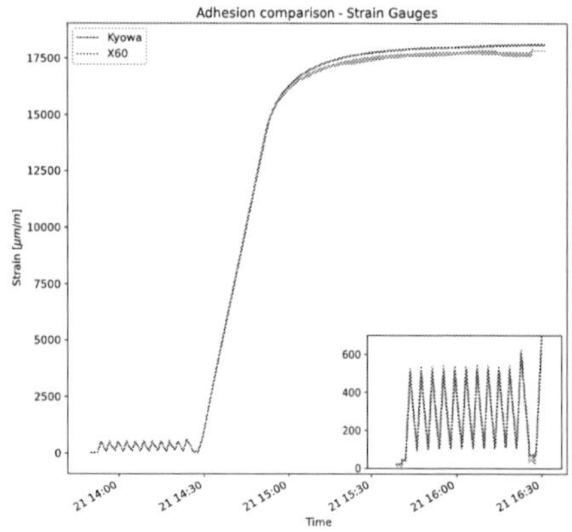

Fig. 10. Comparison of the strain gauges with different adhesives.

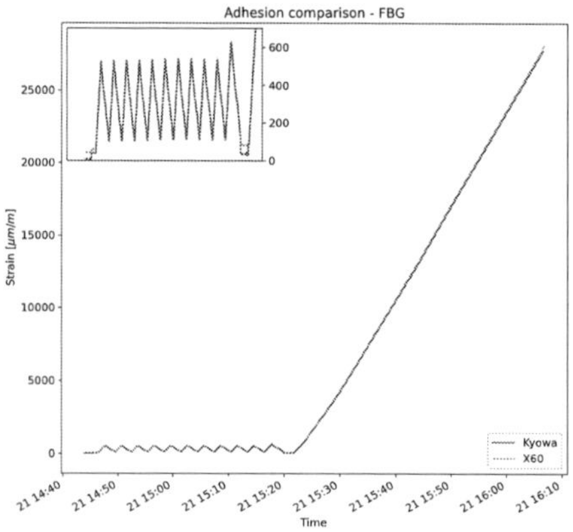

Fig. 11. Comparison of the FBGs with different adhesives.

The results from testing are seen in the Figures below (Fig. 10, Fig. 11, Fig. 12). In all three cases, there is not much difference after load variation on the sensors. After load variation, the difference in slope between the strain gauges is 1.3%, difference between the FBGs is 0.7%, and difference between DFOS sensors is 0.1%.

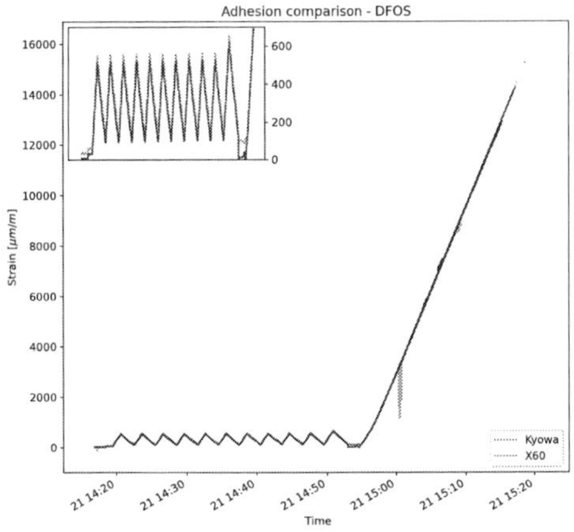

Fig. 12. Comparison of the DFOS with different adhesives.

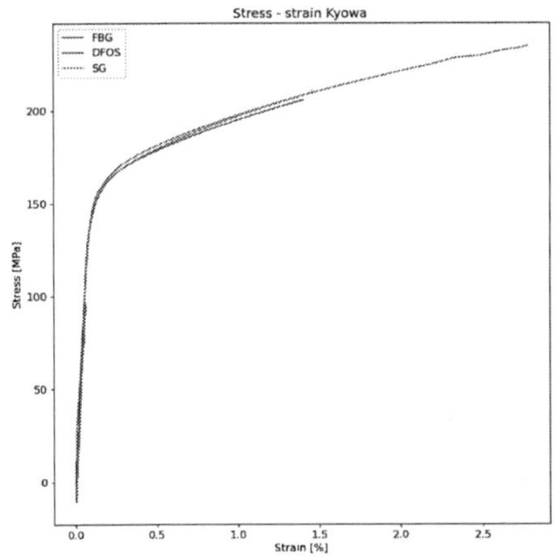

Fig. 13. Sensor comparison by kyowa adhesive.

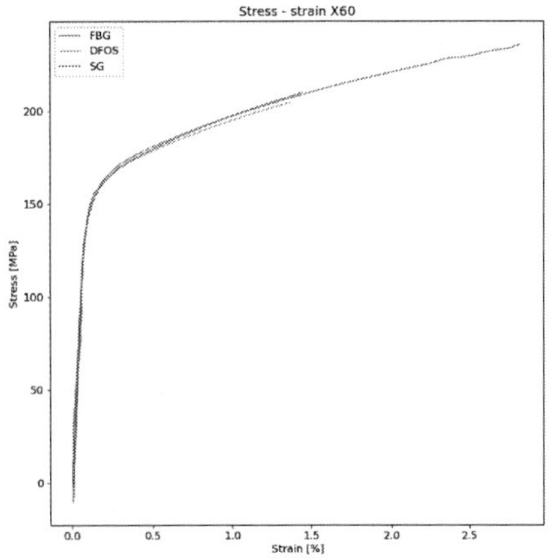

Fig. 14. Sensor comparison by x60 adhesive.

A comparison between the sensors installed with Kyowa adhesive can be seen in the Fig. 13. The strain difference after load variation is not significant between the sensors. The same results can be seen with X60 adhesive, Fig. 14. The difference between sensors are after load variation. The DFOS sensor was installed in area, where the plastic deformation started sooner, therefore there is a greater strain compared to FBG and SG. The DFOS sensor broke first, than the FBG sensor, and the SG measured to the end of testing, but it partially peeled off. The figures below show the dependence of stress and strain for different sensor installations.

3.3. Metal sheet with a hole

The second specimen has the same dimensions as previous one. The only difference is that the current sample has a 10 mm hole in the middle. This specimen has DFOS sensor installed vertically on one side and DIC gray pattern on the other side. The schema can be seen in the Fig. 15.

The strain distribution is monitored around the hole. The specimen was tested on Tira

Test 2300. The first load variation is performed from 0.5 kN to 6 kN, then from 0.5 kN up to the 9 kN, and finally loading the specimen until failure.

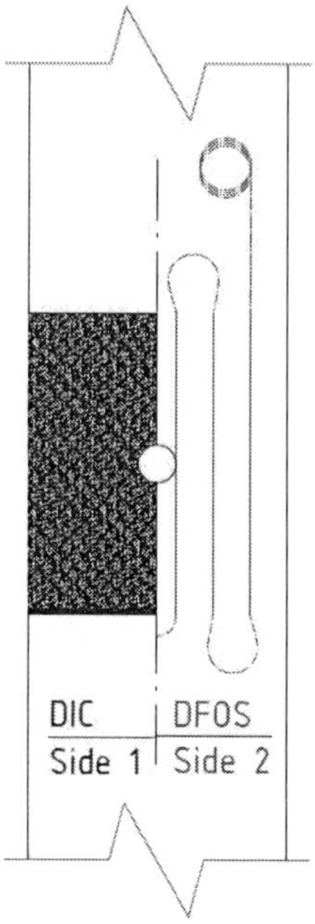

Fig. 15. Metal sheet schema with a hole in the middle.

Fig. 16. Experimental layout with GOM camera.

Fig. 17. Vertical layout of the DFOS sensor on the other side of the specimen.

Experimental layout of the specimen with GOM camera for pattern recording is in the Fig. 16. The camera captured pattern with 2 Hz frequency during testing. Vertical layout of the DFOS sensor is in the Fig. 17.

The recording from the GOM camera was processed in GOM Correlate software. The points on pattern for comparison with the DFOS sensor are picked at a "Hole-section" and at "Nominal-section", see schema in the Fig. 18.

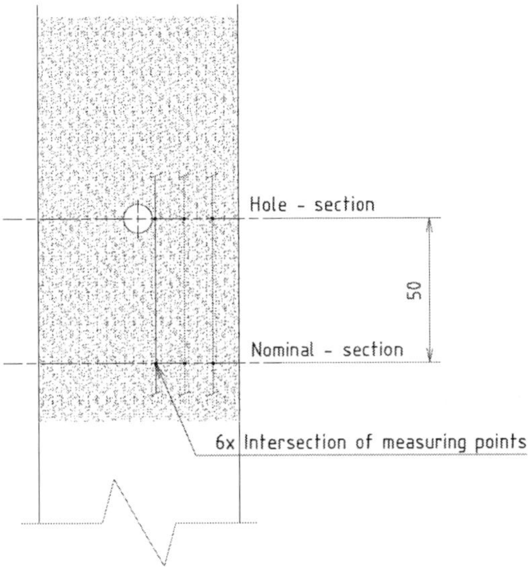

Fig. 18. The points on DIC pattern for comparison with the DFOS sensor.

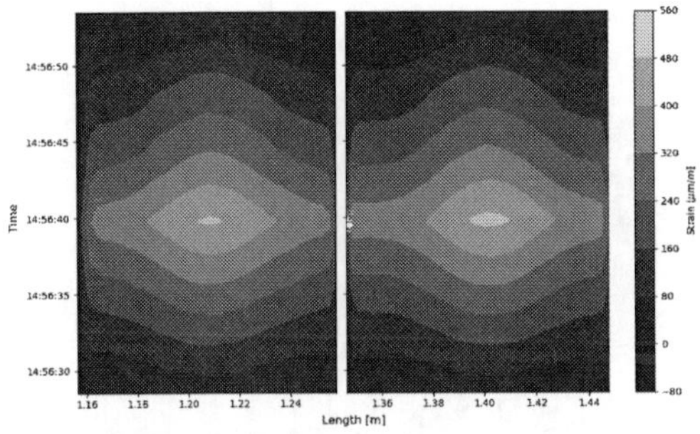

Fig. 19. DFOS sensor along the fiber and near the edge of the specimen during one cycle with a maximum amplitude of 6 kN.

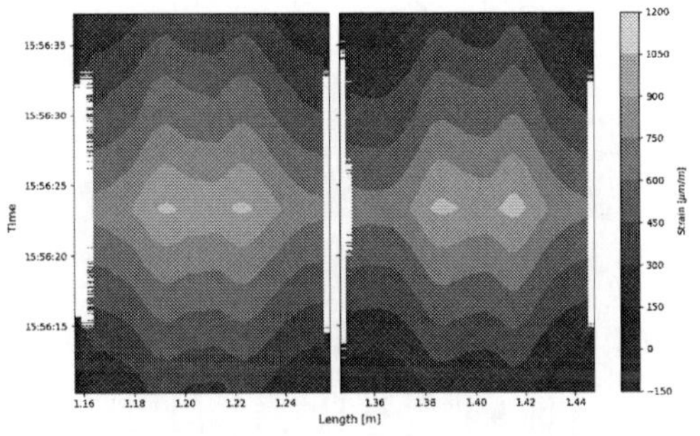

Fig. 20. DFOS sensor along the fiber and near the edge of the specimen during one cycle with a maximum amplitude of 9 kN.

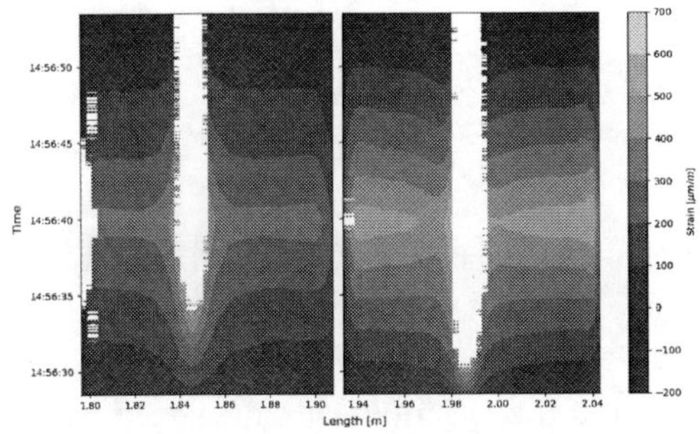

Fig. 21. DFOS sensor along the fiber and close the hole of the specimen during one cycle with a maximum amplitude of 6 kN.

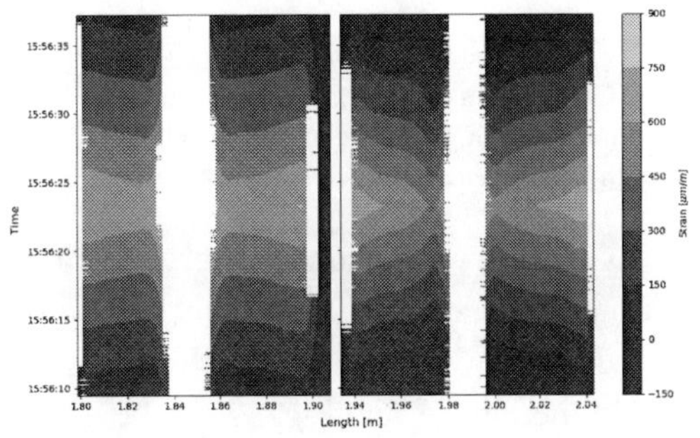

Fig. 22. DFOS sensor along the fiber and close the hole of the specimen during one cycle with a maximum amplitude of 9 kN.

On the first individual results from the DFOS sensor along the fiber (from Nominal to Hole section) show how the sensors respond to the load during one cycle, see Fig. 19–Fig. 22. Both of sensors along the fiber show course of load during 6 kN and 9 kN. On the sensor path is seen large gradient in the middle of the path, especially close to the hole. This confirms the sudden change caused by the hole in the plate, or rather the concentrator. The sensor response near a large gradient exhibits an unmeasurable section, and what's more, this information could be used for defect detection.

In the Fig. 23–Fig. 26 are seen data from the second measurement (up to 9 kN), according to the schema in the Fig. 18. Data from the first measurement (up to 6 kN) are not shown due to low facet resolution during correlation.

A comparison between the closest points near the hole (Hole-section) is not possible due to limitation of the fiber optic sensor, which doesn't respond properly in large strain gradient between adjacent points. Difference between a red line (GOM Correlate) and a green line (DFOS points) is not more than 32%.

The last comparison between GOM Correlate points and the DFOS sensor is in the failure mode. In the Fig. 28 can be seen strain–time dependency during failure with the detail to the strain range up to 1.5% of deformation. The detail is shown due to the DFOS sensor limitation. The fiber optic system provided measurement up to the 1.5% of deformation and then broke.

3.4. Composite rod

The third tensile testing is performed on a carbon/epoxy composite rod. The rod is wounded together with its ending by continuous fiber. Hence, the DFOS sensor was applied along the entire length of the rod.

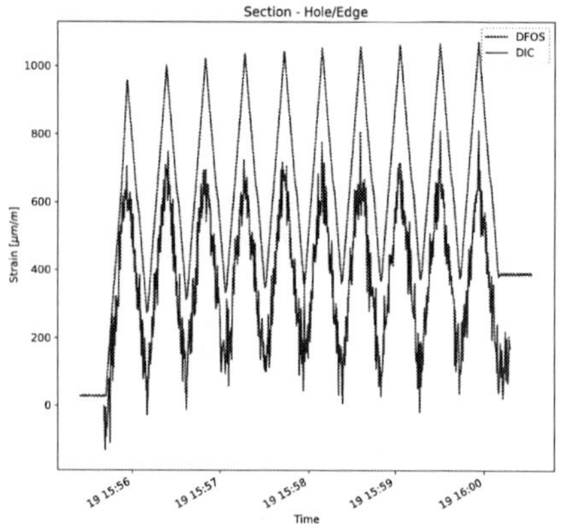

Fig. 23. Comparison between DFOS sensor points and points from GOM Correlate software in the level of the hole.

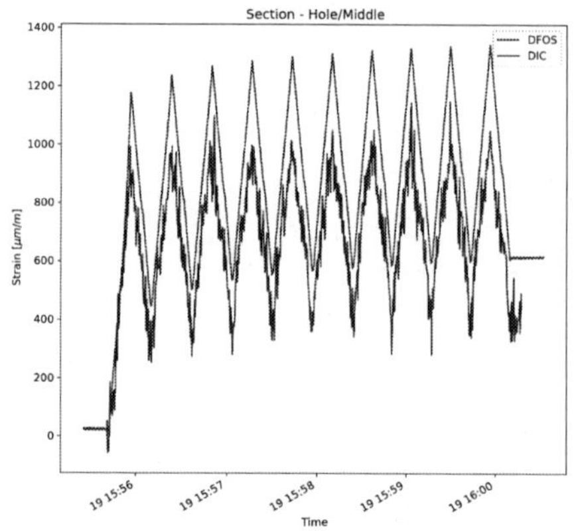

Fig. 24. Comparison between DFOS sensor points and points from GOM Correlate software in the level below the hole.

The specimen was loaded on servo-hydraulic machine Heckert UFP 400 until failure. The DFOS fiber is installed in axial direction to the surface of the composite rod, the strain gauge is installed on the opposite side in the same direction. Both of sensors are compared during the loading. In the Fig. 31 and Fig. 32 can be seen a schema of the specimen and the specimen after loading.

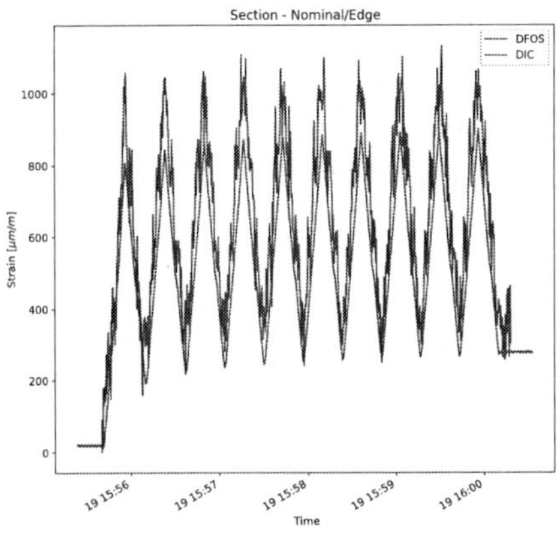

Fig. 25. Comparison between DFOS sensor points and points from GOM Correlate software in the level of the hole.

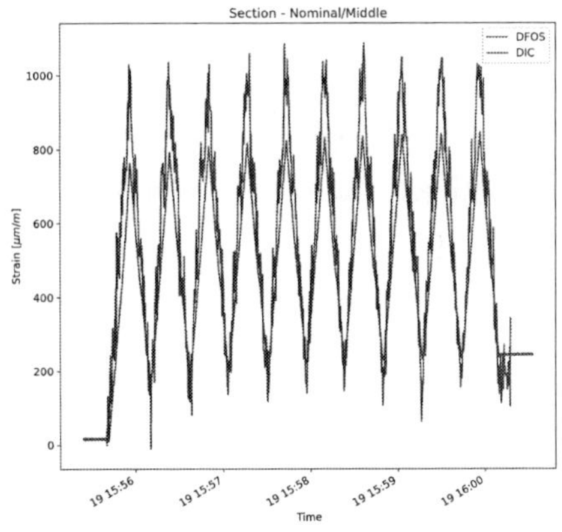

Fig. 26. Comparison between DFOS sensor points and points from GOM Correlate software in the level below the hole.

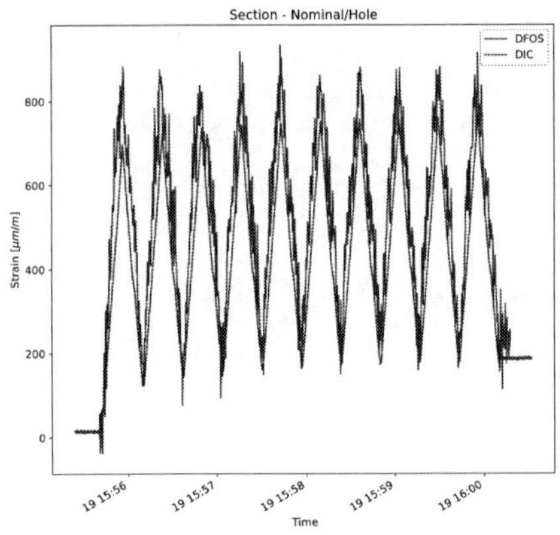

Fig. 27. The points on pattern for comparison with the DFOS sensor in a failure mode.

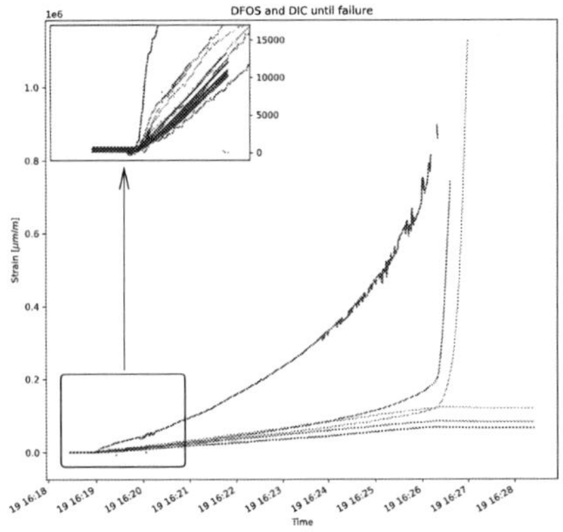

Fig. 28. The lines from GOM Correlate software during the failure mode with a detail to the DFOS measurement limit (DFOS - green, DIC - red).

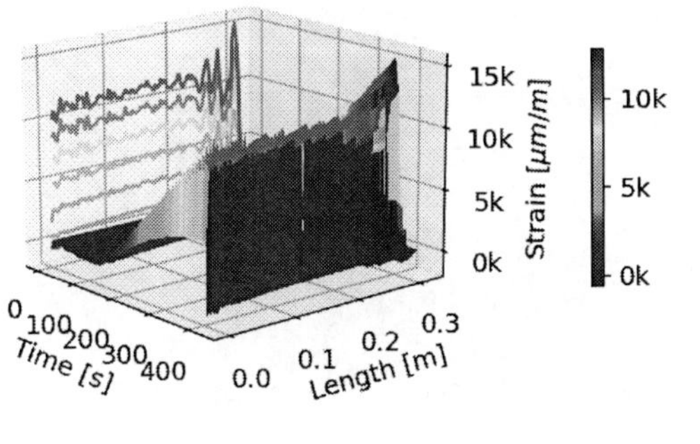

Fig. 29. Strain - length - time data from the DFOS sensor showed in 3D plot.

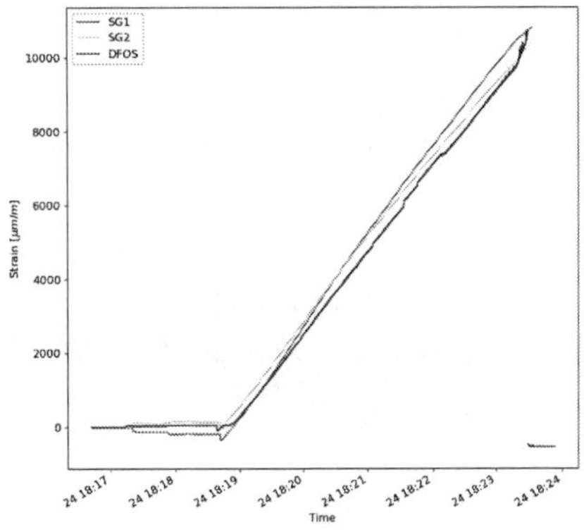

Fig. 30. Comparison between the strain gauges and the DFOS sensor in the same plane, see Fig. 31.

The results from measurement are presented in two Figures. Fig. 29 shows the advantage of the DFOS sensor. All the length of composite rod is measured by single fiber line with the visible strain concentration at the end of the composite rod with wound ending. Fig. 30 shows both strain gauges and the DFOS sensor gauge in the same plane.

The difference between SG1 and the DFOS sensor is 6.4%, and 1.7% between SG2 and the DFOS sensor in the same plane.

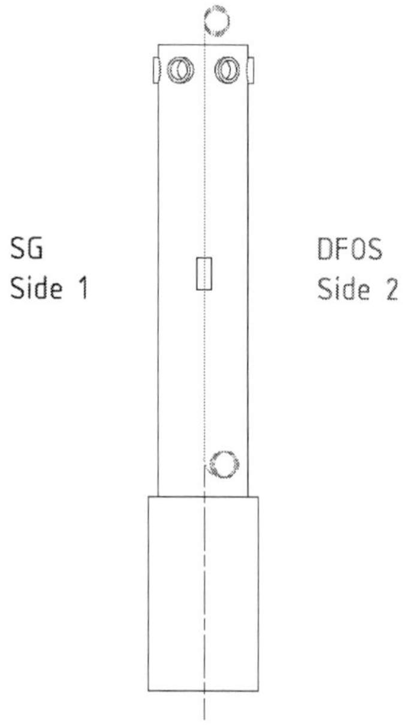

Fig. 31. Composite rod sensor schema.

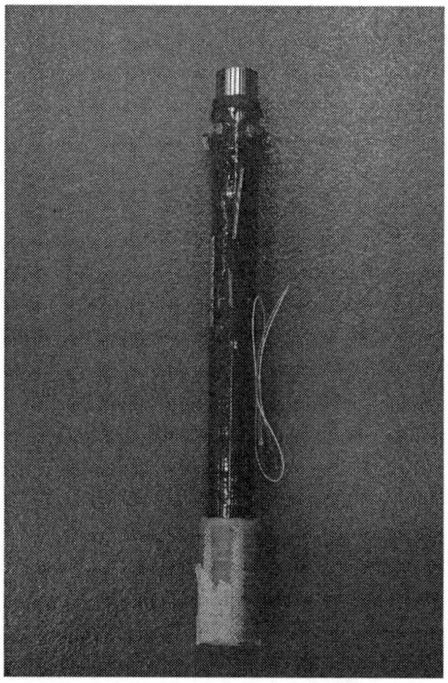

Fig. 32. Composite rod after testing with the DFOS sensor.

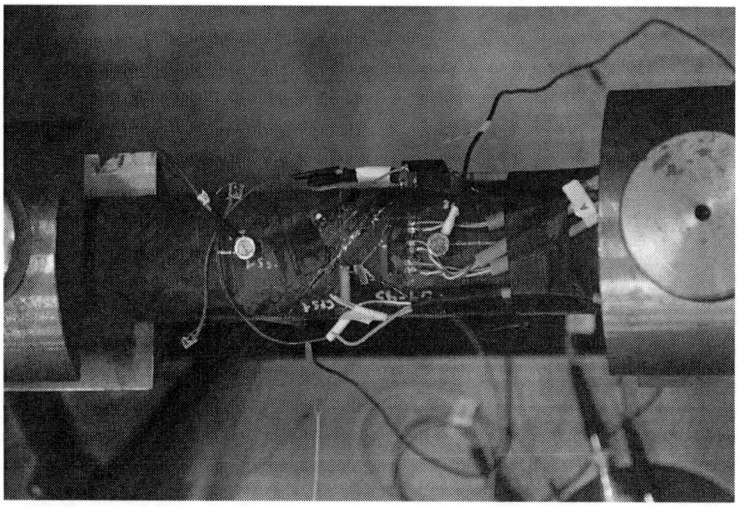

Fig. 33. Sensor layout on the composite bone.

3.5. *Composite bone*

The last type of specimen is the composite bone. The bone is wounded by continuous carbon fiber together with its ending as well. The specimen is subjected to torsional stress.

The torsion testing was provided on hydraulic machine until failure. All the sensors are installed in the direction of the testing, Fig. 33. The DFOS sensor is installed on the surface in 45° from the axial direction, CF sensor and SG are parallel to the DFOS sensor. The DFOS sensors and CF sensors are installed on both sides of the composite bone.

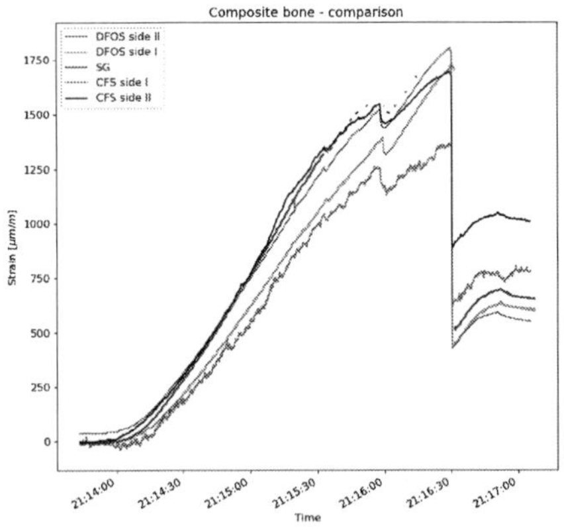

Fig. 34. The composite bone - specimen X0.

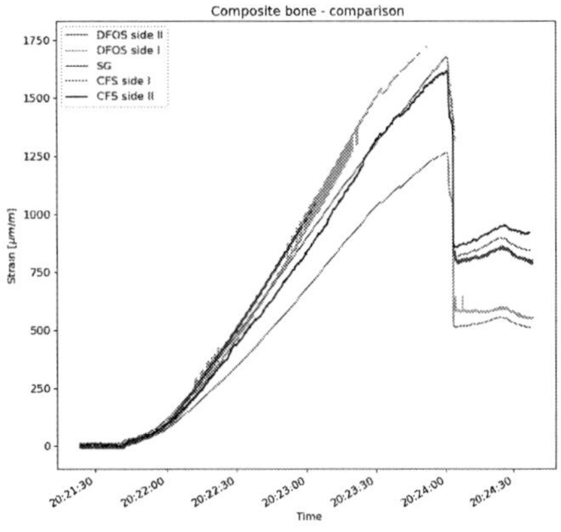

Fig. 35. The composite bone - specimen X1.

The results from testing show comparison between all the sensors, in the Fig. 34 and Fig. 35. The difference between the DFOS sensors and SG are up to 9 % on both sides in the same plane as the SG. The difference between CF sensors and SG varies from the 3% up to 30%. The maximum difference was achieved by the single CF sensor on specimen X1, see the Fig. 35. This can be caused by peeling off the sensor or due to the local damage near the sensor.

4. Conclusions

This article shows limitation of the DFOS sensor in comparison with other type of the strain sensors. Several conclusions can be seen from the measurements on the specimens.

The first conclusion is about the adhesion comparison on optical fibers. Kyowa and HBM glue show similar results during the load variation and the failure testing as well. The difference between the slopes of the FBG curves from each side is 0.7%, between DFOS sensors 0.1%, and between SG 1.3%.

The second conclusion is about the usage of DFOS system. The fiber optic system is advantageous to use for low strain long term monitoring and testing. Measurements until failure are not efficient due to spectral resolution of the method, and the mechanical properties of the fiber.

The third conclusion involves precision of strain measurement by the fiber optic systems. This can be affected by the adhesive installation and the strain transmission from the part to the single fiber core, since the strain has to be transfered to the fiber core through the adhesive and the fiber cladding. Another influence is due to spatial resolution of the method, which averages strain values of each gauge pitch.

The last conclusion suggests a greater focus on defect localization. A metal sample with a hole provides an insight into how a defect can be viewed due to the limits of the measurement method. The large gradients in strain monitoring could be used for a damage detection. Further testing is required to verify and develop further conclusions.

Acknowledgements

Financial support was provided for this study by the Grant Agency of the Czech Technical University in Prague; grant no. SGS21/151/OHK2/3T/12, and by the Technology Agency of the Czech Republic, grant no. TJ02000252, FW01010450.

References

[1] Fu H., Sharif Khodaei Z. and Aliabadi M.H.F., "An event-triggered energy-efficient wireless structural health monitoring system for impact detection in composite airframes," *IEEE Internet of Things Journal*, Vol. 6, Issue 1, pp. 1183–1192 (2019). https://doi.org/10.1109/JIOT.2018.2867722.

[2] Comanducci G., Ubertini F. and Materazzi A-L., "Structural health monitoring of suspension bridges with features affected by changing wind speed," *Journal of Wind Engineering and Industrial Aerodynamics*, Vol. 141, pp. 12–26 (2015). https://doi.org/10.1016/j.jweia.2015.02.007.

[3] Froggatt M. and Moore J., "High-spatial-resolution distributed strain measurement in optical fiber with rayleigh scatter," *Appl. Opt.*, Vol. 37, no. 10, pp. 1735–1740 (1998). http://dx.doi.org/10.1364/AO.37.001735.

[4] Lomperski S., Gerardi C., and Lisowski D., "Fiber optic distributed sensors for high-resolution temperature field mapping," *Journal of Visualized Experiments*, no. 117 (2016). http://dx.doi.org/10.3791/54076-v.

[5] Barrias A., Casas J. and Villalba S., "SHM of reinforced concrete elements by rayleigh backscattering DOFS," *Frontiers in Built Environment*, Vol. 5, (2019). http://dx.doi.org/10.3389/fbuil.2019.00030.

[6] Boller Ch., Chang F.-K. and Fujino Y., editors. *Encyclopedia of structural health monitoring*, (Wiley-Blackwell, Hoboken, NJ, January 2009). http://dx.doi.org/ 10.1002/9780470061626.

[7] Peters W. H. and Ranson W. F., "Digital Imaging Techniques In Experimental Stress Analysis", *Optical Engineering*, Vol. 21, Issue 3, pp. 427 – 431 (1982). https://doi.org/ 10.1117/12.7972925.

[8] Schmidová N., Macken J., Horoschenkoff A., Sedláček R., Kostroun T., Šimota J. and Růžička M., "Impact damage detection of a glass fabric composite using carbon fiber sensors with regard to mechanical loading," *Applied Sciences*, Vol. 12 no.3, Art. no. 1112 (2022). https://doi.org/10.3390/app12031112.

Theoretical and Experimental Analysis of the Dynamic Behavior of an Existing Footbridge

Magdaléna Boháčová[1]

Abstract: The topic of this paper is an experimental and theoretical analysis of the dynamic behavior of an existing footbridge. The examined structure is the cable-stayed footbridge across the Vltava river in Lužec nad Vltavou. The horizontal load-bearing structure is suspended on a 39.85 m high A-shaped steel pylon by means of a system of 17 pairs of stays. The footbridge deck consists of UHPC components, which are prestressed by a pair of external cables. The first part of the paper is dedicated to the experimental analysis of the dynamic behavior of the footbridge. At first, the experimental modal analysis was carried out and natural frequencies and modes of the horizontal load-bearing structure were evaluated. Then a dynamic load test was performed and the dynamic response of the footbridge deck to dynamic pedestrian load (in a scale of acceleration) was monitored in the course of it. The final obtained results were assessed according to current standards, regulations and recommendations and the comfort criteria are then mutually compared. The second part of the paper is dedicated to the theoretical analysis of the dynamic behavior of the footbridge. Within this part a theoretical 3D model of the footbridge was created. Natural vibration frequencies and modes were calculated. Then the validation and subsequently also the identification of the model were performed. In the last part of the paper, the obtained results of the experimental and theoretical analysis of the footbridge dynamic behavior are compared and discussed.

Keywords: Footbridge; UHPC components; Experimental modal analysis; Natural frequencies; Natural modes; Dynamic load test; Pedestrian load; Comfort criterion

1. Introduction

The fundamental reason for choosing the footbridge in Lužec and Vltavou town (see Fig. 1) as the examined structure was the appearance of a problem with vibration of its stays caused by wind. The dynamic experiment focusing on this problem was carried out twice. However, this topic is going to be shown in another paper. The experiment, the result of which are described in this paper, was concurrently performed with that executed on the stays in October 2021.

The presented theoretical and experimental analysis of the footbridge is divided into three substantial parts. The first one is an experimental modal analysis carried on the horizontal load-bearing structure of the footbridge. The second one is a dynamic load test which was focused on the dynamic response of the footbridge deck caused by dynamic pedestrian load. The third part of the paper is dealing with a theoretical 3D model of the footbridge and its validation and subsequently also with its identification based on the experimental results.

Similar experiments which were also carried out on three different cable-stayed footbridges in the Czech Republic are described in references [1–3].

The analogous experimental modal analysis and the similar dynamic load test were carried out on the footbridge in 2020 before putting the structure into operation. One of the natural frequencies of the footbridge evaluated in 2020 was 2.11 Hz, which is substantially susceptible to dynamic pedestrian load. One of the aims of the performed dynamic load test was to focus

[1] *Ing. Magdaléna Boháčová; Department of Mechanics, Czech Technical University in Prague; Thákurova 2077/7, 166 29 Prague 6, Czech Republic;* magdalena.bohacova@fsv.cvut.cz

on the dynamic response of the horizontal bearing structure to the dynamic load with frequency 2.11 Hz. The measured dynamic response of the footbridge deck to the pedestrian load was assessed according to current standards, regulations and recommendations and the pedestrian comfort level was also identified. Some of the applied comfort criteria were then mutually compared.

The basic objective of the experimental modal analysis was to evaluate substantial natural frequencies and modes of the horizontal load-bearing structure of the footbridge and then to compare them with the ones evaluated in 2020 and with the theoretical ones.

As the last part of the described analysis, a theoretical 3D model of the footbridge was created, natural frequencies and modes were calculated and then they were compared with the measured ones. The validation and identification of the model were consequently performed. The identified model can be used for more advance theoretical analysis of the dynamic behavior of the footbridge.

2. Description of the footbridge

The observed footbridge is located in Lužec nad Vltavou town northwards from Prague. It crosses Vltava river and it creates a pedestrian and cycling connection between Lužec nad Vltavou town and Bukol village.

It is a cable-stayed footbridge with two asymmetric spans, which are 99.18 m and 31.90 m long (Fig. 1). The horizontal load-bearing structure is supported by pair of longitudinally sliding bearings on the O1 abutment (see Fig. 1) and frame-connected with the O3 abutment (see Fig. 2). The height of the frame-connection varies from 500 mm to 1000 mm and the shape is parabolic. The width of the horizontal load-bearing structure is 4.50 m and its total length is 131.58 m. The clear width for walking and cycling is 3.00 m. The longitudinal slope of the footbridge deck varies along the length from +8.00% to -8.00%.

Fig. 1. The view of footbridge in Lužec nad Vltavou town.

The substantial part of the horizontal load-bearing structure is assembled of UHPC components (see Fig. 3). The components are made of concrete C 110/130. The footbridge deck consists of 31 typical and 2 atypical components. The length of typical component is 3.998 m and the length of atypical component is 1.998 m. The minimum width of the longitudinal girders, which are a part of the components, is 400 mm and it varies along the hight up to 500 mm. The thickness of the components desk is 60 mm. The width of cross girders, which are located in ¼ and ¾ of the length of the typical component and in ½ of the length of the atypical component, varies symmetrically along the height from 200 mm to 400 mm and there are linear rises between the cross girders and the components desk. The longitudinal slope and transverse slope of the upper surface of the components is ±0.00% and the longitudinal slope of the footbridge is created by the suitable composition of the components. The height of the cross girder above the abutment O1 varies from 750 mm to 500 mm (see Fig. 1).

The UHPC components are prestressed by a pair of external cables (see Fig. 2 and Fig. 3). The diameter of the cables is 15.7 mm and each cable consists of 19 strands. The type of strands is St 1640/1860 MPa manufactured by VSL.

Fig. 2. The view on the abutment O3.

Fig. 3. The bottom view on the footbridge deck assembled from UHPC components.

The horizontal load-bearing structure is suspended on a 39.85 m high A-shaped steel pylon by means of a system of 17 pairs of stays, 5 pair of stays are reverse (see Fig. 1). The stays are anchored into the massive head of the pylon and into certain components of the footbridge deck or into the massive anchor block behind the O3 abutment. The type of stays is "Full Locked Coil Strands" manufactured by Redaelli. The diameter of stays varies from 32 mm to 64 mm.

The vertical load-bearing structure is made of pair of pillar struts. The pillar struts are rectangular-shaped and the heel dimensions are 1000×600 mm, the head dimensions are 600×600 mm. The material of the pillar struts is steel S355J2+N and sheet metal thickness is 20 mm. The axial distance of the pillar struts in the heel is 7060 mm, the struts converge into the massive head. There are 12 stays anchoring levels towards Lužec nad Vltavou and 5 stays anchoring levels towards Bukol. The pylon is filled with self-compression concrete C 30/37 up to the level of the footbridge deck.

3. Experimental modal analysis

As was said above, a dynamic experiment was carried out on the footbridge before putting it into operation in 2020. The aim of the experimental modal analysis, which was the first part of the described experiment, was to find out the current natural frequencies and modes of vibration of the footbridge, to verify the compliance between the current results and the results obtained in 2020 and to compare the actual results with the theoretical ones.

3.1. Measurement line

The footbridge vibration was measured by piezoelectric acceleration transducers Brüel & Kjær Type 8344 (see Fig. 5). The working range of these sensors is from 0.2 Hz to 3 kHz, their sensitivity is very high such as approximately 2500 mV/g and their natural frequency is about 10 kHz. During the measurement, the sensors were magnetically attached to steel weights (see Fig. 4) and via cables connected to the 4-channel data acquisition station SIRIUS Type 6ACC-2ACC+ made by DEWESoft with maximum frequency range 51.2 kHz and with unique 2 × 24-bit A/D converter. During the experimental modal analysis, the data were recorded by DEWESoft software. Subsequent analysis of the records was performed with ME'scope VES software Vibrant Technology Inc.

Fig. 4. Three acceleration transducers Brüel & Kjær attached to the steel weight.

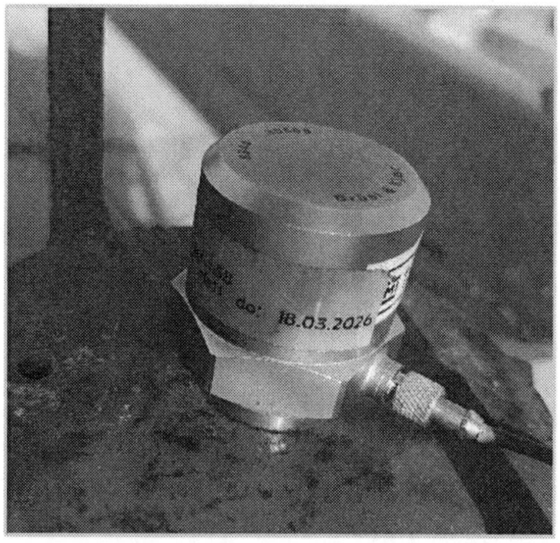

Fig. 5. Detail view on the acceleration transducer Brüel & Kjær Type 8344.

3.2. Measurement system

In the preparation phase of the experiment, the low damping of the examined footbridge was supposed as it is a subtle prestressed concrete structure. Thus, the Ambient Vibration Testing (AVT) system was chosen for the experimental modal analysis. During this method, the excitation forces are not measured and therefore the recorded dynamic responses were normalized to the one observed in the reference point. The reference point should be placed so that none of the searched natural mode of vibration has an oscillation node there and the magnitude of all searched natural modes should be the greatest possible in this point. During the AVT variant of the experiment the pedestrian and cycling traffic on the footbridge does not have to be suspended or restricted and the possible use of an exciter within the Forced Vibration Testing (FVT) method would not be suitable also from economical view.

Nineteen cross-sections on the footbridge deck were selected for placing steel weights with acceleration transducers (see Fig. 6). The number of selected cross-sections was chosen to evaluate a sufficient number of natural modes and also to be able to carry out the whole experiment in one day. There were two points measured in each cross-section and they were located approximately 0.75 m from the edge of the horizontal load-bearing structure right next to the handrail. The placement of the cross-sections on the footbridge deck was identical to the location of the joint-lines of the UHPC components because of obtaining the magnitude of searched natural modes in the same points as in the theoretical model. The numbering of the cross-sections was from 01 to 35 according to the direction of stationing. The selected cross-sections for placing the acceleration transducers were every second numbered cross-section except the area near the O3 abutment where the accelerometers were placed in every numbered cross-section to obtain the influence of the frame-connection between the horizontal load-bearing structure and the O3 abutment on the shape of the searched natural modes. The measured points were marked with the number of selected cross-section and with the number 1 or 2 for distinction of the position in the footbridge cross-section (1 means the left side of the deck and 2 means the right side). The positions of the measured cross-sections with marked observed points are evident from the schematic ground plan shown in Fig. 6.

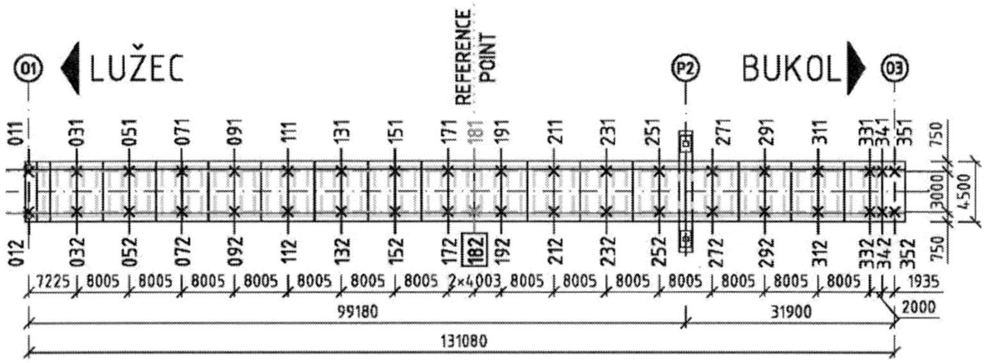

Fig. 6. The schematic ground plan of the observed points on the footbridge in the course of the experimental modal analysis.

The shapes of the natural modes measured and evaluated in 2020 were taken into consideration during searching the best possible position for the reference point. The best position for the reference point was evaluated as the point marked 182. Only the first vertical bending mode of vibration might be problematic to measure, but the magnitude of the mode was still sufficient according to the test measurement.

The dynamic response was observed by six acceleration transducers attached by a neodym magnet to steel weights placed on the deck. It was measured vertically and horizontally in both directions perpendicular and parallel to the longitudinal axis of the footbridge in both points in the selected cross-section. The position of the reference point did not change during the whole experiment. The dynamic response in the reference point was measured vertically and horizontally only in direction perpendicular to the longitudinal axis of the footbridge. The vibration in each cross-section was measured for five minutes to acquire the data sufficient for averaging in FFT with appropriate precision.

The excitation of the structure caused by wind and microseismicity was sufficient to be able to evaluate the required data. It was not possible to place the acceleration transducers on the pylon, therefore it was not possible to measure its natural modes. It means the evaluated natural modes (Fig. 8) are related only to the horizontal load-bearing structure of the footbridge.

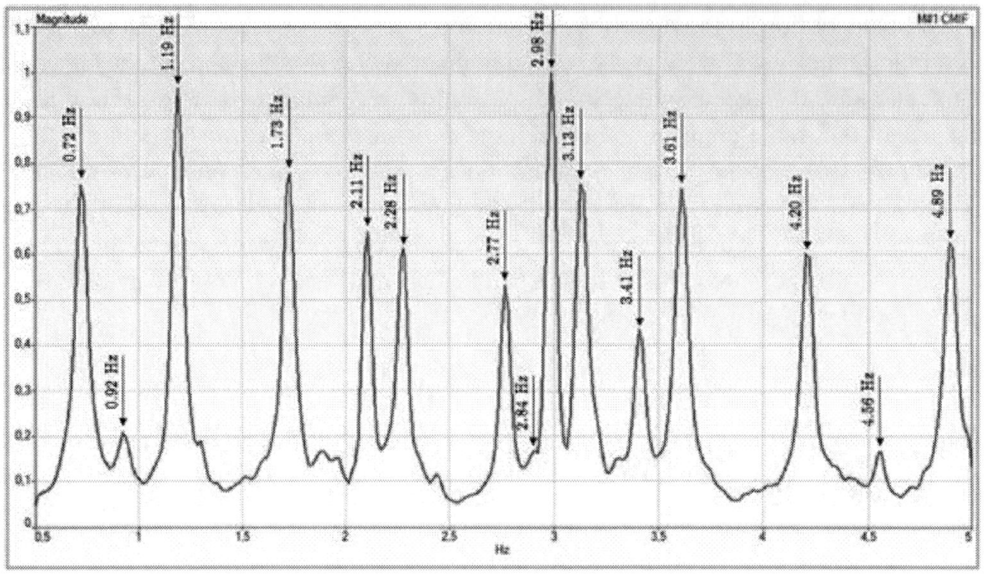

Fig. 7. The evaluated Complex Mode Indicator Function.

Table 1. Natural frequencies of the horizontal load-bearing structure of the footbridge evaluated from the experimental modal analysis.

No.:	$f_{(k)}$ [Hz]	Character of the natural mode
(1)	0.72	1st vertical bending mode of vibration
(2)	0.92	1st horizontal bending mode of vibration
(3)	1.19	2nd vertical bending mode of vibration
(4)	1.73	3rd vertical bending mode of vibration
(5)	2.11	1st torsional bending mode of vibration
(6)	2.28	4th vertical bending mode of vibration
(7)	2.77	5th vertical bending mode of vibration
(8)	2.84	2nd horizontal bending mode of vibration
(9)	2.98	6th vertical bending mode of vibration
(10)	3.13	7th vertical bending mode of vibration
(11)	3.41	2nd torsional bending mode of vibration
(12)	3.61	8th vertical bending mode of vibration
(13)	4.20	9th vertical bending mode of vibration
(14)	4.56	3rd torsional bending mode of vibration
(15)	4.89	10th vertical bending mode of vibration

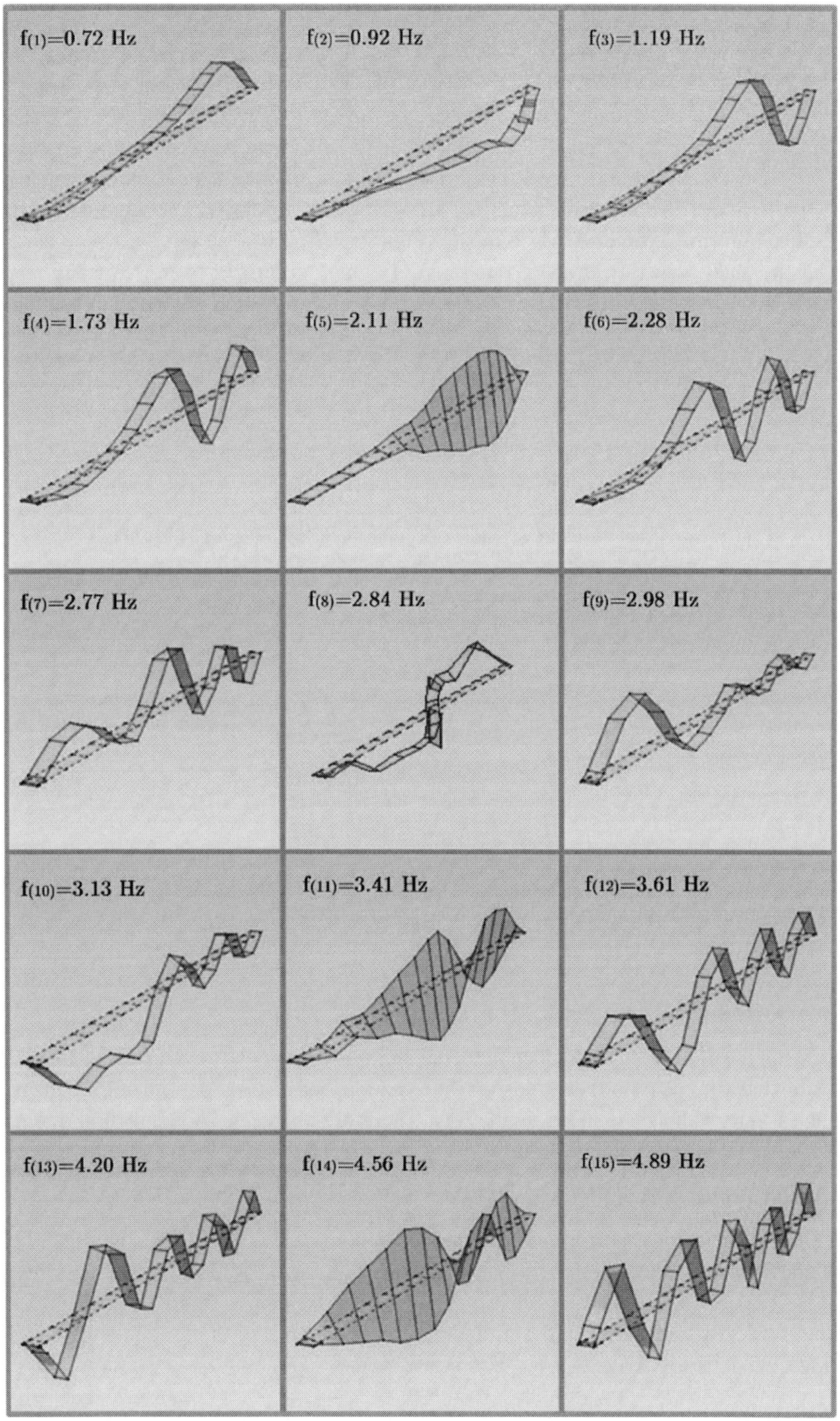

Fig. 8. Natural modes of the horizontal load-bearing structure of the footbridge.

3.3. Evaluation of the natural frequencies and modes

Measured records of the dynamic response were processed in ME'scopeVES. At first, the data from all channels were transformed using Fast Fourier Transform (FFT) from the time domain to the frequency domain. Frequency transfer function of the operational deflection shapes (ODS FRF) were evaluated from these signals for all 38 measured points. The natural frequencies were evaluated from Complex Mode Indicator Function (CMIF) (see Fig. 7), which is defined as the eigen values solved from the normal matrix, which is formed from frequency response function (FRF) matrix. CMIF can be calculated from multiplication of normal matrix at each spectral line. The first fifteen natural frequencies (see Table 1) and appropriate natural modes (see Fig. 8) were evaluated.

3.4. Comparison of the natural frequencies measured in 2020 and 2021

The first experimental modal analysis was carried out on the footbridge before putting it into operation in 2020. The new set of measurement was performed in October 2021.

Table 2. Comparison of the natural frequencies of the footbridge measured in 2020 and 2021.

Order	Character of the natural mode	2020	2021
(1)	Vertical bending mode	0.74	0.72
(2)	Horizontal bending mode	0.94	0.92
(3)	Vertical bending mode	1.19	1.19
(4)	Vertical bending mode	1.73	1.73
(5)	Torsional mode	2.11	2.11
(6)	Vertical bending mode	2.23	2.28
(7)	Vertical bending mode	—	2.77
(8)	Horizontal bending mode	—	2.84
(9)	Vertical bending mode	2.98	2.98
(10)	Vertical bending mode	—	3.13
(11)	Torsional mode	3.37	3.41
(12)	Vertical bending mode	3.60	3.61

4. Dynamic load test

The results of the performed experimental modal analysis could not be evaluated on-line. Therefore, testing measurement for the purpose of the dynamic load test was carried out and the natural frequencies were evaluated in more simple way. Therefore, the values of the natural frequencies in Table 2 and Table 3 may not be identical.

4.1. Measurement system

The dynamic load test was performed with the same equipment as during the experiment modal analysis described in subsection 3.1. The selected points for measuring the dynamic response were 072, 112, 132 and 182 (see Fig. 9). The dynamic response in all selected points was measured vertically and horizontally in direction perpendicular to the longitudinal axis of the footbridge.

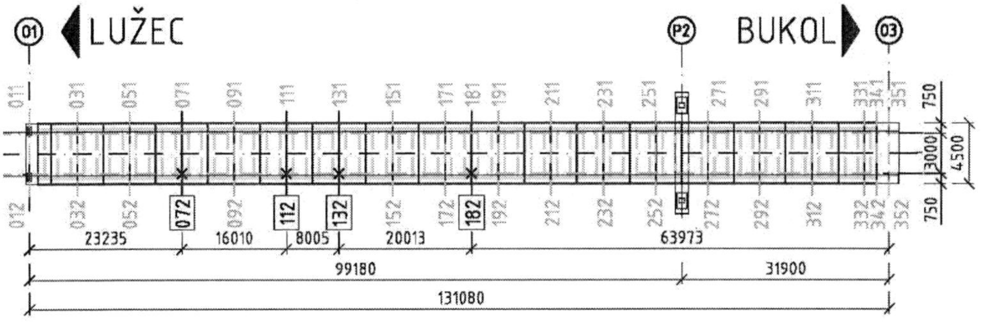

Fig. 9. The schematic ground plan of the observed points on the footbridge deck in the course of the dynamic load test.

Four types of footbridge loading were used during the dynamic load test (Table 3). The first one was random pedestrian traffic. People were crossing the footbridge with various walking speeds without any synchronization. The second type was a synchronous walk of two people behind each other with pace frequency equal to some natural frequency from the range 0.97–2.28 Hz. The synchronization was achieved by using a metronome. The position of pedestrians behind each other (not next to each other) is more effective in exciting torsional modes and not less effective in exciting bending modes (than the position next to each other), therefore it was used. The first and second type of the loading are considered as normal pedestrian traffic according to the Czech standard ČSN 73 6209 [4]. The third type was a synchronous run of two people behind each other with the natural frequencies from the range 3.00–3.42 Hz. The fourth type of footbridge loading was vandalism. In the load cases focused on vandalism, various positions of a group of people were chosen. The choose of the position was dependent on particular natural frequency is to be excited and where does the appropriate natural mode have the largest possible magnitude. According to the greatest possible magnitude of the particular natural mode one of the points 072, 112, 132 and 182 (see Fig. 9) was chosen and then the structure was excited by group of four people swaying in the knees synchronously in one of the evaluated natural frequencies. It is necessary to note that the third and fourth type of the loading are not considered as normal pedestrian traffic according to the Czech standard ČSN 73 6209 [4].

Table 3. Types of loading applied during the dynamic load test.

No.:	Type of loading	Frequency range [Hz]
001-002	Random pedestrian crossings	—
003-012	Synchronous walk of two people behind each other	0.97-2.28
013-016	Synchronous run of two people behind each other	3.00-3.42
017-032	Vandalism	1.19-3.42

4.2. Evaluation of the dynamic response

The measured vibration of the horizontal-bearing structure of the footbridge was processed (see Fig. 10 and Fig. 12) and the maximum peak values and maximum RMS values of measured acceleration were evaluated (see Fig. 11). Time interval of 5 s was used for determination of the RMS values of acceleration.

Table 4. Maximum peak values of acceleration evaluated during the dynamic load test analysis.

Type of loading	Horizontal direction [m·s^{-2}]	Vertical direction [m·s^{-2}]
Random pedestrian crossings	0.039	0.270
Synchronous walk of two people in frequency 2.11 Hz	-0.124	-0.688
Synchronous run of two people in frequency 3.42 Hz	-0.192	0.918
Vandalism	0.283	1.568

Table 5. Maximum RMS values of acceleration evaluated during the dynamic load test analysis.

Type of loading	Horizontal direction [m·s^{-2}]	Vertical direction [m·s^{-2}]
Random pedestrian crossings	0.017	0.123
Synchronous walk of two people in frequency 2.11 Hz	0.069	0.376
Synchronous run of two people in frequency 3.42 Hz	0.069	0.397
Vandalism	0.188	0.994

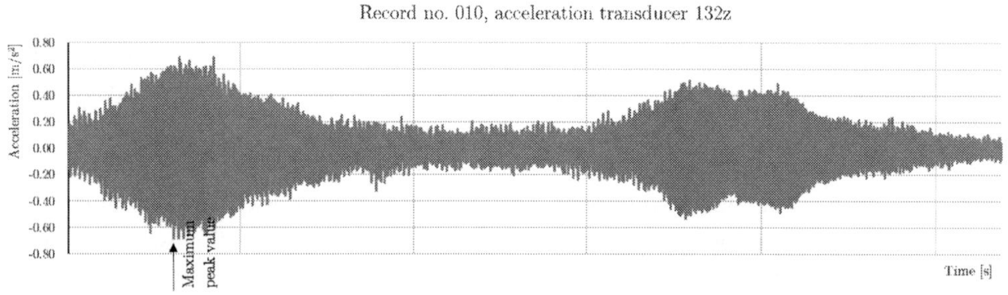

Fig. 10. The whole record no. 010, acceleration transducer 132z, type of loading: Synchronous walk of two people in frequency 2.11 Hz.

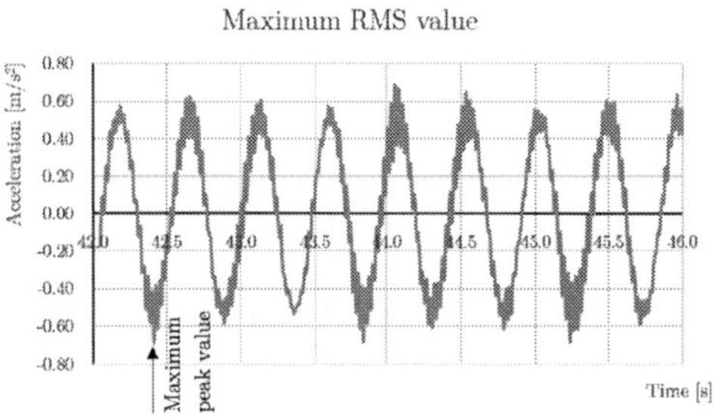

Fig. 11. The 5 s time interval of record no. 010, which maximum RMS value of acceleration was evaluated from.

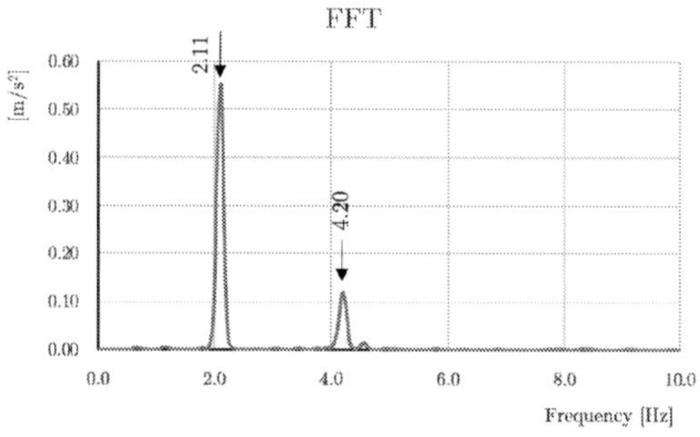

Fig. 12. The record no. 010 in the frequency domain.

4.3. Evaluation of damping

The logarithmic decrement of the damping was evaluated using FFT multispectrum for suitable record of the footbridge vibration. Since the dynamic load test did not suspend traffic on the footbridge and also the excitation by wind and microseismicity was significant, most vibration records were not suitable for damping evaluation. Therefore, only the record no. 022 focused on vandalism at 2.11 Hz was selected as suitable (see Fig. 13) and the determined value of logarithmic decrement (Fig. 14) was $\delta = 0.021$.

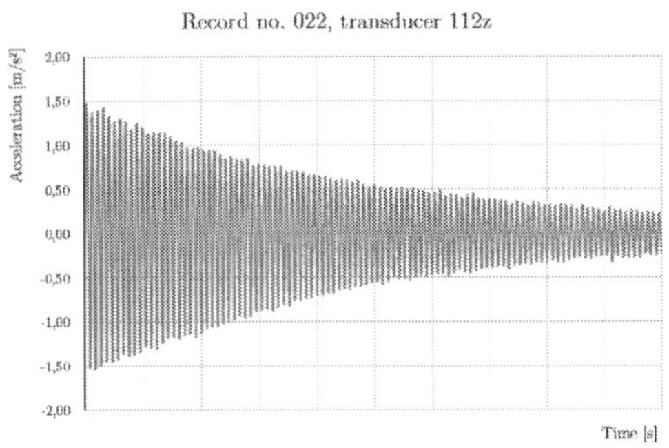

Fig. 13. Selected part of the record no. 022 (vertical acceleration in the point 112) focused on vandalism on 2.11 Hz which was used for damping evaluation.

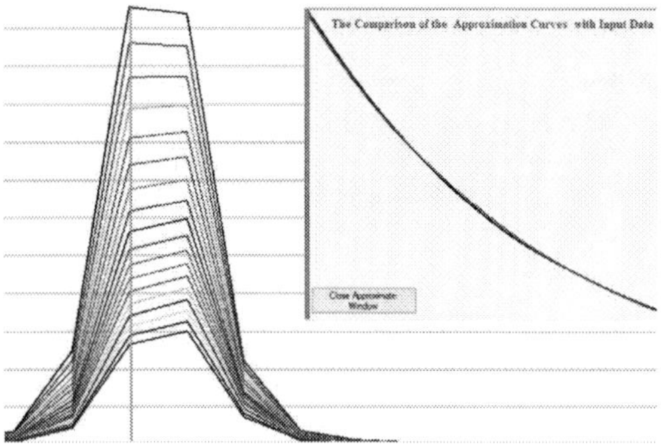

Fig. 14. The FFT Multispectra determined from the vertical acceleration shown in Fig. 13.

4.4. Assessment of the dynamic response

Evaluated maximum peak and RMS values of acceleration were assessed according to the pedestrian comfort criteria defined in Czech standard ČSN 73 6209:2019 [4], National Annex of ČSN EN 1991-2 ed.2 [5], Eurocode ČSN EN 1990:2002 [6], French document SÉTRA [7] and International standard ISO 10137:2007 [8]. Only the values from the records with type of loading, which is considered as normal pedestrian traffic according to the Czech standard ČSN 73 6209 [4], were assessed.

4.4.1. Assessment according to the Eurocode ČSN EN 1990:2002 [6]

In Eurocode ČSN EN 1990:2002 [6], the recommended pedestrian comfort criteria for footbridges are defined as maximum admissible peak acceleration values.

Table 6. Assessment of the dynamic response according to ČSN EN 1990:2002 [6].

Criterion: a [m/s^2]	Type of vibrations	Fulfils the crit.	Not fulfils the crit.
<0.7	Vertical vibrations	100.0%	0.0%
<0.2	Horizontal vibrations caused by normal traffic	100.0%	0.0%
<0.4	Horizontal vibrations caused by exceptional crowd loading	100.0%	0.0%

4.4.2. Assessment according to the National Annex of ČSN EN 1991-2 ed.2 [5]

There are four pedestrian comfort levels defined by maximum admissible peak acceleration values in National Annex of ČSN EN 1991-2 ed.2. The pedestrian comfort level in our case would comply the average comfort level.

Table 7. Assessment of the dynamic response of the footbridge deck according to the National Annex of ČSN EN;1991-2 ed.2 [5].

Pedestrian comfort level	Vertical acceleration a [m·s^{-2}]	Fulfils the criteria	Horizontal acceleration a [m·s^{-2}]	Fulfils the criteria
Maximum	≤ 0.50	85.4%	≤ 0.10	95.8%
Average	≤ 0.70	14.6%	≤ 0.20	4.2%
Minimum	≤ 1.00	0.0%	≤ 0.40	0.0%
Inadequate	< 1.00	0.0%	< 0.40	0.0%

4.4.3. Assessment according to SÉTRA [7]

There are four pedestrian comfort levels defined by maximum admissible peak acceleration values in SÉTRA [7]. The pedestrian comfort level in case of the investigated footbridge complies the average comfort level. There is also a criterion for horizontal vibration set by a maximum peak acceleration value 0.1 $m \cdot s^{-2}$ for avoiding a possibility of "lock-in effect" occurrence, only 4.2 % of evaluated values are over this criterion. It implies that the footbridge is not susceptible to appearing "lock-in effect" during the type of loading, which is considered as normal pedestrian traffic according to the Czech standard ČSN 73 6209 [4].

Table 8. Assessment of the dynamic response of the footbridge deck according to SÉTRA [7].

Pedestrian comfort level	Vertical acceleration a [$m \cdot s^{-2}$]	Fulfils the criteria	Horizontal acceleration a [$m \cdot s^{-2}$]	Fulfils the criteria
Maximum	≤ 0.50	85.4%	≤ 0.15	100.0%
Average	≤ 1.00	14.6%	≤ 0.30	0.0%
Minimum	≤ 2.50	0.0%	≤ 0.80	0.0%
Inadequate	< 2.50	0.0%	< 0.80	0.0%

4.4.4. Assessment according to the ČSN 73 6209:2019 [4]

In Czech standard ČSN 73 6209:2019 [4], the recommended pedestrian comfort criteria for footbridges are defined as maximum admissible weighted RMS acceleration values using the weight function W_m. Since the values of the weight function W_m is for the whole frequency range <1, the weight function was not used. If the evaluated RMS acceleration values fulfil the criteria without using the weight function W_m, they would surely fulfil the criteria with using the weight function.

Table 9. Assessment of the dynamic response of the footbridge deck according to ČSN 73 6209:2019 [4].

Criterion: a_{RMS} [$m \cdot s^{-2}$]	Type of vibrations	Fulfils the crit.	Not fulfils the crit.
<0.5	Vertical vibrations	100.0%	0.0%
<0.14	Horizontal vibrations caused by normal traffic	100.0%	0.0%
<0.28	Horizontal vibrations caused by exceptional crowd loading	100.0%	0.0%

4.4.5. Assessment according to ISO 10137:2007 [8]

In international standard ISO 10137:2007 [8], the recommended pedestrian comfort criteria for footbridges are defined as maximum admissible weighted RMS acceleration values. No weight function was used for the same reason as was said above (in subsubsection 4.4.4). The maximum admissible weighted RMS acceleration values are set by the graph depending on the dominant frequency in the record.

Table 10. Assessment of the dynamic response according to ISO 10137:2007.

Type of vibrations	Fulfils the crit.	Not fulfils the crit.
Vertical vibrations	100.0%	0.0%
Horizontal vibrations	100.0%	0.0%

4.5. Comparison of comfort criteria according to selected standards

There are four pedestrian comfort levels defined by maximum admissible peak acceleration values in both SÉTRA [7] and National Annex of ČSN EN 1991-2 ed.2 [5].

Table 11. Comparison of comfort criteria according to SÉTRA [7] and NA of ČSN EN 1991-2 ed.2 [5].

Pedestrian comfort level	Vertical acceleration a [m·s^{-2}]		Horizontal acceleration a [m·s^{-2}]	
	SÉTRA [7]	NA of ČSN EN 1991-2 ed.2 [5]	SÉTRA [7]	NA of ČSN EN 1991-2 ed.2 [5]
Maximum	≤ 0.50	≤ 0.50	≤ 0.15	≤ 0.10
Average	≤ 1.00	≤ 0.70	≤ 0.30	≤ 0.20
Minimum	≤ 2.50	≤ 1.00	≤ 0.80	≤ 0.40
Inadequatet	> 2.50	> 1.00	> 0.80	> 0.40

As is evident from Table 11, the criteria for pedestrian comfort level are more stringent in National Annex of ČSN EN 1991-2 ed.2 [5] in comparison with SÉTRA [7]. The author of this paper considers the criteria in National Annex of ČSN EN 1991-2 ed.2 [5] more corresponding with pedestrian comfort levels, because according to the criteria set in SÉTRA [7], the dynamic response to the type of loading "vandalism" would fulfil the criteria for minimum comfort level. However, the type of loading "vandalism" is not considered as "normal pedestrian traffic" according to the Czech standard ČSN 73 6209 [4], so the response should fit the "inadequate" acceleration range.

5. Theoretical analysis of the dynamic behavior of the footbridge

For purpose of the theoretical analysis of dynamic behavior of the footbridge a 3D model was created (see Fig. 15). The model was compiled in Dlubal RFEM version 6.01. The theoretical natural frequencies and modes were determined and compared with the results of the experimental modal analysis.

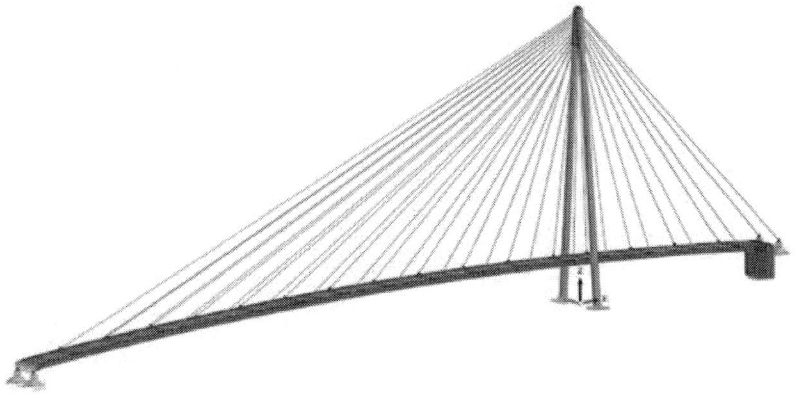

Fig. 15. The 3D model of the footbridge compiled in Dlubal RFEM.

5.1. Theoretical 3D model of the footbridge

The components were modelled as a system of longitudinal and cross girders, corresponding to beam elements, and the slab, which was modelled by the shell elements. The external prestressed cables, the pylon and the O3 abutment were all modelled by the beam elements. The stays were modelled by rod elements. When the shape or the cross-section of some elements were simplified, it was always based on the equality of the areas. All the chosen materials in the model were pickled out in accordance with the description in the realization documentation of the footbridge structure [9]. The structure was modelled without the stiffness parameters of the substructure, only the theoretical supports were used. The support on the O1 abutment consists of two longitudinal roller supports, the support on the pylon and the O3 abutment was fixed. Adding a load state was necessary for calculation of modal analysis. This load state consisted of the permanent action including the prestress of the stays and cables.

5.2. Validation of the model

The comparison of the corresponding calculated and measured natural frequencies is shown in following Table 12. Since it was not possible to measure the vibration of the pylon in the course of the experiment, the calculated natural frequencies with major vibration of the pylon were omitted from the comparison.

Table 12. Validation of the 3D model of the footbridge.

Order	Character of natural mode	Experimental $f_{(k)}$ [Hz]	Theoretical $f_{(j)}$ [Hz]	$\Delta_{(j,k)}$ [%]	Limit deviation [%]
(1)	Vertical bending mode	0.72	0.604	-19.21	-15; +10
(2)	Horizontal bending mode	0.92	0.947	2.85	±15.6
(3)	Vertical bending mode	1.19	1.050	-13.33	±15.7
(4)	Vertical bending mode	1.73	1.504	-15.03	±16.5
(6)	Torsional mode	2.11	1.925	-9.61	±17.2
(7)	Vertical bending mode	2.28	2.020	-12.87	±17.3
(8)	Horizontal bending mode	2.84	2.533	-12.12	±18.2
(9)	Vertical bending mode	2.98	2.548	-16.95	±18.2
(10)	Vertical bending mode	3.13	2.849	-9.86	±18.7
(11)	Torsional mode	3.41	3.028	-12.62	±19.0
(12)	Vertical bending mode	3.61	3.082	-17.13	±19.1
(14)	Vertical bending mode	4.20	3.589	-17.02	±19.9
(15)	Torsional mode	4.56	3.941	-15.71	±20.5
(16)	Vertical bending mode	4.89	4.041	-21.01	±20.7

As can be seen in Table 12, the limit deviation specified in the Czech standard ČSN 73 6209 [4] was not fulfilled for the first natural frequency, therefore the identification of the model was performed.

5.3. Identification of the model

The identification of the model was performed in three steps. The first one was the reduction of the bulk density of the UHPC concrete according to the real results of material experiments performed during its production. The default value of bulk density in the software was 2400 kg·m^{-3} and the real experimental value was 2380 kg·m^{-3}. The values of $\Delta_{(j,k)}$ decreased after changing the bulk density, but the limit deviation was still not fulfilled for the first natural frequency. The second step was more detail modelling of the top of the pylon, but this step did not provide lower values of $\Delta_{(j,k)}$, especially not for the first natural frequency. The third step was the adjustment of the supports. The two longitudinal roller supports were changed to the two pinned ones. This modification caused the maximum reduction of $\Delta_{(j,k)}$ and the limit deviation was already fulfilled for the first natural frequency (see Table 13). The conformity between the vast majority of the fundamental frequencies also substantially improved (see Table 12 and Table 13).

Table 13. The final step of the identification of the model.

Order	Character of natural mode	Experimental $f_{(k)}$ [Hz]	Theoretical $f_{(j)}$ [Hz]	$\Delta_{(j,k)}$ [%]	Limit deviation [%]
(1)	Vertical bending mode	0.72	0.775	7.10	-15; +10
(2)	Horizontal bending mode	0.92	0.976	5.74	±15.3
(3)	Vertical bending mode	1.19	1.118	-6.44	±15.4
(4)	Vertical bending mode	1.73	1.617	-6.99	±16.1
(6)	Torsional mode	2.11	1.955	-7.93	±16.5
(7)	Vertical bending mode	2.28	2.088	-9.20	±16.7
(8)	Horizontal bending mode	2.84	2.596	-9.40	±17.3
(9)	Vertical bending mode	2.98	2.633	-13.18	±17.4
(10)	Vertical bending mode	3.13	2.933	-6.72	±17.8
(11)	Torsional mode	3.41	3.076	-10.86	±18.0
(12)	Vertical bending mode	3.61	3.128	-15.41	±18.0
(14)	Vertical bending mode	4.20	3.672	-14.38	±18.7
(15)	Torsional mode	4.56	3.971	-14.83	±19.1
(16)	Vertical bending mode	4.89	4.126	-18.52	±19.3

5.4. *Calculated natural frequencies and modes*

All the first sixteen natural frequencies calculated after the identification of the model are summarized in previous Table 14 and corresponding theoretical natural modes on Fig. 16.

Table 14. Calculated natural frequencies after the identification of the model.

No.	$f_{(j)}$ [Hz]	Character of the natural mode
(1)	0.775	Vertical bending mode of vibration
(2)	0.976	Horizontal bending mode of vibration
(3)	1.118	Vertical bending mode of vibration
(4)	1.617	Vertical bending mode of vibration
(5)	1.822	Pylon
(6)	1.955	Torsional bending mode of vibration
(7)	2.088	Vertical bending mode of vibration
(8)	2.596	Vertical bending mode of vibration
(9)	2.633	Horizontal/Torsional bending mode of vibration
(10)	2.933	Vertical bending mode of vibration
(11)	3.076	Torsional bending mode of vibration
(12)	3.128	Vertical bending mode of vibration
(13)	3.304	Pylon
(14)	3.672	Vertical bending mode of vibration
(15)	3.971	Torsional bending mode of vibration
(16)	4.126	Vertical bending mode of vibration

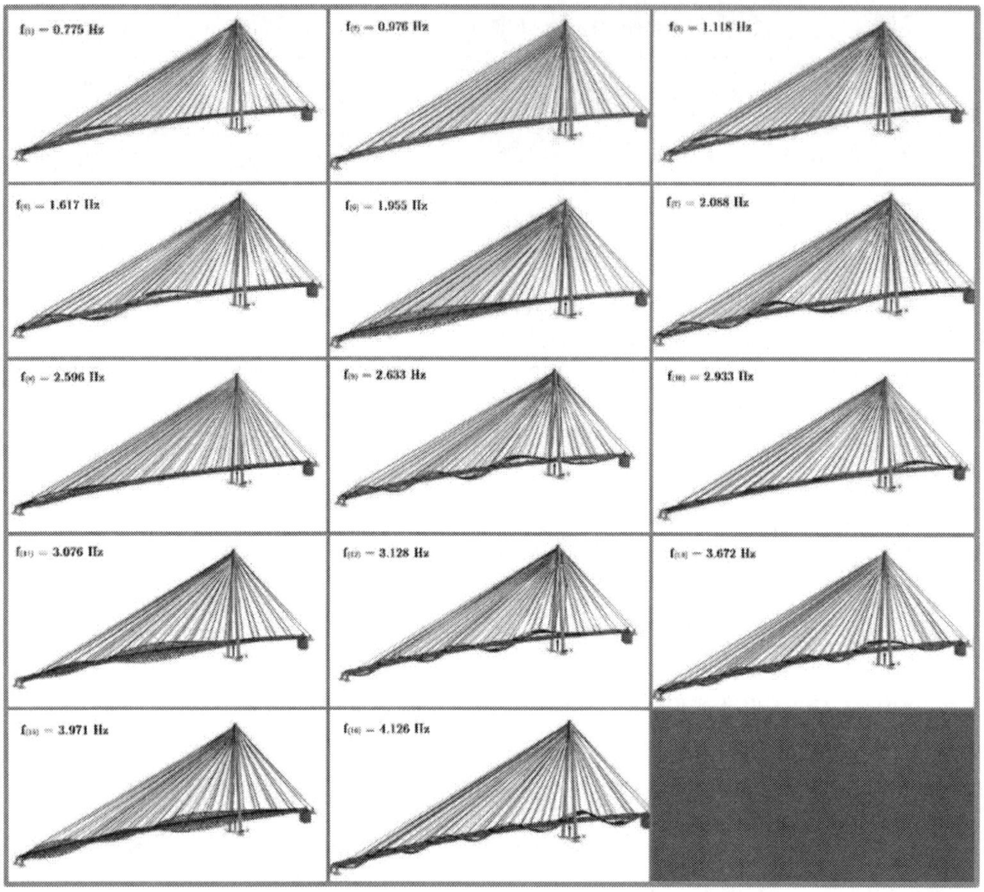

Fig. 16. Calculated natural modes of the footbridge.

5.5. Comparison of measured and calculated natural modes

The measured natural modes of the footbridge were compared with the calculated ones. The mutual assignment of the calculated and measured modes was done using the Modal Assurance Criterion (MAC). The determined values of MAC are shown in Table 15.

According to Table 15, the measured natural mode related to the natural frequency 2.77 Hz was not assigned to any calculated one. This natural mode may have dominant magnitude of vibration on the pylon and since it was not possible to measure on pylon during the experiment, the shape of this natural mode is not complete, so it is not corresponding with any of the theoretical natural mode. The calculated natural frequencies with main magnitude on pylon were omitted from the comparison.

Table 15. Comparison of calculated and measured natural modes using MAC.

k;	$f_{(k)}$[Hz]	1	2	3	4	5	6	7	8	9	10	11	12	13	14	15
j	$f_{(j)}$[Hz]	0.72	0.92	1.19	1.73	2.11	2.28	2.77	2.84	2.98	3.13	3.41	3.61	4.20	4.56	4.89
1	0.78	0.99	0.01	0.00	0.01	0.00	0.00	0.02	0.03	0.00	0.00	0.00	0.00	0.00	0.00	0.00
2	0.98	0.00	0.92	0.00	0.00	0.01	0.00	0.00	0.03	0.00	0.00	0.01	0.00	0.00	0.00	0.00
3	1.12	0.00	0.00	0.99	0.02	0.00	0.00	0.02	0.03	0.01	0.00	0.00	0.00	0.00	0.00	0.00
4	1.62	0.00	0.00	0.00	0.95	0.00	0.06	0.04	0.00	0.00	0.00	0.00	0.00	0.00	0.00	0.00
6	1.96	0.00	0.01	0.00	0.00	0.98	0.00	0.00	0.00	0.00	0.00	0.01	0.00	0.00	0.00	0.00
7	2.09	0.00	0.01	0.00	0.02	0.00	0.89	0.43	0.04	0.00	0.00	0.00	0.00	0.00	0.00	0.00
8	2.60	0.00	0.03	0.00	0.00	0.01	0.00	0.00	0.64	0.00	0.00	0.01	0.00	0.00	0.00	0.00
9	2.63	0.00	0.00	0.00	0.00	0.00	0.03	0.45	0.04	0.64	0.17	0.00	0.06	0.00	0.00	0.00
10	2.93	0.00	0.00	0.00	0.00	0.00	0.00	0.04	0.00	0.30	0.75	0.00	0.05	0.00	0.00	0.00
11	3.08	0.00	0.00	0.00	0.00	0.00	0.00	0.00	0.03	0.00	0.00	0.93	0.00	0.00	0.00	0.00
12	3.13	0.00	0.00	0.00	0.01	0.00	0.00	0.03	0.00	0.00	0.06	0.00	0.79	0.11	0.00	0.01
14	3.67	0.00	0.00	0.00	0.00	0.00	0.00	0.00	0.03	0.00	0.00	0.00	0.09	0.81	0.00	0.11
15	3.97	0.00	0.00	0.00	0.00	0.00	0.00	0.00	0.00	0.00	0.00	0.03	0.00	0.00	0.96	0.00
16	4.13	0.00	0.00	0.00	0.00	0.00	0.00	0.00	0.02	0.00	0.00	0.00	0.00	0.08	0.00	0.83

6. Conclusions

The results of both experimental and theoretical analysis of the dynamic behavior of the footbridge in Lužec nad Vltavou town can be summarised to the following conclusions.

The differences between compared natural frequencies measured in 2020 and in 2021 are very low (see Table 2), but in 2020 some natural frequencies of the footbridge structure were not captured or evaluated, e. g. the frequency $f_{(8)}$ = 2.84 Hz. The minimum value of the differences between natural frequencies measured in 2020 and in 2021 is indicative of the fact that the important properties of the horizontal load bearing structure did not change substantially between 2020 and 2021.

The evaluated damping of the horizontal load-bearing structure of the footbridge is surprisingly small (logarithmic decrement $\delta = 0.021$). However, the damping was successfully evaluated from measured data for natural frequency 2.11 Hz only and for other natural frequencies it may be different. Though, the natural frequency 2.11 Hz is the most important for the level of the footbridge deck response to the dynamic pedestrian load. The fact that the damping is small, does not influence negatively the level of usual vibration of the footbridge deck, as the criteria of comfort level of pedestrians during the "normal pedestrian traffic" type of loading were fulfilled.

The dynamic response was assessed according to current standards, regulations and recommendations. The criteria for pedestrian comfort level were fulfilled for 100% evaluated values within the "normal pedestrian traffic" type of loading. The pedestrian comfort level was identified as "average" according to [5, 7].

The theoretical 3D model was created and natural frequencies and modes were calculated. The validation of the model was done. The limit deviation was not fulfilled for the first natural frequency. Therefore, the identification of the model was consequently performed. Two crucial steps of the identification process of the model were the change of the bulk density of the footbridge deck and the adjustment of the support.

The measured and calculated natural frequencies were compared mutually using the limit deviation by [4]. After the identification of the model the limit deviation was fulfilled for all evaluated natural frequencies. The natural modes were compared using the Modal Assurance Criterion (MAC). The identified model can be used for more advance theoretical analysis of the dynamic behavior of the footbridge.

Acknowledgements

The results presented in this paper are outputs of the research project No. SGS22/089/OHK1/2T/11 supported by the Czech Technical University in Prague.

References

[1] Polák M., Plachý T., Čítek A., Berková K., Hájková P. and Čáp M., "Experimental dynamic analysis of the existing footbridge in Dobřichovice town," *MATEC Web of Conferences*, Vol. 313, p. 2 (2020). ISSN 2261-236X. http://dx.doi.org/10.1051/matecconf/202031300002.

[2] Šána V., Polák M. and Plachý T., "A Dynamic Analysis of the Cable-Stayed Footbridge in Čelákovice Town," *Procedia Engineer.*, Vil. 199, pp. 28772882 (2017). ISSN 1877-7058. http://dx.doi.org/10.1016/j.proeng.2017.09.582.

[3] Karásek P., Polák M., Plachý T., Němec P., Kalný M. and Studničková M., "Dynamic Behaviour of a Steel Cable-stayed Footbridge Damped by TMD," in Proceedings of the 6th International Conference on Structural Dynamics EURODYN 2005, Paris, 2005 (Millpress Science Publishers, Rotterdam, 2005), pp. 521–526. ISBN 90-5966-033-1.

[4] ČSN 73 6209, "Zatěžovací zkoušky mostních objektů," Prague: Úřad pro technickou normalizaci, metrologii a státní zkušebnictví, (2019).

[5] National Annex of ČSN EN 1991-2 ed. 2 "Eurokód 1: Zatížení konstrukcí – Část 2: Zatížení mostů dopravou," Prague: Úřad pro technickou normalizaci, metrologii a státní zkušebnictví, (2019).

[6] ČSN EN 1990, "Eurokód: Zásady navrhování konstrukcí," Prague: Český normalizační institut, (2004).

[7] SÉTRA, "Assessment of vibrational behaviour of footbridges under pedestrian loading, Technical guide", Paris: Ministry of Transport and Infrastructure, The Technical Department for Transport, Roads and Bridges Engineering and Road Safety, (October 2006).

[8] ISO 10137:2007, "Bases for design of structures - Serviceability of buildings and walkways against vibrations," International Organization for Standardization, (2004).

[9] Novák & Partner, s. r. o. "Realization documentation of the structure (RDS): The footbridge across the river Vltava in Lužec and Vltavou," (Prague, 2020).

Methods and Means in Teaching of Experimental Methods

Eliška Cézová[1]

Abstract: This paper describes how the computer technology can be used for preparation and processing of various technical reports and protocols. Such computer aided process can for example be used in the engineering design, processing of experiments, or in the technical education. The approach chosen in this work suggests the use of a combination of specialized software tools for data processing and typesetting of technical reports. The freely available open-source software Python and LaTeX is presented here as a viable choice offering numerous advantages over the widely adopted general purpose software tools, using standard office spreadsheet and text editors. The use of this software is demonstrated here on a simple mechanical engineering design test case, taught in the elementary machine design courses at technical universities. The paper shows how the software tools can be used to simplify and automate the repetitive (routine) activities using simple templates and automatically generated data files. This approach leads to significant time saving and reduction of technical personnel workload and overall costs.

Keywords: Teaching; Experimental Methods; Python; LaTeX

1. Introduction

The data processing and preparation of related technical reports is equally important in engineering design as well as in various experiments. This is one of the main reasons why technical reporting and data processing also plays an important role in technical education at specialized high schools and universities. The applied methods and tools are mostly identical technical praxis and education. This is why most of the material presented below, related to teaching, can also be used in practical engineering.

The aim of this paper is to present the use of freely available software in the preparation of technical documentation or methodological aids for teaching technical subjects.

In teaching of technical subjects, the preparation of courses and lessons is rather complex and time-consuming. Individual lessons must be internally interrelated and the whole course has to meet certain set of objectives. When preparing such a course, the following questions must be considered:

- What to include in the topic plan, i.e. what will be the lesson plan?
- How to structure the lesson logically?
- What are the objectives of each lesson?
- What should the students learn?
- How are the students supposed to learn it?
- What didactic aids can/will be used?
- How to verify the achievement of the learning objective?
- What homework assignment to give?
- What will be the continuation of the lesson?

[1] *Department of Designing and Machine Components, Faculty of Mechanical Engineering, Czech Technical University in Prague; Technická 4, 166 07 Prague, Czech Republic;* eliska.cezova@fs.cvut.cz

- Is the teaching supported by computer (e-learning)?

This list shows how complicated and time consuming the preparation of a course and individual lectures usually is. This is why any tools and methods that will help to reduce lecturers workload are of crucial importance, allowing to focus on the substance of the subject, leaving the repetitive, mechanically performed tasks to automatic devices and computers.

As an example, consider the subject of Mechanical Engineering Design, which is typically taught in the elementary course at many technical universities. The aim of this course is to introduce students to the principles of engineering design and the creation of technical documentation. Machine designing is naturally connected with appropriate calculations, proving that the essential design objectives were met and technical parameters achieved. In addition to the mastering the theoretical and computational apparatus of the design process, students are also introduced to the computational software tools that can facilitate design process. An integral part of the design process is the technical reporting, summarizing and recording all steps of the design procedure. Also at this stage, the appropriate software and IT tools can significantly facilitate the preparation, processing and presentation of the final report. For both, the computational and reporting tasks, there exists a wide choice of available software tools, from which it is advisable to choose those that will effectively meet the essential requirements (cost, laboriousness, possibility of automation,...).

General purpose commercial products from the MS Office family (Word and Excel) are often used to process simple computations and technical reports. For the evaluation of a larger sample of statistical data, it is preferable to use more sophisticated tools, like for example Minitab or Statistica for larger data sets and more sophisticated statistical processing.

The choice of software tools can also be approached from the point of view of the possible automation of the processing of the document and the calculations contained therein. This is particularly important if it is clear in advance that the calculation report or the teaching material will be further modified/extended in different variants. In this case it may be of interest to use the combination of Python for data processing and LaTeX for report preparation. This software may initially appear complex, but once mastered, its robustness and efficiency will outweigh the initial complications. This combination of software tools becomes even more efficient for repetitive processing of similar tasks, which can easily be scripted and automated on almost arbitrary computer platform. There are of course other possible software combinations that are commonly used for SW support and some level of "automated" processing:

- Word + Excel,
- Matlab + LaTeX,
- R + LaTeX,
- Python + LaTeX.

2. Methods of teaching

Teaching is the most important part of all primary, secondary and higher education. Its cornerstone is the transmission of theoretical knowledge and practical experience. From them, students are supposed to understand the laws that they subsequently encounter in everyday practice. In the context of vocational training, it is appropriate to extend the interpretation of the theoretical basis with practical demonstrations. Using the same procedures, it is also possible to save time in the preparation and evaluation of laboratory tasks from the position of the teacher. It is possible to entrust the control of the results of the tasks to a specialized program that is able to evaluate them. Currently, the Moodle platform is often used for this purpose. The weakness of this approach, however, is that students may share test results with each other and, without active knowledge of the subject matter, simply hand in the results. In this respect, the use of customized individual tests with their generation also can help the presented approach. Another area for its use is also the generation of a solution to the assignment, which reveals to the students the recommended procedure including its numerical evaluation. Such "help" can be provided to students, e.g. after failing a test. Thus, students can find out on their own where they have made mistakes. This approach also took on a whole new dimension at a time of massive deployment of distance learning, when the space for individual instruction was significantly limited.

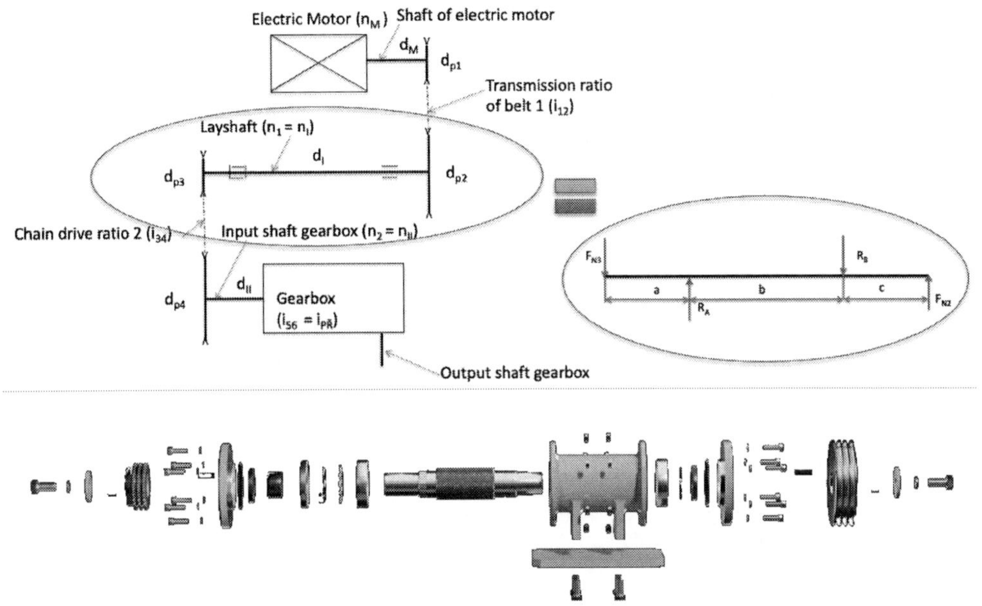

Fig. 1. Schematic of the transmission including an example of a layshaft.

3. Example of using LaTeX and Python

Python is a modern, versatile and high-level programming language. It can be learned very quickly. It has a simple and clean syntax. It can be used for all common platforms (Windows, Linux, Unix, MacOS) and is available for free. There are many examples published on the Internet, which you can just take and add your own formulas. Python can also display images that can be easily inserted into LaTeX. In this way, the above mentioned automation will occur with all the benefits cited. This can also be used to generate customized assignments and term paper assignments for students to complete.

In the following text, we will describe in detail the general procedure for creating documentation or documentation in Python and LaTeX software.

- Preparation of the documentation (what we want to teach the students and what should be the result)
- Preparation of a template in LaTeX - technical report including formulas and their completion
- Preparation of calculations in Python

- Merging LaTeX and Python
- Checking the results

The presented example comes from the design of a transmission involving a layshaft, see Fig. 1, for which the bending moment must be calculated.

```
\documentclass{report}
\usepackage[utf8]{inputenc}
\pagenumbering{arabic}
\usepackage{graphicx}
\usepackage[czech]{babel}
\usepackage{amsmath,systeme}
\usepackage{mathtools}
\usepackage{a4wide}

\include{config}

\begin{document}
........
.....
\end{document}
```

```
For field II. $x_2 = \langle a; a+b \rangle$, $x_2 =
\langle \aI; \aI+\bI \rangle$ (procedure from the left)
\\
Calculation of shear force  in field II \\
$T_{II}(x_2) = F_{RET} - R_A$\\
Calculation of shear force at $0$\\
$T_{II}(a) = F_{RET} - R_A= \FxxxRET - \RAI =
\TIIxxxa\, [N]$\\
Calculation of shear force at $a+b$\\
$T_{II}(a+b) = F_{RET} - R_A = \FxxxRET - \RAI =
\TIIxxxab \,[N]$\\
\\
Calculation of the bending moment at the general
location $x$ (at the cut point)\\
$Mo_{II}(x_2) = - F_{RET} \cdot x_2 + R_A (x_2 - a) =
$ \\
Calculation of the bending moment at $a$ \\
$Mo_{II}(a) = - F_{RET} \cdot a + R_A (a - a)
= - \FxxxRET \cdot \aI + \RAI (\aI - \aI) = \MoIIxxxa
\;[Nmm]$ \\
Calculation of the bending moment at $a+b$ \\
$Mo_{II}(a+b) = - F_{RET} \cdot (a+b) + R_A (a+b - a) =
- \FxxxRET \cdot (\aI+\bI) + \RAI (\aI+\bI - \aI) =\\
=\MoIIxxxab \;
[Nmm]$ \\
```

Fig. 2. Extract from LaTeX.

For field II. $x_2 = \langle a; a+b \rangle$, $x_2 = \langle 40; 40+60 \rangle$ (procedure from the left)
Calculation of shear force in field II
$T_{II}(x_2) = F_{RET} - R_A$
Calculation of shear force at 0
$T_{II}(a) = F_{RET} - R_A = 100 - 233.33 = -133.33\,[N]$
Calculation of shear force at $a+b$
$T_{II}(a+b) = F_{RET} - R_A = 100 - 233.33 = -133.33\,[N]$

Calculation of the bending moment at the general location x (at the cut point)
$Mo_{II}(x_2) = -F_{RET} \cdot x_2 + R_A(x_2 - a) =$
Calculation of the bending moment at a
$Mo_{II}(a) = -F_{RET} \cdot a + R_A(a - a) = -100 \cdot 40 + 233.33(40 - 40) = -4000.00\,[Nmm]$
Calculation of the bending moment at $a+b$
$Mo_{II}(a+b) = -F_{RET} \cdot (a+b) + R_A(a+b-a) = -100 \cdot (40+60) + 233.33(40+60-40) =$
$= 4000.00\,[Nmm]$

For field III. $\overline{x}_3 = \langle \overline{0}; \overline{c} \rangle$, tj. $\overline{x}_3 = \langle \overline{0}; \overline{40} \rangle$ (procedure from the right)
Calculation of shear force at the general location $\overline{x}$ (at the cut point)
$T_{III}(\overline{x}_3) = -F_{REM}$
Calculation of shear force at the site $\overline{0}$
$T_{III}(\overline{0}) = -F_{REM} = -100\,[N]$
Calculation of shear force at the site $\overline{c}$
$T_{III}(\overline{c}) = -F_{REM} = -100\,[N]$

Fig. 3. Compiled PDF document from LaTeX (calculation of shear force and moments).

Fig. 2 shows the preparation of the LaTeX report. On the left is the basic template used for writing the report. This is followed by `\begin{document}`, where the technical message is written. The values of variables will be computed and inserted by Python into LaTeX source file via the included text file `config.tex`. In this way, the variables such as aI, FxxxRET or RAI are replaced in the report by their numerical values. Then the `\end{document}` is written at the end of the main LaTeXsource file. More details regarding the LaTeXlanguage structure, syntax and code processing can be found in numerous resources on the internet. Most of the documentation, templates and examples are freely available online, including installation packages of the associated compilers and editors. Fig. 3 shows a PDF document created by compiling the LaTeXsource code, which is part of the technical report. The complete technical report for the calculation of the layshaft is not shown here.

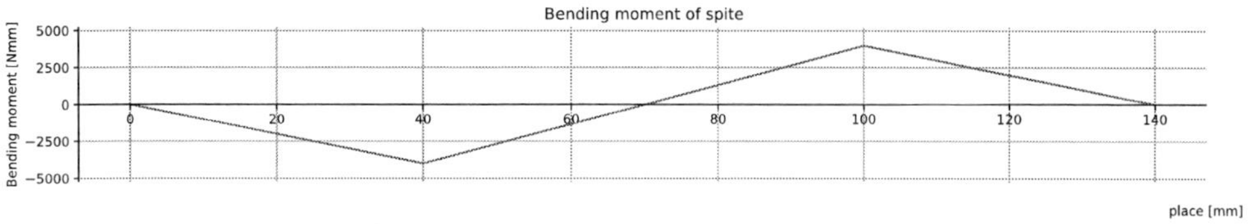

Fig. 4. Graph of the bending moment along the layshaft.

```python
import numpy as np

def zaokrouhleni(x,m):
    multiplier = 10**m
    a = x * multiplier
    a = np.rint(a)
    vysledek = str(a/multiplier)
    if '.0' in vysledek and m <= 0:
        return (vysledek[0:-2])
    elif vysledek[-2] == "." and m > 1:
        vysledek = vysledek + (m-1)*'0'
        return vysledek
    else:
        return vysledek

def nameStripper(x):
    x = x.strip()
    x = x[6:]
    x = x[:-7]
    return x

puvodniSoubor = open("Vypis_Promennych.txt", 'r', encoding="utf-8")
koncovySoubor = open("config.tex", 'w', encoding="utf-8")
puvodniSoubor.readline()
puvodniSoubor.readline()
```

Fig. 5. Sample of a Python code.

```latex
\newcommand{\DxxxIIn}{25}
\newcommand{\DxxxIIp}{24.95}
\newcommand{\DxxxIL}{30}
\newcommand{\DxxxIn}{20}
\newcommand{\DxxxIp}{19.43}
\newcommand{\Email}{Eliska c@email.cz}
\newcommand{\FxxxREM}{100}
\newcommand{\FxxxRET}{100}
\newcommand{\Lxxhod}{20000}
\newcommand{\MKI}{35980.97}
\newcommand{\MKII}{51812.59}
\newcommand{\Mkm}{19554.87}
\newcommand{\MoIIIxxx0}{0}
\newcommand{\MoIIIxxxc}{4000}
\newcommand{\MoIIxxxa}{-4000.00}
\newcommand{\MoIIxxxab}{4000.00}
\newcommand{\MoIxxx0}{0}
\newcommand{\MoIxxxa}{-4000}
\newcommand{\Momax}{4000}
\newcommand{\PI}{2760.00}
\newcommand{\PII}{2649.60}
\newcommand{\Pm}{3000}
\newcommand{\RAI}{233.33}
\newcommand{\RBI}{233.33}
```

Fig. 6. List of variables from config.tex.

Fig. 4 shows the bending moment diagram created using the Python for the considered layshaft. Fig. 5 shows a sample Python output. Here, a variable list, the config.tex file, is generated, which will further be inserted into the main LaTeX source code template shown above. Fig. 6 shows a listing of variables from the config.tex file that is loaded into the main LaTeX source file. The added variables in LaTeX from the config.tex file are shown in Fig. 3.

4. Conclusion

Although all the necessary know-how and software is freely available, the above procedure is not commonly used in schools and companies. The presented approach could inspire other lecturers and technicians and make their work easier, thus contributing to the overall improvement of teaching.

Acknowledgements

This work was partly funded by the Czech Ministry of Education, under the Institutional support for the development of research organization, No.:RVO12000 for Faculty of Mechanical Engineering of the Czech Technical University in Prague.

References

[1] Goossens M., Rahtz S. and Mittelbach F., *The LATEX Companion*, (Addison- Wesley, 1994).

[2] Cézová E., "Design concept of sandwich beams deflection,"in Experimental Stress Analysis - 57th International Scientific Conference, EAN 2019 - Conference Proceedings, (Česká společnost pro mechaniku, Prague, 2019), p. 19-23. ISBN 9788021457669.

[3] Pecinovský R., *Python - Kompletni příručka jazyka pro verzi 3.10*, (Grada, Prague 2021). ISBN 978-80-271-3442-7.

Residual Stresses and Microstructural Parameters of Laser Hardened Railway Axles

Jiří Čapek[1], Karel Trojan[2], Jan Kec[3], Nikolaj Ganev[4] & Ivo Černý[5]

Abstract: The paper describes the influence of laser hardening on the state of residual stress, microstructural parameters, and fatigue life of laser hardened railway axles made of EA1N steel. The difference between single- and multi-track hardened surfaces has been found. Tensile residual stresses and low dislocation density are located on the overlapping tracks and at the boundary of the heat-affected zone and the bulk surface. These critical zones, accompanied by a decrease in hardness, can be a potential area for crack initialization. Therefore, it is important to have these critical zones outside the area typical for the initiation of fretting fatigue caused by operation of press fitted railway wheels.

Keywords: Steel for railway axles; Laser hardening; X-ray diffraction; Microstructure; Dislocation density; Hardness; Fatigue

1. Introduction

Railway wheels are usually connected to axles by press fitting, and thus failure with fatal consequences can occur on the axle seats due to ongoing mechanical processes during operation [1]. Surface treatment of axles using laser hardening could improve the resistance to fatigue failure. Therefore, it is important to achieve optimal characteristics of the surface integrity of the axle seat in critical zones, including fatigue resistance, which is positively affected by compressive residual stresses, sufficient hardness, and a suitable microstructure [2].

2. Experimental details

For experimental tests, a 1:3 axle seat model forged from EA1N steel was used. Three samples were prepared with different surface treatments: ground, single-, and multi-track hardened. The laser hardening was carried out circumferentially, with a controlled hardening temperature of 1150 °C and a beam speed of 3 mm/s. The area of the inner edge of the axle seat (side of the axle shaft) was analysed; see Fig. 1. Metallographic specimens were hot embedded into conductive bakelite resin with carbon filler and prepared by machine grinding and polishing. Structure was revealed by etching in 2% Nital (98 ml ethanol + 2 ml HNO_3). The microstructure was investigated using the *Zeiss Axio Observer Z1M* optical microscope.

Macroscopic residual stresses (RS) were determined using the *X'Pert Pro MPD* diffractometer by X-ray diffraction (XRD) and chromium radiation. The *Rachinger's* method was used to separate the diffraction lines $K\alpha_1$ and $K\alpha_2$ of the planes *{211}* of the ferrite phase and diffrac-

[1] *Ing. Jiří Čapek, Ph.D.; Department of Solid State Engineering, Faculty of Nuclear Sciences and Physical Engineering, Czech Technical University in Prague; Trojanova 13, 120 00 Prague, Czech Republic;* jiri.capek@fjfi.cvut.cz
[2] *Ing. Karel Trojan; Department of Solid State Engineering, Faculty of Nuclear Sciences and Physical Engineering, Czech Technical University in Prague; Trojanova 13, 120 00 Prague, Czech Republic;* karel.trojan@fjfi.cvut.cz
[3] *Ing. Jan Kec; SVÚM a.s., Tovární 2053, 250 88 Čelákovice, Czech Republic;* kec@svum.cz
[4] *prof. Ing. Nikolaj Ganev, CSc.; Department of Solid State Engineering, Faculty of Nuclear Sciences and Physical Engineering, Czech Technical University in Prague; Trojanova 13, 120 00 Prague, Czech Republic;* nikolaj.ganev@fjfi.cvut.cz
[5] *Ing. Ivo Černý, Ph.D.; SVÚM a.s., Tovární 2053, 250 88 Čelákovice, Czech Republic;* ivo.cerny@seznam.cz

tion lines $K\alpha_1$ were fitted by the *Pearson VII* function. The *Winholtz & Cohen* method and the X-ray elastic constants s_1 = -1.25 TPa^{-1}, $^1/_2 s_2$ = 5.76 TPa^{-1} were used for the stress calculation.

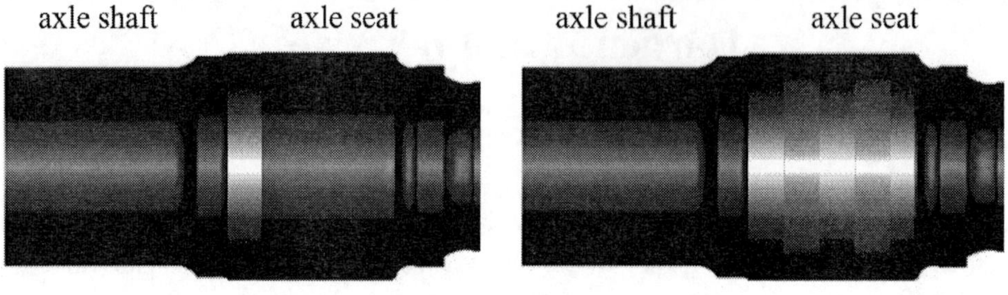

Fig. 1. Axle seat with marked single-track (left) and multi-track (right) hardening.

The dislocation density was calculated according to the *Williamson and Smallman* method [3] based on the crystallite size (size of coherently diffracted domains) and microstrain (inhomogeneous distribution of elasto-plastic deformation on a microscale) which were determined from the XRD patterns obtained using cobalt radiation and the Rietveld refinement, performed in *MStruct* software [4]. The irradiated volume, where the information about the discussed data came from, was defined by the geometry of the experiment, the effective penetration depth of the X-ray radiation (approx. 5 and 10 μm for chromium and cobalt radiation, respectively), and the pinhole size (4×0.5 mm^2).

Hardness was evaluated as Vickers hardness HV10 using the *Vickers Limited HTM*. A sample with multi-track hardening was selected for hardness measurements.

The rotational bending fatigue tests were carried out on the *Schenck UMBI 750* test machine. The tests were carried out at a stress ratio R = -1 (tension-compression) and a load frequency of 23.8 Hz.

3. Results and Discussions

Detailed XRD analyses of the surface state of residual stress and dislocation density of laser-hardened and ground samples are depicted in Fig. 2. Figures show the correlation with the previous work [5], where the compressive residual stresses were determined on the hardened surface, high tensile RS on the ground surface, and a sharp maximum of crystallite size at the boundary of the heat-affected zone and the bulk (non-treated) surface. Generally, this area shows worse mechanical properties such as hardness and mainly fatigue strength.

In the case of multi-track hardening, the correlation with the work [6] was found too. The distributions of the investigated surface integrity parameters are the worst at the boundary between two laser tracks, namely, the preceding track and the subsequent overlapping track. In this area, the tensile RS and a sharp crystallite size maximum were observed. However, contrary to [6], the tensile RS were found on the hardened surface. This inconsistency could be caused by different hardening conditions, e.g., the proportion of laser power and the amount of material for heat-conducting.

The dislocation density of the laser hardened surface is higher compared to the ground surface. Moreover, Figs. 2 show a lower dislocation density localized approx. at the distance of 11 mm and 18 mm from the origin for multi- and single-track samples, respectively, which is caused by tempering for both cases. Strain hardening is an important strengthening process that involves plastically deforming the material during manufacturing to greatly increase the number of dislocations. Therefore, zones with lower dislocation density are usually accompanied by a decrease in hardness [7] and are prone to the initialization of surface fatigue cracks.

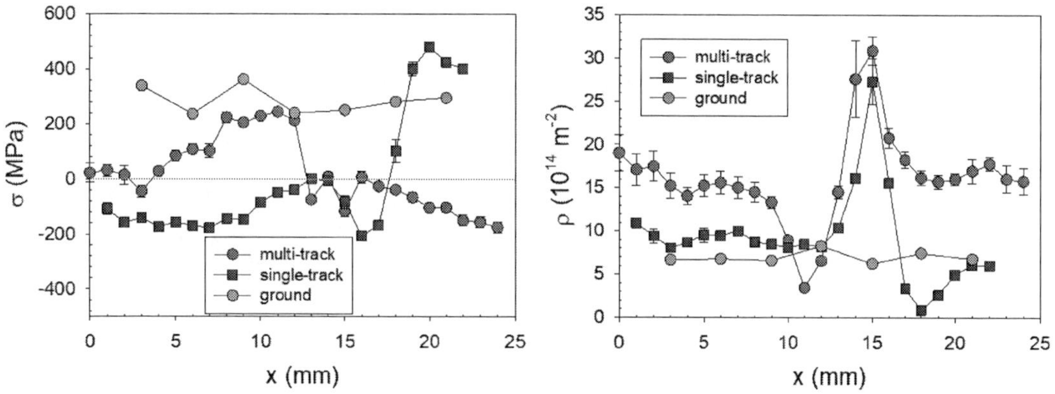

Fig. 2. Comparison of residual stresses σ and dislocation density ρ for ground, single-track, and multi-track hardened samples, where x = 0 mm denotes the inner edge of the axle seat — side of the axle shaft.

The course of hardness (Fig. 3) is consistent with the XRD multi-track measurements shown in Fig. 2, where the highest hardness 525 HV10 was observed at a distance of 4 mm. Subsequently, there is a slight decrease in the hardness to values around 440 HV10. In the overlapping area (at the distance of 11 mm), the hardness dropped to 270 HV10 which correlates with the area of the dislocation density minimum. As mentioned in previous work [6], the hardness drop in the overlapping layers cannot be avoided, however, even the lowest measured hardness values exceed the values of the bulk material, which is approximately 170 HV.

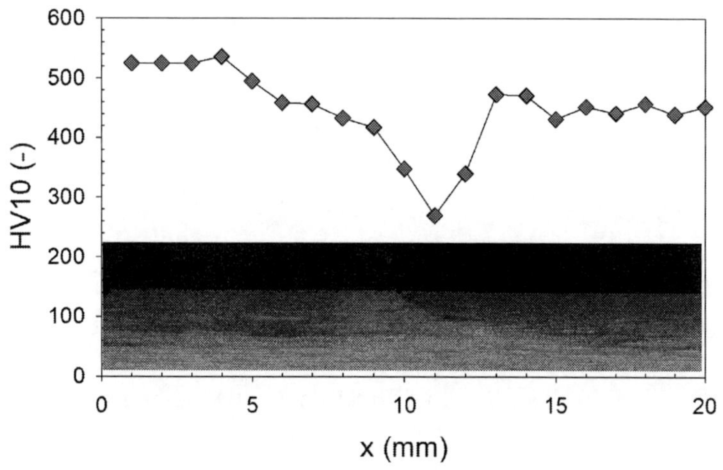

Fig. 3. The course of hardness in comparison with macrostructure image.

The macrostructure of two laser tracks can be observed in Fig. 4. The depth of the laser-hardened tracks is about 1.5 mm. A slight inhomogeneity in the carbon distribution can be observed in the hardened region, which was detected optically. Typical microstructures of the individual observed areas can be observed in Fig. 5. The hardened track typically shows a needle-shaped martensitic structure in which undissolved pearlite or Widmanstätten ferrite can be occasionally observed. The martensitic grains are occasionally quite large, as a consequence of the large grains of perlite in the parent material. In the tempered region, tempered cubic martensite can be seen with a smaller proportion of retained austenite. At the beginning of the 2nd laser track, i.e., where the structure is re-hardened (overlapped region), it can be seen that the martensitic structure was dissolved in the mixture of ferrite and complex carbides.

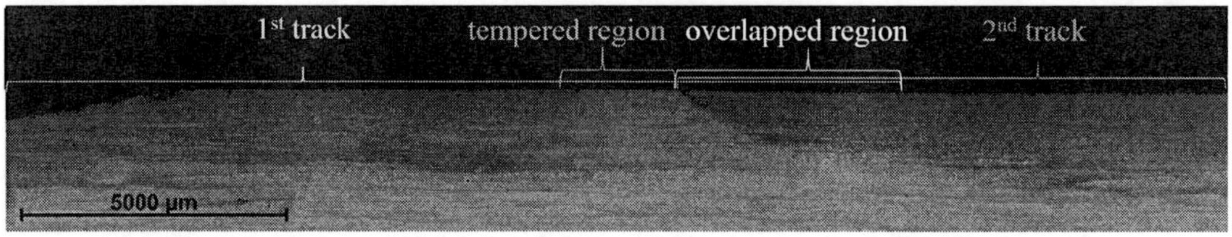

Fig. 4. Macrostructure image of laser tracks, overlapped and tempered regions.

Fig. 5. Microstructures of laser track, overlapped and tempered region.

The standard minimum stress amplitude for axle rotational bending tests is 120 MPa and the axle must withstand at least 10^7 cycles without failure. This was not achieved for the ground specimen; without laser hardening. For the single-track hardened specimen, 10^7 cycles were achieved at a stress level of 135 MPa and failure occurred at a stress level of 150 MPa. For the multi-track specimen, failure occurred at a stress level of 165 MPa. Typically, fractures occur under the press fit at a distance between 4.5 and 7.5 mm from the inner edge of the axle seat. This is a typical initiation of the fretting fatigue mechanism.

4. Conclusions

Hardened and ground surfaces were analysed and compared to find the best combination of properties to delay surface crack initialization. Both the compressive and tensile residual stresses are present on the hardened track surface in a direction parallel to the axial axis of the axle shaft, which is more critical from a technological point of view. However, high values of tensile residual stresses and/or dislocation density (related to crystallite size and microstrains) were located on the overlapped track and at the boundary between the track and bulk material of single-track. These zones are accompanied by a decrease in hardness and can be a potential area for crack initialization. Therefore, these critical zones would be suitable for a detailed study and for analysing which of these two treatment options is the most appropriate from a fatigue strength point of view. Both laser-hardened samples have better properties from the residual stresses and dislocation density points of view than the ground one.

The high-cycle fatigue tests showed that the press fitting operation of the railway wheel caused the maximum load initialization in the hardened area. The fatigue life of hardened axle seats with press fitted wheels is higher in comparison with non-treated axle seat. It is of utmost importance that the localization of the mentioned critical zones is outside the area, typical for fretting fatigue initiation.

Acknowledgements

This work was supported by the project TH04010475 of the Technology Agency of the Czech Republic. K.T.'s work was supported by the Grant Agency of the Czech Technical University in Prague, grant No. SGS22/183/OHK4/3T/14.

References

[1] Gürer G. and Gür C.H., "Failure Analysis of Fretting Fatigue Initiation and Growth on Railway Axle Press-Fits," in *Engineering Failure Analysis*, Vol. 84, pp. 151–166 (2018). https://doi.org/10.1016/j.engfailanal.2017.06.054.

[2] Linhard V. and Černý I., "An Effect of Strength of Railway Axle Steels on Fatigue Resistance Under Press Fit," in *Engineering Fracture Mechanics*, Vol. 78, pp. 731–741 (2011). https://doi.org/10.1016/j.engfracmech.2010.11.023.

[3] Williamson G.K. and Smallman R.E., "Dislocation densities in some annealed and cold-worked metals from measurements on the X-ray Debye-Scherrer spectrum," *Philosophical Magazine*, Vol. 1, pp. 34–46 (1955).

[4] Matěj Z., Kadlecová A., Janeček M., Matějová L., Dopita M. and Kužel R., "Refining bimodal microstructure of materials with MSTRUCT," *Powder Diffraction*, Vol. 29(S2), pp. S35–S41 (2014). https://doi.org/10.1017/S0885715614000852.

[5] Trojan K., Čapek J., Ganev N., Černý I., Kec J. and Němeček S., "Residual stresses of laser-hardened railway axles", in Proceedings of Experimental Stress Analysis Conference, online, 2020 (VŠB — Technical University of Ostrava, 2020), pp. 547–554. ISBN 978-80-248-4451-0.

[6] Čapek J., Trojan K., Černý I., Kec J., Ganev N. and Němeček S., "Influence of Overlapping on Surface Integrity Parameters of Laser-Hardened Railway Axles", in Proceedings of Experimental Stress Analysis Conference, Litomyšl, 2021 (CTU, Faculty of Civil Engineering, Prague, 2021), pp. 1–4. ISBN 978-80-01-06884.

[7] Martínez S., Lesyk D., Lamikiz A., Ukar E. and Dzhemelinsky V., "Hardness simulation of over-tempered area during laser hardening treatment," in *Physics Procedia*, Vol. 83, pp. 1357–1366 (2016). https://doi.org/10.1016/j.phpro.2016.08.143.

Non Standard Experimental Procedures for Design and Verification of Thermoplastic Structure Made of Hot Forming Process

Radek Doubrava[1], Roman Růžek[2], Jakub Šedek[3], Ivan Mlch[4] & Adam Karkulín[5]

Abstract: A hot forming process of thermoplastics requires a row of special material characteristics for correct design and optimization of the structure part. Paper describes hot forming process modelling and design optimization. Selected special material characteristics relevant for hot forming process and FE design optimization of thermoplastic composite structure parts are introduced. Behavior of the real structure part (gear box) for current metal solution compared with optimized new composite solution is presented.

Keywords: Composite gear box; FEM; Hot forming proces; Thermoplastic composite structure parts

1. Introduction

Thermoplastic composites are more and more exploited in the aerospace application during last decades. The goal of the series works and investigations that were carried out for several years has been performed with the goal to produce aerospace components with complex curvature. These investigations were focused mainly to assisted/automatic fiber placement [1–3] and hot forming [4–7] technology owing to persistent high prices of thermoplastics and possibility to apply these materials in a serial production. The main advantage of thermoforming is a possibility to apply forming and stamping process like metal forming during manufacturing process. This technology can significantly decrease manufacturing costs in a serial production. Although the price of material is still higher compared with thermosets, the hot forming process is very competitive, and perspective compared with other manufacturing technologies

The design and optimization of any structural part are very risky and time and cost consuming. The cost of proving this novel concept on large components increases exponentially with the size of the part. These factors imply that the return on the research investment carries a higher risk. The only way how the risk and costs can be reduced is through virtual try-out technology with simulation to develop and verify the process before applying it to the structural part.

Using of an appropriate software for simulation of the moulding process of composites can reduce the occurrence above mentioned defects and can make possible to perform the strain and stress analysis and the shape optimization in design stage [8]. However, any software requires several different material characteristics. These characteristics are depended mainly on the fi-

[1] *Ing. Radek Doubrava, Ph.D.; Czech Aerospace Research Centre; Beranových 130, 199 05 Prague - Letňany, Czech Republic;* doubrava@vzlu.cz

[2] *Dr. Ing. Roman Růžek.; Czech Aerospace Research Centre; Beranových 130, 199 05 Prague - Letňany, Czech Republic;* ruzek@vzlu.cz

[3] *Ing. Jakub Šedek, Ph.D.; Czech Aerospace Research Centre; Beranových 130, 199 05 Prague - Letňany, Czech Republic;* sedek@vzlu.cz

[4] *Ivan Mlch; Czech Aerospace Research Centre; Beranových 130, 199 05 Prague - Letňany, Czech Republic;* mlch@vzlu.cz

[5] *Ing. Adam Karkulín; Czech Aerospace Research Centre; Beranových 130, 199 05 Prague - Letňany, Czech Republic;* karkulin@vzlu.cz

bre and matrix of a composite used in the structure. Properties published for the semi-products are aimed mainly for structure design, not for the technologist and simulation of moulding process at 300–320 °C temperature. For these elevated temperatures the strength of the matrix is practically zero and composite properties are related mainly to viscosity. Unfortunately, the special material characteristics are unavailable in generally. All information regarding to special material behaviour are know-how of each user. Without knowledge of the special material characteristics, any reliable analysis cannot be performed. This paper shows some of these special characteristics, and following newly designed composite part designed and manufactured based on these characteristics.

2. Hot forming process modelling and design optimization

Process simulation techniques were supported using of virtual prototyping softwares [8] to achieve good correlation in simulation to the physical process that causes different types of faults (wrinkles, lack of consolidation, excessive in-plane shearing). The process simulation through those software's can facilitates the development of new composite parts that would be considered far too risky to be developed by experimental means. The software's updates the fibre orientations throughout the forming process, and so enables the shear angles and corresponding changes in permeability to be calculated. These local permeability changes and the overall geometry of the pre-form are essential to ensuring an accurate infusion prediction. Those software's packages have great possibilities, but without knowledge of some special material characteristics the results cannot be comfortable. These special material properties must be determined based on experiments. These tests enable to specify material behaviour under higher temperature (shear stress, warping point, slip of fabric layers, etc.).

The second part of proper part design is its optimization from a viewpoint of the functionality, stress-strain and strength properties. Design optimization in structural redesign is an application of the operations research field of applied mathematics to problems engineering design. The optimization algorithms in MSC/NASTRAN [9, 10] which use the function gradients to assist in the numerical search for an optimum have been used for finding of appropriate thickness of thermoplastics composite material.

3. Material characterisation

From the point of view above mentioned requirements two basic groups of tests for material characterisation must support the process simulation and design optimisation. While for design optimisation the standard methods for material characterisation (such as tensile and compression strength, shear strength, ILLS, curved beam strength, [11–17]) are sufficient, for the process simulation a row of special tests must be carried out. Such special tests were designed and conducted so that the main factors of hot forming process were defined. The typical time temperature dependence of forming process is shown in Fig. 1. The main factors that have a strong influence on the thermomechanical behaviour of composites are:

- Temperature (range between 240 °C and 320 °C).
- Velocity (2–5 mm/sec).
- Forming pressure.

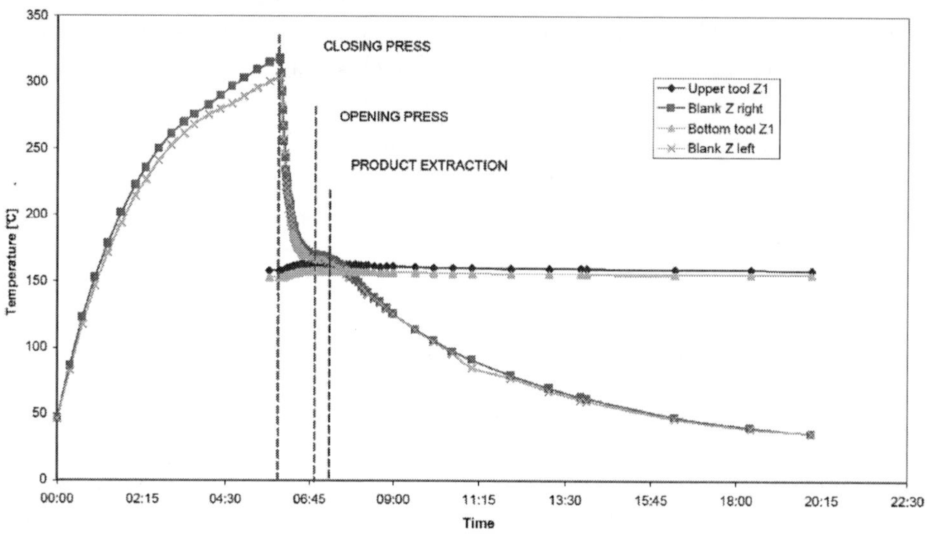

Fig. 1. Typical time temperature dependence for hot forming process of composites.

While for metal materials the ductile behaviour of material is crucial, for composites the ductile is negligible and interlaminar shear deformation between layers and interlaminar shear deformation in the plane of each layer is important. Therefore, three basic experiments for material characterisation were designed and performed:

- Interlaminate shear test in tension.
- Interlaminar shear test in bending.
- Interlaminar test of shear strain.

Two types of test coupons were designed for interlaminar shear test in tension. The first represents a coupon with one shear plain and the second coupon design represents a tensile specimen with two shear plains – see the scheme of critical part of both coupons in Fig. 2. The shear between plies is induced by tension load. Test coupons were tested under different test conditions (temperature varies in range from RT to 310 °C, loading rate 0.01 and 3 mm/sec). The splice plates for coupon geometry stabilization were used. The reason why the splice plates were used is the fact that the individual layers have tendency to lose under higher temperature. The displacement during the loading was measured by digital image correlation method. The symmetric set up of the test coupon induce the symmetric displacement. Under RT condition the splice plates using is not necessary. Fig. 3 shows the typical force vs. time and cross head displacement vs. time dependences for interlaminar shear test in tension under different temperature and 3 mm/sec load rate.

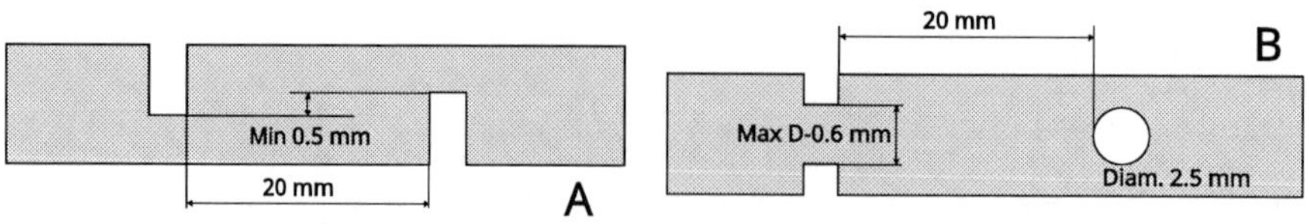

Fig. 2. Two types of coupons used for interlaminar shear test in tension (one shear plain specimen on the left, two shear planes specimen on the right).

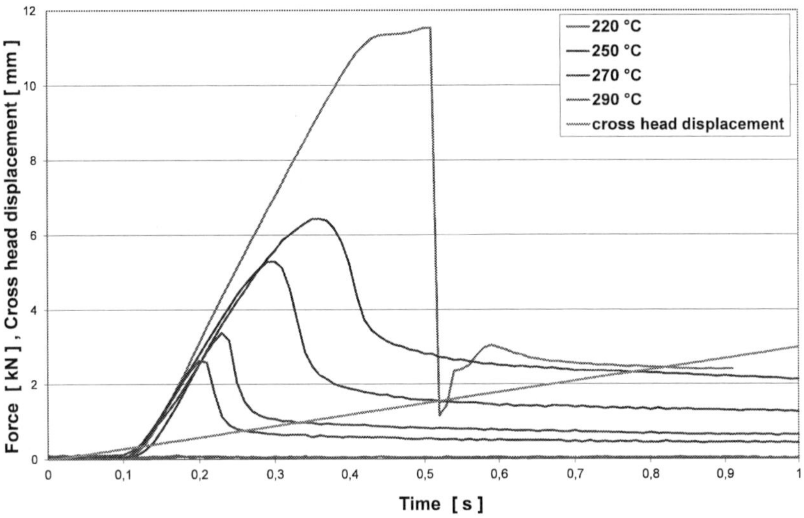

Fig. 3. Typical force vs. time and cross head displacement vs. time dependences for interlaminar shear test in tension under different temperature and 3 mm/sec load rate.

Interlaminar shear test in bending derive benefit from the fact that:

- tension strength in layer plane is in order higher than shear strength between plies;
- deformation of composite bar beam under the temperature when the matrix becomes plastics is determined by interlaminar shear deformation.

The principle of the test follows: the beam (composite plate) on two rigid supports is placed into the chamber. During the test temperature is continuously increased with a low rate. The real temperature is continuously measured on the beam surfaces. The deflection vs. temperature dependence is recorded. The origination of permanent deformation and thickness increasing are determined. Fig. 4 shows the beam configuration with thermocouples after the test termination.

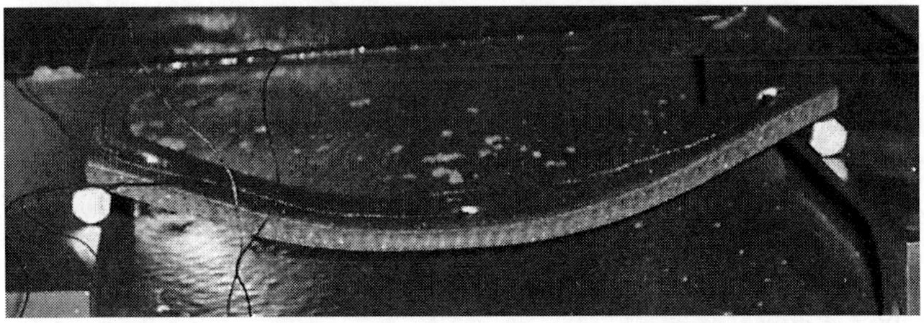

Fig. 4. Interlaminar shear test in bending set up with thermocouples after test termination.

The angle of latching shear deflection of fabrics is the next important parameter which is entering into the numerical models. This fabric characteristic is influence by weave type and ply tightening. The bevel angle at the moment of the shear stiffness of the layer rapidly increases is experimentally verified. One ply was separated from the composite plate under 300 °C temperatures. The shear deformation of the ply was evoked by mechanical loading. The bevel angel of ply was determined. Fig. 5 shows the composite ply before and after interlaminar test of shear strain.

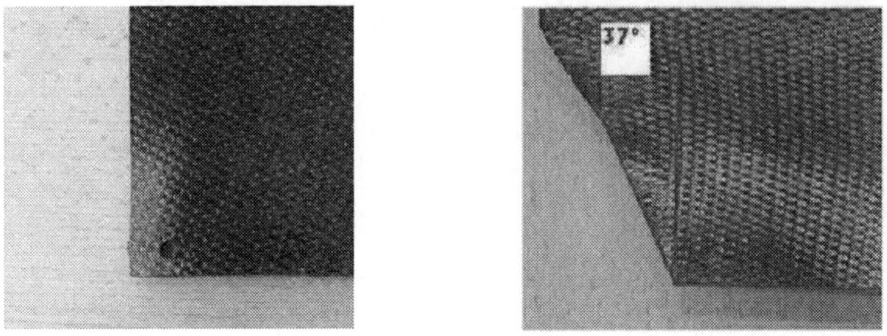

Fig. 5. Composite ply before and after interlaminar test of shear strain.

4. Gear box design optimization

The special material characteristics were included into the process simulation software and used for design and optimization of gear box structure parts. The FE model was set-up in the FEMAP pre-processor.

The composite gear box body was designed for production by press moulding technology from C-PPS (Carbon fibre fabrics reinforced Polyphenylene Sulphid resin) thermoplastic material [10]. For calculation was used isotropic material model with Young modulus 37 GPa. Fig. 6 shows the three various proposals of the gear thermoplastic box body shape designed for the press moulding technology. The objective of optimization was finding appropriate thickness of thermoplastics composite material with equivalent stiffness in comparison with original metal design. The inner cabin pressure load case was used for the process optimization.

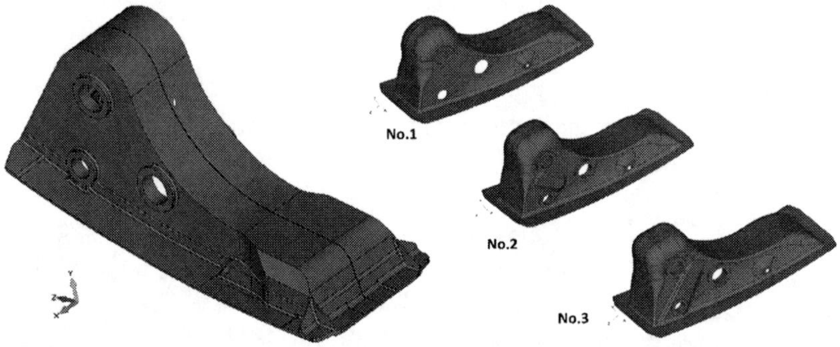

Fig. 6. Original metal gear box (left) and various stages of the thermoplastic gear box body designed for press moulding technology (right).

5. Gear box experimental verification

The deformation measurements for two representative cases of service loading were carried out with aim of numerical solution verification:

- Pressure loading.
- Single force loading.

The special testing chamber and test benches were designed and manufactured for experimental evaluation of mechanical properties. The real deformations of composite gear box were measured by means of optical ARAMIS HS system. The full field measurements of two areas of the gear box were performed for each load case. The first one the side wall in the surrounding of shafts aperture was investigated and then behaviour of the middle section with connection of both parts of gear box with mechanical joints was evaluated.

Fig. 7 shows the setup of the pressure loading case – gear box, pressurized chamber and ARAMIS HS system. Fig. 8 display comparison between calculated and measured displacement of the shafts aperture area of the composite gear box body.

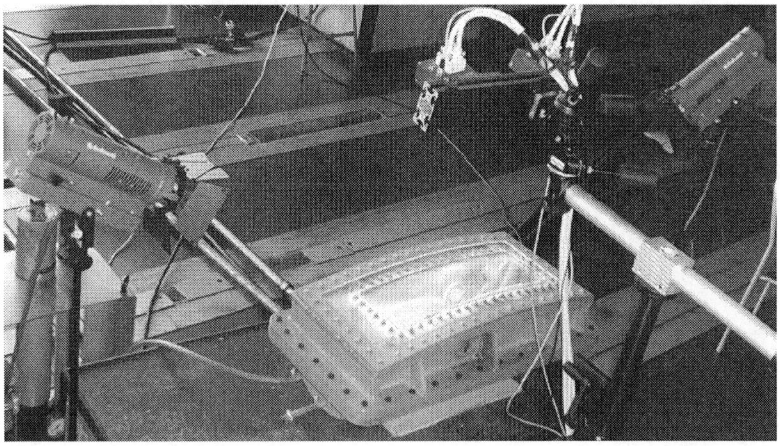

Fig. 7. Setup of the pressure load case.

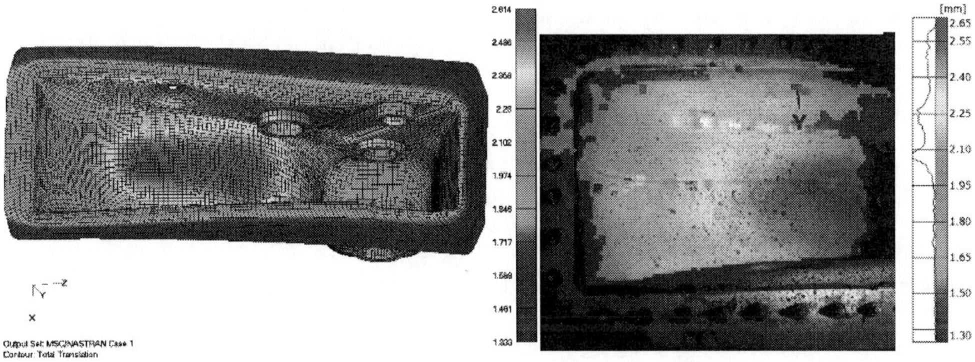

Fig. 8. Comparison between calculated and measured displacement of the gear box.

6. Conclusion

Overall metal structural part replacement by thermoplastic composite part is discussed. Paper describes selected special material tests relevant for hot forming process characterisation and FE design optimization of thermoplastics composite structure part. Behaviour of the real structure part (gear box) for current metal solution in comparison to the new composite solution is presented. Good agreement between numerical simulations and experimental results was obtained. Proposed technique, including stiffness and technology process numerical simulation, improves the credibility of the design phase and reduces the amount of time and costs needed for both prototype manufacturing and tests.

Acknowledgements

This research was funded by the Ministry of Industry and Trade of the Czech Republic (MPO), program TRIO, grant number FV30033 "Structural and technological development of primary aircraft parts made by composite with thermoplastic matrix" and with the help of institutional funding of the MPO in the DKRV01 program dedicated to the development of research organizations.

References

[1] Ursenbach D., Merotte J. and Bahramshahi N., "In-Situ Consolidation of Complex Curvature Thermoplastic Composite Part Manufactured by Automated Fiber Placement," *SAMPE 2020 Virtual Series, Manufacturing Technology*, p.12 (2020). SKU/Code: TP20-0000000201. https://doi.org/10.33599/nasampe/s.20.0201.

[2] Mallon P.J., O'Brádaigh C.M. and Pipes R.B., "Polymeric diaphragm forming of complex-curvature thermoplastic composite parts," *Composites*, 20(1), pp. 48-56 (1989). ISSN 0010-4361. https://doi.org/10.1016/0010-4361(89)90682-4.

[3] Peeters D.M.J., Irisarri F.X., Groenendijk C. and Růžek R., "Optimal design, manufacturing and testing of non-conventional laminates," *Composite Structures*, Vol. 210, pp. 29-40 (2019). ISSN 0263-8223. https://doi.org/10.1016/j.compstruct.2018.10.062.

[4] Dutta A., Niemeyer M. and Cakmak M., "Thermoforming of advanced thermoplastic composites. I: Single curvature parts," *Polymer Composites*, Vol. 12, pp. 257-272 (1991). ISSN:1548-0569. https://doi.org/10.1002/pc.750120407.

[5] Akkerman R. and Haanappel S.P., "Thermoplastic composites manufacturing by thermoforming" in *Advances in Composites Manufacturing and Process Design*,(Woodhead Publishing, 2015) pp. 111-129. ISBN 9781782423072, https://doi.org/10.1016/B978-1-78242-307-2.00006-3.

[6] Růžek R., Křena J., Doubrava R., Tkadlec J., Kadlec M. and Bělský P., "Optimal Design and Testing of a Thermoplastic Pressurized Passenger Door Manufactured Using Thermoforming," *Polymers*, Vol. 13(19), pp. 3394. (2021), https://doi.org/10.3390/polym13193394

[7] Růžek R., Šedek J., Kadlec M. and Kucharský P., "Mechanical behavior of thermoplastic rib under loading representing real structure conditions," in EAN 2016 - 54th International Conference on Experimental Stress Analysis. Srní (CZ), 30 May – 2 June 2016. EID: 2-s2.0-84986296610.

[8] Dörr D., Brymerski W., Ropers S., Leutz D., Joppich T., Kärger L., and Henning F., "A benchmark study of finite element codes for forming simulation of thermoplastic UD-tapes," *Procedia CIRP*, Vol 66, pp. 101-106 (2017). https://doi.org/10.1016/j.procir.2017.03.223.

[9] Moore, G.J. 1992. MSC/NASTRAN. Design Sensitivity and Optimization, MacNeal-Schwendler Corporation.

[10] Kim K.-P. and D. Hale R. D., in "Design of Directionally Uniform Extensional and Flexural Stiffness in Quasi-Isotropic Laminates," 49th AIAA/ASME/ASCE/AHS/ASC Structures, Structural Dynamics, and Materials Conference, ..., Schaumburg(IL), (AIAA, 2008). https://doi.org/10.2514/6.2008-1749.

[11] Hron R. Kadlec M. and Růžek R., "Effect of the Test Procedure and Thermoplastic Composite Resin Type on the Curved Beam Strength," *Materials* Vol.14, no. 2, pp. 352 (2021). https://doi.org/10.3390/ma14020352.

[12] ASTM D 3410/D 3410M – 95. Standard Test Method for Compressive Properties of Polymer Matrix Composite Materials with Unsupported Gage Section by Shear Loading. ASTM International.

[13] ASTM D 3039/D 3039M – 08. Standard Test Method for Tensile Properties of Polymer Matrix Composite Materials. ASTM International.

[14] ASTM D 695-96. Standard Test method for Compressive Properties of Rigid Plastics. ASTM International.

[15] ASTM D 2344/D 2344M – 00. Standard Test Method for Short-Beam Strength of Polymer Matrix Composite Materials and Their Laminates. ASTM International.

[16] ASTM D6415-99 - Standard test method for measuring the curved beam strength of a fiber reinforced polymer matrix composite. ASTM International.

[17] ABS 5045 A 09 A 2G : Carbon fibre fabrics reinforced Polyphenylene Sulphide resin (PPS), Laminates (consolidated plates) with or without lightning protection, with or without anti-corrosion protection - Structural material.

The Analysis of Fatigue Life Prediction of the Metallic Bone Plates for Osteosynthesis

Veronika Drátovská[1], Radek Sedláček[2], Zdeněk Padovec[3],
Pavel Růžička[4] & Adam Kratochvíl[5]

Abstract: This paper deals with fatigue life prediction of metallic bone plates performed by SIMULIA fe-safe software. Two materials commonly used for surgical implants were subjected to the investigation: stainless steel for implants 316 L (ASTM F138) and titanium alloy Ti6Al4V ELI (ASTM F136). Material properties necessary for the prediction were determined from uniaxial tensile tests and four-point bending fatigue tests. Several types of bone plates were subjected to four-point bending fatigue testing in order to allow for the comparison of the fatigue life prediction results with the measured data from the experiments.

Keywords: Bone plates; Fatigue prediction; Four-point bending

1. Introduction

Metallic bone plates are used to fixate broken bones. During their application in the human body, the bone plates are typically cyclically bent. Therefore, their fatigue properties in bending need to be examined. However, fatigue testing is quite time-consuming and expensive. Nowadays the finite element method (FEM) or other computational approaches are usually employed to reduce the amount of necessary experimental testing during product development.

This paper is focused on performing the fatigue life prediction of metallic bone plates using SIMULIA fe-safe software to address this problem. Material parameters necessary for the prediction were determined experimentally. Four types of bone plates were subjected to fatigue testing in order to allow for a comparison of the fatigue life prediction results to the results from experiments.

2. Methods

In the experimental part of the work two materials commonly used for surgical implants were subjected to the investigation: stainless steel for implants 316 L (ASTM F138) and titanium alloy Ti6Al4V ELI (ASTM F136). The uniaxial tensile test according to ČSN EN ISO 6892-1 standard [1] was performed to determine the elastic modulus E, the ultimate tensile strength R_m and the tensile yield strength R_p.

The fatigue material parameters were determined from four-point fatigue tests with prismatic samples. Basquin's model $\sigma_a = \sigma'_f(2N)^b$ was used to mathematically describe the measured

[1] *Ing. Veronika Drátovská; Department of Mechanics, Biomechanics and Mechatronics, Fac. of Mech. Engineering, Czech Technical University in Prague; Technická 4, 166 07 Prague, Czech Republic; veronika.dratovska@fs.cvut.cz*
[2] *Ing. Radek Sedláček, Ph.D.; Department of Mechanics, Biomechanics and Mechatronics, Fac. of Mech. Engineering, Czech Technical University in Prague; Technická 4, 166 07 Prague, Czech Republic*
[3] *Ing. Zdeněk Padovec, Ph.D.; Department of Mechanics, Biomechanics and Mechatronics, Fac. of Mech. Engineering, Czech Technical University in Prague; Technická 4, 166 07 Prague, Czech Republic*
[4] *Ing. Pavel Růžička, Ph.D.; Department of Mechanics, Biomechanics and Mechatronics, Fac. of Mech. Engineering, Czech Technical University in Prague; Technická 4, 166 07 Prague, Czech Republic*
[5] *Ing. Adam Kratochvíl; Department of Mechanics, Biomechanics and Mechatronics, Fac. of Mech. Engineering, Czech Technical University in Prague; Technická 4, 166 07 Prague, Czech Republic*

$S - N$ curves, its parameters the fatigue strength coefficient σ_f' and the fatigue strength exponent b were obtained using linear regression. The fatigue tests were conducted applying an R ratio (minimum to maximum loading force ratio) equal to 0.1; however, the fe-safe requires material parameters for an R ratio equal to -1 (fully-reversed cycle). Therefore, Walker's equation [2] was used to convert Basquin's model parameters to an equivalent fully-reversed cycle. Seeger's method [3] was employed to estimate other fatigue parameters, which were not determined from experiments.

Besides gathering the material parameters, the experimental part was also focused on measuring the fatigue properties of four types of bone plates. Humeral proximal bone plates of two generations, fibular distal, and tibial distal medial bone plates were subjected to load-controlled high-cycle four-point bending fatigue testing (see Fig. 1). The methodology of the tests was based on the ASTM F382 standard [4].

The experimental part of the work was followed by the computational part. Abaqus was employed to compute the stress in the bone plates for a given static loading force. Simulation of the four-point bending test included a model of the bone plate and four cylinders which represented the loading and support rollers (see Fig. 2). The diameters of the rollers and the distances between them were the same as in the experiments. Planar symmetry of the models enabled using only half of the model for the computations. Results of the Abaqus computations were used as an input to fe-safe, where the loading cycle was defined, and the material definition was created using the measured parameters. The prediction in fe-safe was performed on several loading levels to get the load vs the number of cycles curves which were subsequently compared to the experimental results.

Fig. 1. The set up for the four-point bending test with steel humeral proximal bone plate of the 1. generation.

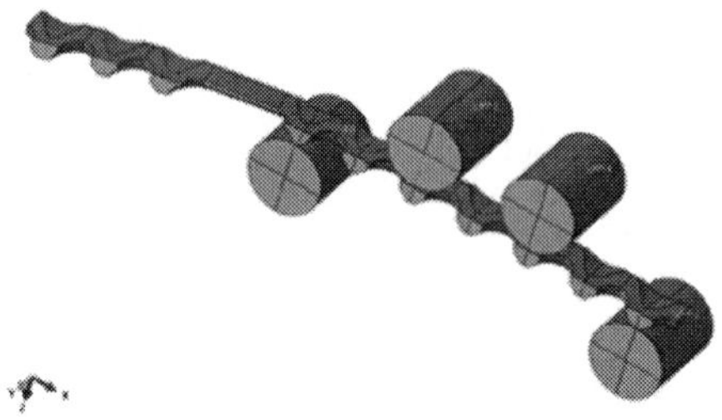

Fig. 2. Half model of the four-point bending test with the fibular distal bone plate in Abaqus.

3. Results

Material parameters gathered from the experiments and subsequently used in the fe-safe analysis are listed in the table below (see Table 1). The table also contains the parameters estimated by the Seeger's method: the strain hardening coefficient K', the strain hardening exponent n', the fatigue ductility coefficient ϵ_f', and the fatigue ductility exponent c.

Table 1. Material parameters used in the fe-safe.

Material	E [GPa]	R_m [MPa]	R_p [MPa]	σ_f' [MPa]	b [-]	K' [MPa]	n' [-]	ϵ_f' [-]	c [-]
Steel	168.6	935	887	9 755	-0.23	14 965	0.40	0.34	-0.58
Ti6Al4V	112.1	925	822	1 159	-0.03	1 220	0.05	0.35	-0.69

The output of the fe-safe analysis is the number of cycles, which the most loaded node of the model endures for a given loading scenario, in other words, it is the lowest lifetime calculated on the whole part. The results can also be displayed in Abaqus as a contour plot, which shows the calculated lifetime for the whole bone plate (see Fig. 3). The figure shows that the lowest lifetime was calculated on the edge of the screw holes between the loading rollers, which corresponds with the areas where the highest stress occurred. It was confirmed by the experiments that the bone plates crack in the predicted area.

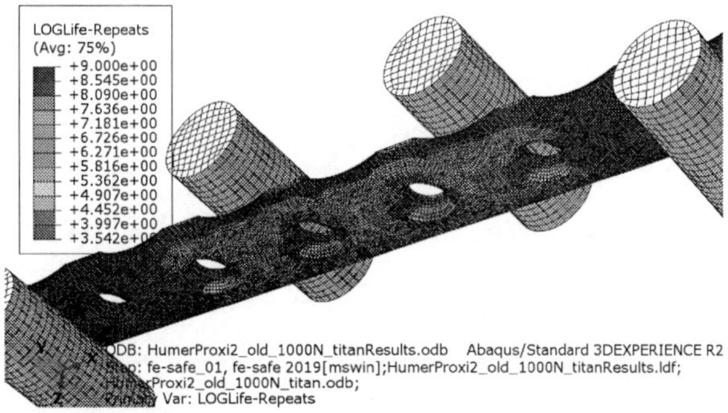

Fig. 3. Fe-safe analysis results for the humeral proximal bone plate of the first generation displayed in Abaqus.

The fe-safe analysis was performed for each type of bone plate on several loading levels to get the predicted load vs the number of cycles curves. The graphs below compare the fe-safe fatigue life prediction results to experimental results for the steel (see Fig. 4) and titanium alloy (see Fig. 5) bone plates. The curves show dependency between the loading force amplitudes and the number of cycles. The abbreviations in the legends of the graphs stand for the types of the bone plates: humeral proximal bone plates of the first and second generation (HP I, HP II), fibular distal (FD), and tibial distal medial (TDM) bones plates. The load of the bone plates is represented by force amplitude; therefore, the results show differences between the types of bone plates due to their different geometries. The comparison shows that the predictions are more accurate for the steel bone plates, for the titanium alloy plates the calculations predict a higher lifetime than was achieved during experiments.

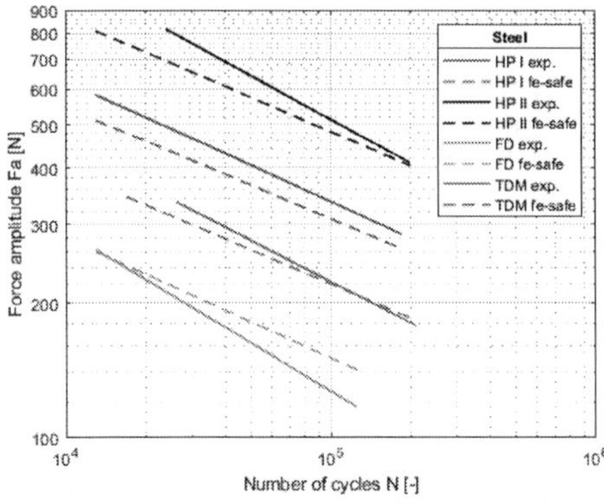

Fig. 4. Comparison of fe-safe and experimental results in force amplitudes for the steel bone plates.

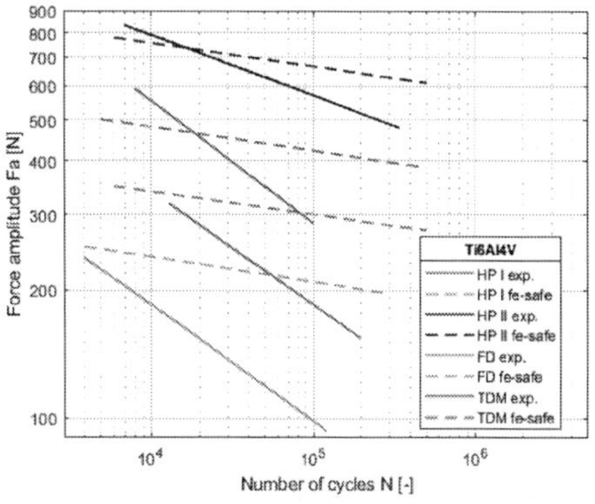

Fig. 5. Comparison of fe-safe and experimental results in force amplitudes for the titanium alloy bone plates.

The following graphs (see Fig. 6 and Fig. 7) shows the same results as the graphs above but in stress amplitudes. The highest stress on the bone plate for a given force was considered. Therefore, the presented curves are related to the edge of screw holes in the bone plates, where the highest stress occurred. The stress for a given force was computed in Abaqus.

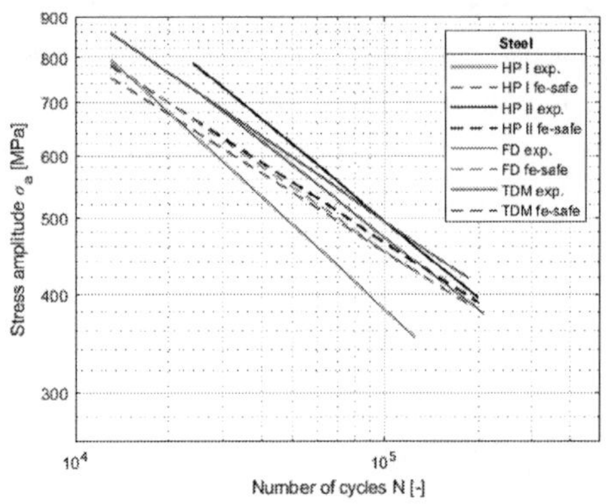

Fig. 6. Comparison of fe-safe and experimental results in stress amplitudes for the steel bone plates.

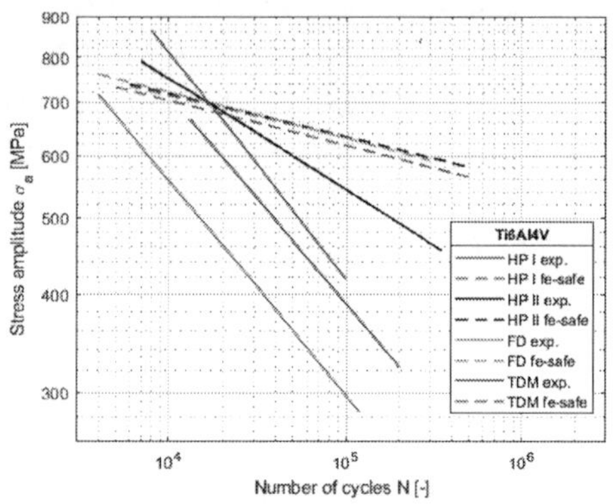

Fig. 7. Comparison of fe-safe and experimental results in stress amplitudes for the titanium alloy bone plates.

The influence of different geometries of the bone plates is eliminated while the stress amplitudes are considered instead of the force amplitudes. Theoretically, the curves should overlap for the same material. It holds well for the predictions and also quite well for the experiments with the steel bone plates; however, the experiments for titanium alloy show more significant differences. The lines should be at least parallel since the vertical location of the curves is given by the computed stress, which might not correspond perfectly with the actual stress during the experiments, but the different slope of the line points to different material properties. Therefore, for example, the experimental results for titanium alloy HP I and II bone plates cannot be satisfyingly described by one material model due to their different slopes.

4. Conclusions

Fatigue life prediction for four types of bone plates was performed using SIMULIA fe-safe software. The results of the prediction were compared to the results of fatigue tests with the bone plates. The prediction achieved better accuracy for the steel bone plates. The introduced methods might be applied in the development process of new types of bone plates.

Acknowledgements

This study was supported via a grant project awarded by the Technology Agency of the Czech Republic, No. TACR TH03010418 and Czech Technical University in Prague, grant number SGS22/149/OHK2/3T/12.

References

[1] ČSN EN ISO 6892-1. Kovové materiály – Zkoušení tahem – Část 1: Zkušební metoda za pokojové teploty. Praha: Úřad pro technickou normalizaci, metrologii a státní zkušebnictví, 2017, 76 s.

[2] Dowling N.E., Calhoun C.A. and Arcari A., " Mean stress effects in stress-life fatigue and the Walker equation," *Fatigue & Fracture of Engineering Materials & Structures*, Vol. 32, Issue 3, pp. 163–179 (2009). ISSN 8756758X. doi:10.1111/j.1460-2695.2008.01322.x.

[3] Strain Based Fatigue Analysis. Autodesk help [online]. [cit. 2020-12-09]. Available from: http://help.autodesk.com/view/ASMECH/2014/ENU/?guid=GUID-ECDCD9AB-BE65-43A3-9ECB-D25C38606A3E.

[4] ASTM F382-17. Standard Specification and Test Method for Metallic Bone Plates. (West Conshohocken, PA: ASTM International, 2017).

Experimental Analysis of Construction Hoist Operating Loads

Martin Dub[1], Marek Šťádler[2], Martin Havlíček[3] & František Lopot[4]

Abstract: The contribution introduces the process of determining the operating loads of the construction hoist during various operational states and hoist mast configurations. The operational loads are expressed by means of the set of loading coefficients which include the influence of dynamics, excentric loading and roughness of the rails. FEA is used to analyse the structure and determine convenient test points for experimental measurement. Complex strain gauge measurement is desined and carried out both on the hoist mast and cage guide arms.

Keywords: Construction hoist; Operating loads; Load coefficient; Experiment; Strain gauge; FEA

1. Introduction

Construction hoist is the device used for transport material and personnel on construction sites. It consists of the hoist mast, which can be build up to 350 m height, and the elevator cage. The drive of the cage is realized by means of the electric motor in combination with the pinions and rack. The transversal position of the cage is defined by means of guide arms with two rolling pulleys, [1].

The loading, strength and stability of the hoist mast and guide arms are very important areas, which has to be solved precisely, [2, 3]. The construction hoist can be built in the single, double or triple cage configuration where one or two hoist masts are used. The double or triple configuration generates different loads to the hoists mast then the single configuration. That has to be taken into account. Next not negligible factor is the influence of dynamics. Dynamic loading is caused by means of acceleration and deceleration of the cage, emergency breaking and roughness of the rails, i.e. the vertical tubes of the hoist mast.

The main goal of this contribution is to describe the process of determination of the operating loads during various test regimes.

2. Methods

To obtain information about operating loads of the construction hoist, the experimental measurement was designed. Foil strain gauges are used in various arrangement to identify appropriate strain of the hoist mast and guide arms. A finite element model of the hoist mast and guide arm is created to find suitable points for strain gauge installation.

[1] *Ing. Martin Dub, Ph.D.; Department of Designing and Machine Components, Faculty of Mechanical Engineering, Czech Technical University in Prague; Technická 4, 166 07 Prague, Czech Republic;* martin.dub@fs.cvut.cz

[2] *Ing. Marek Šťádler; Department of Designing and Machine Components, Faculty of Mechanical Engineering, Czech Technical University in Prague; Technická 4, 166 07 Prague, Czech Republic;* marek.stadler@fs.cvut.cz

[3] *Ing. Martin Havlíček; Department of Designing and Machine Components, Faculty of Mechanical Engineering, Czech Technical University in Prague; Technická 4, 166 07 Prague, Czech Republic;* martin.havlicek@fs.cvut.cz

[4] *Ing. František Lopot, Ph.D.; Department of Designing and Machine Components, Faculty of Mechanical Engineering, Czech Technical University in Prague; Technická 4, 166 07 Prague, Czech Republic;* frantisek.lopot@fs.cvut.cz

2.1. *Finite element models*

The complex beam model of the hoist mast created in RFEM 5 software is used to simulate different positions of the cages as well as the different loading including loading eccentricity. The whole beam model is in Fig. 1. Detailed view on the measured segment with indication of strain gauges positions can be seen in Fig. 2. Strain gauge arrangement on the hoist mast is used to measure tensile strains and stresses in chosen beams.

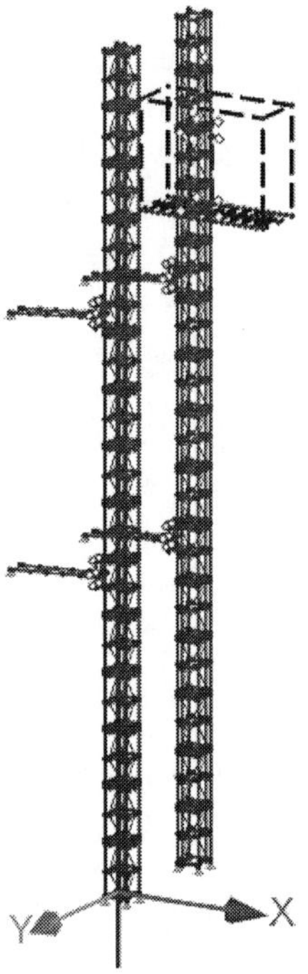

Fig. 1. Beam model of the double mast hoist.

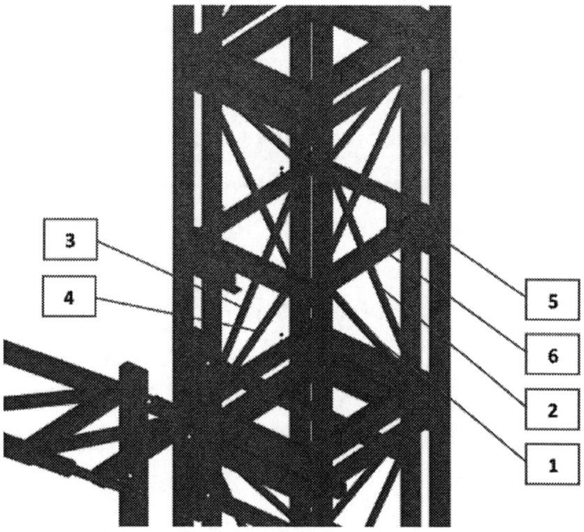

Fig. 2. Detail view on the hoist mast segment with positions of strain gauges.

The guide arm is modelled as a solid element model in Abaqus software Fig. 3. The guide arms are used to identify reaction forces. The strain gauge arrangement is used to measure bending strains. For this purpose two chamfers are manufactured on the guide arm edge, Fig. 4.

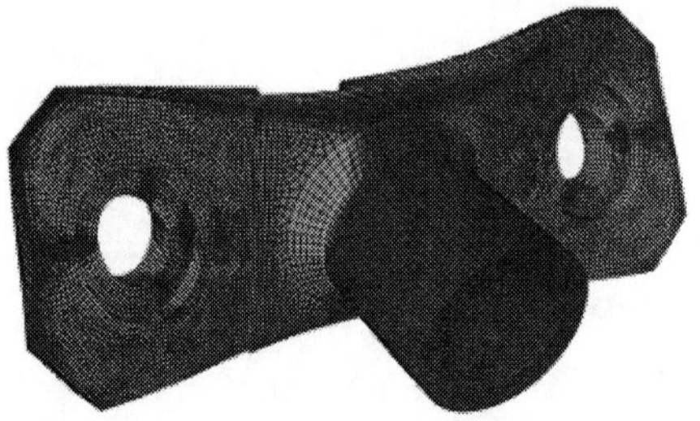

Fig. 3. Guide arm welding stress field.

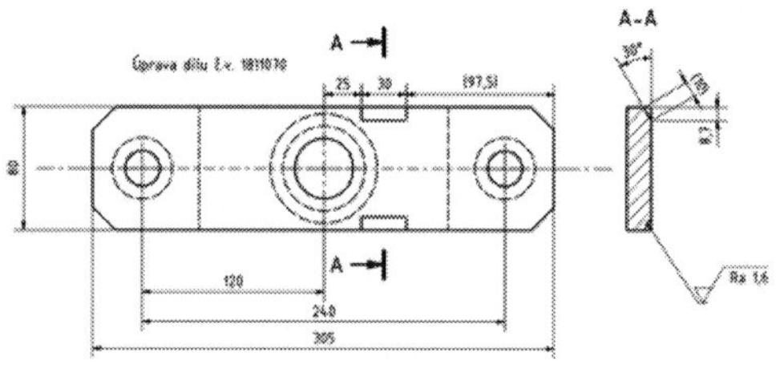

Fig. 4. Modification of the current guide arm design – chamfers for strain gauge installation.

2.2. *Experiment*

For each modified guide arm, calibration was carried out in order to determine the dependence between the load and the strain at the place of the installed strain gauges. Knowing these dependencies, modified guide arms serve as force sensors. Calibration took place in a research laboratory. Each guide arm was loaded in a specially prepared tool, see Fig. 5. The device simulates the load of the guide arms through the mast pipe. The load force was checked by a HBM U10M force sensor. The maximum load force was 15 kN due to the loading machine. All guide arms showed a very good linear relationship between the load force and strain gauge deformation. Calibration was carried out during loading and unloading. The guide arm showed negligible hysteresis.

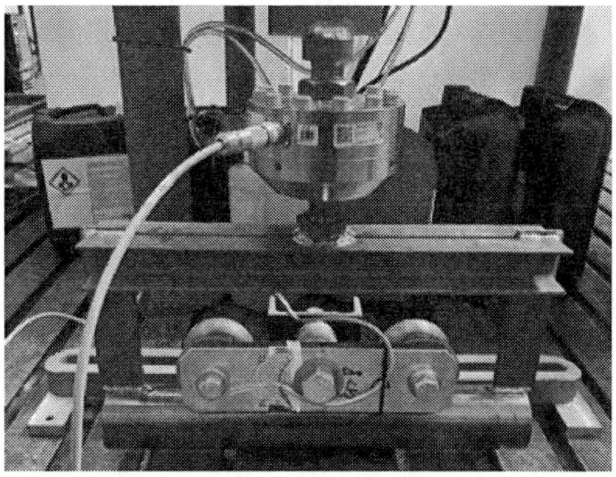

Fig. 5. Guide arm calibration using the special tool and HBM U10M sensor.

Experimental measurement was carried out on the test rig of STROS company in Sedlčany, Fig. 6. Totally 32 test modes in two different hoist mast arrangement were carried out. Data from six measuring points on the hoist mast segment, six guide arms of single cage and six guide arms of the twin cage were acquired by means of the Dewetron autonomous stations with the sampling frequency 10 kHz.

Fig. 6. Test rig in STROS company in Sedlčany, hoist mast with strain gauges is in red box.

3. Results and evaluation

All the acquired data were processed and evaluated for each test mode. The result for each test mode and measuring point is the loading coefficient, that includes the influence of the dynamics and also the influence of the static deflection of the hoist mast. Test mode where only single cage rides up is taken as a reference (test mode 16), because it is common and validated state. The rest of the modes is compared to the reference mode.

The dynamic coefficients on the guide arms of the single cage were determined as follows. From the measured data in the given mode, a section in which dynamic loading of the structure occurs is always selected. In the case of a single cage, the effect of static loading is filtered out of the data. A histogram of individual dynamic reaction forces is compiled from the obtained dynamic oscillations to verify the normal distribution, and then the value of the given dynamic load is determined as the standard deviation of the dynamic oscillations. The value of the dynamic coefficient k_{dyn} is illustrated in Fig. 7 and described in following equation, where R_{xydyn} is the dynamic reaction force in the guide arm in arbitrary mode and R_{16dyn} is the reaction force in the test mode.

$$k_{dyn} = \frac{R_{xydyn}}{R_{16dyn}} \tag{1}$$

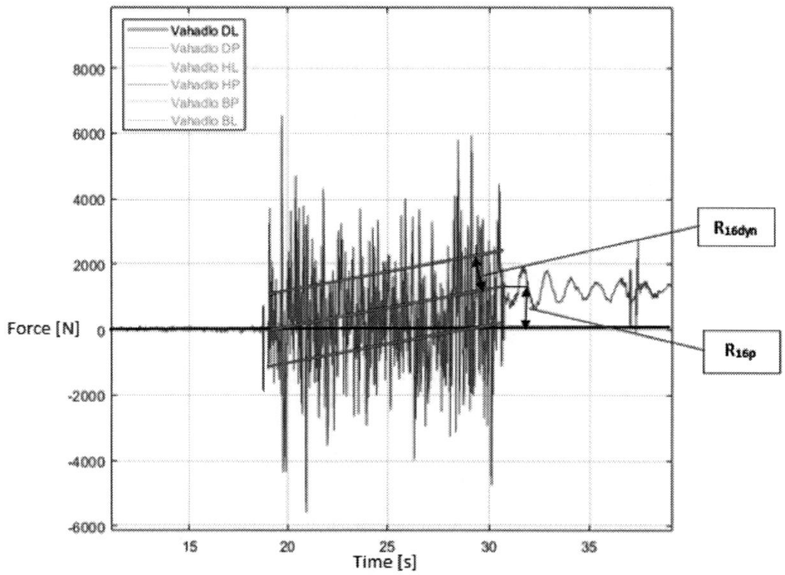

Fig. 7. Typical time course of the reaction force of the guide arm on the single cage.

If we want to be on the side of safety, the overload of the structure should also be evaluated in the reference mode 16. This evaluation is again quantified by coefficients. This time let's call them the overload coefficients k_p, because they also respect the value of the static load of the guide arms of the single cage in the reference test mode - reaction force R_{16p} according Fig. 7. The methodology for determining the coefficients is designed to be on the side of safety as much as possible, i.e. the highest possible coefficients. These coefficients also take into account the theoretically calculated reaction force of the guide arms R_s for symmetrical loading of the single cage, which forms a comparative value. The equation for the overload coefficient k_p follows.

$$k_p = \frac{R_s + R_{16dyn} + R_{16p}}{R_s} \tag{2}$$

By means of the overload coefficient k_p the reference test mode 16 is evaluated. The overall overload and dynamics is taken into account. To evaluate the total overload in various test modes, the total coefficient k_c is used.

$$k_c = k_{dyn} k_p \tag{3}$$

The loading coefficients calculated according data acquired from the measuring segment are determined as a simple ratio of the maximum values during arbitrary test mode and the reference test mode 16. The time course of the normal stress at the measuring points of the mast segment during the reference test mode can be seen in Fig. 8. There are significant stress peaks during the ride up and down.

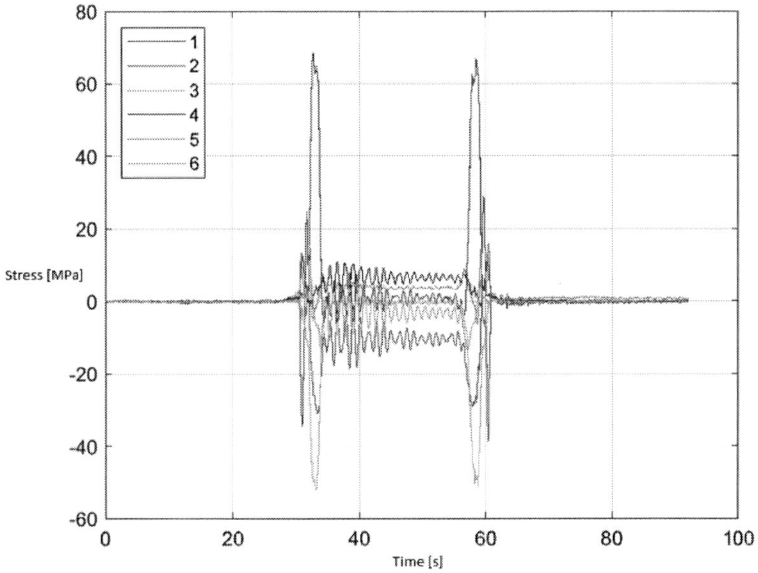

Fig. 8. Typical time course of the normal stress obtained from the measiring segment test points.

Calculated coefficients of the structure overloading are not further presented due to the business secret of the STROS company.

4. Conclusions

The contribution describes the finite element analysis of the hoist mast and the guide arms of the construction host. The result of these analysis is identification of suitable measuring points for strain gauges. After that, the experimental measurement is carried out. Acquired data are used to determine the loading coefficients for each test regime. These coefficients can be used during designing and static control of the structure. Obtained coefficient should be also compared with the overload coefficients provided by appropriate standards.

Acknowledgements

The experimental measurement was carried out in the cooperation with the company STROS-Sedlčanské strojírny a.s. on their test rig.

References

[1] Kim T. et al., "Inclined construction hoist for efficient resource transportation in irregularly shaped tall buildings," *Automation in Construction*,Vol. 62, pp. 124–132 (2016). ISSN 0926-5805. https://doi.org/10.1016/j.autcon.2015.11.008.

[2] Petr K. et al., "Measurement of response distribution on mass crossbeams in dependence on load position," in Proceedings of 56th International Scientific Conference on Experimental Stress Analysis, EAN 2018, Harrachov, 2018, pp. 332–337. ISBN 978-802704062-9.

[3] Starý F. et al., "Comparison of different construction hoist masts by FEM analysis," in Proceedings of 58th International Conference of Machine Design Departments (ICMD), Prague, 2018, pp. 344–347. ISBN 978-80-213-2769-6.

Measurement of the Axial Load of the Tapered Pin Pivot

Milan Dvořák[1], Ivan Jeřábek[2] & Daniel Pleschner[3]

Abstract: The purpose of the test described in this article was to verify whether optical Fiber Bragg Grating (FBG) strain sensors are usable for axial load measurement of the tapered pin pivot. One piece of the pin pivot was instrumented with the four FBG sensors and underwent an axial tensile load testing in order to identify the sensitivity of such sensor system design. Part of the experimental work was also the design of a special loading fixture. During the experimental testing, the FBG sensor technology proved to be useful for determining the local strain at the notch on the pin. It allows to measure in places where it is not possible to install a conventional electric strain gauge. However, shortcomings have also emerged, such as the problematic quality control of the adhesion of the optical fiber in the drilled hole, or the quality control of the drilled hole at all. Furthermore, it is obvious that for other than laboratory measurements it is necessary to use a robust secondary protection of the optical fiber, especially in the place where the fiber exits the pin.

Keywords: Experimental Analysis; Pin Pivot; FBG Sensor; Fiber Optic Sensor; Embedded Sensor; Axial Load Measurement

1. Introduction

The purpose of the test described in this article was to verify whether optical Fiber Bragg Grating (FBG) strain sensors are usable for axial strain measurement of the tapered pin pivot. An area, where the threaded cylindrical end of the pin changes to a conical shape together with the step increase of the pin diameter, was our area of interest (see Fig. 1).

Theoretically, this place is accessible from the inside, because the pin has a through axial hole with a diameter of about 5 mm. However, the place where we want to experimentally determine the strain is more than 30 mm away from the beginning of the hole, so the installation of a traditional electric strain gauges would be very difficult.

If it is proven that the FBG sensor can be reliably installed in the hole in the body od the pin pivot, or bolt, it would mean that it would be possible to measure the axial preload of the joint during tightening and subsequently during its operation. It would thus be possible to detect loosening of the bolt joints during operational life. In addition, with a suitable configuration of FBG sensors, it would be possible to indicate any additional stress on the screw, for example from bending.

Designs based on fiber-optic strain sensors, that are used for strain monitoring of large elements of building structures, are described in the literature. For example, in [1,2] concrete girders with measurement of the steel reinforcement prestress using an FBG sensors are described. Axial load measurement in steel beams with FBG sensors is described in [3]. Regarding the dimensions corresponding to the specimen in our assignment, the so-called Smart Bolt is described in [4]. Due to the used configuration of the FBG sensors, the bolt shaft is weakened

[1] *Ing. Milan Dvořák, Ph.D.; Department of Mechanics, Biomechanics and Mechatronics, Fac. of Mech. Engineering, Czech Technical University in Prague; Technická 4, 160 00 Prague, Czech Republic;* milan.dvorak@fs.cvut.cz

[2] *Ing. Ivan Jeřábek, Ph.D.; Doosan Bobcat EMEA, s.r.o., U Kodetky 1810, 263 01 Dobříš, Czech Republic;* ivan.jerabek@doosan.com

[3] *Ing. Daniel Pleschner, Ph.D.; Doosan Bobcat EMEA, s.r.o., U Kodetky 1810, 263 01 Dobříš, Czech Republic;* daniel.pleschner@doosan.com

by holes with a diameter of 2 mm. In addition, the sensor is not glued in its entire length and actually works as an extensometer. This does not allow evaluating local strain changes. In [5], an optical FBG sensor is temporarily glued along its entire length into a blind hole with a diameter of 1.8 mm. Two adhesive systems are compared: paraffin wax and thermoplastic, and as with the previous solution, the weakening of the bolt shaft by the hole for the FBG sensor is the limiting factor. A possible solution to avoid this weakening is the use of a measuring bolt washer. Designs using FBG sensor are described, for example, in [6,7], but the washer naturally does not allow monitoring the local strain behavior of the bolt, only the preload of the entire bolt joint.

Our strain measurement design of the bolt/pin joint uses a hole with a diameter only slightly larger than the outer diameter of the FBG sensor, so that there is only a minimal layer of an adhesive between the primary coating of the fiber and the surface of the hole. Among other things, this makes it possible to use the effect of capillary rise for gluing. Another advantage is the transfer of strain from the bolt shaft to the FBG sensor with minimal losses caused by the deformation of the adhesive layer.

2. Experimental Method

Verification of the functionality of the proposed experimental configuration was carried out using a quasi-static tensile load test. Using a custom-designed fixture, a tensile load was introduced into the steel pin to such a level that no plastic deformation of the pin material occurred.

2.1. Experimental Specimen

One piece of the pin pivot was instrumented with the FBG sensors. A total of 4 FBG sensors were used, each on a separate optical fiber, with the following parameters: single-mode optical fiber with cladding diameter: 0.125 mm, outer diameter of the fiber: 0.155 mm; primary coating material: polyimide; central wavelength of the FBG: 1530/1550 nm; FBG length: 3.3 mm. The optical fibers were placed in the holes with diameter of 0.3 mm, that were oriented parallelly to the longitudinal axis of the pin. The measuring grid of the FBG sensor was placed in the transition between the thread and the conical part of the pin (see Fig. 1). HBM Z70 cyanoacrylate adhesive was used for gluing the optical fibers into the holes.

2.2. Experimental Procedure

Safibra FBGuard 1550 FAST Mini optical measurement device was used to evaluate and store strain readings from the FBG sensors. The pin was placed inside the custom-made fixture (see Fig. 2) to ensure proper tensile loading of the area of interest.

The TiraTEST 2300 electromechanical testing device (see Fig. 3) was used to load the pin by tensile force. The pin specimen was loaded at a controlled feed rate of 2 mm/min (see Fig. 4). At first, a force level of 50 kN was reached, followed by a gradual relief to 40–30–20–10–0 kN. At each load level, the steady state value of the mechanical strain was evaluated, indicated by the appropriate FBG sensor. The FBG03 sensor was damaged during clamping. The FBG01 sensor did not respond to the load. This was probably due to its poor adhesion to the walls of the hole. The data were therefore evaluated from the FBG02 and FBG04 sensors, placed on a pitch circle facing each other.

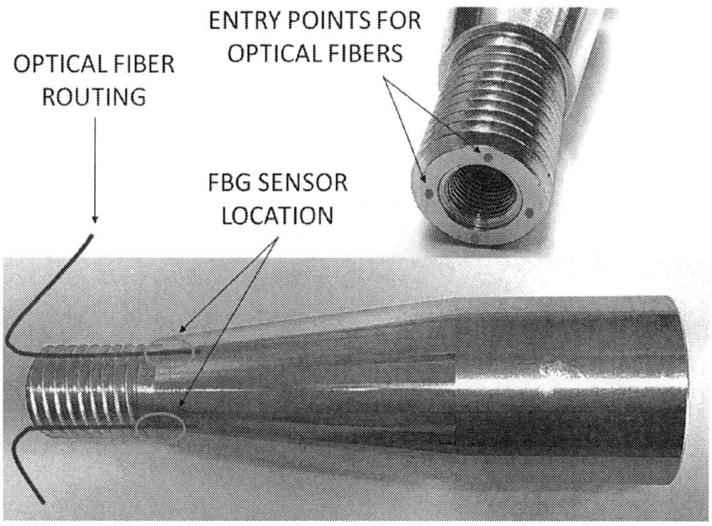

Fig. 1. Tapered pin pivot with FBG sensors locations.

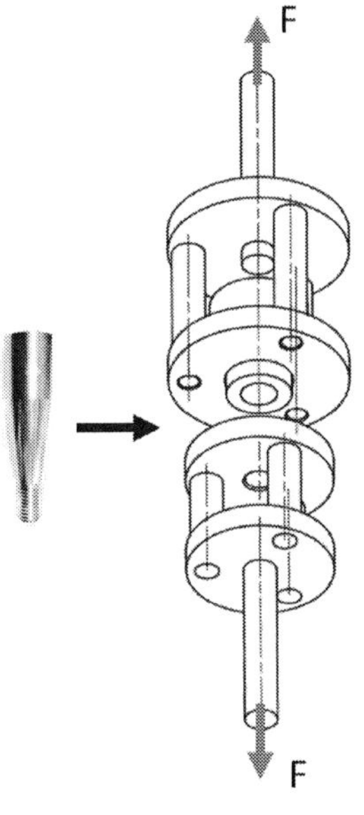

Fig. 2. Custom-made loading fixture.

Fig. 3. Specimen placed in the tensile test machine.

3. Results and Discussion

A total of four load tests were performed, where the pin pivot was removed after each test and remounted in the fixture. A scatter plot of the dependence of the indicated mechanical strain on the axial loading force was evaluated from the time records of the loading tensile force from the load cell and the strain indicated by the FBG sensor. The aim was to evaluate the strain sensitivity of the proposed design of the measuring bolt/pin in the operational load range. And secondly, to verify influence of the changes in its placement in the fixture after repeated assembly. An example of the data obtained during the first load test is in the graph in Fig. 5. The linearity of the dependence of the strain response of the FBG sensor on the axial load force was also evaluated.

The results are summarized in Table 1. Very good linearity can be seen in the entire measured range up to 50 kN. The change in strain sensitivity responds to re-assembly in the fixture. Sensitivity between 15.28 and 20.65 (µm/m)/kN is sufficient in the operating range. It can therefore be assumed that with a higher number of sensors it would be possible to indicate, in addition to the axial strain of the bolt, the bending component of the strain, however, this has not been confirmed experimentally yet.

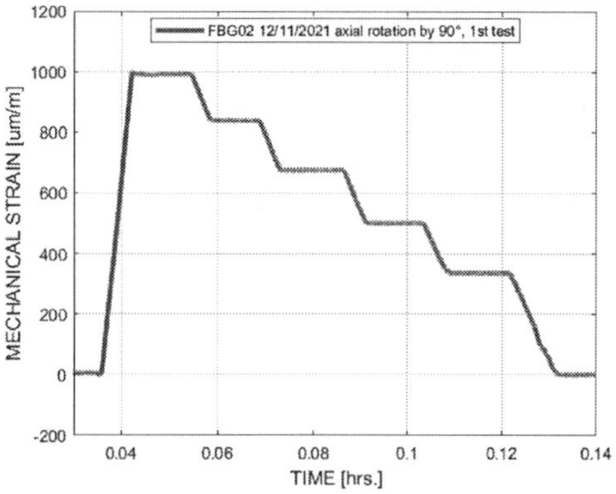

Fig. 4. Time record of mechanical strain measured by FBG sensor FBG02 during one load cycle.

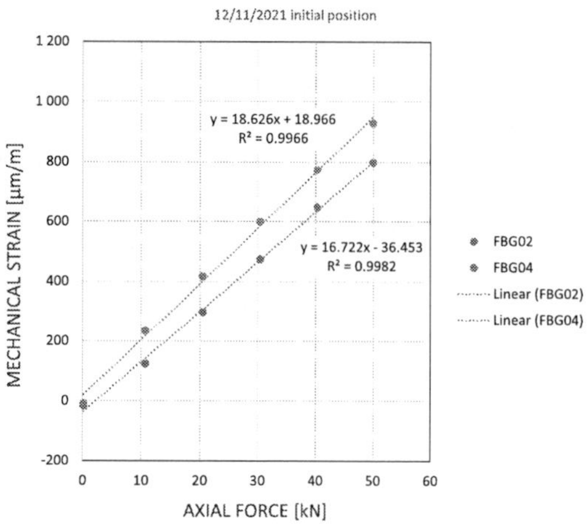

Fig. 5. Strain response of the FBG sensors to the tensile loading force.

Table 1. Linearity measurements results.

Measurement no.	Strain response to axial load [(µm/m)/kN]		R-squared	
	FBG02	FBG04	FBG02	FBG04
1.	16.72	18.63	0.9982	0.9966
2.	16.92	19.14	0.9916	0.9952
3.	15.51	20.42	0.9921	0.9966
4.	15.28	20.65	0.9928	0.9972

4. Conclusion

During the experimental testing, the FBG sensor technology proved to be useful for determining the mechanical strain at the notch on the pin. It allows to measure in places where it is not possible to install a conventional electric strain gauge. However, shortcomings have also emerged, such as the problematic quality control of the adhesion of the optical fiber in the drilled hole, or the quality control of the drilled hole at all. Furthermore, it is obvious that for other than laboratory measurements it is necessary to use a robust secondary protection of the optical fiber, especially in the place where the fiber exits the pin.

References

[1] Cho K., Kim S.T., Cho J.R. and Park Y.H., "Estimation of tendon force distribution in prestressed concrete girders using smart strand," *Appl. Sci.*, Vol. 7, Art. no. 1319 (2017). eISSN 2076-3417. https://doi.org/10.3390/app7121319.

[2] Sung H.J., Do T.M., Kim J.M. and Kim Y.S., "Long-term monitoring of ground anchor tensile forces by FBG sensors embedded tendon," *Smart. Struct. Syst.*, Vol. 19, pp. 269–277 (2017). eISSN 1738-1991. https://doi.org/10.12989/sss.2017.19.3.269.

[3] Bonopera M., Chang K.-Ch., Chen C.-C., Lee Z.-K. and Tullini N., "Axial load detection in compressed steel beams using FBG-DSM sensors," *Smart Structures and Systems*, Vol. 21., pp. 53–64 (2018). https://doi.org/10.12989/sss.2018.21.1.053.

[4] Ren L., Feng T., Ho M., Jiang T. and Song G., "A smart "shear sensing" bolt based on FBG sensors," *Measurement*, Vol 122, pp. 240–246 (2018). https://doi.org/10.1016/j.measurement.2018.03.026.

[5] Khomenko A., Koricho E. G., Haq M. and Cloud G. L., "Bolt tension monitoring with reusable fiber Bragg-grating sensors," *The Journal of Strain Analysis for Engineering Design*, Vol. 51, Issue 2 (2015). https://doi.org/10.1177/0309324715598265.

[6] Yeager M., Whitaker A. and Todd M., "A method for monitoring bolt torque in a composite connection using an embedded fiber Bragg grating sensor," *Journal of Intelligent Material Systems and Structures*, Vol. 29, Issue 3 (2017). https://doi.org/10.1177/1045389X17705217.

[7] Chen D., Huo L., Li H. and Song G., "A Fiber Bragg Grating (FBG)-Enabled SmartWasher for Bolt Pre-Load Measurement: Design, Analysis, Calibration, and Experimental Validation," *Sensors*, Vol. 18, Art. no. 2586 (2018). https://doi.org/10.3390/s18082586.

Experimental and Computational Modal Analysis of a Newly Designed Rotor Rig

Petr Ferfecki[1], Michal Molčan[2], Jakub Cienciala[3], Jaroslav Zapoměl[4], František Fojtík[5] & Miroslav Trochta[6]

Abstract: Machinery rotors are frequently supported by hydrodynamical bearings. To improve vibration response, magnetically sensitive fluids such as ferrofluids, magnetorheological oils, and nano-micro composite magnetic fluids are commonly used as bearing lubricants. A new experimental rotor rig including a hydrodynamic bearing lubricated with a magnetically sensitive fluid was designed. The preliminary characteristics of the bearing of the rotor were experimentally measured. This paper presents an experimental and computational modal analysis of the selected components and assemblies of the designed experimental rotor rig. Modal parameters such as natural frequencies, mode shapes, and damping ratios are measured. The computed and measured results show that they are in good agreement. Moreover, it has been confirmed that the natural frequencies of the rotor rig are far enough from the operating speed.

Keywords: Rotor rig; Magnetically sensitive hydrodynamic bearing; Prestressed modal analysis; Experimental validation of the computational model

1. Introduction

Stiffness and damping of the lubrication layer have a strong effect on the dynamical performance of the hydrodynamical bearing. Recent research in the improvement of the bearing and damper performance characteristics is focused on lubrication with magnetically sensitive fluids [1–6]. Using these lubricants allows the apparent viscosity to change with the action of a magnetic field. Therefore, the stiffness and damping properties of the hydrodynamical bearing can be altered according to the current operating conditions of the rotating machine.

A new experimental rotor rig with a magnetically sensitive hydrodynamic bearing for the testing of the rotor-bearing characteristics was designed at the Department of Applied Mechanics of the VSB - Technical University of Ostrava. The experimental rotor rig and the magnetic bearing have a modular arrangement, and their main parts are shown in Fig. 1.

The experimental test rig and the magnetically sensitive bearing (1) are proposed for a rotational speed of up to 6000 rpm. To drive and control the rotor speed, a low-voltage 2-pole asynchronous electric motor (2) with a rated power of 2.2 kW with a frequency converter (3)

[1] *Ing. Petr Ferfecki, Ph.D.; IT4I, VSB-TUO; Department of Applied Mechanics, Fac. of Mech. Eng., VSB-TUO; 17. listopadu 2172/15, 708 00 Ostrava, Czech Republic;* petr.ferfecki@vsb.cz

[2] *Ing. Michal Molčan; IT4I, VSB-TUO; Department of Applied Mechanics, Fac. of Mech. Eng., VSB-TUO; 17. listopadu 2172/15, 708 00 Ostrava, Czech Republic;* michal.molcan@vsb.cz

[3] *Ing. Jakub Cienciala; Department of Applied Mechanics, Fac. of Mech. Eng., VSB-TUO; 17. listopadu 2172/15, 708 00 Ostrava, Czech Republic;* jakub.cienciala@vsb.cz

[4] *prof. Ing. Jaroslav Zapoměl, DrSc.; Department of Dynamics and Vibration, Inst. of Thermomechanics; Dolejškova 1402/5, 182 00 Prague 8, Czech Republic; Department of Applied Mechanics, Fac. of Mech. Eng., VSB-TUO; 17. listopadu 2172/15, 708 00 Ostrava, Czech Republic;* jaroslav.zapomel@vsb.cz

[5] *doc. Ing. František Fojtík, Ph.D., Department of Applied Mechanics, Fac. of Mech. Eng., VSB-TUO; 17. listopadu 2172/15, 708 00 Ostrava, Czech Republic;* frantisek.fojtik@vsb.cz

[6] *Ing. Miroslav Trochta, Ph.D.; Department of Machine Parts and Mechanisms, Fac. of Mech. Eng., VSB-TUO; 17. listopadu 2172/15, 708 00 Ostrava, Czech Republic;* miroslav.trochta@vsb.cz

is used. A cone clamping element (4) and a jaw coupling (5) are employed for the shaft (6) of 15 mm in diameter to connect the drive to the proposed bearing. The rotor is loaded with three discs (7) with a weight of 4.6 kg in total. The magnetically sensitive bearing and a single row deep groove ball bearing on the drive side end are employed for its support. The rotor displacements are measured by the eddy current proximity probes. A custom application was developed in the MATLAB software for a real-time visualization of the rotor orbits and measured time-dependent signals. The electric motor, rotor, and bearing are mounted on the base element (8) of the hot-rolled U profile that is bolted to the foundation desk (9). The stands for the vibration protection (10) and for the probes can be easily adjusted to the desired position as can be seen in Fig. 1.

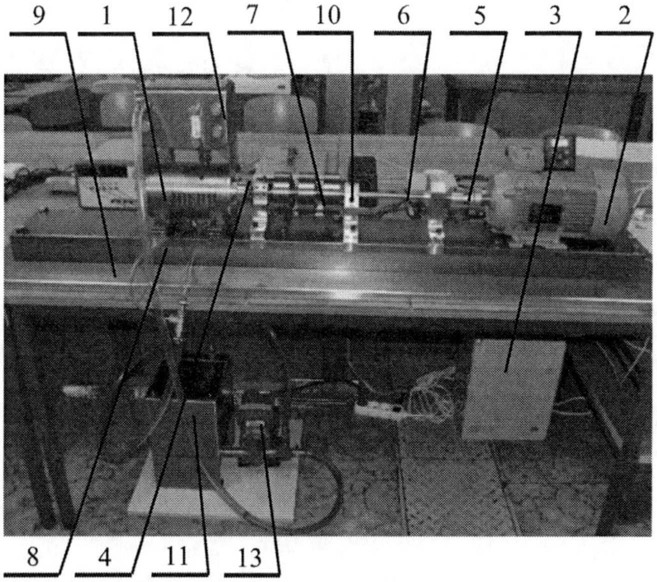

Fig. 1. The experimental rotor rig with the magnetic hydrodynamic bearing.

The proposed hydrodynamic bearing [1, 2] consists of the housing with the bearing shell in which the hollow journal is inserted onto the stationary pin. The source of the magnetic field in the bearing provides the electric coil. The apparent viscosity of the thin layer of the magnetically sensitive lubricant between the inner surface of the bearing shell and the outer surface of the hollow journal is affected by a magnetic field and changes the magnitude of the hydraulic force. The bearing can be lubricated with a ferrofluid, a magnetorheological oil, a nano-micro composite magnetic fluid [7, 8] as well as a turbine oil.

The bottom reservoir (11) serves as storage of the magnetic fluid, and it also cools the fluid to the desired temperature. The fluid is fed into the upper reservoir (12) from the bottom reservoir by the gear pump (13). The bearing inlet is not pressurised, and the fluid flow is controlled by the gravity force from the upper reservoir.

The results of the experimental and computational modal analysis for the foundation desk (9), the base element (8), and the proximity probe stand are presented. The shaft (6) with the disc (7) and the foundation desk (9) with the base element (8) and the electric motor (2) are bolted assemblies. Hence, this paper mainly draws attention to some subjects of the computational modal analysis for prestressed structures.

2. The methodology of the experimental modal analysis

Modal properties such as natural frequencies, mode shapes, and damping ratios are determined with the roving impact hammer method. During the measurements, an approximation of the free-free boundary condition is performed by suspension on belts or by placement on soft foam. The Brüel & Kjær (B&K) data acquisition device is utilized for processing signals of the impact hammer and a piezoelectric three-axis acceleration sensor. The B&K Connect software is used.

Fig. 2. Photography of the realisation of free-free suspension for an experimental modal analysis of the base element.

For modal measurement, the foundation desk and the base element are suspended on the belts and the proximity probe stand is placed on the soft foam to approximate free-free boundary conditions. The realisation of suspension on the belts for the base element with the grid lines and with the vibration accelerometer mounted by a thin layer of wax is depicted in Fig. 2.

3. The comparison of the experimental and computational modal analysis results for the rotor rig components

Natural frequencies and mode shapes of the investigated components are numerically computed in the Ansys Workbench software. In computational models, the geometry of the components is without simplifications and further refinement of the finite element mesh has negligible influence on the results. The solution is performed with the undamped assumption.

Table 1. Physical properties of the tested components.

Component name	Material	Young's modulus [GPa]	Poisson's ratio [–]	Weight [kg]	Density [kg·m^{-3}]
Foundation desk	EN AW-7075 T6	72	0.33	47.5	2810 / 2848.4
Base element	S235JR (11375)	210	0.30	30.4	7800 / 7877.20
Proximity probe stand	EN AW-7075 T6	72	0.33	0.75	2810 / 2798.1
Shaft	C45 (12050)	210	0.30	0.76	7800 / 8017.5
Disc	St60-2 (11600)	210	0.30	1.54	7800 / 8017.5
Bolt M14×100	8.8 zinc-plated steel	210	0.30	0.13	7850 / 7850

The linear elastic isotropic material behaviour is assumed for all components. For the material type, Young's modulus, Poisson's ratio, and the density from data sheets are stated in Table 1. The data sheet densities are slightly modified according to the measured weight of the components. See the last column behind the slash in Table 1.

The measured natural frequencies with corresponding mode shapes of the simple components such as the foundation desk, base element, and proximity probe stand are compared with the computed ones and the achieved results are summarised in Table 2, Table 3 and Table 4.

Table 2. Comparison of the measured and computed modal properties of the foundation desk.

Natural frequency order / Dominant shape	Measured natural frequency [Hz] / Damping ratio [%]	Computed natural frequency [Hz] for 72.0 GPa / 75.2 GPa	Relative error [%]
1 / 1st bending	46.6 / 5	44.8 / 45.8	+3.9 / +1.7
2 / 1st torsional	112.2 / 0.2	111.3 / 113.7	+0.8 / -1.3
3 / 2nd bending	125.1 / 0.2	123.9 / 126.6	+1.0 / -1.2
4 / 2nd torsional	237.0 / 0.2	229.9 / 234.8	+3.0 / +0.9
5 / 3rd bending	247.9 / 0.1	243.4 / 248.7	+1.8 / -0.3
6 / 3rd torsional	368.2 / 0.1	362.7 / 370.5	+1.5 / -0.6
7 / 4th bending	412.6 / 0.2	403.7 / 412.4	+2.2 / +0.1

The results of the modal parameters for the foundation desk in a free-free boundary condition are given in Table 2. The values of 72.0 / 75.2 GPa Young's modulus are used in the computation of the natural frequencies and relative errors in the fourth column. The higher value of Young's modulus than the data sheet is determined by an optimization procedure based on minimizing the sum of the squared difference between the first five measured and computed natural frequencies. The relative errors between the computed and measured natural frequencies are up to 4 or 2%, respectively.

Table 3. Comparison of the measured and computed modal properties of the base element.

Natural frequency order / Dominant shape	Measured natural frequency [Hz] / Damping ratio [%]	Computed natural frequency [Hz] for 210.0 GPa / 216.2 GPa	Relative error [%]
1 / 1st torsional	80.0 / 0.1	77.6 / 78.7	+3.0 / +1.6
2 / 1st bending	149.7 / 0.1	146.9 / 149.1	+1.9 / +0.4
3 / 2nd torsional	187.1 / 0.2	182.0 / 184.6	+2.7 / +1.3
4 / 2nd bending	387.5 / 0.1	382.8 / 388.2	+1.2 / -0.2
5 / 3rd torsional	395.6 / 0.1	391.0 / 396.6	+1.2 / -0.3
6 / 1st lateral bending	427.3 / 0.1	443.7 / 450.3	-3.8 / -5.4
7 / 2nd lateral bending	455.9 / 0.1	473.0 / 479.7	-3.6 / -5.2

The measured and computed modal parameters for the base element and the proximity probe stand are summarized in Table 3 and Table 4. The suspension of the base element was on the belts and the proximity probe stand was placed on the soft foam. The measured higher value of the damping ratio of the first natural frequency of the proximity probe stand in Table 4 is due to the attenuation of the suspension foam. Most of the measured natural frequencies are higher than the computed ones for the case of the data sheet Young's modulus. This is due to neglecting the material damping and due to an inaccurate model of suspension in the computational model. The optimization procedure with the variation of up to 10% of Young's modulus value was performed for both components. The relative errors for the computations with the modified Young's modulus were reduced only for the several lowest natural frequencies of both components, see Table 3 and Table 4. The measured and computed mode shapes for both components are visually almost identical.

Table 4. Comparison of the measured and computed modal properties of the proximity probe stand.

Natural frequency order	Measured natural frequency [Hz] / Damping ratio [%]	Computed natural frequency [Hz] for 72.0 GPa / 74.3 GPa	Relative error [%]
1	156.6 / 1.9	148.5 / 150.9	+5.2 / +1.7
2	593.0 / 0.1	589.7 / 599.1	+0.6 / -1.0
3	806.8 / 0.1	812.2 / 825.2	-0.7 / -5.6
4	1133.2 / 0.1	1098.4 / 1116.1	+3.1 / +1.4

The 1st bending mode shape for the natural frequency of 150 Hz (left) and 149 Hz (right) of the base element is shown in Fig. 3.

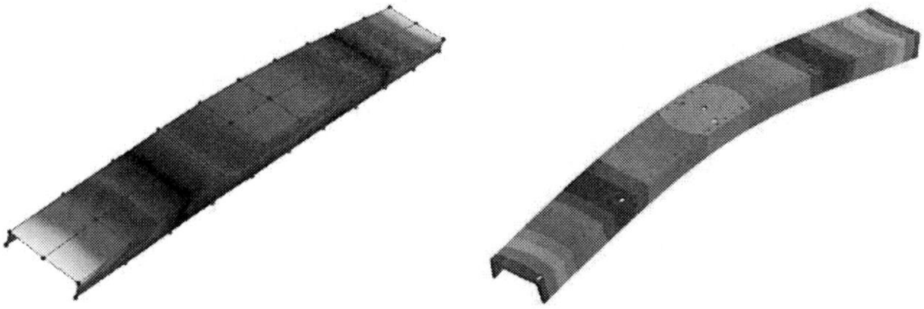

Fig. 3. Measured (left) and computed (right) 1st bending mode shape of the base element.

In Fig. 4 the mode shape of the proximity probe stand for the measured natural frequency of 157 Hz (left) and computed one 151 Hz (right) is depicted.

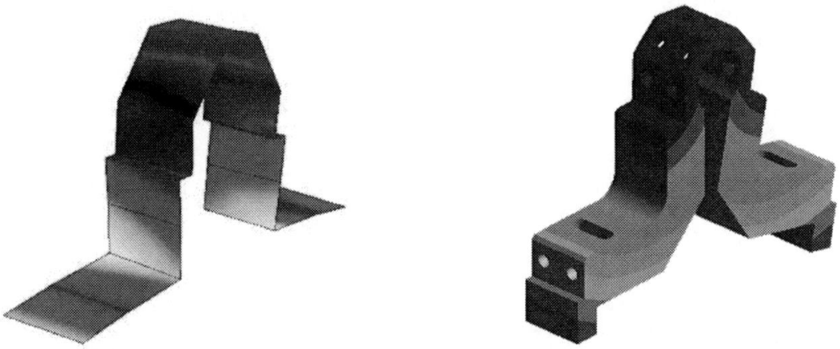

Fig. 4. Measured (left) and computed (right) 1st mode shape of the proximity probe stand.

4. Results of the experimental and numerical modal analysis for the rotor rig assemblies

The assembly of shaft with disc and the assembly of the foundation desk, base element and electric motor are experimentally and numerically investigated.

The disc is connected to the shaft by the grub screws and its modal analysis is solved without the preload. In the computational model, a contact surface of the shaft is connected to the tip of the grub screw by the bonded contact type [9]. The results of numerical simulation show that the proper selection of the contact pair region is crucial to obtain the correct natural mode shapes and frequencies. The experimentally measured and numerically computed natural frequencies are presented in Table 5. In the third column, two values of the bending natural frequencies are stated, each for one of the two main axes of stiffness of the shaft with disc. Relative error is determined between the measured natural frequency and the mean value of the computed bending natural frequencies. Its value is approximately equal to 1%.

Table 5. Comparison of the measured and computed modal properties of the shaft with disc.

Natural frequency order / Dominant shape	Measured natural frequency [Hz] / Damping ratio [%]	Computed natural frequency [Hz] 1st / 2nd axis	Relative error [%]
1 / 1st bending	172.9 / 1.3	171.3 / 171.3	+0.9
2 / 2nd bending	479.9 / 0.2	476.6 / 476.7	+0.7
3 / partial bending	972.3/ 1.2	983.2 / 983.3	-1.1
4 / 3rd bending	1245.1 / 0.8	1259.2 / 1259.7	-1.2
5 / partial bending	1268.4 / 0.8	– / –	–
6 / partial bending	– / –	2542.2 / 2542.6	–
7 / partial torsional	– / –	2885.4	–

The assembly of the foundation desk, base element and electric motor is connected by the preloaded bolted joints. The effect of prestress is achieved in a two-step computational analysis. First, the bolt preload is imposed in the nonlinear static analysis, and then, based on the resulting stress field, the natural frequencies and mode shapes are solved using linear modal analysis.

Fig. 5. Geometry assembly of the foundation desk, base element, and electric motor.

In the computational model, the electric motor is considered as a rigid body, see Fig. 5. The moment of inertia of the motor was measured through a bifilar pendulum swing experiment. The bolts are discretized with beam-like bodies and their masses are added as mass point elements. Between the foundation desk and the base element and between the frame bracket of the motor and the base element, there is the frictional contact type with a friction coefficient of 0.61. The bolts between the foundation desk and the base element are prestressed by the axial force determined according to the tightening torque. The various definitions of the prestressed modal analysis by the contact status parameter [9] defined in the ANSYS Workbench environment were tested.

The comparison of the measured and numerically computed natural frequencies is summarized in Table 6. For such a complex assembly, the relative errors of the natural frequencies are less than 8%.

Table 6. **Comparison of the modal properties of the foundation desk, element base, and motor.**

Natural frequency order / Dominant shape	Measured natural frequency [Hz]	Computed natural frequency [Hz]	Relative error [%]
1 / 1st bending	148.3	151.1	+1.9
2 / 1st torsional	179.2	185.2	+3.3
3 / partial bending	232.5	250.5	+7.7
4 / 2nd torsional	358.7	343.9	-4.1
5 / 2nd bending	411.2	389.8	-5.2
6 / partial lateral bending	437.5	413.0	-5.6
7 / 3rd torsional	455.6	455.2	-0.1

The measured (148 Hz) and computed (151 Hz) first bending mode shape of the foundation desk with base element and with electric motor is shown in Fig. 6. The first bending mode shape is not symmetric in the longitudinal direction due to the arrangement of the bolt joints and the motor body. The influence of the sticking and sliding contact zones on the mode shapes is obvious in the higher natural frequencies.

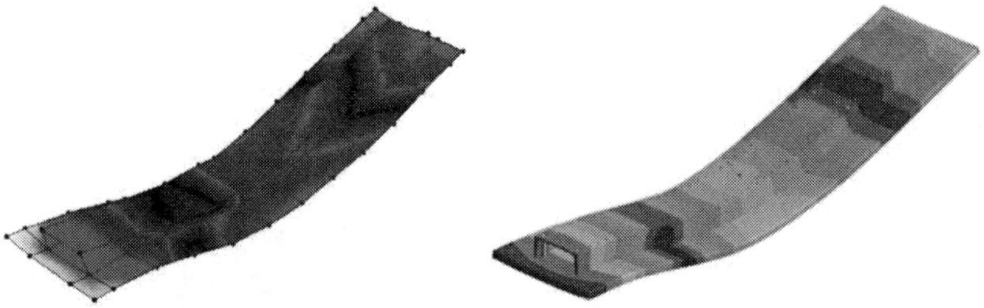

Fig. 6. Measured (left) and computed (right) 1st bending mode shape of the foundation desk with the base element and the electric motor.

5. Conclusions

The experimental rotor rig with the hydrodynamic bearing intended to work with the magnetically sensitive fluids was designed, manufactured, and tested. The preliminary measurement results demonstrated that the bearing load capacity is sensitive to variations in the applied electric current. The experimental and numerical results confirmed that the natural frequencies of the rotor rig are far enough from the operating speed of the designed magnetically sensitive hydrodynamic bearing. The relative errors between the lowest measured and computed natural frequencies were found to be insignificant. However, for the assembly of the foundation desk with the base element and electric motor, small discrepancies were observed between the measured and computed natural frequencies and mode shapes. The computational simulations with assemblies showed that the modelling of the contact interface between the components is crucial to the accuracy of the results.

Acknowledgements

The authors gratefully acknowledge the support of the Doctoral Grant Competition of VSB-TUO (No. CZ.02.2.69/0.0./0.0/19_073/0016945) within the Operational Programme Research, Development and Education (No. DGS/TEAM/2020-033), the Support for Science and Research in the Moravian-Silesian Region (No. RRC/02/2020), and the Czech Science Foundation (No. 19-06666S).

References

[1] Zapomĕl J. and Ferfecki P., "Vibration control of rotors mounted in hydrodynamic bearings lubricated with magnetically sensitive oil by changing their load capacity," *Bulletin of the Polish Academy of Sciences: Technical Sciences*, Vol. 69 (6), pp. 1–8 (2021). ISSN 2300-1917. http://dx.doi.org/10.24425/bpasts.2021.137988.

[2] Ferfecki P., Zapomĕl J. and Molčan M., "Design of the magnetically sensitive hydrodynamic bearing for the experimental rotor rig," *in Proceedings of 27/28th International Conference Engineering Mechanics 2022*, Milovy, Institute of Theoretical and Applied Mechanics of the Czech Academy of Sciences, Prague, 2022, pp. 105–108. ISSN 1805-8256. http://dx.doi.org/10.21495/51-2-105.

[3] Zapomĕl J., Ferfecki P. and Forte P., "Vibrations attenuation of a Jeffcott rotor by application of a new mathematical model of a magnetorheological squeeze film damper based on a bilinear oil representation," *Acta Mechanica*, Vol. 230, pp. 1625–1640 (2019). ISSN 1619-6937. https://doi.org/10.1007/s00707-018-2343-8.

[4] Zapomĕl J. and Ferfecki P., "The oscillation attenuation of an accelerating Jeffcott rotor damped by magnetorheological dampers affected by the delayed yielding phenomenon in the lubricating oil," *Journal of Vibration and Acoustics*, Vol. 140 (1), pp. 1–10 (2018). ISSN 1048-9002. http://dx.doi.org/10.1115/1.4037512.

[5] Urreta H., Leicht Z., Sanchez A., Agirre A., Kuzhir P. and Magnac G., "Hydrodynamic bearing lubricated with magnetic fluids," *Journal of Intelligent Material Systems and Structures*, Vol. 21 (15), pp. 1491–1499 (2010). ISSN 1045-389X. http://dx.doi.org/10.1088/1742-6596/149/1/012113.

[6] Wang X., Li H., Li M., Bai H., Meng G. and Zhang H., "Dynamic characteristics of magnetorheological fluid lubricated journal bearing and its application to rotor vibration control," *Journal of Vibroengineering*, Vol. 17 (4), pp. 1912–1928 (2015). ISSN 1392-8716.

[7] Susan-Resiga D. and Vékás L., "From high magnetization ferrofluids to nano-micro composite magnetorheological fluid: properties and applications," *Romanian Reports in Physics*, Vol. 70 (501), pp. 1–29 (2018). ISSN 1221-1451 43 822.

[8] Zapomĕl J. and Ferfecki P., "A new concept of a hydrodynamic bearing lubricated by composite magnetic fluid for controlling the bearing load capacity," *Mechanical Systems and Signal Processing*, Vol. 168 (1), pp. 1–13 (2022), ISSN 0888-3270. https://doi.org/10.1016/j.ymssp.2021.108678

[9] ANSYS INC., "Theory Reference for the Mechanical APDL and Mechanical Application," (2022).

Strength Analysis of the Spreader Carrier

Jan Flek[1], Martin Dub[2], Martin Havlíček[3] & František Lopot[4]

Abstract: This contribution is focused on experimental determination of dynamic effects acting on a construction of a spreader carrier. During the design, it is necessary to respect the dynamic load, which is related to the operating mode of agricultural machines. The aim of the contribution is to experimentally determine the dynamic load coefficients for the subsequent strength calculation of the spreader carrier structure. The dynamic effects acting on the structure are determined by measuring the acceleration on the spreader and on the carrier. Due to the strength analysis of the structure performed using the FEM, it is possible to analyse the structure with a load respecting the dynamic operation of the agricultural equipment.

Keywords: Agricultural Machines; Dynamic Effects Measurement; Strength Analysis

1. Introduction

Agricultural machines requires the design of structures that meet specific operating loads. When operating agricultural machines, dynamic forces and impact loads act on the structure, which can damage the machine [1]. For this reason, it is necessary to determine the dynamic load coefficients that correspond to the specific operating conditions of the machines and to take these coefficients into account in the strength calculations of the proposed structures. Worldwide research is devoted to this area, which includes analytical models of agricultural machines loading [2, 3]. Furthermore, the finite element method is used as an alternative to analytical tools. FEM analysis for specific uses of agricultural machinery is important for comparing theoretical, empirical and experimental data [4]. Nowadays, experimental detection of loads acting on specific constructions of agricultural machines is also very widespread. The dynamic effects must be determined and included in the calculations and design of these machines with regard to the operation and quality of the terrain, which causes a significant increase in force effects on the structure. Due to the possibilities of measurements an experimental analysis of the dynamic effects on the structures was performed in this case on the structure of the spreader carrier. Based on the identified dynamic effects, which cause a significant load on the structure it was possible to perform a subsequent strength analysis of the structure using the finite element method and subsequently evaluate critical points.

2. Methods

In order to determine the magnitudes of the dynamic effects during the normal operation of the spreader, the acceleration was measured on the carrier and the spreader. The obtained accelerations are the basis for determining the dynamic coefficients that increase the overall force effect on the investigated structure. The experiment was preceded by partial steps that led to the selec-

[1] *Ing. Jan Flek; Dep. of Designing and Machine Components , Fac. of Mech. Engineering, Czech Technical University in Prague; Technická 4, 166 07 Prague, Czech Republic; jan.flek@fs.cvut.cz*
[2] *Ing. Martin Dub, Ph.D.; Dep. of Designing and Machine Components , Fac. of Mech. Engineering, Czech Technical University in Prague; Technická 4, 166 07 Prague, Czech Republic; martin.dub@fs.cvut.cz*
[3] *Ing. Martin Havlíček; Dep. of Designing and Machine Components , Fac. of Mech. Engineering, Czech Technical University in Prague; Technická 4, 166 07 Prague, Czech Republic; martin.havlicek@fs.cvut.cz*
[4] *Ing. František Lopot, Ph.D.; Fac. of Mech. Engineering, Czech Technical University in Prague; Technická 4, 166 07 Prague, Czech Republic; frantisek.lopot@fs.cvut.cz*

tion of a measuring apparatus. These steps are: Identification of the construction of the spreader carrier, selection of places for placement of accelerometers, selection of suitable accelerometers, selection of equipment for data recording, testing of equipment functionality, determination of driving modes (driving on the road, field road and crossing an obstacle), data processing, and defining the FEM model. These methodological steps are described in the following paragraphs. The resulting detected overload was used for strength analysis using the finite element method.

2.1. Identification of the construction of the spreader carrier

A complete load condition analysis was performed on a conventional spreader carrier (Fig. 1) with a specified maximum load capacity of 6,000 kg in static load mode. The carrier's construction consists of an axle with variable track setting (1800–2500 mm), and a base frame made of square tube rectangular steel profile. There are welded two support arms at the front of the structure and a welded sheet metal structure for placing the pull rod and towing eye. Due to the two arms and the pull rod, attaching the spreader to the carrier is possible. The towing eye secures the connection to the tractor.

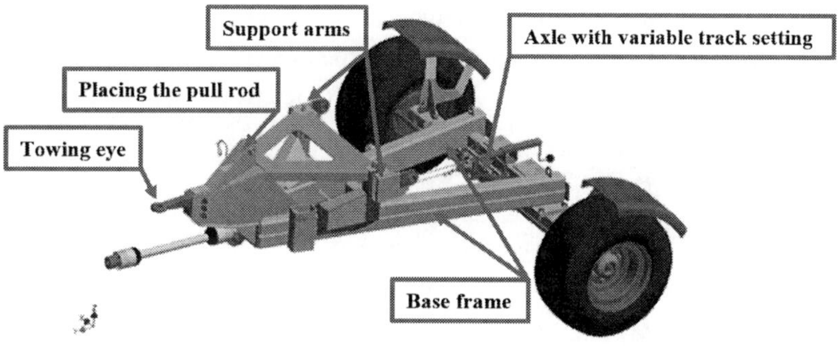

Fig. 1. The spreader carrier.

Fig. 2. Placement of accelerometers.

2.2. *Selection and placement of accelerometers and data recording equipment*

Selected measuring equipment included:

- Kistler accelerometer: type 8766A100BB, measuring range ±100 g
- MMF accelerometer: type KS943B.10, measuring range ±100 g
- Measuring apparatus by Dewetron: Dewesoft with a sampling frequency of 20,000 Hz

Two places for accelerometers were selected on the machine (Fig. 2). The accelerometers were attached to the chassis frame and spreader with strong magnets, which ensured contact of the accelerometers with the structure throughout the measurement. An accelerometer from Kistler was placed on the chassis frame. An MMF accelerometer was placed on the spreader. The accelerometers were connected to the Dewetron measuring apparatus with cables.

2.3. *Determination of driving modes*

The driving modes were set to correspond to the normal operation of the measured machine. For this reason, road and field road were included in the measurements. At the same time, these operating modes were measured with a fully loaded spreader (weight of the spreader contents was 720 kg) and with an empty spreader. An overview of the mentioned driving modes is given in the Table 1 (modes 1–7). Other specific modes were measured outside the field road with a fully loaded spreader, this is the crossing of obstacles (curb, trough). These modes are listed in positions 8–10 in the Table 1.

Table 1. Overview of realized driving modes.

Mode	Measured section	Speed [km/h]	Spreader contents mass [kg]
1	road	33	0
2	field road	accelerating–breaking	0
3	field road	15	0
4	field road	manimulation	0
5	road	40	720
6	field road	accelerating–breaking	720
7	field road	20	720
8	curb	proportionate to the obstacle	720
9	trough (1 wheel)	proportionate to the obstacle	720
10	trough (2 wheels)	proportionate to the obstacle	720

2.4. *Data processing*

The output of experimental measurements using accelerometers are acceleration time courses in three axes of measuring sensors x, y, z (longitudinal, transverse, vertical direction).To obtain the most frequent overload during the measurement time period and to obtain the real overload that acts on the measured construction while driving, histograms of the frequency of individual acceleration components (assumed normal frequency distribution) were created. Due to them, it was possible to determine the resulting and long-term overload that acts on the carrier and the spreader during a given driving mode. From the thus determined additional accelerations of the carrier and the spreader, it was possible to subsequently determine the force effect on the carrier and the spreader during the FEM analyses. Specifically, the FEM analyses include a static mode including the maximum spreader load extended by the dynamic effects of the supported mass during acceleration and a static mode including the maximum spreader load extended by the dynamic effects of the supported mass during braking.

2.5. *Defining the FEM model*

The FEM modeling method of the Dlubal RFEM 5 software using shell elements with the appropriate sheet metal thickness was selected for the calculation. In places such as the plates for connecting the axle to the spreader carrier frame or the towing eye for connecting to the towing machine (tractor), it was not possible to use shell elements in terms of part geometry. Therefore solid elements modeling was used for these parts. On the contrary, the modeling of the drawbar of the spreader hinge was modeled due to its geometry using a beam element, which well respects the real design. The Table 2 shows the statistics of the mesh.

Table 2. Statistics of the mesh.

Number of beam elements (1D)	71
Number of shell elements (2D)	70,731
Number of solid elements (3D)	41,440
Number of mesh nodes	75,507
Finite element size [mm]	10
Finite element size at critical points of the model [mm]	3

The spreader carrier model is loaded by the structure's own weight and a vertical force of 60,000 N at the spreader's center of gravity, which in the static calculation mode represents a fully loaded spreader hopper with 6,000 kg of filling. The calculation with regard to dynamic effects takes into account the actual weight of the structure and the vertical force load at the center of gravity of the spreader, which is the sum of the full hopper weight (6,000 kg) with dynamic overload in the vertical z-axis. The center of gravity is connected with rigid beams to the side connecting eyes of the carrier and to the third point of the suspension - the drawbar eye.

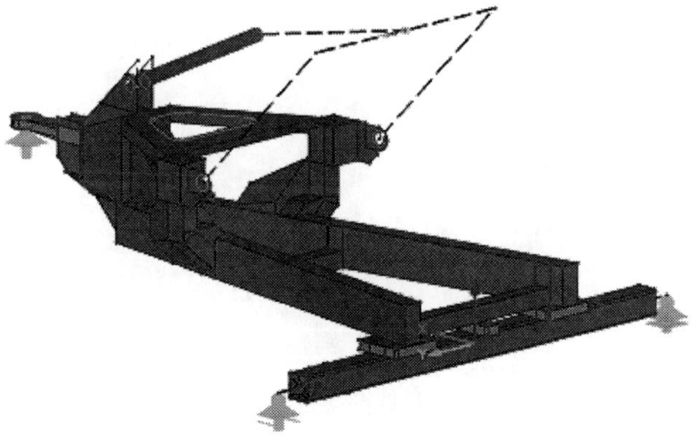

Fig. 3. FEM model of the spreader carrier

The model is mounted on two supports at the axle ends, which allow rotation around all three axes (the left support is also sliding on the x-axis, and the right support is sliding on the x-axis and y-axis) and one support in place of the towing eye, which prevents all displacements and allows rotation in all three axes. The model of spreader carrier is shown in Fig. 3.

3. Results

The procedure of processing the measured data was described in subsection 2.4. The results are therefore divided into a part that deals with the interpretation of the measured acceleratios and a part that deals with the implementation of the load extended by the measured dynamic effects into the static stress analysis of the spreader carrier structure performed using FEM calculation.

3.1. Interpretation of measured acceleratios

The measuring sensors (accelerometers) provided acceleration courses along the x, y, z axes during the measured time on the carrier and on the spreader. An example of the acceleration courses on the carrier and on the spreader is shown in Fig. 4. Fig. 4 shows the accelerations along the x-axis in driving mode 2 described in Table 1.

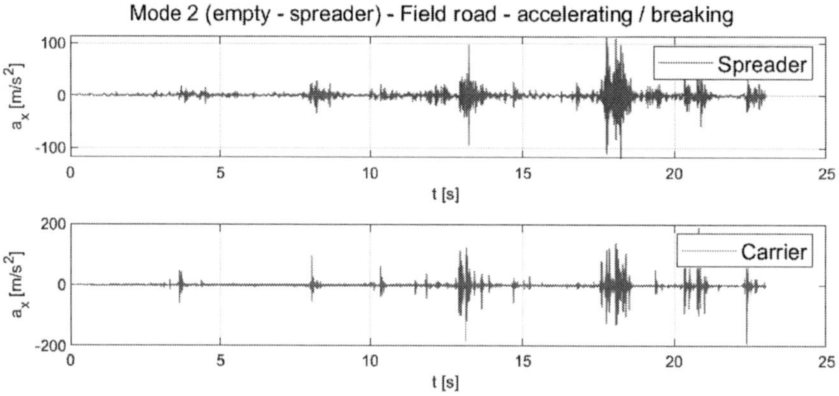

Fig. 4. Acceleration course on carrier and spreader (along the x-axis, driving mode 2).

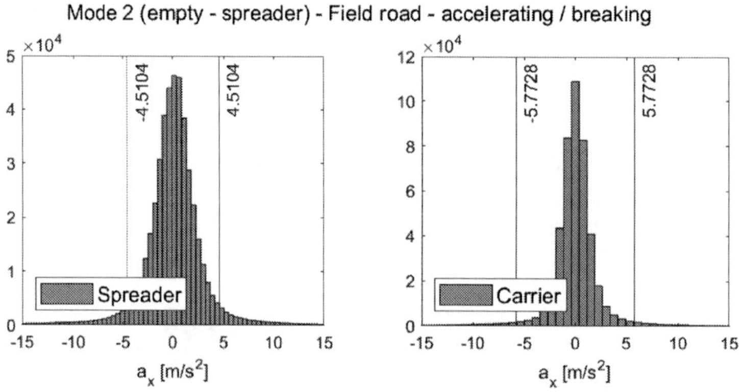

Fig. 5. Histograms of the frequency of accelerations along the x-axis (spreader, carirrier, driving mode 2).

To obtain the most frequent overload during the measurement period and to obtain the real overload acting on the measured system while driving, histograms of the frequency of individual acceleration components were created so it was possible to determine the resulting and long-term overload acting on the carrier and spreader during a given driving mode. For the ilustration of clarity, the case of the mode 2 is given here. The overload frequency histogram in Fig. 5 is therefore based on the measured acceleration curves shown in Fig. 4.

From the histograms in Fig. 5 it is possible to see the individual standard deviations from zero acceleration, which occurred most frequently during the driving mode and therefore have the greatest influence on the load-bearing capacity of the structure. The overall evaluation of the structural overload during all modes is recorded in the Table 3.

Table 3. Overload overview for all driving modes.

Axis	x-axis		y-axis		z-axis	
Mode	Spreader	Carrier	Spreader	Carrier	Spreader	Carrier
1	0.116 g	0.179 g	0.074 g	0.087 g	0.198 g	0.194 g
2	**0.451 g**	**0.577 g**	**0.495 g**	–	**0.687 g**	**0.580 g**
3	0.343 g	0.371 g	0.365 g	0.206 g	0.581 g	0.516 g
4	0.053 g	0.070 g	0.073 g	0.056 g	0.101 g	0.093 g
5	0.120 g	0.341 g	0.145 g	0.104 g	0.159 g	0.181 g
6	0.218 g	0.401 g	0.252 g	0.200 g	0.343 g	0.461 g
7	0.155 g	0.283 g	0.173 g	0.175 g	0.286 g	0.454 g
8	0.082 g	0.159 g	0.130 g	0.107 g	0.136 g	0.135 g
9	0.216 g	0.413 g	0.231 g	**0.208 g**	0.327 g	0.364 g
10	0.183 g	0.384 g	0.217 g	–	0.273 g	0.327 g

The Table 3 shows that the greatest dynamic effects, overload values in the x and z directions, were achieved in driving mode 2, which corresponds to the accelerating and braking of a tractor with a coupled carrier with an empty spreader hopper on a field road. According to the measurement of the greatest overload in the y-direction, the carrier reached during driving mode 9, which corresponds to the passage of an obstacle (trough) by one wheel when the spreader hopper was fully filled.

Due to the additional accelerations of the carrier and spreader determined in this way, it was possible to subsequently determine the force effect on the carrier and the spreader for the FEM calculation.

The following forces (given by Eq. 1) in the longitudinal, transverse, and vertical directions characterize the load of the carrier, including the influence of dynamic effects. Final forces are mentioned in Table 4.

$$F = a \cdot m, \tag{1}$$

where a is overload from Table 3 (Mode 2), and m is required load capacity of the carrier.

Table 4. Force effect on the carrier.

F_x	F_y	F_z
±30 kN	30 kN	102 kN

3.2. FEM analysis

The force effects, which were determined from the measurements in subsection 3.1, were applied to the spreader carrier model, which is described in subsection 2.5. In addition to the load defined from the measurement, the own weight of the structure is applied to the model. The most unfavorable load condition was used to evaluate the strength of the structure for the possibility of correctly evaluating the suitability of the structure. This load condition corresponds to the values in the Table 4. The calculation considers dynamic loading in all three directions at the same time, so it is a simulation of the extreme load of the structure, when the structure is stressed asymmetrically and neglecting the suspension axle and tire. The material of the construction of the spreader carrier is steel S355 with a yield strength of 355 MPa.

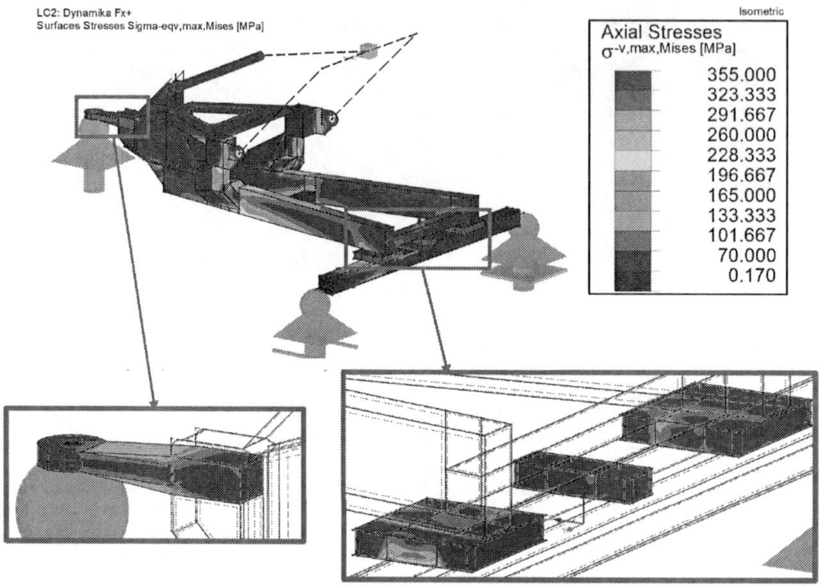

Fig. 6. Global FEM analysis results – Mises stress.

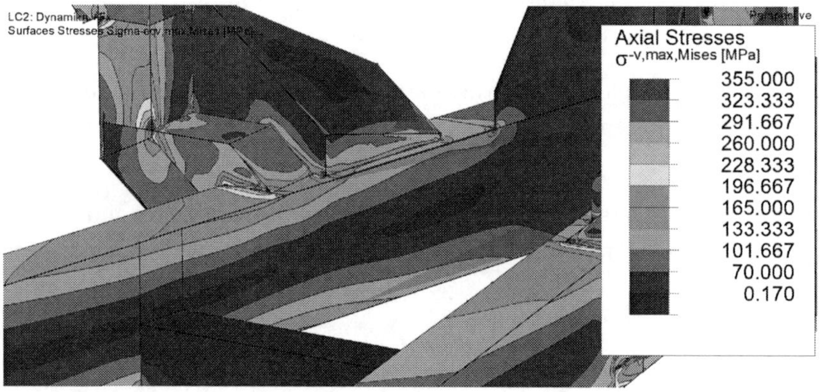

Fig. 7. The most stressed places – Mises stress.

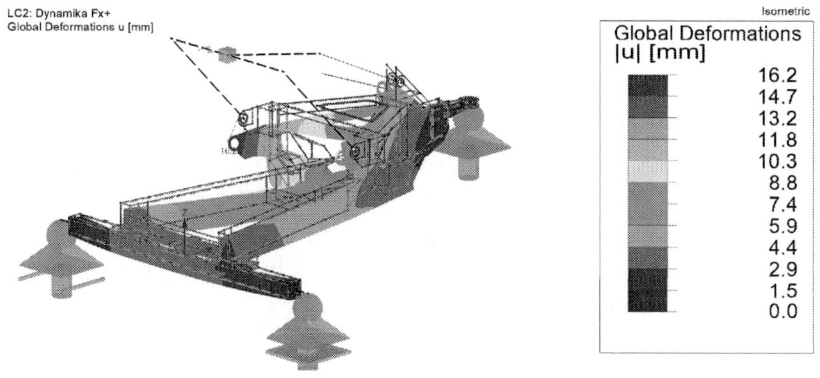

Fig. 8. Global deformation of structure.

The Fig. 6 describes the Mises stress distribution. Final stress range is limited to 355 MPa from above. The most stressed places are located in the area of the weldment of the two attachment arms and in the place of welding the flanges for the connection of the supporting frame with the axle. The most stressed places are shown in the detail Fig. 7.

From a global perspective, the Mises stress in the structure is up to about 165 MPa. This corresponds to a static safety factor of 2.15 against yield strength. The maximum stress in the structure is 320 MPa but it is a local stress concentration caused by the modeling method (sharp surface transitions, absence of rounding and filling of corners by welds, etc.).

The load applied to the structure caused global deformations, which are shown in the Fig. 8. The maximum deformation value is 16.2 mm and can be read on the left attachment arm.

4. Conclusions

This contribution dealt with the issue of experimental determination of the operational dynamic load of agricultural machines structures (specifically spreader carriers), more precisely determination of dynamic coefficients that must be included in the static load of the structure in order to correctly verify the strength of the structure with respect to its operating condition. Due to measurements using accelerometers, dynamic coefficients in three directions x, y and z were determined, and were included in the total static load of the structure. These coefficients were expressed as multiples of the gravitational acceleration g and were defined here as overload. The highest overload values were reached in the tested driving mode 2. Unfortunately, it was not possible to evaluate the overload on the carrier in the direction of the y-axis from the measured data from mode 2, therefore the value in this direction was determined from mode 9. The resulting overload values considered for the strength calculation are 0.577 g for the x-axis, 0.208 g for the y-axis and the maximum overload was determined 0.687 g in the z-axis direction.

The strength calculation was performed by the FEM method and the results show that including the maximum measured overloads, from a global perspective on the structure, a Mises stress of 165 MPa and at critical points of approximately 320 MPa is achieved. Critical points are not entirely relevant, as the model does not include welds and the individual surfaces do not follow each other continuously. In the end, it can be stated that the structure is designed adequately with regard to its operation, as in terms of safety it reaches a value of 2.16 with regard to the yield strength of material (355 MPa). The calculation was performed in a way that the safety criterion of the structure was respected. The total maximum deformation was 16.2 mm and was achieved on the side connecting arm of the structure.
The performed experimental method of determining the overload of the structure is a suitable method that can be used for a wide range of other machines.

References

[1] Flores M.J. et al., "Evaluation of Methods to Calculate Dynamic Load Coefficients," *Revista Ciencias Técnicas Agropecuarias*, Vol. 28, no. 4, (2019). ISSN 2071-0054.

[2] Bulgakov V., Pascuzzi S., Nadykto V. and Ivanovs S., "A Mathematical Model of the Plane-Parallel Movement of an Asymmetric Machine-and-Tractor Aggregate," *Agriculture*, Vol 8(10), no. 151, (2018). https://doi.org/10.3390/agriculture8100151.

[3] Kromulski J., "New method for identification of uncorrelated forces acting in the agricultural machines," *Journal of Research and Applications in Agricultural Engineering*, Vol. 51, (2006). ISSN 1642-686X.

[4] Shen J. and Kushwaha R.L., *Soil-Machine Interactions: A Finite Element Perspective (1st ed.)*. (Routledge, New York, 1998). https://doi.org/10.1201/9780203739228.

Viscoplastic Behaviour of Nylon Reinforced by Short Carbon Fibres under Room Temperature

Radim Halama[1], James Nicholas[2], Bhuvanesh Govindaraj[3] & Marek Pagáč[4]

Abstract: This paper presents mainly experimental determination of the viscoplastic properties of Onyx material used in the FDM technology of Markforged 3D printers. A dog bone specimen was used in the monotonic, creep, and relaxation tests performed under room temperature. First, the sensitivity of stress-strain behaviour to strain rate was investigated by standard tensile tests considering three-position rates realized up to 10 % of axial strain. A biaxial extensometer ensured simultaneous measurement of transverse and the longitudinal strains. For two chosen position rates, stress relaxation was followed with a fixed 10 % axial strain. Creep tests were performed on a dog bone and an hourglass specimen for comparison. The latter option required the application of the digital image correlation method. All experiments were simulated using the finite element method using the Perzyna viscoplastic model combined with the nonlinear kinematic hardening rule of Chaboche. The material model gives good predictions in monotonic loading cases with relaxation of axial stress.

Keywords: Composites; Carbon fiber; Nylon; Relaxation; Creep; FEM

1. Introduction

Recent progress in additive manufacturing brings new challenges to the characterisation of composite materials. The short carbon fiber content can significantly improve the mechanical properties of nylon [1, 2]. An example of such composite is Onyx, used in the FDM technology of Markforged 3D printers. The material is used to manufacture precise parts, which can be directly reinforced by long fibres in the chosen layers [3]. The previous study focused on the measurement of elastic properties of Onyx and carbon fibres (CF) as well as evaluating a numerical model to be able to predict the stiffness of designed structures reinforced in chosen layers or by CF contours. The investigation was limited to the field of elastic behaviour. Significant differences between datasheet values and the optimal Young modulus were discussed also in the master's thesis [4] and the Ph.D. thesis [5].

The purpose of this study is to determine the viscoplastic properties of the Onyx material at room temperature by tensile testing. Short-term creep and relaxation tests were performed. The experimental data are then used for the validation of a viscoplastic material model available in commercial software.

2. Mechanical testing

A dog bone specimen was used in the monotonic, creep and relaxation tests performed under room temperature; see Fig. 1. Specimens were prepared by the Fused Deposition Modelling

[1] *prof. Ing. Ph.D. Radim Halama; Dep. of Applied Mechanics, Fac. of Mech. Engineering, VSB-Technical University of Ostrava; 17.listopadu 2172/15, 70800 Ostrava, Czech Republic;* radim.halama@vsb.cz

[2] *Ing. Rajkumar Arokiaraj James Akshay Nicholas; Dep. of Applied Mechanics, Fac. of Mech. Engineering, VSB-Technical University of Ostrava; 17.listopadu 2172/15, 70800 Ostrava, Czech Republic;* nicholas.james39@yahoo.com

[3] *Ing. Bhuvanesh Govindaraj; Field Application Engineer- EMEIA Region, JPT ELECTRONICS PTE. LTD., 2 Bukit Batok Street 23, Singapore 659527;* bhuvanesh@jptoe.com

[4] *doc. Ing. Ph.D. Marek Pagáč; Dep. of Machining, Assembly and Eng. Metrology, Fac. of Mech. Engineering, VSB-Technical University of Ostrava; 17.listopadu 2172/15, 70800 Ostrava, Czech Republic;* marek.pagac@vsb.cz

(FDM) technology, more specifically on a Markforged X7 3D printer. The thickness of each Onyx layer was 0.125 mm. All specimens were printed in the 90° orientation. This means that the fibres in each even layer are perpendicular to the applied load (Fig. 1). A similar printing strategy was also used for the hourglass specimen visible in Fig. 2.

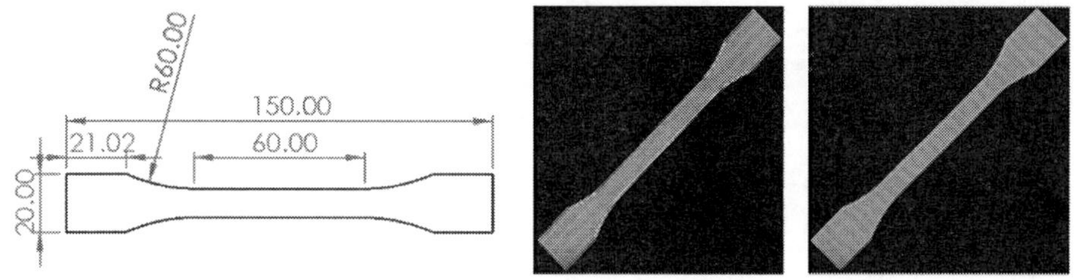

Fig. 1. Dog bone specimen geometry and a design of its structure in the first and the second layer.

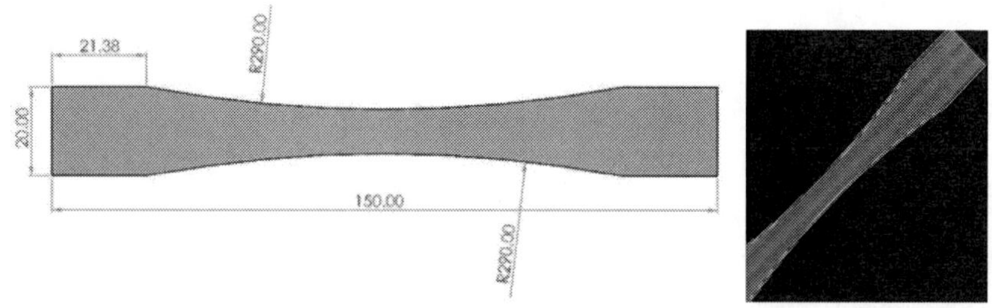

Fig. 2. Hourglass specimen geometry and a design of its structure in the first layer.

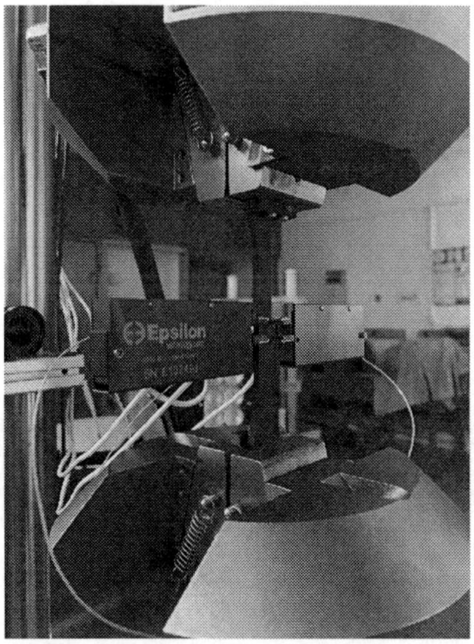

Fig. 3. Biaxial extensometer.

Fig. 4. LabControl 100kN/1kNm and DIC system.

Tensile testing was performed using a LabControl 100kN / 1000Nm hydraulic testing machine available in the Department of the applied mechanics of the FME of VSB-Technical University of Ostrava. Epsilon 3560-BIA-010M-010-HT2 biaxial extensometer (Fig. 3) was used for the simultaneous measurement of longitudinal and transverse strain. With the hourglass specimen, a 2D optical measurement of creep strain has been done, based on Digital Image Correlation (DIC). A mono camera system with a 2.3Mpx CCD camera and Mercury RT software was used to evaluate the gradient of strain along with the specimen. The experimental setup is visible in Fig. 4.

The sensitivity of the stress response of Onyx to the strain rate was investigated by standard tensile tests considering three-position rates (2.5 mm/min, 25 mm/min, 250 mm/min). For maximal and minimal position rates, stress relaxation followed at a fixed 10% of axial strain; see Fig. 5.

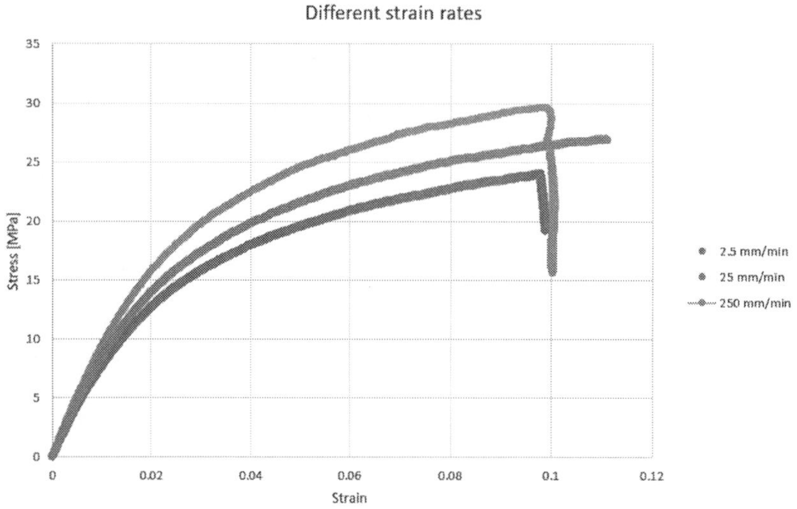

Fig. 5. Resulting stress-strain curves in tensile tests performed for three different position rates.

An hourglass specimen has been subjected to a creep test followed by relaxation in four steps. The specimen was subjected to a uniaxial load of 1000 N for 3600 seconds and then relaxed at four intervals for 1000 seconds, while the force was decreased by 250 N in each of them until the force reaches 0 N.

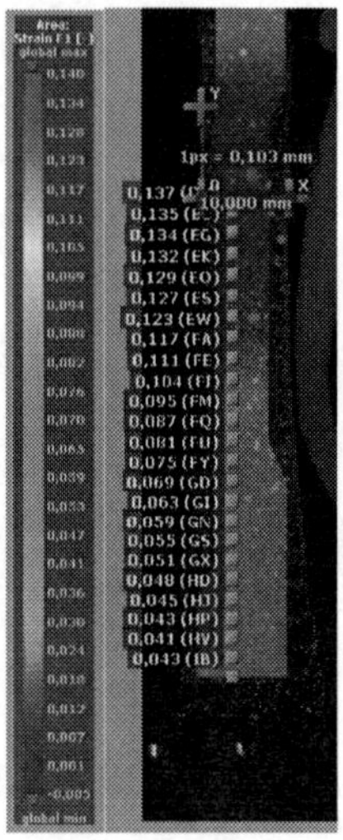

Fig. 6. Contours of the first principal strain – creep/relaxation test [6].

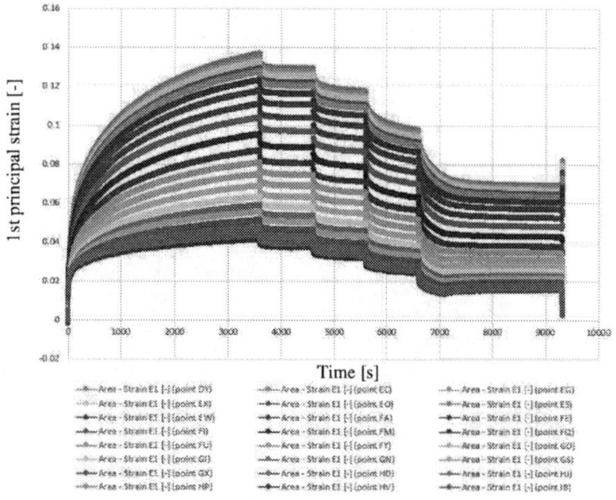

Fig. 7. Creep/relaxation curves evaluated for points considered in the DIC measurement [6].

The contour of the first principal strain was recorded using DIC. The distribution of the quantity for the 3600 s is depicted in Fig. 6. Particular points were selected along the axis of the specimen to be used for the time-history evaluation of the first principal strain in the form of a graph in Fig. 7.

3. FE modelling

The experiments were simulated using the Finite Element Method (FEM) in ANSYS APDL Mechanical. In this contribution, the constitutive equations are written for the uniaxial loading case. The viscoplastic strain is assumed to be composed of the transient plastic strain ε_{tr} and the steady plastic strain ε_{ss}. Then, the total strain can be expressed as

$$\varepsilon_{tot} = \varepsilon_{el} + \varepsilon_{tr} + \varepsilon_{ss'}, \tag{1}$$

where ε_{el} is the elastic strain component, which can be calculated by the Hooke's law

$$\varepsilon_{el} = \frac{\sigma}{E}, \tag{2}$$

where σ is the axial stress and E is Young modulus.

Perzyna viscoplastic model in combination with Chaboche nonlinear kinematic hardening rule [6] is considered for the computation of the transient viscoplastic strain. The flow rule according to Perzyna [7] is defined as

$$\dot{\varepsilon_{tr}} = \nu\left(\frac{\sigma - \alpha}{\sigma_\gamma} - 1\right)^{\frac{1}{m}}, \tag{3}$$

where $\dot{p}$ is equivalent plastic strain rate, α is the backstress, ν is the material viscosity parameter, σ_Y is the yield stress and m is the strain rate hardening parameter. The superposition of the back-stress parts is considered according to Chaboche [8]:

$$\alpha = \sum_{i=1}^{M} \alpha_i, \; \mathrm{d}\alpha_i = C_i \mathrm{d}\varepsilon_p - \gamma_i \alpha_i \mathrm{d}p, \tag{4}$$

where α_i is the i-th backstress part, ϵ_p is the plastic strain, C_i and γ_i are material parameters and M is the number of back-stress parts (in this study M = 5). The steady plastic strain ϵ_{ss} is aimed to describe the steady creep behaviour of material, thus the Norton law is considered

$$\dot{\varepsilon_{ss}} = A\sigma^n, \tag{5}$$

where A and n are material parameters. All the parameters of the material model are stated in Table 1.

Table 1. Parameters of Perzyna/Chaboche model combined with Norton creep law for Onyx.

Parameters	m [–]	ν [s^{-1}]	σ_γ [MPa]	C_1 [MPa]	γ_1 [–]	C_2 [MPa]	γ_2 [–]	C_3 [MPa]	γ_3 [–]	C_4 [MPa]	γ_4 [–]	C_5 [MPa]	γ_5 [–]	A [s^{-1}]	n [–]
Values	0.38	$2.5 \cdot 10^{-4}$	1	1520	400	720	200	380	100	210	70	160	10	$4.4 \cdot 10^{-10}$	3

The elastic properties considered in the analyses are Young modulus E = 1022 MPa and Poisson's ratio $\mu = 0.43$. The simulation of tests performed on the dog-bone specimen was done on the model containing one 20-node brick element. The dog-bone model was created using the same element type and its detailed description is given elsewhere [9].

4. Results comparison

A graphical comparison of numerical predictions with experimental results obtained on the dog-bone specimens are shown in Fig. 8 - Fig. 11. Considered viscoplastic model can describe the strain rate influence in the monotonic tensile tests very well, see Fig. 8. The Norton law was calibrated from the creep test performed on the hourglass specimen considering creep strain rate observed for 3600 s, see Fig. 9.

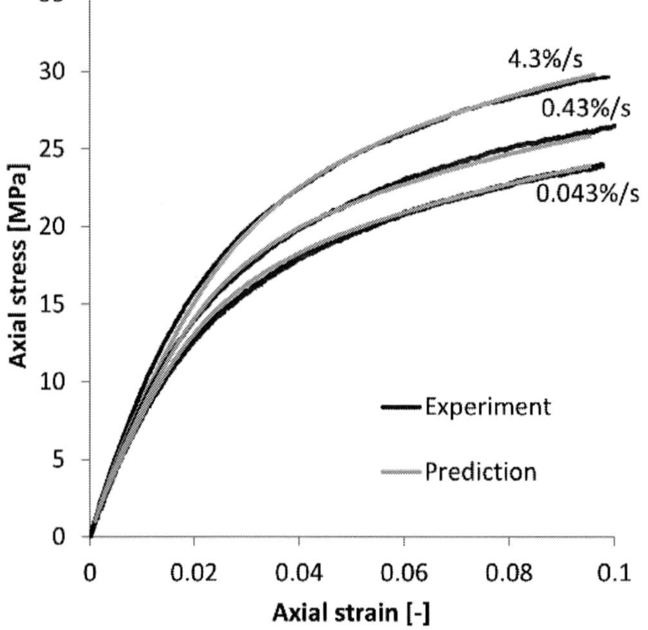

Fig. 8. Comparison of experiment and prediction obtained by the viscoplastic model – tensile tests under 2.5 mm/min, 25 mm/min and 250 mm/min position rates [6].

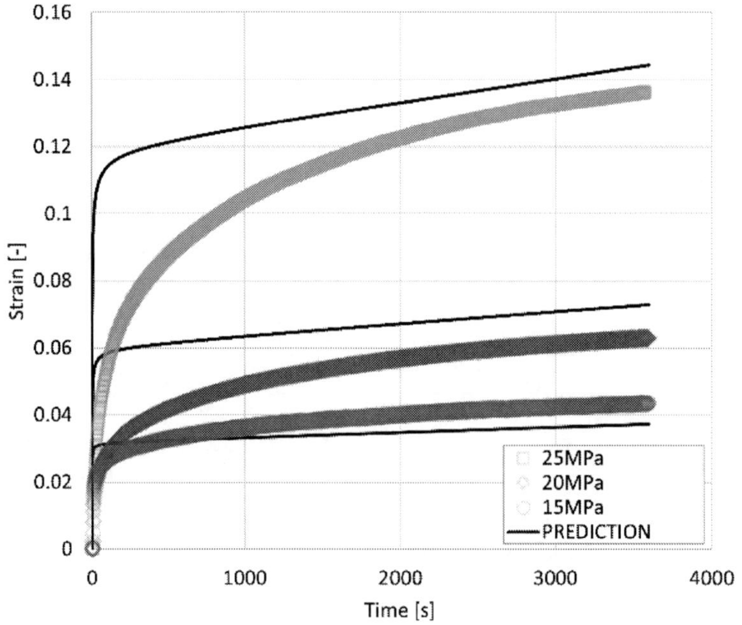

Fig. 9. Comparison of experiment and prediction obtained by the viscoplastic model – the creep test performed on an hourglass specimen (experiment and numerical model taken from [9]).

In the simulation of stress relaxation, two cases of initialization of axial strain are considered: 2.5 mm/min and 250 mm/min. For both cases, the stress relaxation is slower than observed in the mechanical tests (Fig. 10 and Fig. 11).

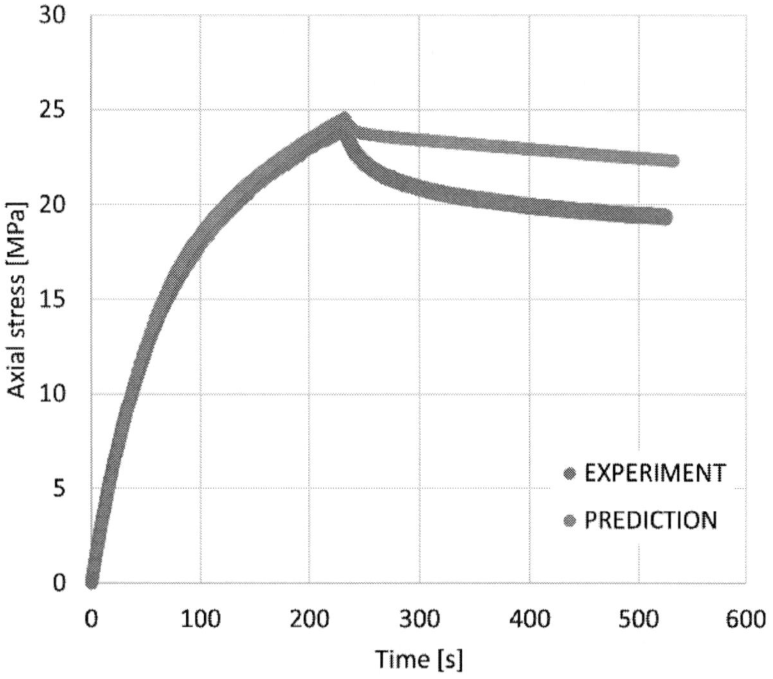

Fig. 10. Comparison of experiment and prediction obtained by the viscoplastic model – test with initialization of 10% axial strain by 2.5 mm/min position rate [6].

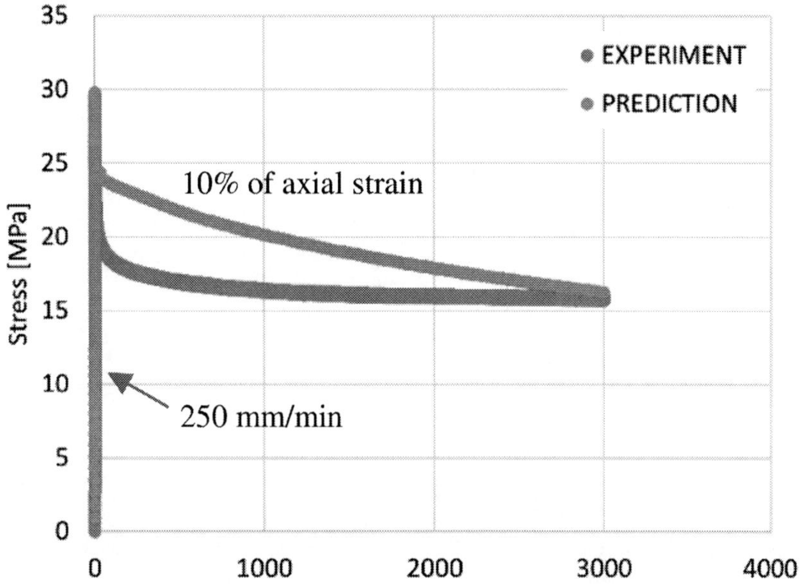

Fig. 11. Comparison of experiment and prediction obtained by the viscoplastic model – test with initialization of 10% axial strain by 250 mm/min position rate [6].

5. Conclusions

The experimental investigations show the sensitivity of Onyx to the strain rate even under room temperature; the higher the strain rate, the higher the stress response. Significant creep behaviour has also been observed above 15 MPa of constant uniaxial load, which was evaluated from a digital image measurement on the newly designed specimen [9]. Tensile tests followed by a relaxation part and the creep test performed on an hourglass specimen were simulated using the finite element method.

The Perzyna flow rule and the Chaboche nonlinear kinematic hardening rule make possible describing the correct stress-strain behaviour of Onyx for a different strain rates under monotonic tensile loading. Norton law is used to capture the creep behaviour more reliably. Unfortunately, the relaxation of Onyx is not described perfectly by the model considered. Therefore, this study will continue by using or developing a more robust material model to describe the behaviour satisfactorily.

Acknowledgements

This work was supported by The Ministry of Education, Youth and Sports from the Specific Research Project (SP2022/26) and has been done in connection with the DMS project reg. no. CZ.02.1.01/0.0/17_049/0008407 financed by Structural Funds of Europe Union.

References

[1] Blok L. G., Longana M. L., Yu H. and Woods B. K. S., "An investigation into 3D printing of fibre reinforced thermoplastic composites," *Additive Manufacturing*, 22, pp. 176-186 (2018). ISSN 2214-8604. https://doi.org/10.1016/j.addma.2018.04.039.

[2] Sága M., Bárnik F., Vaško M., Handrik M. and Kopas P., "Identification of physical characteristic of composite materials produced by additive technology from perspective of selected mechanical properties," *Acta Physica Polonica A*, 138(2), pp. 249 – 252 (2020). ISSN 0587-4246. http://dx.doi.org/10.12693/APhysPolA.138.249.

[3] Horňáček L., Halama R., Krzikalla D., Měsíček J., Pauldoss J. and James N., "Validation of Methodology for Numerical Modeling of 3D Printed Structures Reinforced by Long Carbon Fibers," in *Proceedings of Experimental Stress Analysis - 59th International Scientific Conference*, Litomyšl, 2021, pp. 58-63. ISBN 978-80-01-06885-4.

[4] Pauldoss J.S.D.P., *Modeling of Mechanical Tests on Composite Specimens Produced By 3d Printing*, (VSB - Technical University of Ostrava, Faculty of Mechanical Engineering, Department of Applied Mechanics,Ostrava, 2021), 68 p.

[5] Horňáček L., *Application of ESPI Method for Experimental Analysis of Vehicle Frame on Basis of Composite-Metal*, VSB - Technical University of Ostrava, Faculty of Mechanical Engineering, Department of Applied Mechanics, Ostrava, 2022, 119 p.

[6] Nicholas R.A.J.A., *Investigation of stress-strain behaviour in ONYX material and its modelling*, (VSB - Technical University of Ostrava, Faculty of Mechanical Engineering, Department of Applied Mechanics, Ostrava, 2020), 57 p.

[7] Neto E.A. de S., Perič D. and Owens D.R.J., *Computational methods for plasticity: theory and applications* (John Wiley & Sons, Ltd, 2008). ISBN 978-04-70-69452-7.

[8] Lemaitre J. and Chaboche J.L., *Mechanics of solid materials* (Cambridge Un. Press, 1990). ISBN 978-0-521-32853-1.

[9] Govindaraj B., *Digital Image Correlation and its Application for Accelerated Testing of Specimens Manufactured by 3D Printing* (VSB - Technical University of Ostrava, Faculty of Mechanical Engineering, Department of Applied Mechanics, Brno, 2020), 68 p.

EAN 2022 60th annual conference on experimental stress analysis

Fatigue Life Prediction Using Recurrent Neural Networks and Critical Plane Approach

Jiří Halamka[1] & Michal Bartošák[2]

Abstract: Lately the machine learning methodology has been used in fatigue life prediction models as an attempt to overcome limitations of the classical approach. In this paper, a fatigue life prediction method for components under cyclic plane stress loading is proposed. This method combines the critical plane approach with the capabilities of artificial neural networks. Eleven different cases of cyclic axial-torsion loadings of the 42CrMo4 steel are utilized for the demonstration purpose in this study. The model achieved great prediction accuracy on this data set, however its ability to extrapolate is very unstable.

Keywords: Fatigue life; Multiaxial loading; Neural networks; Critical plane

1. Introduction

The issue of multiaxial material fatigue is complicated and far from solved. Classical models are usually limited to specific materials and loading cases. To overcome these limitations, machine learning based models such as artificial neural networks have started to be used [1, 2].

Recently it has been shown that recurrent neural units like LSTMs are suitable for predicting multiaxial fatigue life even for various nonproportional loadings [3]. Yang et al. [3] proposed that timeseries of all 6 stress components are used as an input for their proposed Artificial Neural Network. However, that would cause the predicted fatigue life values to vary with rotation of the coordinate system.

In this work, an attempt is made by combining the capabilities of artificial neural networks with the property of invariance incorporated within the critical plane approach. The scope of this work is limited to the plane stress case, because there are no available datasets for 3D stress loading conditions.

2. Brief introduction to Artificial Neural Networks

The simplest form of Artificial Neural Networks is so-called Feedforward Neural Network (FNN) (Fig. 1), that constructs a relationship between input vector $\vec{x}$ and output vector $vecy$. The output vector $\vec{y}$ of a single hidden layer is simply computed by (1)

$$\vec{y} = g(\boldsymbol{W}\vec{x} + \vec{b}), \tag{1}$$

where $\boldsymbol{W}$ is the weights matrix, $\vec{b}$ is the bias vector and g is the activation function. The FNN constists of one or more of these layers, where each neuron in one layer is connected to every neuron in the next layer.

Recurrent Neural Networks (RNN) (Fig. 2) are a type of neural networks that are capable of processing history dependent relationship between input sequence $[\vec{x}_1, \vec{x}_2, \cdots, \vec{x}_L]$ and output

[1] *Ing. Jiří Halamka; Division of Strength and Elasticity, Fac. of Mech. Engineering, Czech Technical University in Prague; Technická 4, 166 07 Prague, Czech Republic;* jiri.halamka@fs.cvut.cz

[2] *Ing. Michal Bartošák, Ph.D.; Division of Strength and Elasticity, Fac. of Mech. Engineering, Czech Technical University in Prague; Technická 4, 166 07 Prague, Czech Republic;* michal.bartosak@fs.cvut.cz

sequence $[\vec{h}_1, \vec{h}_2, \cdots, \vec{h}_L]$. That is possible by adding the concept of internal states to FNNs i.e., using vector $\vec{h}_{t-1}$ as a cell input in time t alongside the vector $\vec{x}_t$.

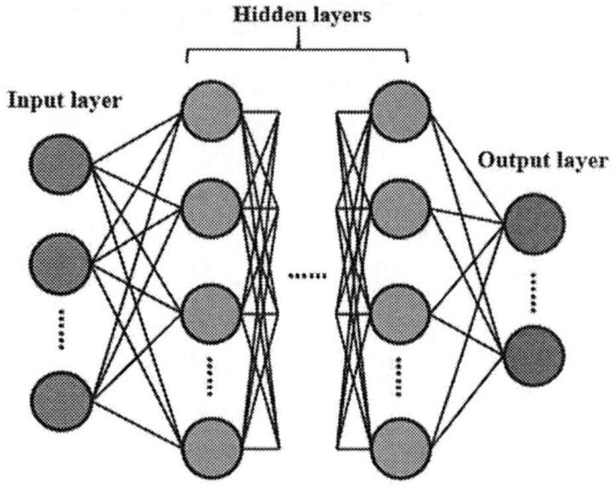

Fig. 1. Diagram of the FFN [3].

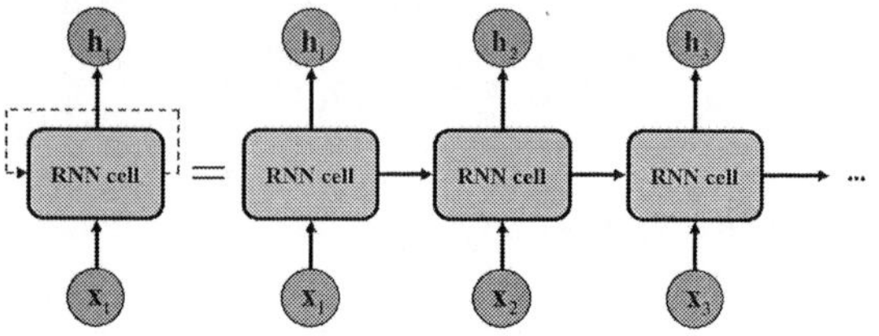

Fig. 2. Diagram of the RNN [3].

During the learning process of these classical RNNs a problem called vanishing/exploding gradient can occur, especially when utilized with longer sequences. This happens because the gradients computed by the chain rule end up having too small/large value and the network is unable to learn effectively.

The Long Short-Term Memory network (LSTM) (Fig. 3) proposed in [4] is a special type of recurrent neural network, which was designed to overcome this problem. The current state of this network is defined by internal variables called cell state $\boldsymbol{C_t}$ and hidden state $\boldsymbol{h_t}$. The data flow between these variables is conveyed through the forget gate $\boldsymbol{f_t}$, input gate $\boldsymbol{i_t}$, cell candidate $\tilde{\boldsymbol{C}}_t$ and output gate $\boldsymbol{o_t}$. The Gated Recurrent Unit (GRU) (Fig. 4) proposed in [5] is a similar network to LSTM, but it has no cell state and output gate. It has fewer parameters, therefore it is faster to train, although it still usually matches the performance of a LSTM cell.

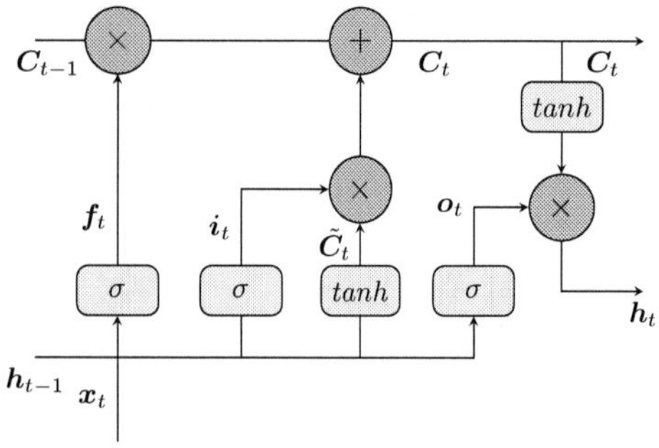

Fig. 3. Diagram of the LSTM cell.

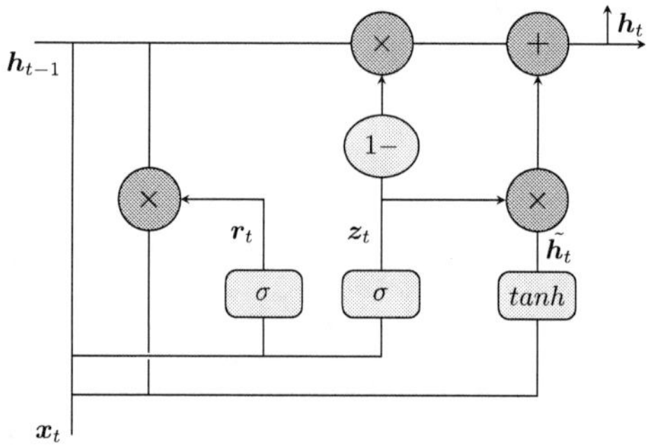

Fig. 4. Diagram of the GRU cell.

3. Data pre-processing

Eleven different loading path cases of proportional and non-proportional cyclic axial-torsion loadings (Fig. 5) of the 42CrMo4 steel are utilized for the demonstration purpose in this study. The raw data is taken from literature and consists of 63 tests, where an hourglass specimen is loaded by timeseries of stresses $\sigma_x(t)$ and $\tau_{xy}(t)$ following a scaled chosen loading path. The test ended when the specimen was totally seperated, marking the fatigue life cycle count N_f [6–9].

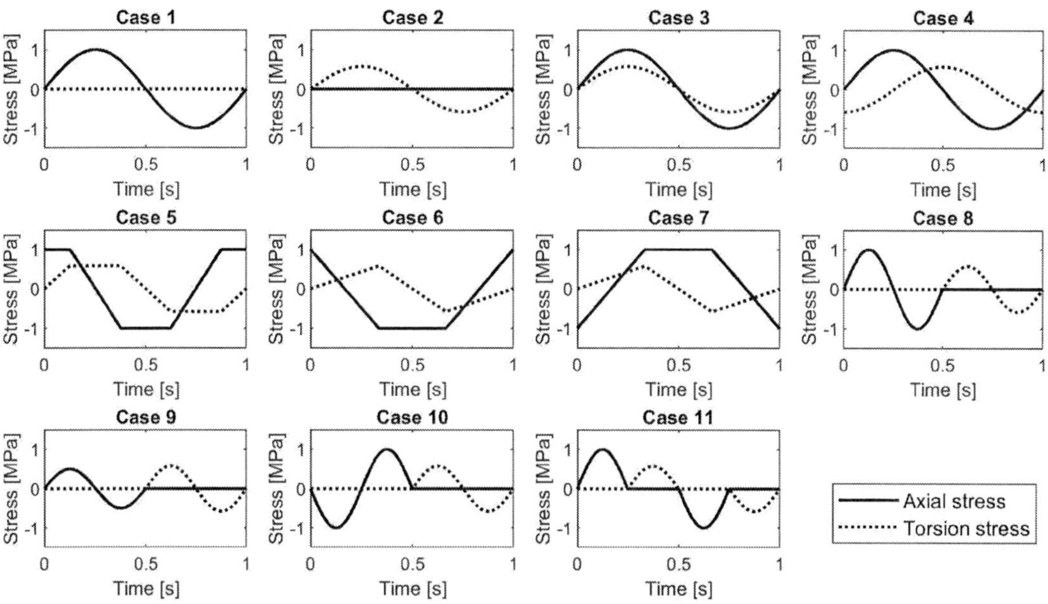

Fig. 5. Axial and torsion stress time series.

In the case of plane stress, the critical plane is always perpendicular to the plane with non-zero stress components, thus the stresses on any potential critical plane can be calculated via Mohr's circle i.e., (2) and (3).

$$\sigma_n(t,\theta) = \frac{\sigma_x(t)+\sigma_y(t)}{2} + \frac{\sigma_x(t)-\sigma_y(t)}{2} \cdot \cos(2\theta) + \tau_{xy}(t) \cdot \sin(2\theta), \quad (2)$$

$$\tau_n(t,\theta) = -\frac{\sigma_x(t)-\sigma_y(t)}{2} \cdot \sin(2\theta) + \tau_{xy}(t) \cdot \cos(2\theta), \quad (3)$$

where σ_n and τ_n are normal and shear stresses on a plane defined by normal that is rotated counterclockwise by angle θ from x axis.

The critical plane is selected to be the plane with the maximum shear stress range (4) as defined in [10]

$$\theta_{cr.} = \arg\max_{\theta}(\max_{t}(\tau_n(t,\theta)) - \min_{t}(\tau_n(t,\theta))). \quad (4)$$

If multiple planes share the same shear stress range, the one with larger maximum value of normal stress is chosen. This definition has been successfully used to predict fatigue life of the 42CrMo4 steel in [11].

The data samples are split into train dataset of 43 samples and test dataset of 20 samples. Stratified random sampling was used, so every loading case has 1-2 samples in the testing dataset.

4. Proposed Artificial Neural Network

The input of the Artificial Neural Network is a sequence of vectors $\vec{x}_i$ with length of L, where $\vec{x}_i$ is defined by (5)

$$\vec{x}_i = [\sigma_n(t_i,\theta_{cr.}), \tau_n(t_i,\theta_{cr.})], \quad (5)$$

where t_i is discretized time. These vector sequences are then normalized, so that the largest absolute value across all samples and time indexes is 1.

The first layer of the ANN is a LSTM or a GRU layer with hidden vector length of m. The last output vector $\vec{h}(L)$ is then connected to a fully connected layer of n neurons with activation function g being either ReLU, Sigmoid or PureLin function. This layer is finally connected to a single output neuron, which should return logarithm of the expected fatigue life $\log(N_f)$.

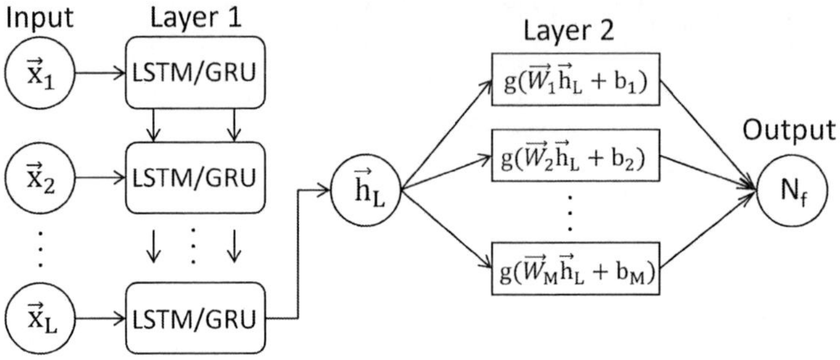

Fig. 6. Diagram of proposed artificial neural network.

Hyperparameter optimalization has been performed using grid search. Training of proposed neural network with every hyperparameter combination has been repeated 300 times with random parameter initialization using the mean squared error (MSE) loss function.

Table 1. Range of examined hyperparameters.

Symbol	Hyperparameter descriptioon	Tested values
L	Input vector length	120; 200
	Layer 1 type	LSTM; GRU
m	Length of hidden vector $\vec{h}(i)$	1–6
n	Number of neurons	1–3
g	Activation function	ReLU; Sigmoid; PureLin

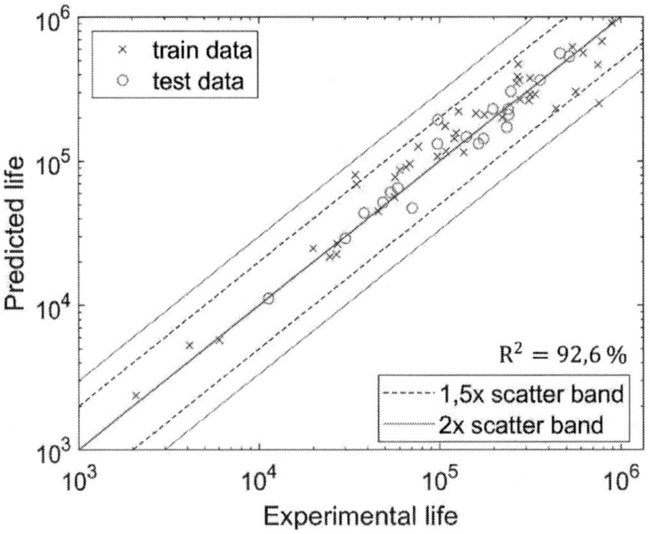

Fig. 7. Predicted life using network with GRU layer (m = 4) and 1 neuron with ReLU activation function.

5. Results

The metric chosen for quantitative comparison of individual trained networks is the root mean squared error (RMSE) on the testing dataset. The lowest RMSE was achieved with the hyperparameters being $L = 200$, GRU layer with $m = 4$ and a single ($n = 1$) neuron with the ReLU activation function (Fig. 7).

More generally the RMSE of networks trained on input vectors with $L = 120$ is on average 20% higher than RMSE with $L = 200$. The difference between using LSTM or GRU layer and between the 3 tested activation functions is insignificant.

It is known that neural networks are unreliable when used for an extrapolation. In this study, it has been observed, that the networks using GRU layer with relatively small values of hyperparameters m and n tend to predict smooth and monotonic curves outside the training data range (Fig. 8). This does not however apply for the networks with the LSTM layer (Fig. 9) and/or for networks with higher values of hyperparameters m and n.

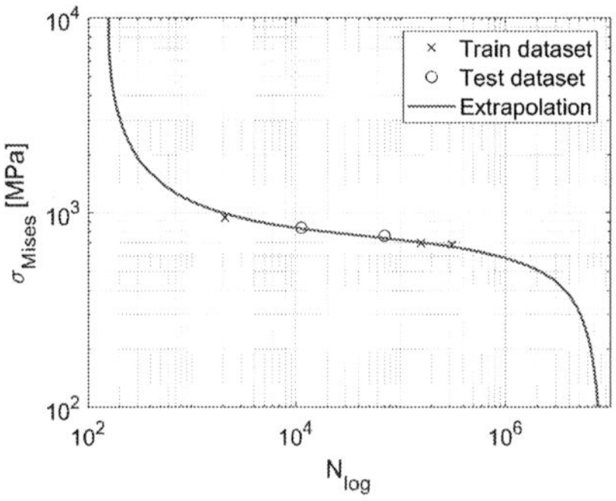

Fig. 8. S-N curve of case 3 predicted by network with GRU layer (m =3) and a single (n = 1) neuron using Sigmoid activation function.

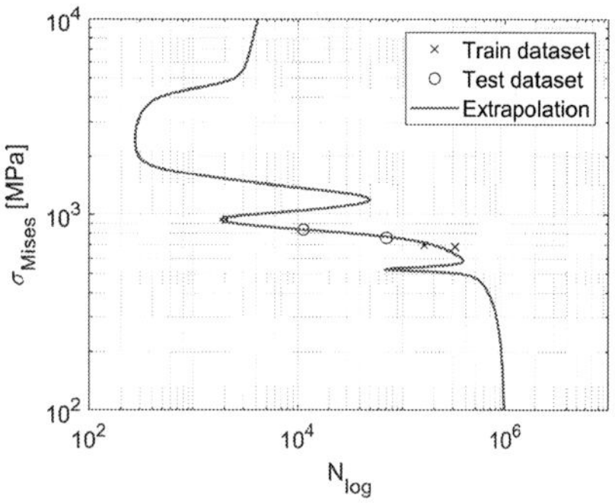

Fig. 9. S-N curve of case 3 predicted by network with LSTM layer (m = 3) and a single (n = 1) neuron using Sigmoid activation function.

6. Conclusions

In this study, the critical plane data pre-processing method for applications with Recurrent Neural Networks has been proposed for fatigue life prediction of components under cyclic plane stress. Implementation of this method brings the desired invariance trait to the rapidly evolving Artificial Neural Networks.

A dataset (taken from [6–9]) of 63 specimens made of 42CrMo4 steel loaded by various axial-torsion loading paths was used for validation purposes. An Artificial Neural Network utilizing the LSTM/GRU cells has been proposed and its hyperparameters and parameters have been optimized for this dataset. A great value of $R^2 = 92.6\%$ of the coefficient of determination has been achieved with this method. However it has been shown that proposed ANN tends to perform poorly when used outside the range of the dataset used for training. It's recommended to use the network with the GRU cell and low number of total learnable parameters to overcome this overfitting problem.

Validation of the method on other materials and more loading paths presents itself as convenient future work. Further development of the method can be adjusting the ANN and imposing physics knowledge by constraining its weights and biases.

Acknowledgements

The authors acknowledge support from Grant project GA21-06645S of the Czech Science Foundation (CSF).

References

[1] Zhang X.-C., Gong J.-G. and Xuan F.-Z., "A deep learning based life prediction method for components under creep, fatigue and creep-fatigue conditions," *International Journal of Fatigue*, Vol. 148, Art. no. 106236 (2021). ISSN 0142-1123. https://doi.org/10.1016/j.ijfatigue.2021.106236

[2] Chen J. and Liu, Y., "Probabilistic physics-guided machine learning for fatigue data analysis," *Expert Systems with Applications*, Vol. 168, Art. no. 114316 (2021). ISSN 0957-4174. https://doi.org/10.1016/j.eswa.2020.114316.

[3] Yang J., Kang G., Liu Y. and Kan Q., "A novel method of multiaxial fatigue life prediction based on deep learning," *International Journal of Fatigue*, Vol. 151, Art. no. 106356 (2021). ISSN 0142-1123. https://doi.org/10.1016/j.ijfatigue.2021.106356.

[4] Hochreiter S. and Schmidhuber J., "Long Short-Term Memory," *Neural Computation*, Vol. 9, no. 8, pp. 1735–1780 (1997). ISSN 0899-7667. http://dx.doi.org/10.1162/neco.1997.9.8.1735.

[5] Cho K., van Merriënboer B., Bahdanau D. and Bengio Y.,"On the Properties of Neural Machine Translation: Encoder-Decoder Approaches," Available from https://arxiv.org/pdf/1409.1259.pdf Accessed: 2022-05-22

[6] Anes V., Reis L., Li B. and de Freitas M., "Crack path evaluation on HC and BCC microstructures under multiaxial cyclic loading," *International Journal of Fatigue*, Vol. 58, pp. 102–113 (2014). ISSN 0142-1123. https://doi.org/10.1016/j.ijfatigue.2013.03.014.

[7] Anes V., Reis L., Li B. and de Freitas M., "New approach to evaluate non-proportionality in multiaxial loading conditions," *Fatigue & Fracture of Engineering Materials & Structures*, Vol. 37, Issue 12, pp. 1338–1354 (2014). ISSN 1460-2695. https://doi.org/10.1111/ffe.12192.

[8] Anes V., Reis L., Li B. and de Freitas M., "New cycle counting method for multiaxial fatigue," *International Journal of Fatigue*, Vol. 67, pp. 78–94 (2014). ISSN 0142-1123. https://doi.org/10.1016/j.ijfatigue.2014.02.010.

[9] Reis L., Anes V., Li B., and M. de Freitas, "The influence of tension / torsion load sequences on fatigue life and early crack propagation of 42crmo4 steel," *Anales de Mecánica de la Fractura 28*, Vol. 1, pp. 341–346 (2011). ISSN 0213-3725.

[10] Brown M. and Miller K., "A theory for fatigue failure under multiaxial stress-strain conditions," *Proceedings of the Institution of Mechanical engineers*, Vol. 187 no. 1, pp. 745–755 (1973). https://doi.org/10.1243/PIME_PROC_1973_187_161_02.

[11] Bartošák M., Horváth J. and Španiel M., "Multiaxial low-cycle thermo-mechanical fatigue of a low-alloy martensitic steel: Cyclic mechanical behaviour, damage mechanisms and life prediction," *International Journal of Fatigue*, Vol. 151, Art. no. 106383 (2021). ISSN 0142-1123. https://doi.org/10.1016/j.ijfatigue.2021.106383.

Design, Computational and Experimental Analysis of Tree Shear

Martin Havlíček[1], Jiří Jakubský[2], Martin Dub[3] & Michal Šperňák[4]

Abstract: This paper is focused on an experimental measurement of acting forces during cutting trees. For this measurement is designed a new tree shear as an attachment for different excavators and this design is supported by FEM analysis. This frame consists of a main unmovable frame with a cutting edge which is connected to an excavator arm and a movable frame which is mounted to a main frame by hinges. Movable frame is moved by hydraulic cylinder. After assembling, experimental measurement took place in a forest in South Bohemia where several different trees are cut. The main goal of this measurement is to measure pressure emerging in a hydraulic cylinder during cutting of trees by pressure gauge. Thanks to this acting pressure and geometry knowledge of used hydraulic cylinder, forces needed to cut trees according to a tree diameter are calculated. From previous research are chosen three different methods how to calculate this cutting force analytically depending on tree diameter and all obtained data are compared in graphs with measured data.

Keywords: Tree shear; Force; Analysis; Excavator

1. Introduction

Agriculture and forestry are two of the most important industries in the world. Both industries are based on maintaining the balance on landscape. Both of these industries have also other similarities like for examples changing hand work for machine work. Thanks to this innovation it is necessary to design machines which can gently care of our lanscape. Because of this reason, a tree shear is designed and tested by experimental measurement. Main goal of this experimental measurement is to compare acting force to the cutting edge of tree shear with three hypothesis. This experimental measurement could help with design of new machines working with trees.

2. Instrumentation

For this experiment is designed a tree shear (Fig. 1), which can be connected to different types of excavators through a special coupling head on a tree shear. This tree shears consists of two main frames. First and main welded frame is connected to an excavator arm by special coupling head and contains a cutting edge. Second welded frame is connected to the main frame through hinges, so this frame is movable. Motion of this second frame is needed because it is used to push the trees against the cutting head. This motion is ensured thanks to the hydraulic circuit which contain hydraulic cylinder.

Design of this tree shear is supported by FEM analysis where the frame is loaded by calculated force obtained from Lisican's hypothesis for the worst type of measured wood with

[1] *Ing. Martin Havlíček; Dep. of Designing and Machine Components , Fac. of Mech. Engineering, Czech Technical University in Prague; Technická 4, 166 07 Prague, Czech Republic; martin.havlicek@fs.cvut.cz*

[2] *Bc. Jiří Jakubský; Dep. of Designing and Machine Components , Fac. of Mech. Engineering, Czech Technical University in Prague; Technická 4, 166 07 Prague, Czech Republic; jiri.jakubsky@fs.cvut.cz*

[3] *Ing. Martin Dub, Ph.D.; Dep. of Designing and Machine Components , Fac. of Mech. Engineering, Czech Technical University in Prague; Technická 4, 166 07 Prague, Czech Republic; martin.dub@fs.cvut.cz*

[4] *Bc. Michal Šperňák; Dep. Of Engineering Technology, Fac. of Mech. Engineering, Czech Technical University in Prague; Technická 4, 166 07 Prague, Czech Republic; michal.spernak@fs.cvut.cz*

diameter equal to 25 centimeters. FEM analysis is made by RFEM software [4]. Designed tree shear is remodeled to this software for FEM analysis, where all parts of this model are modeled as surface models. Tree shear is supported by hinges in place where tree shear is connected to an arm of an excavator and hydraulic cylinder is replaced by prestressed beam with internal acting force. From Fig. 2 is seen that in part of the movable frame is higher stress than yield stress.

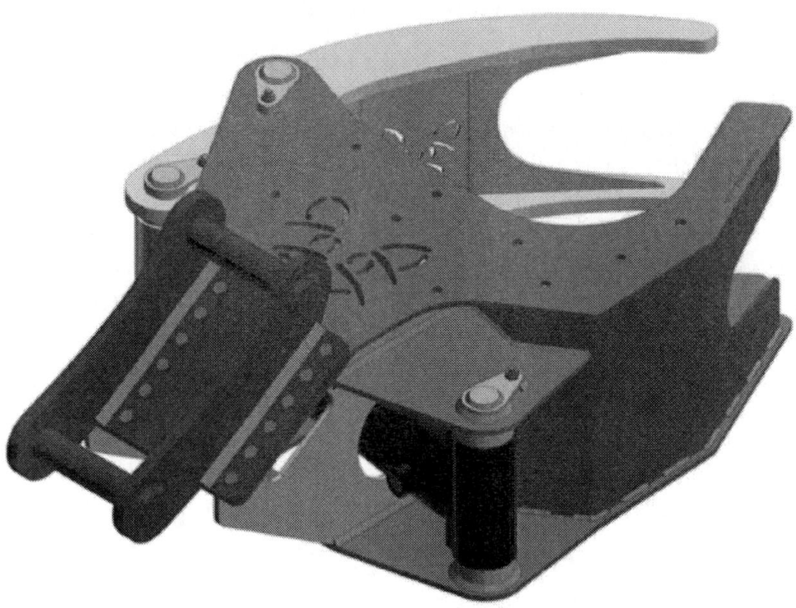

Fig. 1. Design of tree shears.

Fig. 2. Model in RFEM.

3. Experimental measurement

During winter period were made several experimental measurements with designed tree shear to find out forces acting to a tree shear's cutting edge during cutting of trees. These measuremnets are made to verify analytical calculations because several methods, how to calculate this forces, are found [1–3]. One of the method, how to calculate this force, is settled by Lisican and other methods are calculated from shear stress. One of the method is calculated according to Kohler's tables and second one is calculated with half humidity of trees. During experiments are substracted several important parameters (Fig. 3). These important parameters are: diameter of trees, position of trees againts the cutting edge and maximal pressure developed during cutting. Then the pressure is recalculated to a force thanks to the knowledge of an area of used hydraulic cylinder, where the pressure acts.

Fig. 3. Substraction of diameter and position of trees.

After these measurements, all the data are processed and plotted in graphs. Cutting force of oak depending on the diameter of trees is shown in Fig. 4. During this experiment is cut oak and the range of diameter of oak is from 7 cm to diameter 14 cm. Orange line represents data obtained during experimental measurement, blue line represents data calculated according to Lisican's hypothesis. Yellow line represents data according to shear stress from Kohler's tables and grey line obtain data gained from shear stress with half of humidity. It can be seen that during the smaller diameter measured data correlate with shear stress hypothesis, but from diameter 9.5 cm measured data have similar trend like Lisican's hypothesis.

In Fig. 5 is shown cutting force of spruce depending on the diameter of trees. During this experiment is cut spruce and diameter of spruce is from 6 cm to diameter 15 cm. Orange line represents data obtained during experimental measurement, blue line represents data calculated according to Lisican's hypothesis. Yellow line represents data according to shear stress from Kohler's tables and grey line obtain data gained from shear stress with half of humidity. It can be seen that from small diameter measured data correlate with Lisican's hypothesis.

In Fig. 6 is shown cutting force of pine depending on the diameter of trees. During this experiment is cut pine and diameter of pine is from 7 cm to diameter 23 cm. Orange line represents data obtained during experimental measurement, blue line represents data calculated according to Lisican's hypothesis. Yellow line represents data according to shear stress from Kohler's tables and grey line obtain data gained from shear stress with half of humidity. It can be seen that all measured data correlate with Lisican's hypothesis.

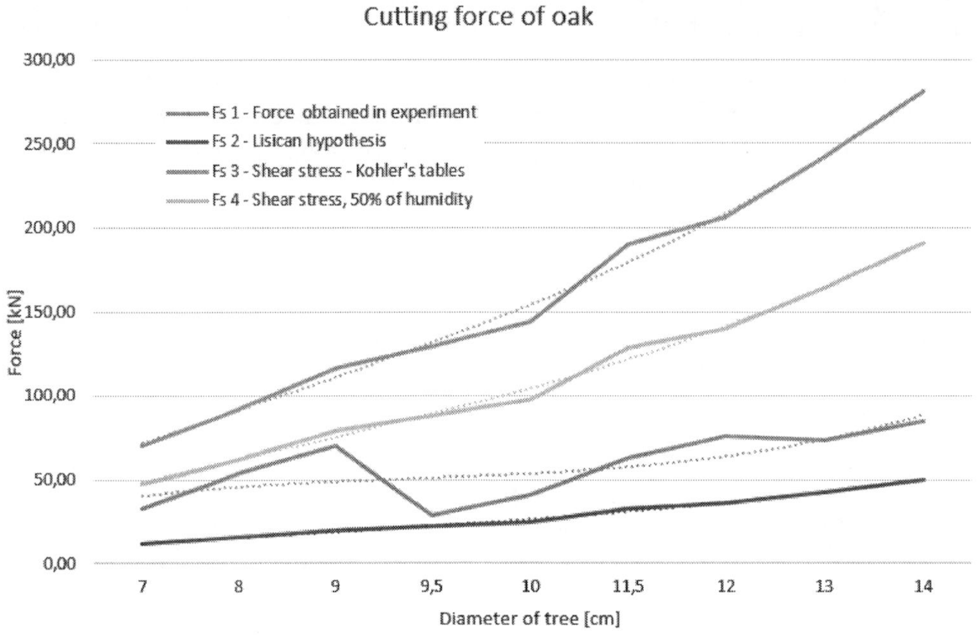

Fig. 4. Cutting force of oak.

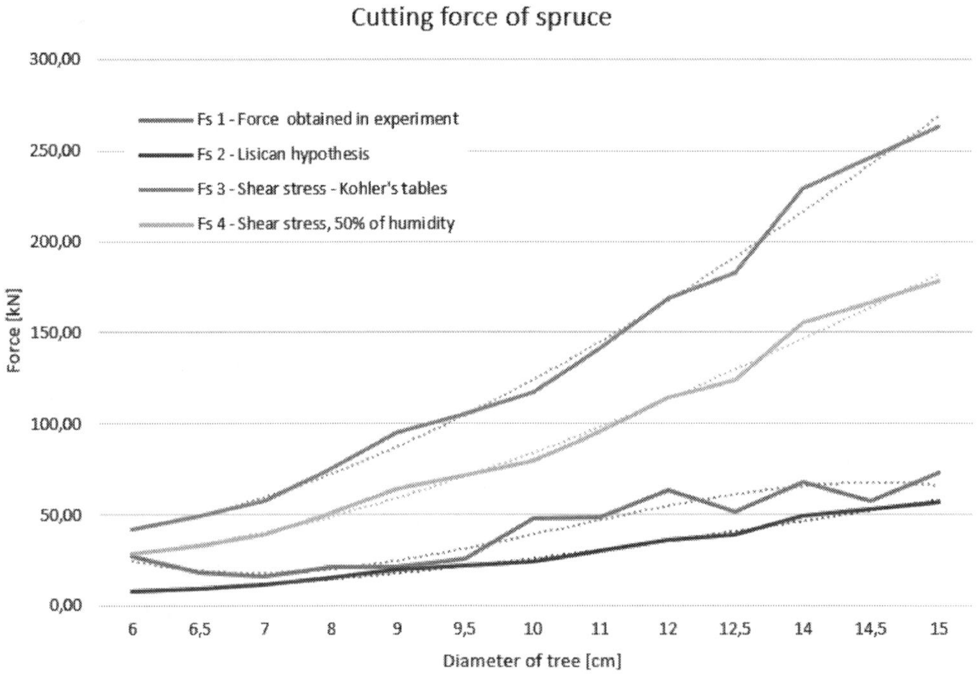

Fig. 5. Cutting force of spruce.

From all these graphs can be summarized that all measured data from all trees with different diameter have similar trend with Lisican's hypothesis. Only data measured during cutting of oak with small diameter do not correlate. This can be caused by reading wrong value from pressure gauge.

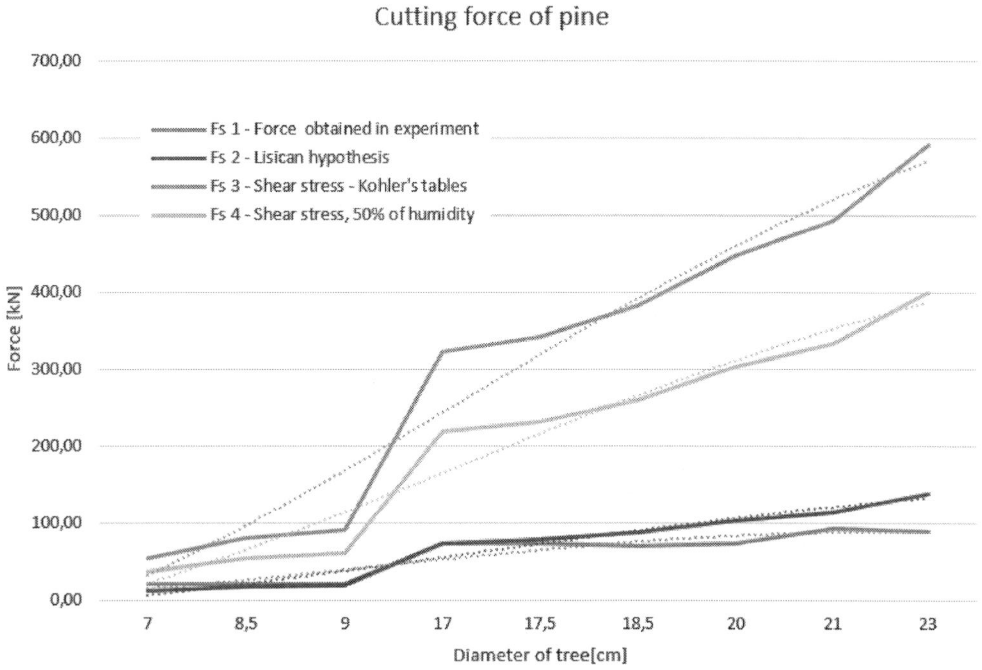

Fig. 6. Cutting force of pine.

4. Conclusions

Thanks to this designed tree shear frame with a new and special coupling head, acting forces during cutting trees were found out. This special frame consist of two welded frames where main one of them is connected to an arm of an excavator and second frame is coupled with a main frame through hinges. Thanks to this hinges, second frame is movable and this motion is mediated by hydraulic circuit with hydraulic cylinder. Movable frame is needed because this frame pushes trees against cutting edge of a tree shear. Force which is needed to cut the trees is measured thanks to an emerging pressure in hydraulic circuit. Pressure of the oil is measured by pressure gauge. This tree shear was then tested in a forest to find out forces needed to cut the trees. Lots of data were obtain from all measurements, all of them were plotted in a graph and compared with existing analytical hypothesis. Thanks to these graphs can be made a conclusion, that hypothesis bases on shear stress do not correlate with experimental measurement. On the other hand, Lisican's hypothesis correlate with experimental measurement. It shows that this hypothesis should be used for designing of new cutting machines.

References

[1] Lisican J., *Teória a technika spracovania dreva* (MATCENTRUM, Zvolen, 1996). ISBN 80-967315-6-4.

[2] Lysý F. and Jírů P. *Nauka o dřevě* (Státní nakladatelství technické literatury, Praha, 1954).

[3] Gandelová L., Horáček P. and Šlezingerová J. *Nauka o dřevě.*, Vyd.3. nezměn. (Mendelova zemědělská a lesnická univerzita, Brno, 2009). ISBN 978-80-7375-312-2.

[4] Dlubal Software GmbH, RFEM 5.26.02 (2020).

Finite Element Analysis of the Large-strain Viscoelastic Model for Magnetorheological Elastomer

Mohammad Hdaib[1], Iva Petríková[2] & Tran Huu Nam[3]

Abstract: Magnetorheological elastomers (MREs) are smart materials that consist of an elastomer matrix filled with iron particles. The distinguished functionality of this type of material is its response to magnetic fields by changing its mechanical properties. This reliable functionality makes MREs suitable for various applications. In this study, the MRE samples were fabricated using liquid silicon rubber (LSR) and Iron particles. Then, the hyper-viscoelastic parameters of MRE are estimated by using static and dynamic uniaxial compression tests. Moreover, the experimentally determined parameters were verified using finite element analysis with a hyper-viscoelastic model at low and high strain rates.

Keywords: MRE; FEA; Hyper-viscoelastic model; Large-strain viscoelasticity

1. Introduction

Rubber MRE composites are viscoelastic materials that exhibit a lag between force and displacement, resulting in stress relaxation, displacement creep, and energy dissipation. These properties can be investigated by employing a linear or a nonlinear viscoelastic model. For the linear viscoelastic model, a combination of springs and dashpots is used to predict the viscoelastic behavior [1]. While for the nonlinear model, springs and spring-pots elements are used [2]. The nonlinear model achieves better accuracy as it has the sprig-pot element, which acts like a spring, dashpot, or a hybrid between both elements depending on the value of the fractional order [1]. On the other hand, the rate-independent properties are estimated by employing a spring-only branch that can be a linear elastic modulus or a hyperelastic model. In the case of using a hyperelastic model, then the overall generalized Maxwell model is called a large-strain viscoelastic model [2].

In this study, the isotropic MRE composite was fabricated using silicon rubber with iron particles. Then the material model parameters were identified by performing two types of uniaxial compression tests: the first one was at a relatively slow strain rate (0.000154, 0.0077, and 0.077 mm/(mm·s)). In contrast, the other was a frequency sweep test starting from 10 Hz to 50 Hz with a 2 Hz step. The acquired parameters were verified using finite element analysis with a large-strain viscoelastic model at high and low strain rates. Moreover, the viscoelastic parameters were determined using Prony-series in the frequency and time domains. Then the FE results were illustrated and compared using the experimentally identified parameters.

2. Theory

According to the strain rate dependency and the generalized Maxwell model, the strain energy function is divided into hyperelastic and viscoelastic parts, as shown in equation (1) and

[1] *Mohammad Yousef Hdaib, Faculty of Mechanical Engineering, Technical University of Liberec, Studentská 1402/2, 461 17 Liberec 1, Czech Republic;* Mohammad.yousef.hdaib@tul.cz

[2] *doc. Ing. Iva Petríková, Ph.D., Faculty of Mechanical Engineering, Technical University of Liberec, Studentská 1402/2, 461 17 Liberec 1, Czech Republic;* iva.petrikova@tul.cz

[3] *doc. Ing. Tran Huu Nam, Ph.D.,Faculty of Mechanical Engineering, Technical University of Liberec, Studentská 1402/2, 461 17 Liberec 1, Czech Republic;* nam.tran.huu@tul.cz

Fig. 1 [2].

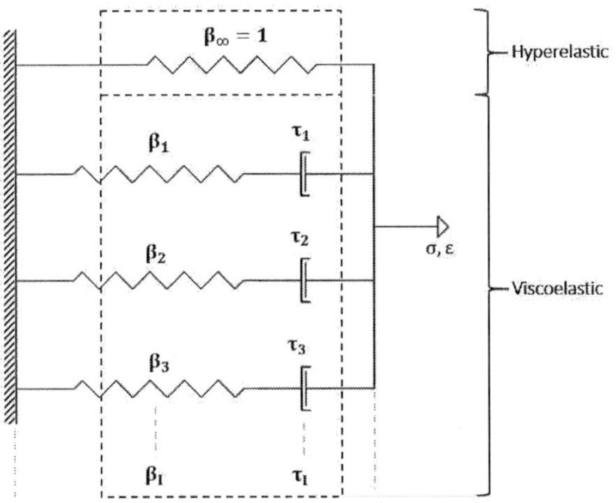

Fig. 1. Generalized Maxwell model.

$$W = W_\infty + \sum_{i=1}^{I} W_i \tag{1}$$

Where the W_∞ represents the isochoric and volumetric strain energies of the hyperelastic branch. While the W_n represent the strain energy corresponding to each viscoelastic branch. Also, the 2nd Piola Kirchhoff stress (S) can be expressed as [2]:

$$S = 2\frac{\partial W}{\partial C} + \sum_{i=1}^{I} 2\frac{\partial W_i}{\partial C} = S_\infty + \sum_{i=1}^{I} q_I \tag{2}$$

Where the q_n is the stress tensor for each viscoelastic branch which its time evolution ($\dot{q}_n$) is expressed as [2]:

$$\dot{q}_i + \frac{1}{\tau_i} q_i = \dot{S}_i \tag{3}$$

Where S$_n$ is the 2nd Piola Kirchhoff stress from each viscoelastic branch, which can be estimated by multiplying the hyperelastic stress (S_∞) by the energy factor (β_n) of that branch as [2]:

$$S_i = \beta_i S_\infty \tag{4}$$

Then the energy factors can be determined by fitting the experimental data from stress relaxation or dynamic tests to a viscoelastic model.

3. Experimental

Liquid silicon rubber was mixed with iron particles to create the MRE samples. Three samples were formed with a 27% volume fraction of iron particles, a height of 13 mm, and a diameter of 27 mm. The formed samples agree with the standard ASTM D575-91.

Two types of uniaxial compression tests were performed to identify the model parameters: low strain rate and high strain rate tests. The former was used to estimate the hyperelastic parameters at a loading velocity of 0.002 mm·s^{-1} (0.00015 s^{-1}). In contrast, the latter was used to predict the viscoelastic parameters with a frequency sweep starting from 10 Hz to 50 Hz with a step of 2 Hz and 10 minutes of holding time before each step.

A very slow loading velocity was chosen to determine the hyperelastic parameters to estimate them independently of the viscoelastic stress [3]. In contrast, the frequency sweep test was used because it estimates the viscoelastic parameters more accurately than the stress relaxation test, especially at high strain rates [4].

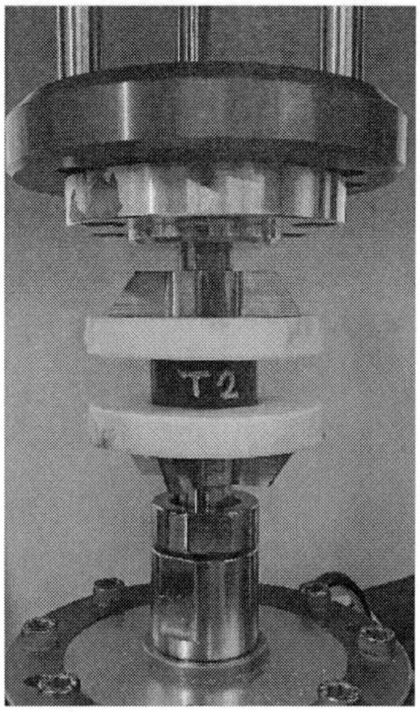

Fig. 2. Experimental setup for the parameters identification tests.

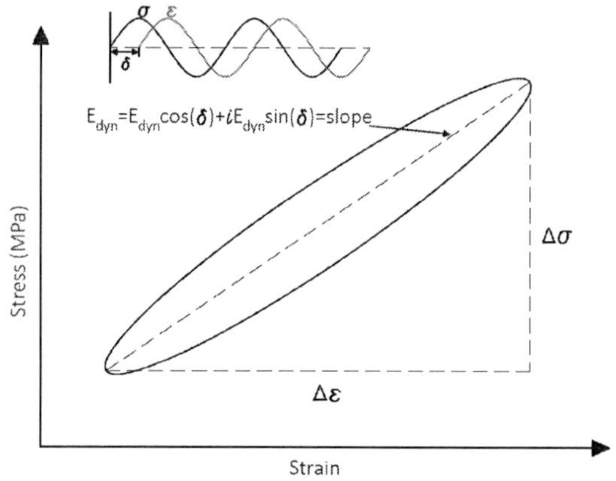

Fig. 3. Phase lag between the stress and strain signals.

The previously mentioned uniaxial compression tests were performed using the Instron Electropuls E3000 testing system (Instron Corp., MA, US). Teflon parts were attached to the steel grips to reduce the static friction between the sample and the grips, as shown in Fig. 2. Then the results were processed using the Fast Fourier Transform (FFT) to determine the phase lag between the force and displacement. Consequently, the experimental dynamic moduli were calculated as follows:

$$E_L = E_{dyn} \sin(\delta) \tag{5}$$

$$E_S = E_{dyn} \cos(\delta) \tag{6}$$

Where the E_L, and the E_S represents the loss and storage moduli, respectively. While E_{dyn}, and δ are the dynamic modulus and lag between the stress and strain signals, respectively, as shown in Fig. 3. Then the optimization module on Comsol Multiphysics was used to fit the experimental data from the quasi-static and dynamic tests to the following equations:

$$\sigma_1^{eng} = 2C_{10}\left(\lambda - \frac{1}{\lambda^2}\right) + 2C_{01}\left(1 - \frac{1}{\lambda^3}\right) \tag{7}$$

$$E_S = E_\infty\left(1 + \sum_{i=1}^{I} \beta_i \frac{\omega^2\tau_i^2}{1+\omega^2\tau_i^2}\right) \tag{8}$$

$$E_L = E_\infty\left(\sum_{i=1}^{I} \beta_i \frac{\omega\tau_i}{1+\omega^2\tau_i^2}\right) \tag{9}$$

$$E_{ins} = E_\infty\left(1 + \sum_{i=1}^{I} \beta_i e^{\frac{-t}{\tau_i}}\right) \tag{10}$$

Equation (7) is the two-parameters Mooney-Rivlin hyperelastic model for a uniaxial test, where C_{10} and C_{01} are the hyperelastic model parameters, while $\lambda$is the stretch in the loading direction. In contrast, equations (7) and (8) represent the Prony-series in the frequency domain for the storage and loss moduli, where ω and τ_n are the frequency and the relaxation time, respectively. On the other hand, equation (10) represents the Prony-series in the time domain.

4. Results

4.1. Parameters identification

Fig. 4 and Fig. 5 represents the fitting results of the hyperelastic and viscoelastic models, respectively. Comsol multiphysics optimization module was used to fit the experimental data using the least-squares method on equations (7–10) .

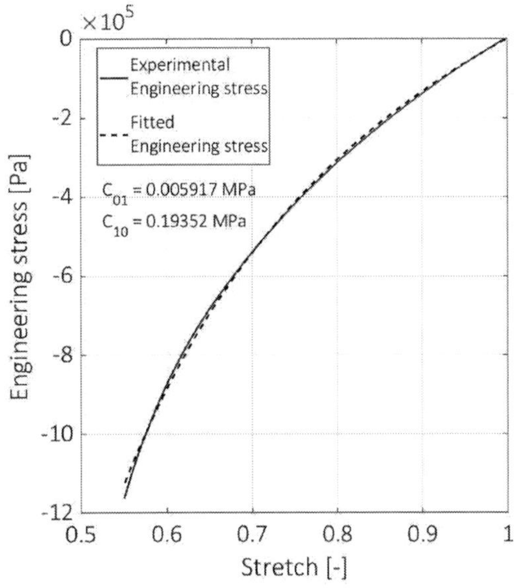

Fig. 4. Hyperelastic model fitting results: Experiment (solid line), Fitting (dashed line).

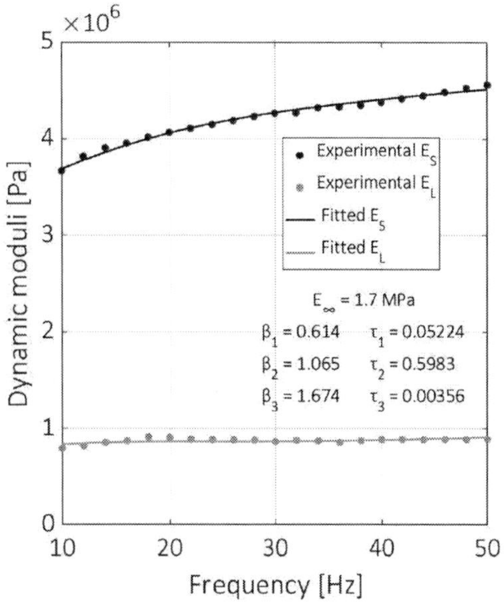

Fig. 5. Viscoelastic model fitting results: Experiment (dotted line), Fitting (solid line). Black (storage modulus), Silver (Loss modulus).

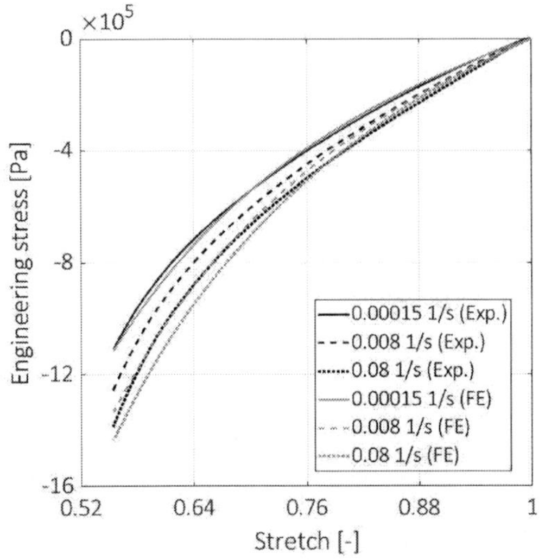

Fig. 6. FE results at relatively low strain rates using the viscoelastic parameters fitted in the time domain Prony-series.

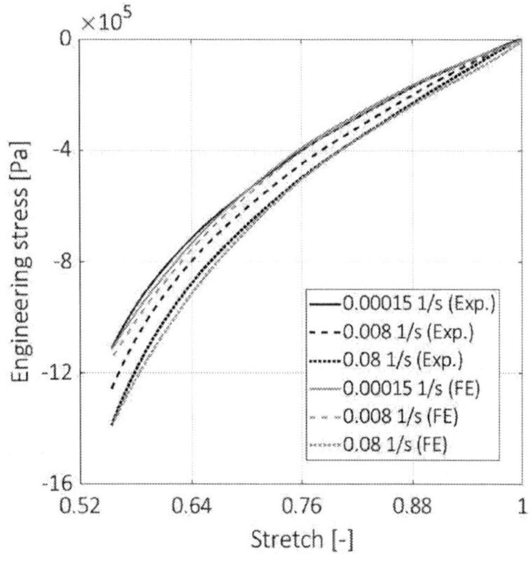

Fig. 7. FE results at relatively low strain rates using the viscoelastic parameters fitted in the frequency domain Prony-series.

4.2. FEA of the low strain rate tests

To verify the fitting results of the hyperelastic model, a FEA was performed on the low strain-rate tests. The analysis was performed using the generalized Maxwell model in both the frequency and time domains. Fig. 6 and Fig. 7 demonstrate the FE results at low strain rates (0.00015, 0.008 and 0.08 s^{-1}). The FE results accurately predicted the engineering stress with a maximum relative error of 13% and 11% using the generalized Maxwell model in frequency and time domains, respectively.

4.3. FEA of the dynamic test.

The material model was verified by comparing the hysteresis loops with the experiment done by Nam et al. [5]. The loops were estimated by FEA using Comsol Multiphysics software. Then their areas were compared with experimental ones after two minutes of dynamic loading. As shown in Fig. 8–Fig. 10, the hysteresis-loops were predicted accurately only using the viscoelastic parameters fitted in the frequency domains; because they represent the dynamic test more accurately and can be evaluated at variable pre-strains.The maximum error observed for the area estimation was 15%, while the minimum error was less than 1%.

While for the stress relaxation test, a high-velocity loading is required to detect the viscoelastic behavior at small relaxation times (< 0.01 s). In some cases, the high-velocity loading causes displacement overshoot, which leads to inaccurate results. Therefore, the parameters determined from the frequency sweep test are better suited for such type of FEA.

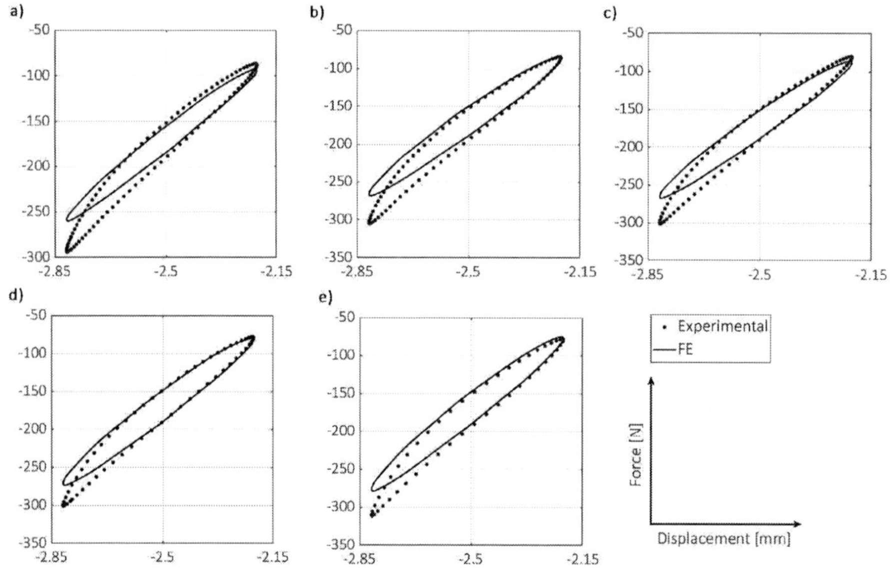

Fig. 8. Simulation results at 3% strain amplitude, 20% pre-strain, and variable frequencies: a) 10 Hz, b) 20 Hz, c) 30 Hz, d) 40 Hz and e) 50 Hz.

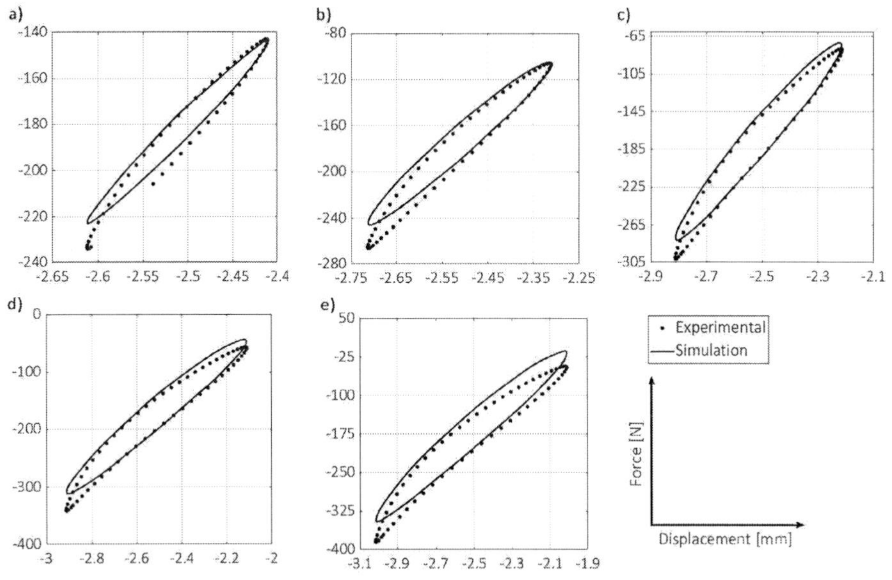

Fig. 9. Simulation results at 40 Hz Frequency, 20% pre-strain, and variable strain amplitudes: a) 1%, b) 2%, c) 3%, d) 4% and e) 5%.

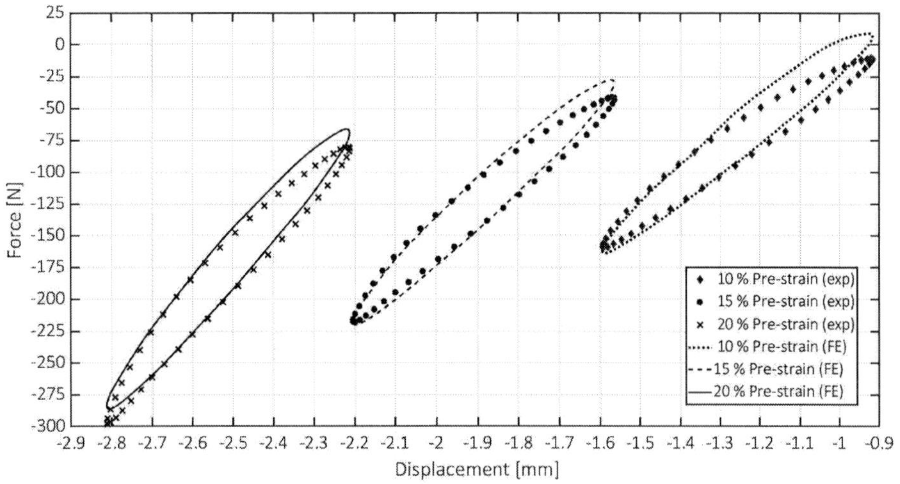

Fig. 10. Simulation results at 50 Hz Frequency, 3% strain amplitude, and variable pre-strains: 10%, 15%, and 20%.

5. Conclusion

In this study, a FEA was performed on MRE under uniaxial compression tests. A large-strain viscoelastic model was employed for the FEA, and the model parameters were fitted experimentally.

According to the achieved results, the following remarks were concluded:

- According to the low strain tests, the engineering stress can be predicted accurately using the Prony-series in the frequency or time domains. Implementing the static friction model is recommended for better results in future analysis.
- At high strain rates (Frequencies of 10 Hz, 20 Hz, 30 Hz, 40 Hz and 50 Hz), the force-

displacement strain loops were predicted solely using the Prony-series fitted in the frequency domain. The achieved results can estimate phenomena involved with energy dissipation, such as self-heating. Moreover, the results can be enhanced by considering the stress softening due to temperature increase and the utilization of a nonlinear viscoelastic model.

Acknowledgements

This work was supported by the Student Grant Competition of the Technical University of Liberec under project No. SGS-2022-5003. This work was partly supported by the project Hybrid Materials for Hierarchical Structures (HyHi, Reg. No. CZ.02.1.01/0.0/0.0/ 16_019/0000843).

References

[1] Zhang H., Li S., Zhang Z., Luo H. and Y. Wang, "A five-parameter fractional derivative temperature spectrum model for polymeric damping materials," *Polym. Test.*, Vol. 89, p. 106654 (2020). doi: 10.1016/j.polymertesting.2020.106654.

[2] Holzapfel G., "Nonlinear Solid Mechanics: A Continuum Approach for Engineering Science," *Meccanica*, Vol. 37, no. 4–5, pp. 489–490 (2002). doi: 10.1023/A:1020843529530.

[3] Haldar K., "Constitutive modeling of magneto-viscoelastic polymers, demagnetization correction, and field-induced Poynting effect," *Int. J. Eng. Sci.*, Vol. 165, p. 103488 (2021). doi: 10.1016/j.ijengsci.2021.103488.

[4] Zhi J., Wang Q., Zhang M., Zhou Z., Liu A. and Jia Y., "Coupled analysis on hyper-viscoelastic mechanical behavior and macromolecular network alteration of rubber during thermo-oxidative aging process," *Polymer (Guildf).*, Vol. 171, no. March, pp. 15–24 (2019). doi: 10.1016/j.polymer.2019.03.029.

[5] Nam T. H., Petríková I., Marvalová B. and Hdaib M. Y., "Self-heating and dynamic mechanical behavior of silicone rubber composite filled with carbonyl iron particles under cyclic compressive loading," *J. Compos. Mater.*, Vol. 55, no. 28, pp. 4273–4292 (2021). doi: 10.1177/00219983211037055.

Glued Joint Strength of Hybrid Structure

Marek Hejman[1] & Jaroslav Václavík[2]

Abstract: The hybrid structures are subjected to fatigue loading in operation. There is level of loading which doesn't initiate debonding of the glued CFRP plate even for 8e6 loading cycles. Above this loading level the debonding appeared for all tested specimens in the beginning of the test. Using the FEM calculation, the ratio of the strain decreasing due to debonding of 5 mm from the end of CFRP plate was evaluated. This ratio was applied for one test specimen to estimate corresponding number of cycles.

Keywords: Hybrid structure; FEM calculation; Debonding

1. Introduction

The hybrid construction consists of steel profiles with glued CFRP plates. To study its properties, experiments were performed on samples of the corresponding steel plate of standard shape with glued CFRP plate. These specimens were subjected to static and fatigue loads. Both types of loading caused a debonding of the glued joint.

2. The debonding of glued CFRP plate

The static tests of the glued joint were broken for the stress level in the steel plate above the yield point, see [1]. The moment of failure of the glued joint was determined by measurement of the stress response on the strain gauges placed on the CFRP plate and on the steel plate. The criterion was a sudden change of measured strain on the strain gauges in relation to the nominal stress of the steel plate.

Fatigue tests were performed on standard fatigue flat specimens with sides formed by radius. The load was repeated with stress ratio $R = 0.3$. The moment of failure of the glued joint was deduced as changes of measured strains in strain gauges. Almost zero decrease of strain was achieved for magnitude of mean stress and stress amplitude approximately 100 MPa. In the cases of higher loads, debonding of glued hybrid joint always occurred.

3. FEM calculations

3.1. Load flow in CFRP

A linear static 3D FEM model of the test specimen with one side CFRP plate gives load flow information see diagram courses of strain and shear see Fig. 1.

On the surface of the CFRP tape, there is practically zero strain at a distance of approximately 2 mm from the edge. That is in contrast of the shear in the interface glue and CFRP plate, where magnitude is highest close to CFRP end and in distance about 20 mm the shear stress decreases almost to zero.

[1] *Ing. Marek Hejman, Ph.D.; Fatigue Analysis RI s.r.o.; V Zátiší, 423, 330 08 Zruč-Senec, Czech Republic;* hejman@fatari.cz

[2] *Ing. Jaroslav Václavík; Výzkumný a zkušební ústav Plzeň s.r.o.; Tylova 1581/46, 301 00 Plzeň, Czech Republic;* vaclavik@vzuplzen.cz

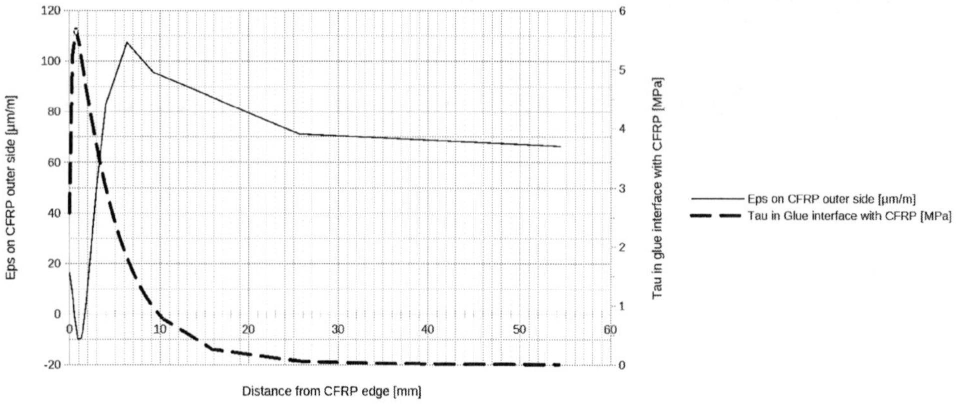

Fig. 1. Comparison of the strain [µm/m] on CFRP plate and shear [MPa] in interface of the glue with CFRP plate.

3.2. Comparison calculations and experiments

Static tests of the steel sample with glued CFRP plate on one side were performed. Strain gauges were installed on the CFRP tape, see Fig. 2. During loading, the values of relative deformation were measured on strain gauges, see Fig. 3. Evaluated stress curves showed the increasing in relative deformation at the measured points on the CFRP tape is not linearly dependent on the magnitude of the loading force at Václavík [1].

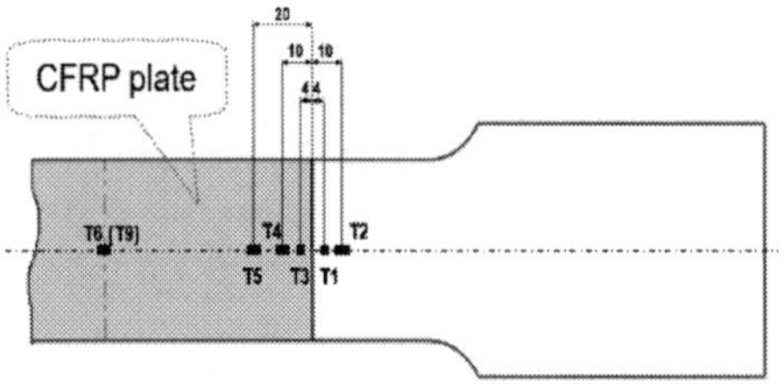

Fig. 2. Scheme of test specimen with one side CFRP plate and installed strain gauges.

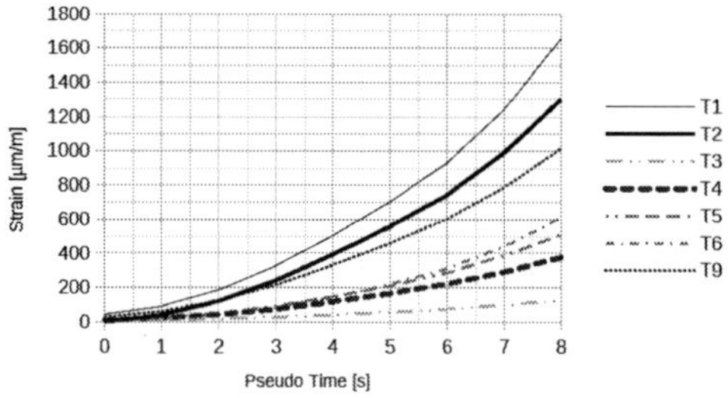

Fig. 3. The courses of strain on strain gauges during the tensile test.

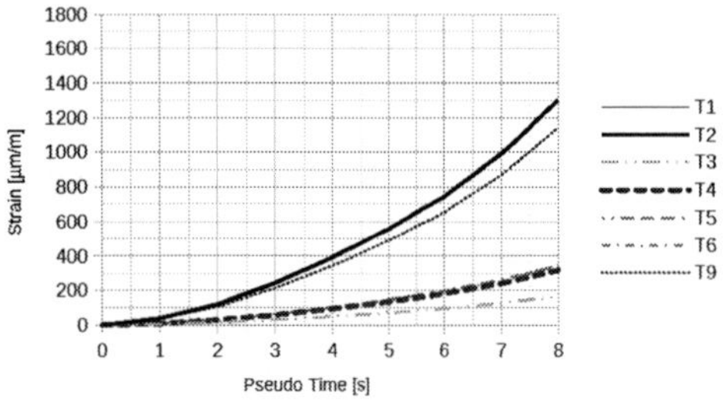

Fig. 4. The courses of strain on strain gauges in linear FEM calculations of the tensile test.

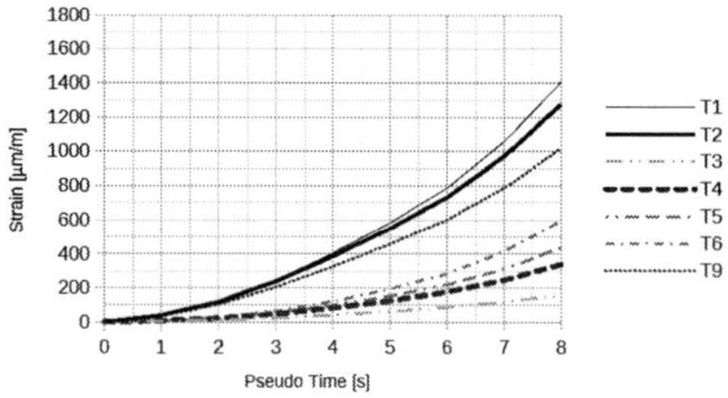

Fig. 5. The courses of strain on strain gauges in geometrically nonlinear FEM calculations of the tensile test.

Calculations using 2D model explain these phenomena. Magnitudes of strain in place of strain gauge T2 correspond to strain during the tensile test see Fig. 4 and Fig. 5. The closer results to experiment are given by geometrically nonlinear calculation with respect of nonlinear properties of glue. The presented range of results cover linear range of steel and CFRP plates and corresponding yield point are not exceed. It is evident that respect of these nonlinearities is very important. Used nonlinear properties of glue were measured experimentally see Fig. 6.

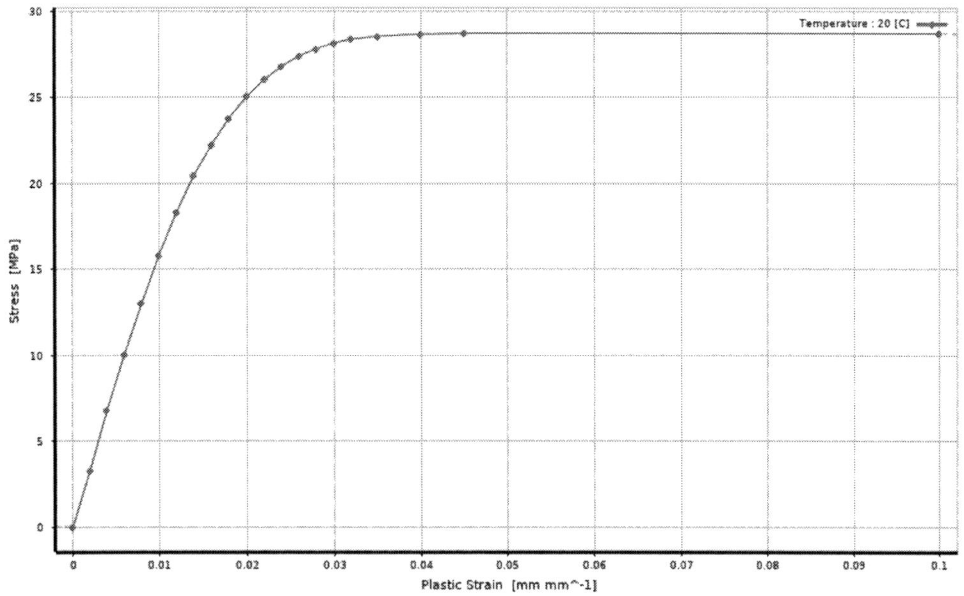

Fig. 6. Nonlinear properties of used glue identified by tensile test.

3.3. FEM simulations of debonding growth

The simplified 2D FEM model can be used to simulate debonding growth. In each step of calculation, a part of the modeled glue 1 mm long was removed up to a total length of 25 mm. The location of strain gauges T3, T4, T5 and T6 is in the middle of the CFRP plate at distances 4, 10, 20 and 70 mm from the edge of the CFRP plate, strain gauges T1 and T2 are on the steel plate at distances 4 and 20 mm and strain gauge T9 is on the opposite side in the center of the sample see Fig. 7. The debonding growth is simulated by elements death in 25 steps, when area without glue is larger by one mm in each step.

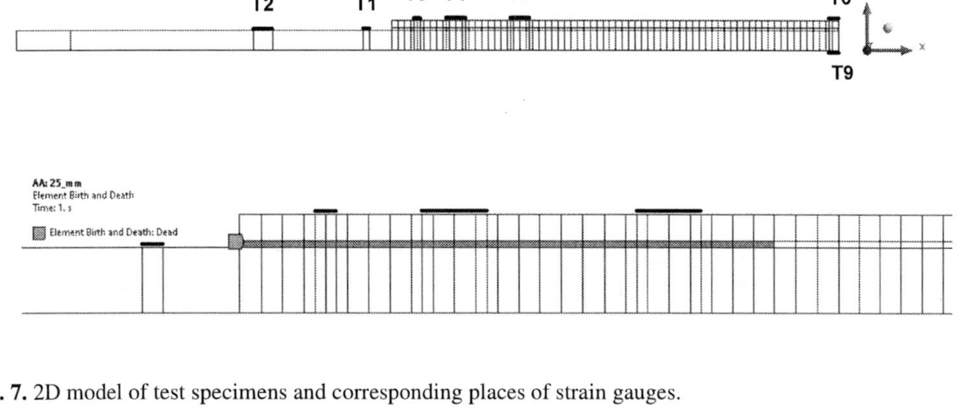

Fig. 7. 2D model of test specimens and corresponding places of strain gauges.

The calculated strain courses at the strain gauges are shown in Fig. 8 and Fig. 9. The presented diagram in Fig. 9 is calculated for level of the loading 300 MPa used for debonding growth study in the section 4.

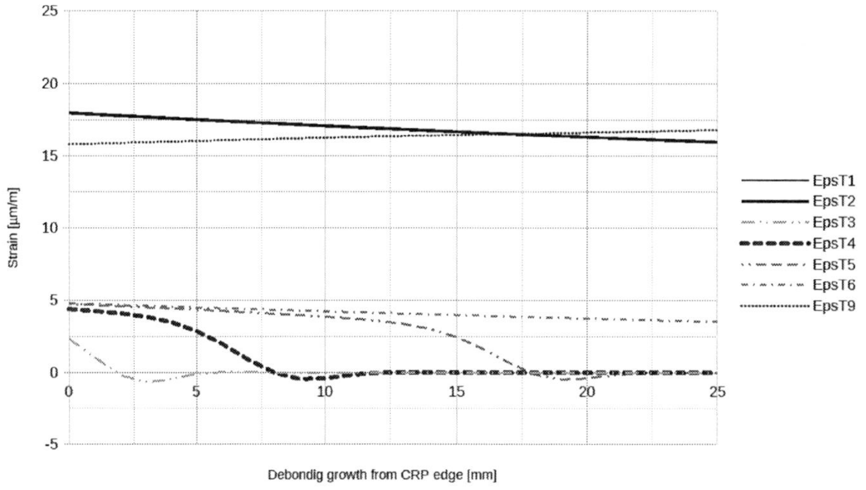

Fig. 8. Comparison courses of the strain [µm/m] in place of strain gauges during the debonding growth evaluated in linear calculation (or small level of load).

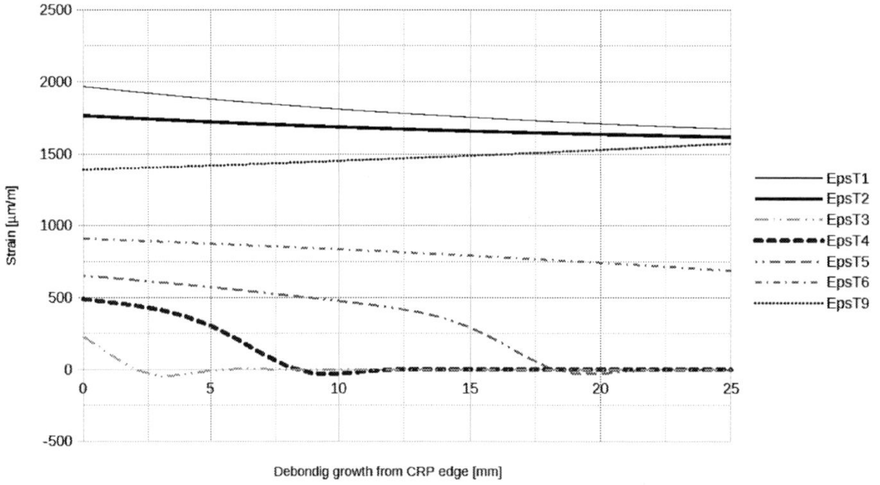

Fig. 9. Comparison courses of the strain [µm/m] in place of strain gauges during the debonding growth evaluated in nonlinear calculation for nominal loading 300 MPa.

The ratio r of the epsilon on strain gauge T4 represents debonding 5 mm and 0 mm is case of linear calculation $r = 1.53$ and in case of nonlinear calculation is $r = 1.61$. It represent scatter of 5%. Evaluated ratio r on strain-gauge T4 is less sensitive to level of loading and nonlinear behaviour of test specimen and it is possible to use it for debonding of 5 mm length.

4. Applying the calculations to the experiments

4.1. Performed experiments

The available fatigue experimental data are test specimens with variable cross-section, side radius respectively, according to ČSN 420363 standard. The loading was realized by repeated force with 8e6 asymmetry cycles of ratio $R = 0.3$. There were no significant changes of strain on installed strain gauges at range of loading 200 MPa. On this stress level was tested only one piece of test specimen. The other provided tests were subjected to the higher load level and changes of strain on strain gauges always appear. That means the separation in these cases were initiate early and were extend by influence of next cycling. It is impossible to identify moment of deboning initiation, only the beginning of cyclic loading is assumed as that moment. But the initiated debonding doesn't means the complete losing of its influence in structural part were the CFRP plate is installed. As consider the courses of stress in Fig. 1, Fig. 8 and Fig. 9 the debonding of 5 mm is appropriate distance to identify by strain-gauge T4 and such damage minimally influence the properties of the hybrid structural part. Scheme of the test specimen and placing of strain gauges are in Fig. 10.

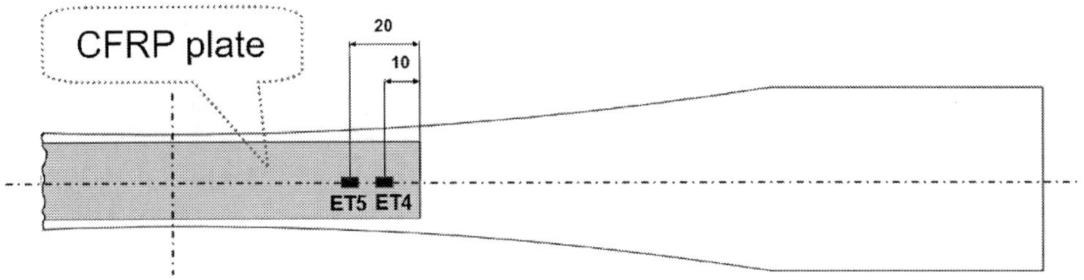

Fig. 10. Scheme of the test specimen and placing of strain gauges.

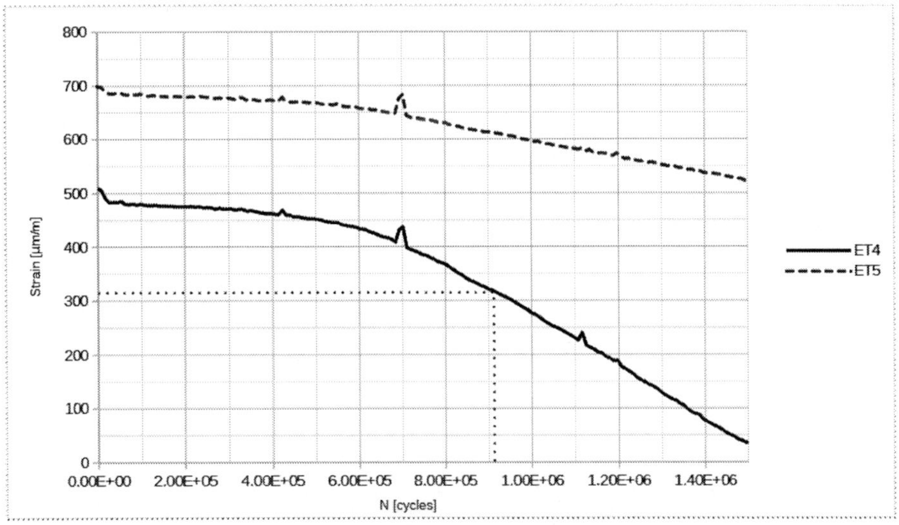

Fig. 11. Comparison of the strain [μm/m] in place of strain gauges during the debonding growth.

4.2. *Number of cycle Estimation at debonding length 5 mm*

Maximum measured strain response during loading by cyclic force on strain gauges is shown in the Fig. 11. Ratio $r = 1.53$ decreasing of the strain on strain gauge ET4 is used for identification of debonding growth 5 mm. The estimated number of cycles at moment of debonding is $N\_5mm = 918\,000$ cycles.

4.3. *Reasons of possible measurement scatter*

The problem of this approach of debonding identification using strain gauges installed on CFRP plates is effect of rest damaged glue which depend on contact with CFRP plate steel or other fragments of glue. This uncertainty increases mainly for highest distances of debonding. Next the thickness of glue and inaccuracy of CFRP plate installation are causes of small strain variations on strain gauges. These factors should be considered during the manufacturing of test specimens.

5. Conclusion

Hybrid joints made from steel plates with glued CFRP plate for one test specimen was subjected to $8{\cdot}10^6$ cycles of repeated loading 200 MPa without visible and measurable damage. The test specimens loaded above this limit were damaged in all tested configurations. The damage appears from beginning of the loading. An appropriate criterion of debonding 5 mm from the end of CFRP plate has small influence to the stiffness of the hybrid joint and it is possible to detect it by strain gauge measurement in laboratory or operation e.g.

Acknowledgements

The article has originated in the framework of M-ERA.NET call 2019 and was supported with Technological Agency of Czech Republic under the No. TH71020003.

References

[1] Václavík J., Dopierała L., Walczak B. and Hejman M., "Load Carrying Capacity of the Hybrid Beam, " in Proceedings of 59 conference on "Experimental Stress Analysis", 2021 (Czech Technical University in Prague, Faculty of Civil Engineering, Litomyšl. 2021). ISBN 978-80-01-06884-7.

[2] Jiao H., Mashiri F. and Zhao X.-L., "Comparative study on fatigue behaviour of steel beams retrofitted with welding, pultruded CFRP plates and wet layup CFRP sheets," *Thin-Walled Structures*, Vol. 59, pp. 144–152 (2012). https://doi.org/10.1016/j.tws.2012.06.002.

Simulations of Robotic Ultrasonic Inspections of Composite Parts in RoboDK Software

Eva Horká[1], Petr Bělský[2] & Michal Podstawka[3]

Abstract: This paper is focused on simulations of robotic inspections of composite test parts, using trough transmission and pulse-echo ultrasonic techniques. The main benefit of this application is the improvement of the whole ultrasonic inspection process by using of a collaborative robot, to inspect test parts with complex geometry and sandwich or monolithic composite test parts. A special NDT workplace has been designed using the Fanuc CRX collaborative robot. This paper describes the process of creating simulations in RoboDK on special composite calibration parts and aileron of turbofan-powered military trainer aircraft.

Keywords: Robotic simulation; Trough-transmission; Water-squirt ultrasonic technique; Composites; Pulse-echo

1. Introduction

Non-destructive testing (NDT) is an essential part of the development and production of new aircraft structure parts from composite materials, which allows for significant optimizing strength and weight parameters. This optimization usually leads to more complex shapes of these structures, where it is inefficient to use contact or standard automated 2D ultrasonic inspections, so new innovative solutions must be available. At VZLU, ultrasonic inspections were initially performed by the classic contact Pulse-echo technique (A-scan), which allows only local NDT inspection of a small number of test pieces and does not provide a sufficiently detailed overview of inspected parts properties. A better overview of inspected parts provides an ultrasonic system using a 3-axis manipulator, which allows performing an ultrasonic C-scan (map) inspection, but only of planar parts. However, flat surfaces are rarely used in aviation and surfaces curved in multiple planes predominate. To obtain C-scan results of parts with complex geometry, more advanced innovative technologies, such as the collaborative robots are required.

Nowadays, there are many applications of collaborative robot across all industries, including aerospace. At VZLU, this technology is integrated into non-destructive testing applications primarily used to detect internal defects in fatigue-loaded aircraft structural parts. Stati and fatigue tests are required to verify the service life of primary structure such as an aileron. Current airworthiness requirements specify the demonstration of a composite structure durability based on the damage tolerance philosophy [1, 2]. These requirements are discussed in detail experimentally in [2].

A Fanuc CRX 10 iA/L collaborative robot is used to demonstrate ultrasonic simulations for two types of ultrasonic water-squirt techniques – pulse-echo and trough transmission. These methods are applied to monolithic, and sandwich composite parts, which are described in detail in section 2.

[1] *Ing. Eva Horká; VZLU – Czech Aerospace Research Centre; Beranových 130; 199 05 Prague; Czech Republic;* eva.horka@vzlu.cz
[2] *Ing. Petr Bělský; VZLU – Czech Aerospace Research Centre; Beranových 130; 199 05 Prague; Czech Republic;* belsky@vzlu.cz
[3] *Ing. Michal Podstawka, Ph.D.; VZLU – Czech Aerospace Research Centre; Beranových 130; 199 05 Prague; Czech Republic;* podstawka@vzlu.cz

2. Methods

2.1. Description of test parts

Monolithic laminates and sandwich structures with a foam or honeycomb core are often used in aircraft structures due to their strength and weight parameters. Internal defects, such as voids, delaminations, cracks, or foreign objects are commonly found in these materials. Ultrasonic technology is used to detect these defects because it provides detailed information about the character of the defect [3]. The following test parts were used to demonstrate the simulation of robotic ultrasonic inspection:

- Aileron of turbofan-powered military aircraft - thin-walled composite shell made of carbon strands (roving) with binder system in combination with a hardener (Fig. 1)
- Sandwich IRP (inspection reference panel) - panel with honeycomb structure (Fig. 2)
- Monolithic IRP panel - prepreg with epoxide matrix reinforced with carbon fabric (Fig. 3)

Fig. 1. Aileron of turbofan-powered military aircraft.

Fig. 2. Sandwich IRP panel with honeycomb structure.

Fig. 3. Monolithic IRP panel.

Both composite IRP (Inspection reference panel) panels contain artificial internal defects of various sizes that represent real defects, typical for composite parts. These panels are designed to set up and verify the correct function of NDT systems. The geometry of the composite fan blades manufactured at VZLU was assumed for the shape of these panels.

2.2. Non-destructive testing

The main purpose of non-destructive testing (NDT) is to detect the surface, subsurface and internal defects of the material without damaging the component. There are several different methods, that are used to monitor the current condition of equipment and evaluate the properties of materials and components. The ultrasonic method is used for this application.

The ultrasonic inspection method is generally based on determination changes in the transmittance and reflectivity of the ultrasonic wave due to imperfections in material. The common frequency range of ultrasonic beam is from 1 to 20 MHz. Two types of ultrasonic water-squirt techniques are used for robotic simulations - trough transmission and pulse-echo. These techniques offer the possibility of ensuring a quality acoustic coupling, good resolution and detection capabilities and easy automatization. To avoid scan results influencing, a special end effector for both methods was designed and manufactured by stereolithographic 3D printing technology. This technology helped to reduce roughness of inner surface of the effector, which improved the laminarity of the water flow, transmitting ultrasonic waves inside the test part. [3–5]

2.2.1. Pulse-echo technique

The pulse-echo method uses only one end effector with one ultrasonic probe, which works as a transmitter and receiver of ultrasonic waves at the same time. This technique provides detailed information about part's material properties and the character of detected defect, such as depth under the surface, its size or orientation. However, it is not suitable for high attenuation materials, such as composite sandwich structures with a honeycomb or foam core [6]. For this reason, this technique is applied to the inspection of monolithic IRP panel (see Fig. 4) and aileron.

For this technique, two types of end effector holders were designed. The first one is aligned with the flange of the collaborative robot and second one is perpendicular to it. Fig. 5 shows the end effector with special holder, aligned with the flange of the robot.

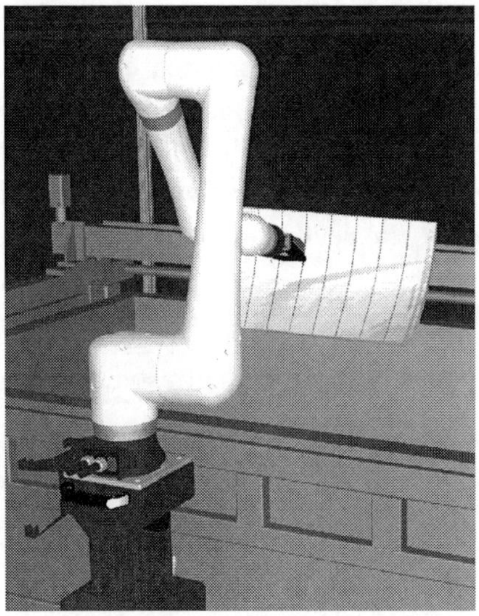

Fig. 4. Pulse-echo ultrasonic technique.

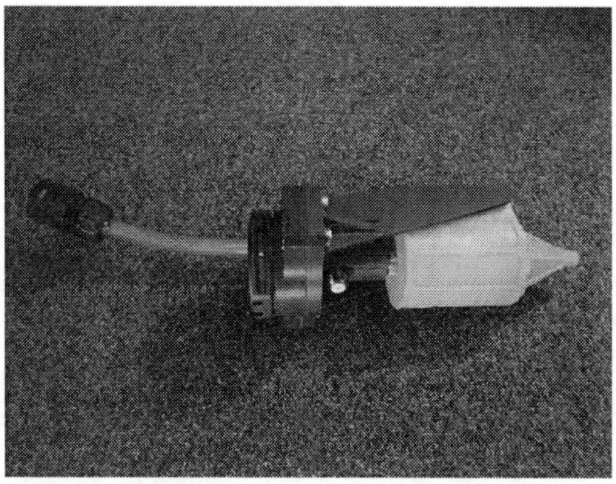

Fig. 5. End effector for pulse-echo ultrasonic technique.

2.2.2. Trough transmission technique

In contrast to the pulse-echo technique, the trough transmission technique is based on using two ultrasonic probes, where the first one works as a transmitter and the second one as a receiver of the ultrasonic wave. This technique requires access to the test part from its both sides and the probes must be aligned, to avoid results influencing. This technique does not provide information about size, orientation, and type of indication, but it is suitable for high attenuation materials, such as sandwich composites. Other advantage of trough transmission technique is the possibility of evaluation in 100% of the material thickness [6]. Fig. 6 shows the trough-transmission method, applied on sandwich IRP panel with honeycomb core. Due to the honeycomb core, it is not possible to inspect this panel by the immersion method to avoid damaging the core, so the trough transmission technique, must be used. For this technique, a special C-frame with two end effectors has been designed (Fig. 7).

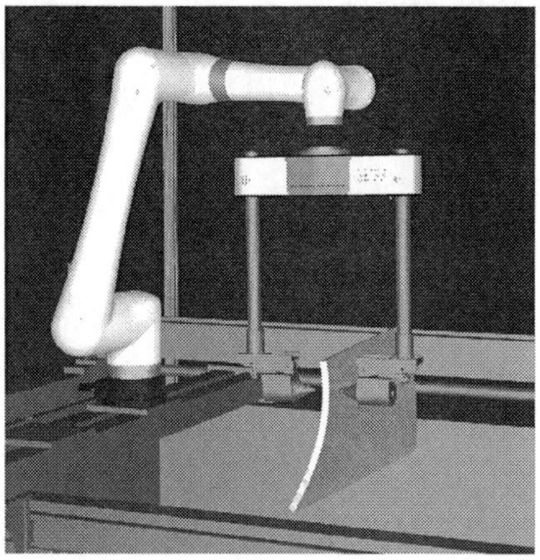

Fig. 6. Trough-transmission ultrasonic technique.

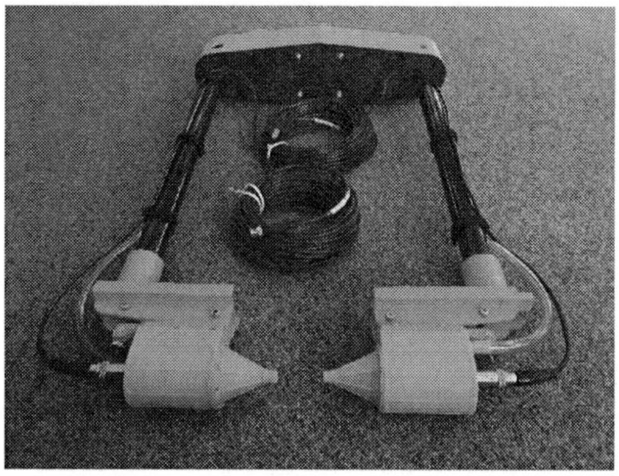

Fig. 7. C-frame tool with two end effectors.

2.3. Workplace design

The whole workplace consists of the collaborative robot, ultrasound system, scanning device with special software for automatic robot trajectory generation, immerse tank with strip curtains and mobile base. Main advantage of the collaborative robot is its mobility, safety, versatility of use and good accessibility to the inspected subjects. Due to its technical parameters, the collaborative robot CRX 10 iA/L of the Fanuc brand was chosen as suitable. The requirements were placed primarily on robot's load-bearing capacity, maximum reach, positioning accuracy, water resistance, number of controlled axes, repeatability, and the price. Due to the maximum reach of this robot, which is 1418 mm, it is possible to inspect structures with a length of approximately 2000 mm from one position [7].

Following figure shows three basic positions (Fig. 8), chosen for the placement of the robot, so the workplace design can adapt to the test part properties and suitable ultrasonic technique.

2.4. Simulations of ultrasonic inspections

To verify the feasibility of automated NDT control, special RoboDK software was used. This software is designed for simulation and offline programming of industrial and collaborative robots. RoboDK has an extensive database of industrial robotic arms, external axes, and effectors from more than 30 different manufacturers. Simultaneously, it is possible to import own tools, models of various objects and to perform a full-fledged simulation of real robotic workstations in a virtual environment.

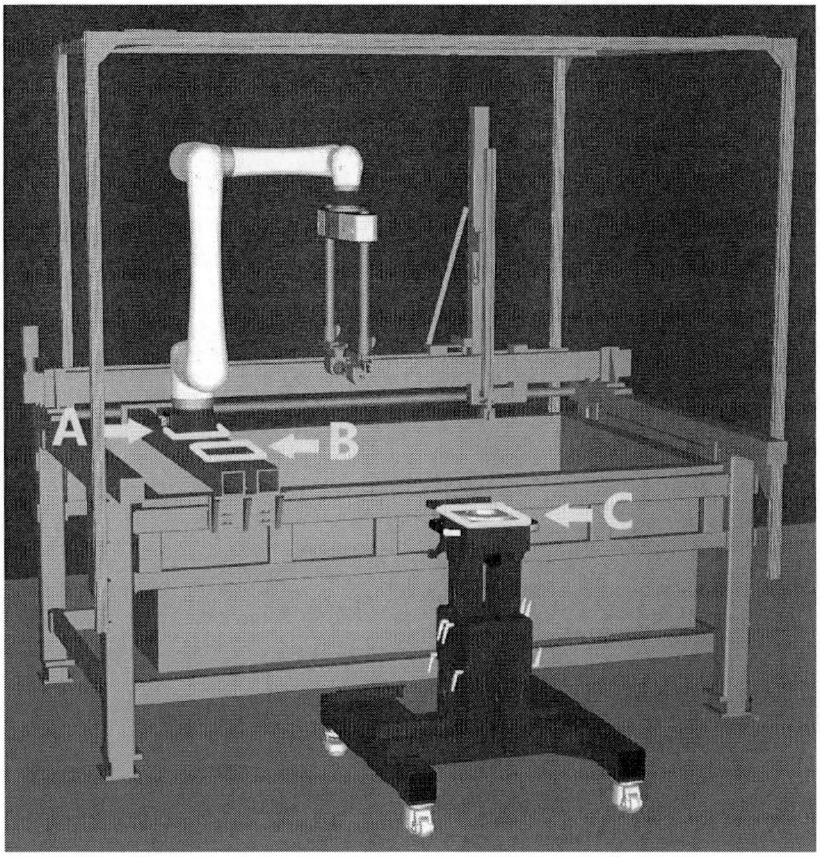

Fig. 8. NDT workplace with three basic positions for robotic arm (A, B, C).

To achieve a successful simulation, the key is to set the relative position trough reference frames of the test part, robotic arm, tool, and its centre point (TCP - Tool Centre Point). TCP is the end point of the robot effector which describes the exact position of the tool and its orientation in that position. In the case of ultrasonic inspection, the distance of the TCP point was set to 20 mm from the end of the effector (see Fig. 9), because at this distance, there is a negligible bending of water flow due to gravity [7]. To avoid inappropriate configuration of the robot or its collisions with the working frame or test piece, the right TCP orientation and suitable position of the robotic arm and the inspected part are the most important steps. That depends on the size of the test piece and the ultrasonic method used. Fig. 10 shows a simulation of the aileron control using the pulse-echo method. Due to the use of a collaborative robot, it is now possible to inspect the outer cover and the leading edge of aileron at the same time and obtain C-scan results of the aileron in the form of an exploded view.

During the simulation, the robot's TCP follows the curve, which was created in Solid-Edge software and then exported to the RoboDK as a toolpath. The tool feed speed was set at 100 mm/s. To avoid result influencing, the main requirement is to maintain the end effector perpendicular to the test surface during the whole inspection. The effector deviation tolerance from its axis is maximum 2–3 angular degrees.

Fig. 9. TCP of end effector.

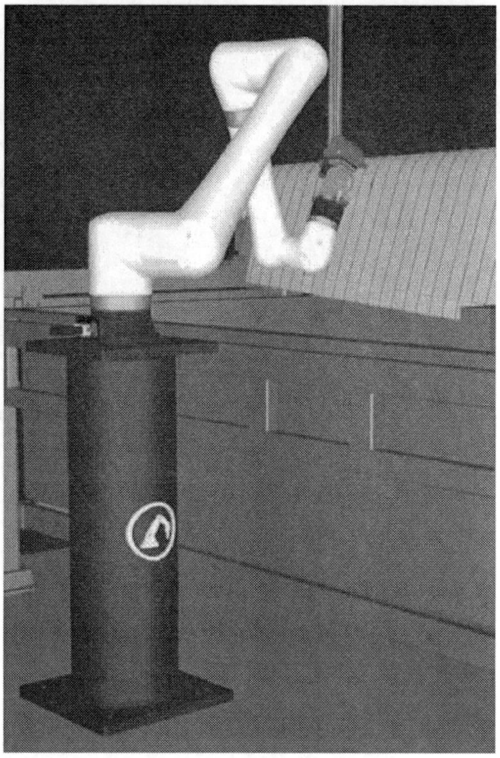

Fig. 10. Inspection of the composite aileron.

The whole simulation consists of three basic parts:

- **Approach** - Safe approach of the end effector to the test part and setting ideal robot configuration for the inspection
- **Inspection** - Inspection of the part, where the TCP follow the path
- **Retract** – Retract of the robotic arm to the home or other safe position, which allows to change the effector or testing part and prepare robot for the new inspection

3. Results

Implementing two water-squirt ultrasonic techniques trough robotics in non-destructive testing applications provides an increase in the level of ultrasonic inspections on composite aircraft structure parts at VZLU. Ultrasonic water-squirt technique is designed to inspect static, and fatigue loaded composite aircraft structure parts, such as the aileron.

The fatigue tests performed to demonstrate the design, to verify the expected or design life, and to detect structural defects that may appear during service. The results of the ultrasonic inspection, after fatigue test, are in the form of C-scans (Fig. 11), which provide comprehensive area information about the defects inside the test part. The results from these inspections are detailed described in [8]. However, results from this inspection are only from outer cover of the aileron without the leading edge. Until now, it has been possible to inspect parts with complex geometry only with contact ultrasonic method, which is not efficient, because it is not time efficient, and it provides only point information of the test part. With new robotic ultrasonic water-squirt techniques, it is possible to inspect various composite parts with complex geometry and obtain result from the inspection in form of C-scans. So, in case of the aileron, this technology allows to perform an inspection of outer cover and leading edge at the same time and thus obtain the results of the C-scan of the aileron in the form of an exploded view [8–10].

Fig. 11. Ultrasonic testing (C-scan) of the aileron with internal defects.

Furthermore, structural modifications of the workplace were made, which led to an increase of its efficiency and adaptability to various types of test parts. This technology allows to perform an inspection of outer cover and leading edge of the aileron at the same time and thus obtain the results of the C-scan of the aileron in the form of an exploded view.

Both ultrasonic water-squirt techniques experiments confirmed a higher effectivity of ultrasonic inspection of test parts with complex geometry, such as the aileron and IRP panels. The main benefit of the trough-transmission technique is the possibility of inspecting materials with high attenuation, such as sandwich composite test parts, that are not suitable for pulse-echo ultrasonic technique or immersion method. The pulse-echo ultrasound technique is suitable for composite monolithic test parts and provides detailed information of detected defects. Also, this technique requires access to the test part surface only from its one side, so it can be applied on more various types of test parts.

4. Conclusions

Simulations of ultrasonic pulse-echo and trough transmission inspections allowed the development and optimization of new robotic ultrasonic workplace at VZLU. The main benefit of this application is the time and cost efficiency of the ultrasonic inspection process due to inspection of the whole part and the results in form of C-scan (maps). Furthermore, it is possible to inspect not only planar test parts, but also parts with complex geometry, that are typical of composite aircraft structures. Other benefits of this application are the repeatability of the inspections and elimination of the influence of the human factors due to use of a collaborative robot.

Acknowledgements

This work was funded by the Ministry of Industry and Trade of the Czech Republic in the DKRV01 program dedicated to the development of research organizations.

References

[1] Růžek R. and Běhal J., "Certification programme of airframe primary structure composite part with environmental simulation," *Int. J. of Fatigue*, Vol. 31(6), pp. 1073-1080 (2009). https://doi.org/10.1016/j.ijfatigue.2008.05.028.

[2] Šedek J., Růžek R., Raška J. and Běhal J., "Comparative study of prediction methods for fatigue life evaluation of an integral skin-stringer panel under variable amplitude loading," *Procedia Engineering*, Vol. 114, pp. 124-131 (2015). https://doi.org/10.1016/j.proeng.2015.08.050.

[3] Bělský P. and Kadlec M., "Capability of non-destructive techniques in evaluating damage to composite sandwich structures," *International Journal of Structural Integrity*, Vol. 10 No. 3, pp. 356-370 (2019). https://doi.org/10.1108/IJSI-10-2018-0067.

[4] Tserpes K., Karachalios E., Růžek R.and Kadlec M., "CFRP Fuselage Panel Behavior Monitoring Using Fibre Optic and Resistance Sensors and Optical Contactless Measurements," *AMM* Vol.827, pp. 51–56 (2016). https://doi.org/10.4028/www.scientific.net/AMM.827.51.

[5] Podstawka M. and Bělský P., "Non-destructive Inspection of L410 NG Aircraft Structure during Full-Scale Fatigue tests", 57th International Scientific Conference on Experimental Stress Analysis, Czech Society for Mechanics, , p 6 (Luhacovice 2019), ISBN 978-80-214-5753-9.

[6] Regazzo R. and Regazzová M., *Ultrazvuk – základy ultrazvukové defektoskopie*, BEN – technická literatura, Praha 2013.

[7] RoboDK, "RoboDK API," *Available from* RoboDK API for Python — RoboDK API Documentation, *Accessed:2015*.

[8] Podstawka M., Snop V., Bělský P. and Kutiová K., "Non-destructive Inspection of Composite Aileron during Fatigue Test," in proceedings of 6th International Conference of Engineering Against Failure (ICEAF-VI 2021), MATEC Web Conf., Vol. 349, Art. No. 03016 (2021). https://doi.org/10.1051/matecconf/202134903016.

[9] Morton H.J., Bishop B.C. and Clark, G. "NDT of Aircraft Structures During Full-Scale Fatigue Testing," National AINDT Conference, Wollongong 1993, 6p. ISBN 0-86418-265-1.

[10] Horák V. and Novotný D., "Full Scale Fatigue Test and Failure Analysis of Advanced Jet Trainer Wing," 5th International Conference of Engineering Against Failure (ICEAF-V 2018), MATEC Web Conf. Volume 188, 9p. (2018). ISSN: 2261-236X. https://doi.org/10.1051/matecconf/201818804023.

Material Characterization of Additively Manufactured Ti-alloy

Jan Horňas[1], Petr Homola[2], Jiří Běhal[3], Sascha Senck[4] & Norica Godja[5]

Abstract: Additive manufacturing (AM) seems to be most perspective candidate for machining of advanced functional lightweight metallic part offering potential benefits in terms of weight, design, and functionality. It is allowing alternative geometric shapes to be produced thereby decreasing the weight of the component without sacrificing component strength and safety. However, AM has not yet been approved for structural components with high safety requirements to date. There are still technology imperfections one has to face with, e.g. material property control, correlation between process and structural properties, effect of defects, etc. The paper is focused on the parameters which can significantly influence mechanical properties of AM parts - defects, surface roughness and microstructure. The work was conducted in the framework of the 3TANIUM Clean Sky project. The main objective of the project is the establishment of quality assessment methods that are capable to provide the detection of process related critical flaws and defects and to understand their effects on material and mechanical properties of Ti-alloy AM parts.

Keywords: Selective laser melting; Additive manufacturing; Non-destructive testing; High cycle fatigue testing; Surface roughness

1. Introduction

Selective laser melting (SLM) is the AM process of bonding materials to make parts from three-dimensional model data layer by layer. The AM offering potential benefits in terms of weight, design, and functionality in the biomedical and aerospace industry. Nevertheless, the AM metallic components containing several types of imperfections.

Various factors have an impact on the mechanical properties of AM metallic parts. Parameters as defects, microstructure, residual stress, and surface roughness were explored in the recent studies. The fatigue performance of SLM Ti-alloy samples was reported to be over 75% lower than wrought material due to surface finish, defects and residual stresses [1]. Mechanical properties of AM components are directly dependent on the in-process factors, such as laser power, laser beam size, scanning speed, scanning pattern, thermal history of the part or build platform and melt-pool dimension [2]. Fatigue life performance is dependent besides on the location, shape and size of the defects [3]. One type of the defect is a lack of fusion (LOF). The LOF is unmelted particle which is caused by fluctuation of laser during the AM process. The fatigue life is more dependent on the LOF than on the layer thickness and build direction during the AM [4]. The fatigue life performance is also influenced by surface morphology of the samples.

[1] *Ing. Jan Horňas; Czech Aerospace Research Centre; Beranových 130, 19905 Prague-Letňany, Czech Republic;* hornas@vzlu.cz
[2] *Ing. Petr Homola, Ph.D.; Czech Aerospace Research Centre; Beranových 130, 19905 Prague-Letňany, Czech Republic;* homola@vzlu.cz
[3] *Ing. Jiří Běhal, CSc.; Czech Aerospace Research Centre; Beranových 130, 19905 Prague-Letňany, Czech Republic;* behal@vzlu.cz
[4] *Dr. Sascha Senck; University of Applied Sciences Upper Austria; Roseggerstraße 15, 4600 Wels, Upper Austria, Austria;* sascha.senck@fh-wels.at
[5] *Dipl.-Ing. Norica Godja; Centre for Electrochemistry and Surface Technology; Viktor-Kaplan-Strasse 2, 2700 Wiener Neustadt, Austria;* norica.godja@cest.at

With higher surface roughness the worse fatigue performance is observed [5]. The most critical ones to the fatigue properties are however the surface defects [6, 7]. These critical factors mentioned above need to be observed and deeply analysed.

This work is carried out based on the 3TANIUM Clean Sky project. The main objective of the 3TANIUM is the establishment of non-destructive testing (NDT) methods that are capable to provide the secure detection of process related critical flaws and defects and to understand their effects on material and mechanical properties of Ti-alloy AM parts.

2. Manufacturing of Ti-alloy Samples

Ti-alloy AM samples were additively manufactured by a commercial Sisma Mysint system using the SLM process, then the stress-relief procedure was adopted. There are 2 types of the samples. First, round sample is shown in Fig. 1, the second, flat sample in Fig. 2.

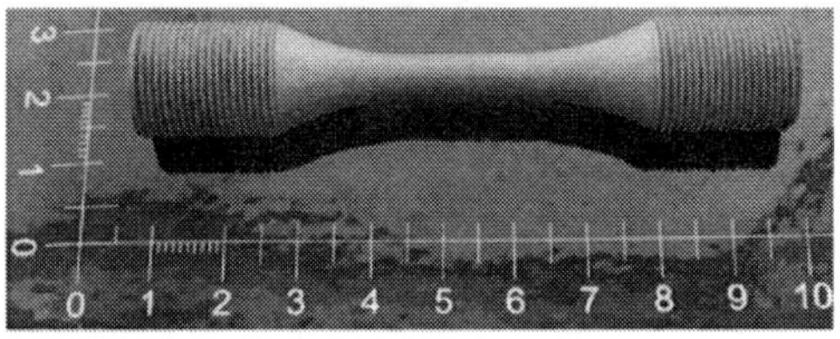

Fig. 1. Round Ti-alloy AM sample.

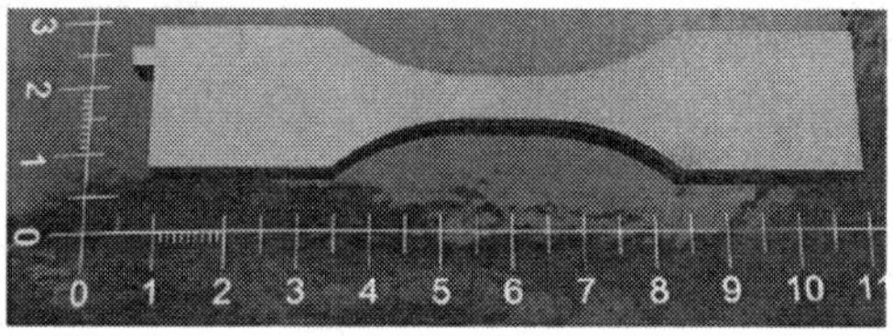

Fig. 2. Flat Ti-alloy AM sample.

3. Characterization of Defects

During the AM small and large voids caused by insufficient energy density lead to irregular shaped pores, which may contain LOF type of defects. Defects are categorized as: gas porosities, LOF and cracks. The major source affecting the formation of the porosities is known to be the trapped gas in the powder material [3, 8]. Characterization of defects was investigated on Phoenix X-Ray Nanotom 180 micro computed tomography (μCT) system with resolution 5 μm. The size and the location of defects in the material were obtained by using the μCT system. In Fig. 3 is shown distribution of pores in one of flat sample in build direction. The projection of distribution of all defects to the cross section of the sample is depicted in Fig. 4.

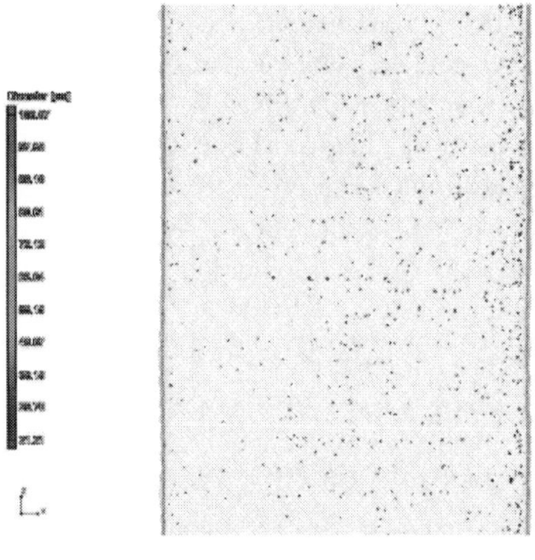

Fig. 3. Distribution of defects in the sample obtained by μCT system.

Fig. 4. Projection of distribution of all defects to the cross section of the sample.

4. Characterization of Microstructure

The microstructure of the AM Ti-alloy is investigated using scanning electron microscopy (SEM) with energy-dispersive X-ray spectroscopy (EDX) electron backscatter diffraction (EBSD) and X-ray diffraction (XRD). Investigations provide information on the texture, present phases, crystallographic orientation, (an)isotropy and crystallite (grain) size where residual stress evaluation is also possible.

5. Characterization of Surface Morphology

Profilometry measurement were performed using the Perthometer S2. The measuring of surface roughness parameters of the samples were done according to DIN ISO 4288. The detailed analysis of the surface morphology is investigated by scanning electron microscope (SEM), see Fig. 5.

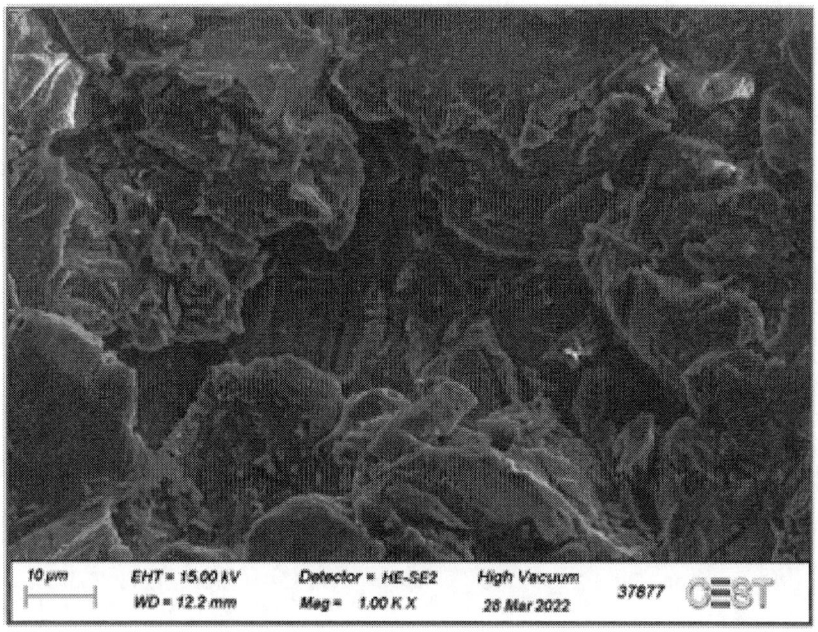

Fig. 5. Surface morphology of AM Ti-alloy as-built sample using SEM.

6. Mechanical Testing

First, quasi-static tests are used to obtain tensile strength and modulus for modelling of both types of specimen. Then, electronic resonance loading machine is used for fatigue testing according to standard ASTM E466. Wöhler curves are established. The interrupt tests are performed at the initiation phase or just before failure and thus obtain specimens with various size of flaws. The specimens after interruption are evaluated by various NDT techniques to measure the flaw size. All loading tests are done in one direction (1D). The fractured sample after high cycle fatigue testing is captured in Fig. 6.

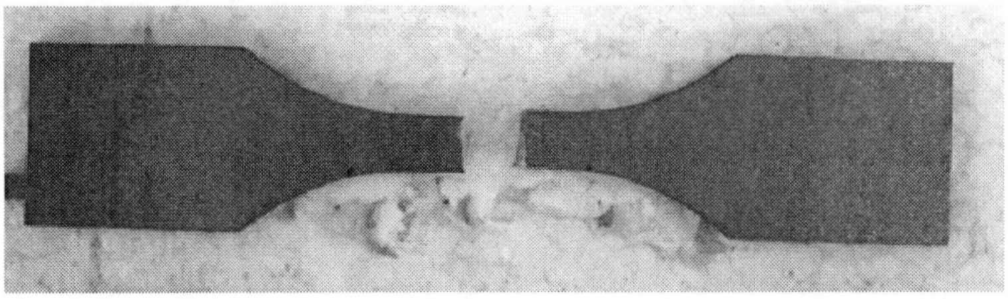

Fig. 6. Fractured flat sample after high cycle fatigue testing.

7. Conclusion

Multiple measurements methods are used for the proper investigation of several factors which have an impact on the final mechanical properties. Obtained data will be used for the modelling of flaws and prediction of fatigue life of AM technology. Measured data also further contribute to better understanding of factors that influence final mechanical properties.

Acknowledgements

The research leading to these results has received funding from the European Community's H2020 for the Clean Sky joint technology initiative under grant agreement No. 101007830.

References

[1] Edwards P. and Ramulu M., "Fatigue performance evaluation of selective laser melted Ti–6Al–4V," *Materials Science and Engineering: A*, Vol. 598, pp. 327–337 (2014). https://doi.org/10.1016/j.msea.2014.01.041.

[2] Fotovvati B., Wayne S., Lewis G. and Asadi E., "A Review on Melt-Pool Characteristics in Laser Welding of Metals," *Advances in Materials Science and Engineering*, Vol. 2018, pp. 1–18 (2018). https://doi.org/10.1155/2018/4920718.

[3] Cunningham R., Narra S., Ozturk T., Beuth J. and Rollett A., "Evaluating the Effect of Processing Parameters on Porosity in Electron Beam Melted Ti-6Al-4V via Synchrotron X-ray Microtomography," *JOM*, Vol. 68, no. 3, pp. 765–771 (2016). https://doi.org/10.1007/s11837-015-1802-0.

[4] Liu Q., Elambasseril J., Sun S., Leary M., Brandt M. and Sharp P., "The Effect of Manufacturing Defects on the Fatigue Behaviour of Ti-6Al-4V Specimens Fabricated Using Selective Laser Melting," *Advanced Materials Research*, Vol. 891–892, pp. 1519–1524 (2014). http://dx.doi.org/10.4028/www.scientific.net/AMR.891-892.1519.

[5] Vayssette B., Saintier N., Brugger C., Elmay M. and Pessard E., "Surface roughness of Ti-6Al-4V parts obtained by SLM and EBM: Effect on the High Cycle Fatigue life", *Procedia Engineering*, Vol. 213, pp. 89–97 (2018). https://doi.org/10.1016/j.proeng.2018.02.010.

[6] Dallago M., Fontanari V., Torresani E., Leoni M., Pederzolli C., Potrich C. and Benedetti M., "Fatigue and biological properties of Ti-6Al-4V ELI cellular structures with variously arranged cubic cells made by selective laser melting", *Journal of the Mechanical Behavior of Biomedical Materials*, Vol. 78, pp. 381–394 (2018). https://doi.org/10.1016/j.jmbbm.2017.11.044.

[7] Chastand V., Quaegebeur P., Maia W. and Charkaluk E., "Comparative study of fatigue properties of Ti-6Al-4V specimens built by electron beam melting (EBM) and selective laser melting (SLM)", Materials Characterization, Vol. 143, pp. 76–81 (2018). https://doi.org/10.1016/j.matchar.2018.03.028.

[8] Tammas-Williams S., Zhao H., Léonard F., Derguti F., Todd I. and Prangnell P., "XCT analysis of the influence of melt strategies on defect population in Ti–6Al–4V components manufactured by Selective Electron Beam Melting," *Materials Characterization*, Vol. 102, pp. 47–61 (2015). https://doi.org/10.1016/j.matchar.2015.02.008.

EAN 2022 60th annual conference on experimental stress analysis

Dynamic Behaviour of 3D Printed Polymers with Respect to the Structure of their Internal Filling

Róbert Huňady[1], Michal Fabian[2] & Peter Palička[3]

Abstract: This paper deals with the experimental assessment of the influence of the internal structure of polymer plates produced by 3D printing on their dynamic properties such as stiffness and damping. The factors considered are the shape and density of the internal infill. The dynamic properties are assessed based on experimental modal analysis for quasi-free-fitting of the test specimens. The paper further discusses the methods of homogenization of material properties and the development of an equivalent numerical model using the FE model updating method. The static stiffness of the plates is considered based on the elastic moduli, the dynamic stiffness based on modal parameters taken from FEM models.

Keywords: 3D print; Modal analysis; Stiffness; Damping; Homogenization; FE model updating

1. Introduction

3D printing is an additive technology that is used to produce parts from a three-dimensional digital model. In the additive process, an object is created by laying down successive thin layers of material until the object is formed. Although the first 3D printing methods were developed in the 1980s, the greatest advances in this field were made in the early 21st century and now it is commonly used to manufacture components from a variety of materials. Additive technology brings a huge benefit to the engineering industry in the form of prototyping parts without the need to produce expensive moulds for injection moulding. Compared to conventional technologies, 3D printing enables the production of parts with complex shapes that generally does not have as much impact on the resulting price. This gives the designer much more design freedom. In addition, the finished products are produced without unnecessary waste, resulting in material savings compared to machining, for example. The resulting product properties depend not only on the material used and the shape of the part itself, but also on the print settings. By varying the layer thickness, changing the density and pattern of the inner filler, selecting the raster, etc., it is possible to obtain a product with different properties. This is why the fabrication of components using additive technologies has also received considerable attention in the field of scientific research, particularly in the investigation of the mechanical properties and behaviour of 3D printed materials.

Chin Ang et al. in a study [1] investigated how air gap, raster width, build orientation, build layer and build profile affect the mechanical properties and porosity of ABS structures made by fused deposition modelling (FDM) method. The Quantachrome Instruments' Ultrapycnometer 1000 was used to measure the porosity. The Quantachrome Instruments' ultrapycnometer (Ultrapycnometer 1000) was used to measure the porosity. The mechanical properties of the samples were investigated on the Instron 5569 machine, based on the standard compressive test

[1] *doc. Ing. Róbert Huňady, PhD.; Fac. of Mech. Engineering, Technical University of Košice, Letná 1/9, 042 00 Košice, Slovakia;* robert.hunady@tuke.sk

[2] *doc. Ing. Michal Fabian, PhD.; Fac. of Mech. Engineering, Technical University of Košice, Letná 1/9, 042 00 Košice, Slovakia;* michal.fabian@tuke.sk

[3] *Ing. Peter Palička; Fac. of Mech. Engineering, Technical University of Košice, Letná 1/9, 042 00 Košice, Slovakia;* peter.palicka@tuke.sk

method specified by ASTM D695-02a. The relationship between structure mechanical properties and porosity was determined to be logarithmic, with the best mechanical properties observed in structures of low porosity.

Medel et al. in study [2] analysed a variety of metrics derived from Laser Doppler Vibrometry (LDV) characterization of 3D printed rectangular prisms. The metrics of interest were natural frequencies and amplitudes of first and second bending vibration modes, equivalent elastic moduli, damping ratios and areas of transmissibility functions. To explore the influence of printing process parameters on the dynamical behaviour and, therefore, on the aforementioned metrics, 48 different combinations were considered, including build orientation, raster angle, nozzle temperature, print speed and layer height as relevant parameters. Based on the equivalent elastic modulus metrics, it was possible to corroborate the influence of printing process parameters on the mechanical performance, being raster angle, build orientation and nozzle temperature the most influential parameters. The analyses of the transfer function areas and vibration mode amplitudes also confirmed better excitation transfer for samples with relatively high equivalent elastic modulus and low damping coefficients. Moreover, damping ratio data confirmed the association between high damping and poor interfilament bonding. In addition, it was possible to identify optimal values of 3D printing process parameters for low damping ratios and, therefore, enhanced interfilament bonding. The possibility of using a laser scanning vibrometry to assess the mechanical behaviour of 3D printed samples fabricated with different combinations of process parameters has been addressed by the same authors in an earlier paper [3], where the equivalent elastic modulus values calculated from the experimental data served to confirm the significant effects and interactions between the process parameters.

A number of papers have also been published that focus on assessing the influence of 3D printing process parameters on the strength and fatigue properties of fabricated samples [4–8].

This paper deals with the experimental assessment of the influence of the internal structure of polymer plates produced by 3D printing on their dynamic properties such as stiffness and damping. The dynamic properties are assessed based on experimental modal analysis for free-support condition of the tested specimens. The static stiffness is considered based on the equivalent elastic moduli of the homogenised plates. Their material constants are determined using the FE model updating method. The updating method uses optimization analysis, in which the natural frequencies of the FE model are fitted to the frequencies determined by experimental modal analysis. The authors successfully used this approach to determine the material constants of the isotropic material [9], the homogenized laminate [10], to determine the fibre orientation of the composite plate [11], and to estimate the stiffness parameters of the bonds [12]. The dynamic stiffness is determined using modal parameters taken from FEM analysis.

2. Object of investigation

The object of investigation was a plastic plates with dimensions 100×160×4 mm made of Z-ULTRAT material [13] using Zortrax M200 Plus printer. Z-ULTRAT filament is an ABS plastic blend designed for durability and excellent surface quality. It can be used for functional prototypes, end-use parts or to 3D print consumer products. It is resistant to high temperatures and impacts. Its exceptional hardness allows you to test models in real life conditions. The printer used is based on Layer Plastic Deposition (LPD®) technology, a technology similar to FDM. Its principle is as follows. A plastic wire filament is melted using a print head and extruded onto a build platform underneath. The movement of the print head against the build platform is automated using electronic motors. The print head extrudes the melted filament to "draw" the model one layer at a time. Each time a single layer has been printed, the height of the print head in relation to the build platform increases by an increment. Because the extruded material is hot, all layers are strongly connected together with the first layer sticking to the build platform.

Nine plates with different structures were analysed. The plates differed in density and the pattern of the infill. The print quality and print type in terms of raster were the same. The plates consisted of 21 layers with a thickness of 0.19 mm. Two patterns were used, a diagonal square (DS) and a honeycomb pattern (HC). The density of the infill varied from 20 to 80%. One plate, named Solid, had 100% fill. The composition of the plates is shown schematically in Fig. 1.

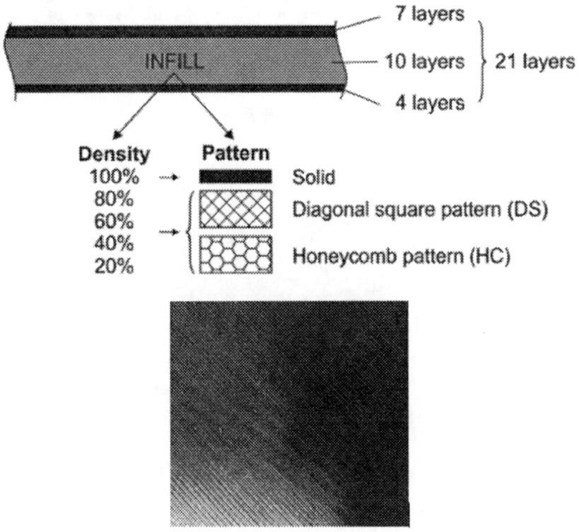

Fig. 1. Plate composition (combination of infill parameters) and surface raster.

Table 1 shows the masses of the plates and their equivalent density determined by calculation. The density will later be used in the FEM analysis in homogenizing the material properties.

Table 1. Mass and equivalent density of the plates.

Plate	Volume [mm^3]	Mass [g]	Equivalent density [kg/m^3]
SOLID		60.56	946.3
DS80		48.27	754.2
DS60		41.60	650.0
DS40		36.13	564.5
DS20	$6.4 \cdot 10^{-5}$	31.12	486.3
HC80		54.67	854.2
HC60		46.00	718.8
HC40		38.60	603.1
HC20		32.67	510.5

3. Experimental modal analysis

Experimental modal analysis was used to assess the effect of plate pattern and infill density on stiffness and damping. Modal parameters, i.e. natural frequencies, mode shapes and damping ratios, were measured and further evaluated.

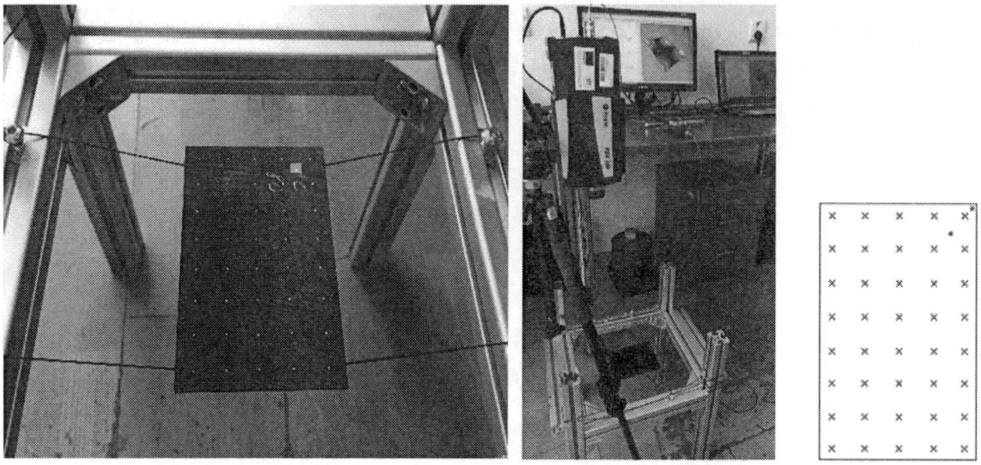

Fig. 2. Experimental setup (left, middle), measurement DOFs (right).

Multi-reference measurements in the range up to 1200 Hz were performed using the Brüel & Kjær Pulse measurement platform. Vibration responses were measured at two points (marked in red in Fig. 2 right) with a Polytec PDV100 single-point laser vibrometer. The plates were quasi-freely supported by soft rubber cords, as shown in Fig. 2 left, and were excited with a Brüel & Kjær 8206 modal hammer at 40 points (marked in green in Fig. 2 right). For a given support to be considered free, the frequency of the highest rigid body mode must be ten times less than the frequency of the first flexible mode. This condition was satisfied for all plates, since the frequency of the rigid body modes was less than 15 Hz, while the frequency of the first mode was more than 200 Hz. See the plot of Complex Mode Indicator Function for the solid plate shown in Fig. 3.

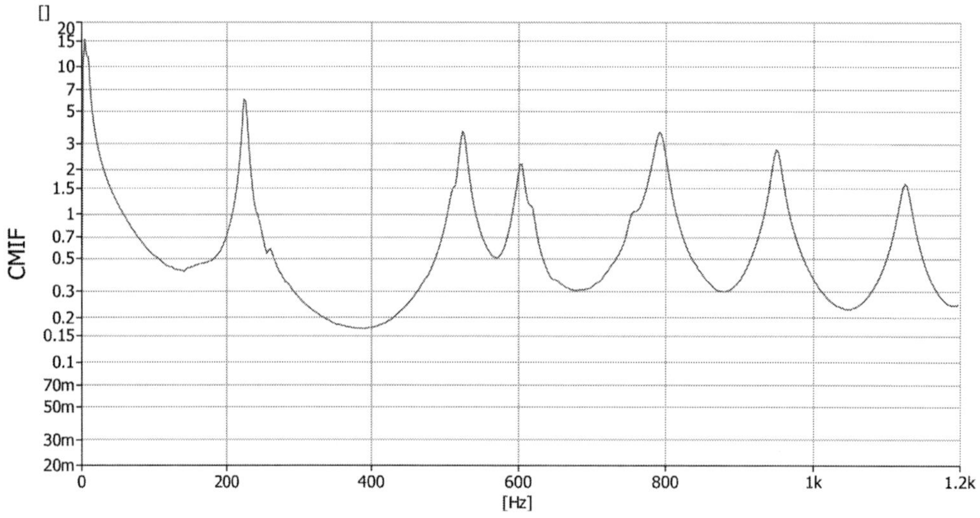

Fig. 3. CMIF plot for the solid plate.

As an example, the first eight mode shapes of the solid plate are shown in Fig. 4. The other plates with reduced fill densities showed similar shapes with only one difference, and that was in the order of the first and second modes (Fig. 5).

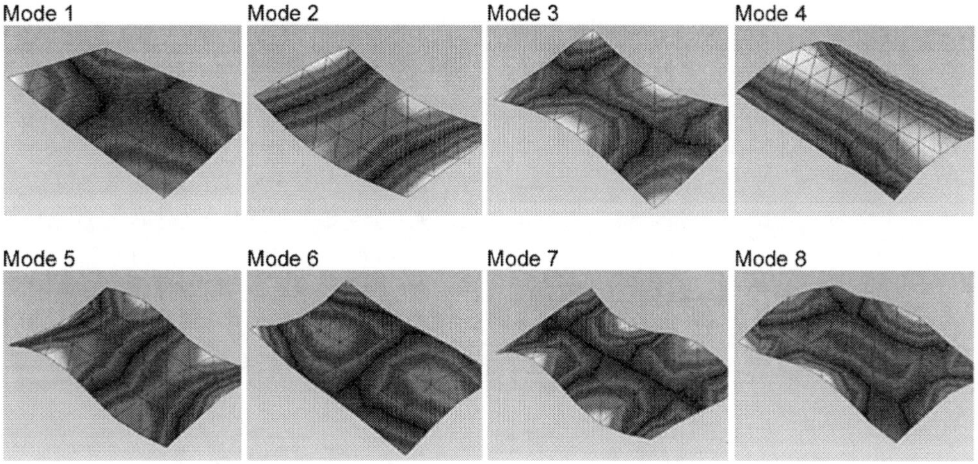

Fig. 4. The first eight mode shapes of the solid plate.

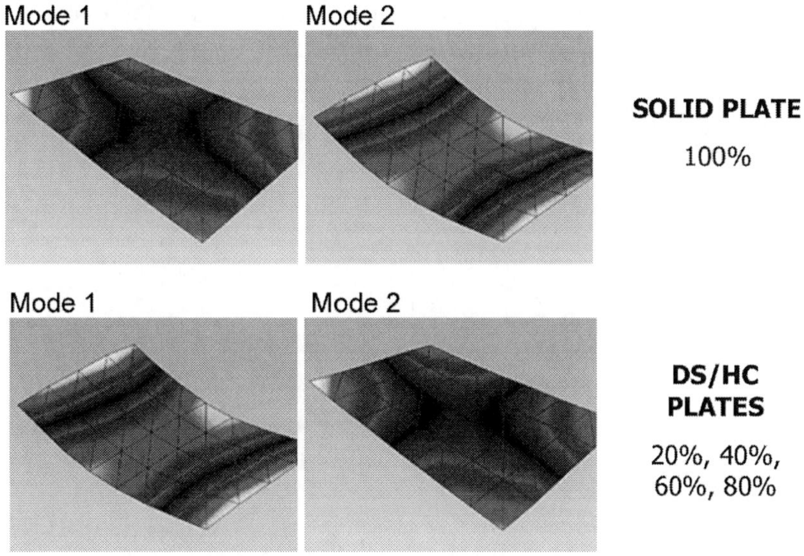

Fig. 5. The difference in the shapes of the plates of the first two modes.

The natural frequencies of all analysed plates are listed in Table 2. Graphically, the frequency changes of the first five modes are shown in Fig. 6. As can be seen, the values vary in the range of units to tens of Hertz. However, the changes are not significant and do not indicate any trend. This is due to the fact that not only the stiffness but also the mass of the plates changes simultaneously with the infill density. For this reason, it will be necessary to assess the effect of infill structure on stiffness of the plates in a different way, which will be described later.

Table 2. Natural frequencies of the plates in Hz.

Mode	Solid	DS80	HC80	DS60	HC60	DS40	HC40	DS20	HC20
1	225.72	214.53	220.54	211.31	214.25	207.85	211.61	204.03	213.71
2	228.37	233.72	228.19	237.62	227.26	242.22	230.57	241.94	241.1
3	525.37	521.52	510.76	526.85	511.32	535.48	515.78	538.69	531.92
4	603.66	560.73	561.76	552.83	543.19	556.86	544.1	538.75	550.34
5	619.04	588.19	602.93	589.45	587.91	590.39	581.24	581.21	587.07
6	793.32	732.65	742.44	737.42	718.91	738.63	724.97	729.94	747.66
7	950.83	927.69	926.83	923.85	908.06	941.98	922.21	916.38	934.65
8	1125.3	1111.81	1090.6	1108.2	1083.1	1109.4	1084.4	1063.9	1094.43

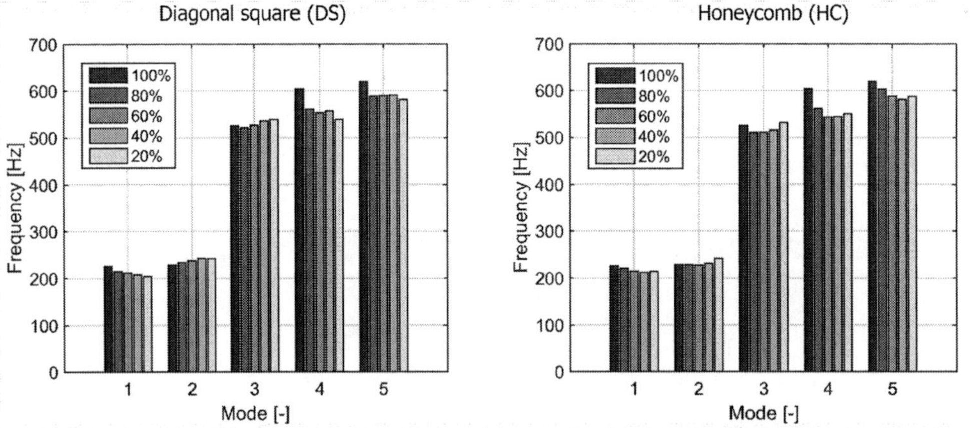

Fig. 6. Natural frequencies of the first five modes.

The pattern and density of the infill has minimal effect on the damping of the boards. As can be seen in Fig. 7, the relative damping does not change significantly. The values range from 0.6% to 1.3%, indicating that the damping is relatively small.

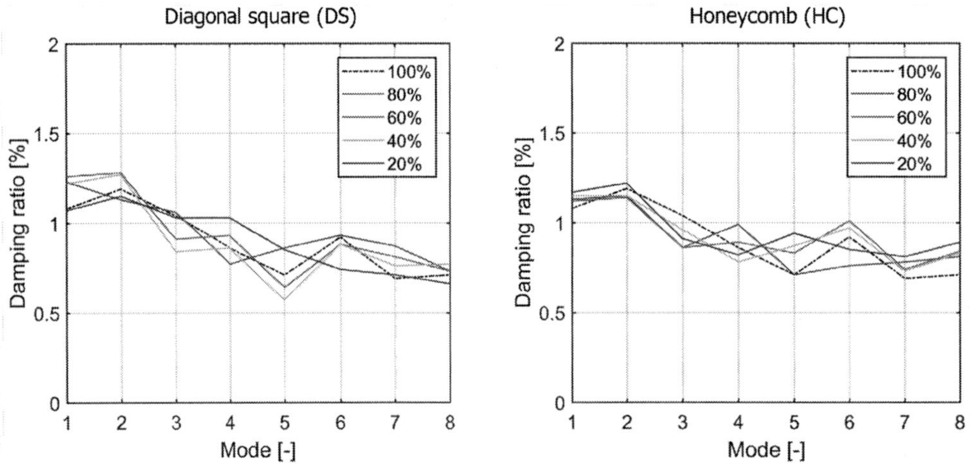

Fig. 7. Damping ratio depending on the infill density of the plates.

4. Homogenisation of the plates

The aim of homogenization is to determine the equivalent material constants, i.e. Young's modulus and Poisson's number, for homogeneous isotropic plates. For this purpose, the FE model updating method was used.

A shell model of the plates with the corresponding dimensions was created in Abaqus/CAE. The material density was defined for each plate according to Table 1. The initial values of Young's modulus and Poisson's ratio were defined. The values of these two design variables were then refined by parametric optimization in Isight. The basis of the optimization was frequency analysis, where the values of the natural frequencies of the first five to seven modes obtained experimentally were used as objective functions. The block diagram of the optimization task is shown in Fig. 8. The NSGA2 genetic algorithm was used to find equivalent material constants. The intervals of the design variables were as follows:

- Young's modulus: <100, 2500> MPa
- Poisson's ratio: <-0.4, +0.4 >

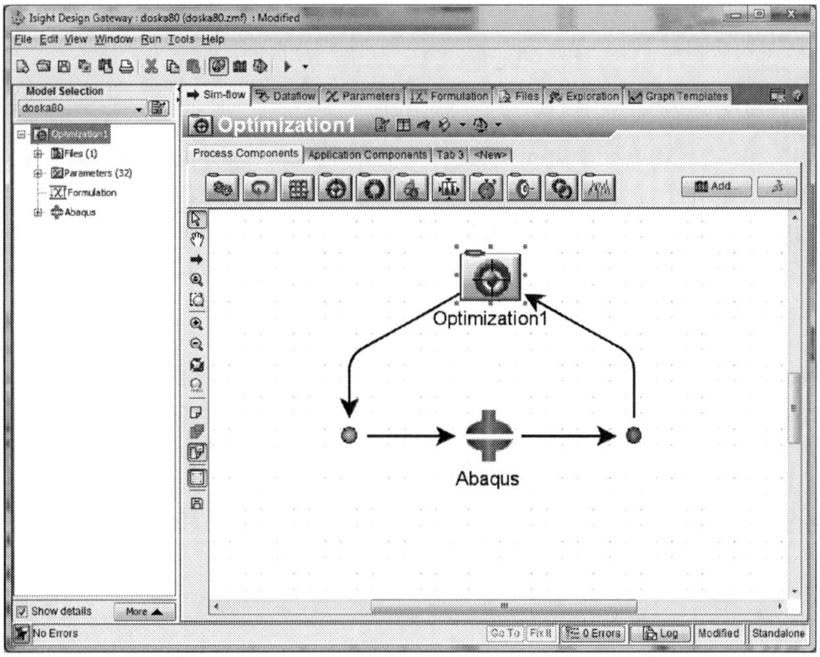

Fig. 8. The block diagram of the optimization task.

Table 3. Equivalent material constants of the homogenised plates.

Plate	Young's modulus [MPa]	Poisson's ratio [–]	Equivalent density [kg/m^3]
SOLID	1928.3	0.291	946.3
DS80	1387.5	0.085	754.2
DS60	1174.1	0.029	650.0
DS40	999.8	-0.069	564.5
DS20	797.3	-0.122	486.3
HC80	1600.7	0.174	854.2
HC60	1257.3	0.056	718.8
HC40	1048.1	0.051	603.1
HC20	910.2	-0.043	510.5

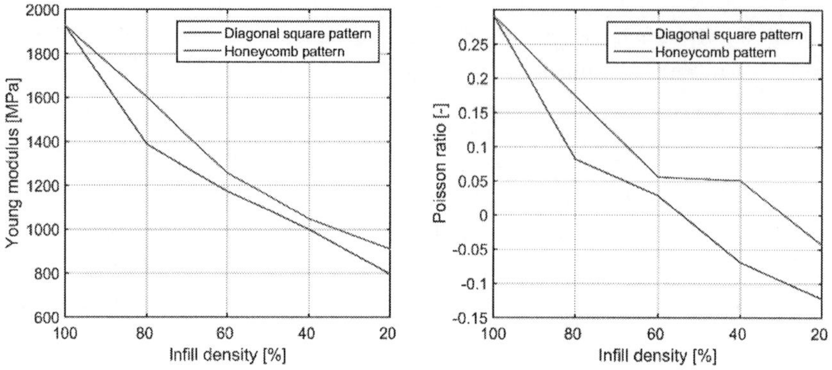

Fig. 9. Young's modulus and Poisson's ratio of the homogenised plates depending on the pattern and density of the infill.

The resulting values of Young's modulus and Poisson's ratio are shown in Table 3. From their plot in Fig. 9, it can be seen that their values decrease with decreasing infill density, while the honeycomb-filled plates are stiffer.

5. Verification experiment

In order to validate the optimization results, a tensile test was performed to determine the Young's modulus and Poisson's number for the 100% fill material. For this purpose, 4 standardized specimens were printed with dimensions as shown in Fig. 10. The specimens were loaded to failure on a FPZ100/1 tensile machine at a speed of 84 mm/min. A Dantec Dynamics Q-450 high-speed 3D correlation system, which also recorded the loading force, was used to measure the surface deformations (Fig. 11 left). The strains in the longitudinal and transverse directions were determined from line virtual gauges determined on the surface contour (Fig. 11 right). The tensile stress as well as the Poisson's ratio were calculated based on well-known formulas. The Young's modulus was determined based on the slope of the stress-strain curve shown in Fig. 12. Since the values agree sufficiently with those determined by optimization, it can be concluded that the method used provides a sufficiently good estimate.

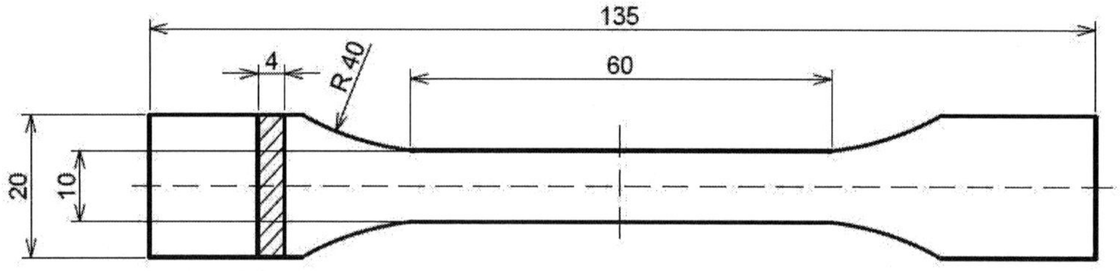

Fig. 10. Specimen dimensions.

Fig. 11. Tensile test (left, middle), virtual strain gauges (right).

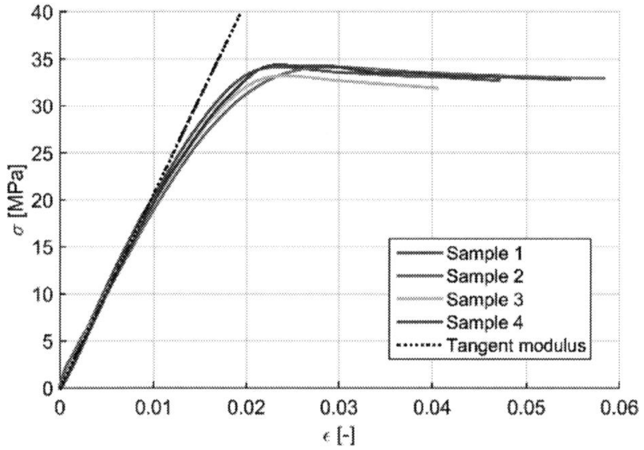

	Poisson's ratio [-]	Young's modulus [MPa]
S_01	0,306	2109
S_02	0,310	1840
S_03	0,302	2019
S_04	0,294	2019
Average	0,303	1996,8
Isight	0.291	1928.3

Fig. 12. Stress-strain curves and resultant material constants of the solid material.

6. Comparison of EMA and FEM

In the case of plates with reduced density of infill, the material constants of the homogenized plates were verified by comparing the natural frequencies of the FEM model with the frequencies determined by the modal test. Table 4 shows the values of the natural frequencies for the three different plates. The comparison shows that the largest difference in the natural frequency values of the FEM model of the homogenized plates and the measured ones is $\pm 3\%$. The mode shapes of the HC60 plate obtained by FEM are shown in Fig. 13.

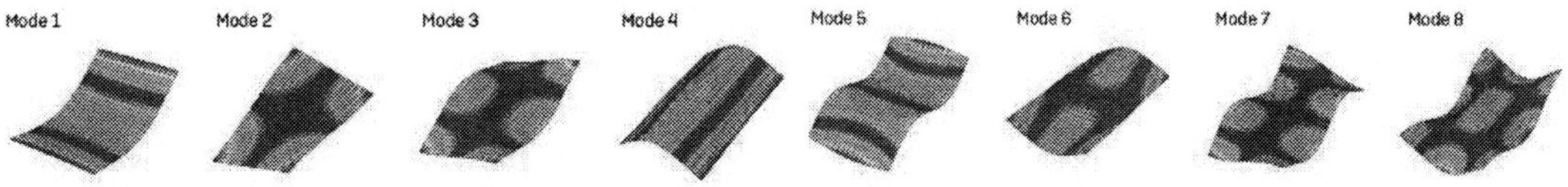

Fig. 13. Mode shapes of the HC60 plates determined by FEM analysis.

Table 4. Comparison of the natural frequencies of the three types of plates.

	Solid plate			DS80			HC20		
Mode	FEM	EMA	Δ[%]	FEM	EMA	Δ[%]	FEM	EMA	Δ[%]
1	227.15	225.72	0.63	219.7	220.54	-0.38	214.06	213.71	0.16
2	229.66	228.37	0.56	227.32	228.19	-0.38	243.16	241.1	0.85
3	515.45	525.37	-1.89	508.73	510.76	-0.40	530.48	531.92	-0.27
4	603.93	603.66	0.04	566.51	561.76	0.85	546.62	550.34	-0.68
5	626.46	619.04	1.20	602.43	602.93	-0.08	588.01	587.07	0.16
6	770.38	793.32	-2.89	729.12	742.44	-1.79	725.75	747.66	-2.93
7	935.89	950.83	-1.57	910.68	926.83	-1.74	925.53	934.65	-0.98
8	1116.1	1125.3	-0.82	1090.1	1090.6	-0.05	1118.1	1094.43	2.16

Based on the knowledge of the eigenvalues and the generalized masses of the modes from the FEM analysis, it was possible to calculate and compare the modal stiffnesses. The values of the first four modes are shown graphically in Fig. 14. From Fig. 14, it can be seen that the stiffness of the plates generally decreases with decreasing infill density. An exception can be observed in the case of the stiffness of the first mode. The reason is that the shape of the first mode of the solid plate differs from that of the reduced density plates. Because of this, there is less mass in motion, which reduces the value of the modal stiffness.

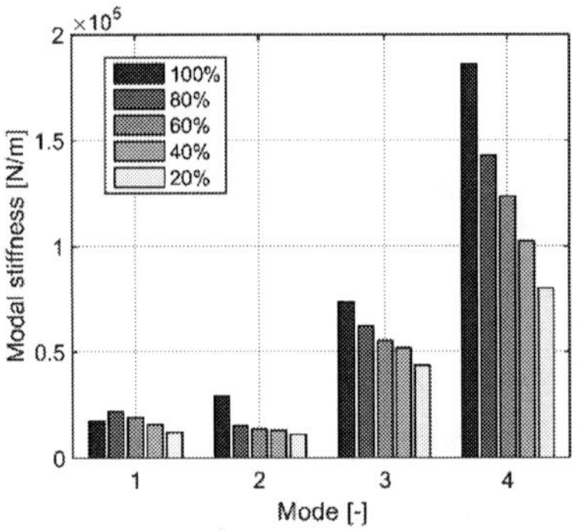

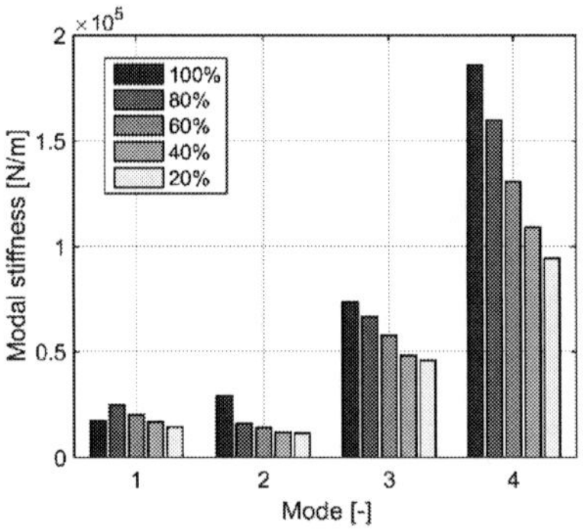

Fig. 14. Modal stiffness of the homogenised plates depending on the pattern and density of the infill.

7. Conclusion

In this paper, attention was paid to the evaluation of the dynamic properties of polymer plates fabricated by 3D printing with respect to the structure of their internal filling. The effect of infill pattern and density on stiffness and damping was assessed using modal analysis tools. The equivalent material constants of the homogenized plates were determined by the FE model updating method. Based on the results obtained, the following can be stated:

- The infill structure does not have a significant effect on the damping of the modes of the tested plates.
- The first and second mode shape of the plates with reduced infill density were interchanged compared to the mode shapes of the solid plate.
- Young's modulus and Poisson's ratio decrease with decreasing fill density.
- The effect of the infill on the dynamic stiffness had to be assessed on the basis of the modal stiffness.

Acknowledgements

This work was supported by the Slovak Research and Development Agency under the grant project VEGA No. 1/0500/20.

References

[1] Chin Ang K., Fai Leong K., Kai Chua C. and Chandrasekaran M., "Investigation of the mechanical properties and porosity relationships in fused deposition modelling-fabricated porous structures," *Rapid Prototyp. J.*, Vol. 12, no. 2, pp. 100–105 (2006). https://doi.org/10.1108/13552540610652447.

[2] Medel F., Abad J. and Esteban V., "Stiffness and damping behavior of 3D printed specimens," *Polymer Testing*, Vol. 109, Art. no. 107529 (2022). http://dx.doi.org/10.1016/j.polymertesting.2022.107529.

[3] Medel F., Esteban V. and Abad J., "On the use of laser-scanning vibrometry for mechanical performance evaluation of 3D printed specimens," *Materials & Design*, Vol. 205, Art. no. 109719 (2021). ISSN 0264-1275. https://doi.org/10.1016/j.matdes.2021.109719.

[4] Song Y., Li Y., Song W., Yee K. et al., "Measurements of the mechanical response of unidirectional 3D-printed PLA," *Materials & Design*, Vol. 123, pp. 154–164 (2017). https://doi.org/10.1016/j.matdes.2017.03.051.

[5] Ziemian S., Okwara M. and Ziemian C.W., "Tensile and fatigue behavior of layered acrylonitrile butadiene styrene," *Rapid Prototyp. J.*, Vol. 21, no. 3, pp. 270–278 (2015). http://dx.doi.org/10.1108/RPJ-09-2013-0086.

[6] Rezaeian P., Ayatollahi M. R., Nabavi-Kivi A. and Javad Razavi S. M., "Effect of printing speed on tensile and fracture behavior of ABS specimens produced by fused deposition modeling," *Engineering Fracture Mechanics*, Vol. 266, Art. no. 108393 (2022). ISSN 0013-7944. https://doi.org/10.1016/j.engfracmech.2022.108393.

[7] Fracio M., Avalle M. and Monti M., "Fatigue strength of plastics components made in additive manufacturing: first experimental results," *Procedia Structural Integrity*, Vol. 12, pp. 32–43 (2018). ISSN 2452-3216. https://doi.org/10.1016/j.prostr.2018.11.109.

[8] Shanmugam V., Das O., Babu K., et al., "Fatigue behaviour of FDM-3D printed polymers, polymeric composites and architected cellular materials," *International Journal of Fatigue*, Vol. 143, Art. no. 106007 (2021). ISSN 0142-1123. https://doi.org/10.1016/j.ijfatigue.2020.106007.

[9] Kaľavský A., Huňady R. and Pavelka, P., "Determining the young's modulus of a specimen using experimental modal analysis," In: *Proceedings of the EAN 2018*, Harrachov, Czech Republic, 5–7 June 2018, pp. 176–183.

[10] Huňady R., Pavelka P. and Lengvarský P., "Homogenization of Material Properties of a Combined rubber/steel Sheet using Modal Analysis," In *Proceedings of the EAN 2019*, Luhacovice, Czech Republic, 3–6 June 2019, pp. 146–150.

[11] Huňady R., Pavelka P., Lengvarský P. and Kaľavský A., "The estimation of elastic material constants of unidirectional fibre-reinforced composite based on modal analysis," In: *Proceedings of the MMaMS 2019*, Sromowce Nižne, Poland, 11–13 September 2019, pp. 1–9.

[12] Huňady R, Lengvarský P, Pavelka P, Kaľavský A, Mlotek J., "Stiffness Estimation and Equivalence of Boundary Conditions in FEM Models," *Applied Sciences*,Vol. 11 no. 4, Art. no. 1482 (2021). https://doi.org/10.3390/app11041482.

[13] Zortrax: Technical Data Sheet, Z-ULTRAT, *Available from* https://zortrax.com/filaments/z-ultrat/ *Accessed:* 2022-04-14.

Experimental Investigation of Defects of Thin 3D-printed Plates

Aleš Jíra[1] , Luboš Řehounek[2] , Gabriela Javorská[3] & Pavel Padevět[4]

Abstract: Contemporary additive technologies push the boundaries of manufacture of intraosseous parts of implants with effective surfaces. Thanks to virtually no limitations in the final geometry, we can design and produce porous structures that support bone ingrowth, stabilizing the implant. Usually, these structures are made of thin beams or walls with widths of ca. 150 – 600 μm. In spite of the advances of 3D printing technologies, 3D-printed products still suffer from micro cracks, discontinuities and other quality issues that negatively influence the mechanical durability of the final implant. The aim of this work is to investigate the mechanical durability in regard to width of 3D-printed titanium plates.

Keywords: 3D printing; Gyroid; Porous implant; Titanium plate

1. Introduction

Intraosseous parts of implants represent the load-bearing pillar of the implant that ensures long-term stability in bone. It is vital to provide a good quality bone-implant connection that is directly proportional to the surface area of the implant [5]. Thanks to the advancement of additive technologies, it is possible to equip the implant with a diverse surface structure. Surface layers decrease the overall bending stiffness of the implant, eliminating the effects of stress shielding [1] while simultaneously creating optimal porosity for good osseointegration of the implant [2].

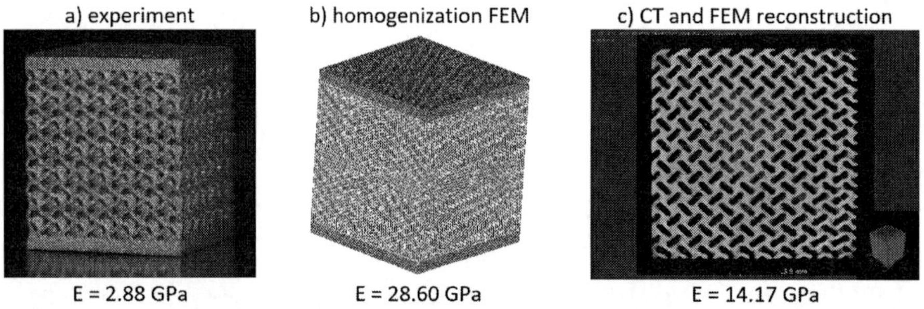

Fig. 1. Comparison of global elastic moduli of the structure created by 3D printing. Values are determined by: a) experiment, b) homogenization methods with FEM and theoretical structure geometry, c) real geometry reconstruction from CT imaging and FEM computation [3].

Achieving an optimal porous surface is possible either with periodically repeating trabecular cells or by using explicitly defined functions, such as the gyroid. The gyroid structure can

[1] *doc. Ing. Aleš Jíra, Ph.D.; Faculty of Civil Engineering, Czech Technical University in Pragu; Thákurova 7, 166 29 Prague, Czech Republic;* jira@fsv.cvut.cz

[2] *Ing. Luboš Řehounek; Faculty of Civil Engineering, Czech Technical University in Pragu; Thákurova 7, 166 29 Prague, Czech Republic;* lubos.rehouneka@fsv.cvut.cz

[3] *Ing. Gabriela Javorská; Faculty of Civil Engineering, Czech Technical University in Pragu; Thákurova 7, 166 29 Prague, Czech Republic;* gabriela.javorska@fsv.cvut.cz

[4] *doc. Ing. Pavel Padevět; Faculty of Civil Engineering, Czech Technical University in Pragu; Thákurova 7, 166 29 Prague, Czech Republic;* pavel.padevet@fsv.cvut.cz

form both trabecular (tubular) and sheet (wall) variants and can eliminate local defects from the manufacturing process [3] due to the absence of singularities in its geometry. Observing the optimal pore sizes for bone ingrowth (0.4–0.8 mm [4]) yields a range for the trabeculae (wall width) of the structure of about 0.15–0.60 mm. By using a width this small, we approach the resolution limits of contemporary 3D printers. Small wall width exacerbates the differences between global moduli determined by experiment and numerical computations (Fig. 1).

This work focuses on comparison between properties of Ti-6Al-4V 3D-printed specimens determined by experiment with values predicted by FEM computations. Recalculated moduli are compared between specimens with thickness of 0.25–2.00 mm during 3-point bending and uniaxial tensile tests.

2. Methedology

Ti-6Al-4V plate specimens with dimensions of $6\times 70\times(0.25\text{–}2.00)$ mm were created using SLM (selective laser melting) 3D printing. Specimens with different thicknesses were mechanically tested using 3-point bending and uniaxial tensile tests on the MTS and LiTeM machines. Recorded values of displacement and load force were used to create the slope of the load curve A (Fig. 2) and subsequently the values of global elastic moduli were recalculated according to expression (Eq. 1).

$$v = \frac{1}{48}\cdot\frac{F\cdot l^3}{E\cdot I} \Rightarrow E = \frac{F}{v}\cdot\frac{l^3}{4\cdot b\cdot h^3} \Rightarrow E = A\cdot\frac{l^3}{4\cdot b\cdot h^3}. \tag{1}$$

Where v is displacement, F is loading force, l is the support span, E is the global elastic modulus, I is the second moment of area, b is specimen width and h is specimen height. The calculation assumes the maximum deflection of a simply supported beam loaded with a single force in the middle of its span.

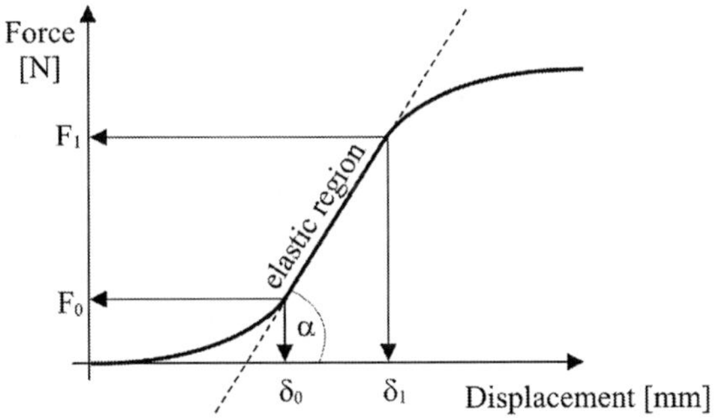

$$slope\ A = \tan\alpha = \frac{F_1 - F_0}{\delta_1 - \delta_0}$$

Fig. 2. Determining the slope A from the laod curve of the 3-point bending test. The slope is determined from the elastic region.

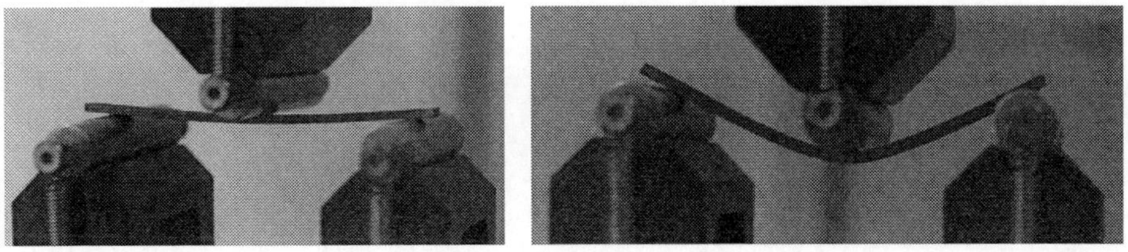

Fig. 3. Deformed shape of the test specimen in the three-point bending test. We can observe great range fo elasticity of the specimens – in this case the thickness is 1500 μm.

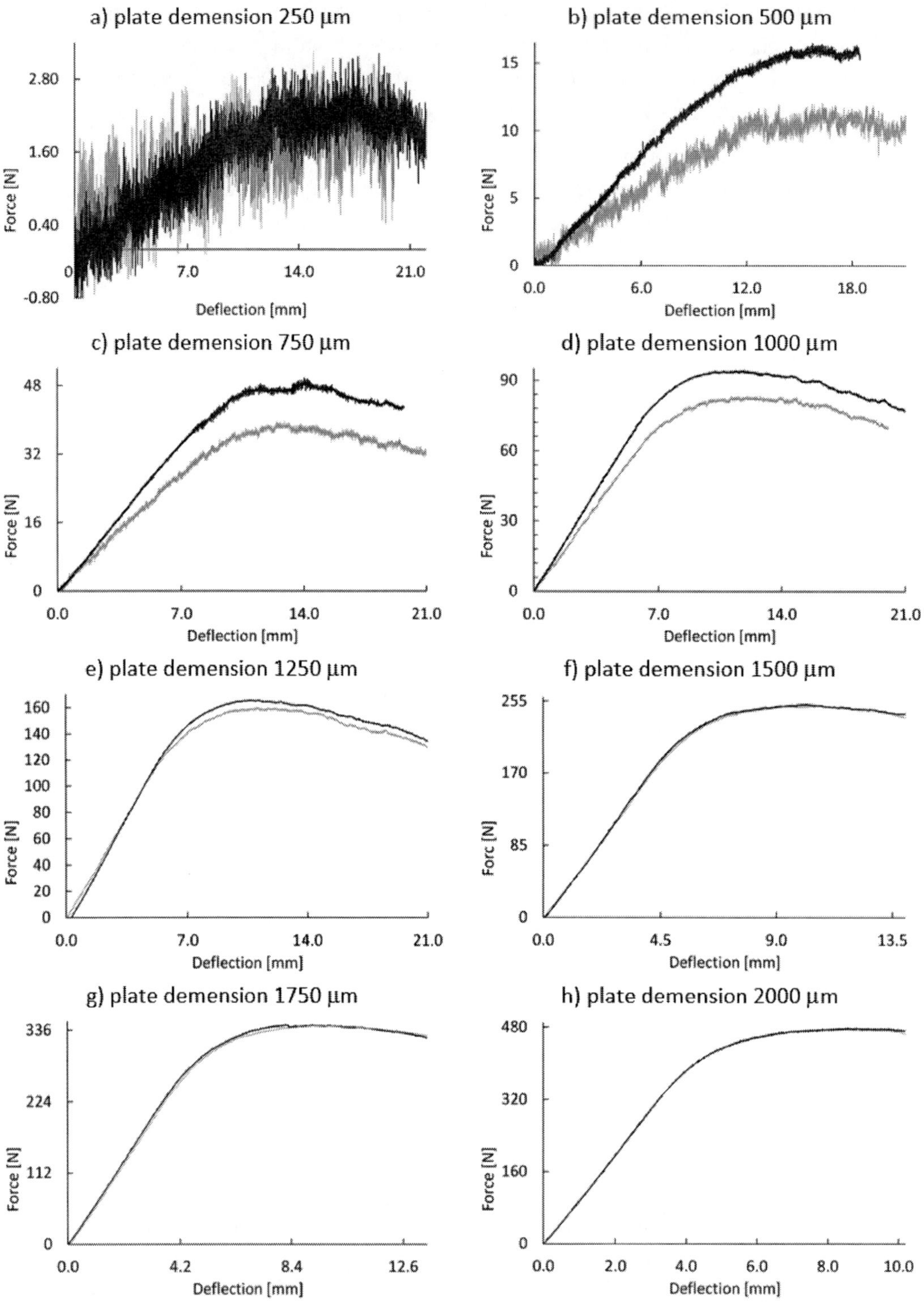

Fig. 4. Dependence of maximum deflection on the load force. Specimens vary in width. Black line are samples without surface etching, orange line are samples with surface etching.

The mechanical tests (Fig. 3) were performed at the laboratory of Faculty of Civil Engineering, CTU Prague with use of the MTS 50 machine according to the methodology for testing of thin plates. All specimens were weighed and measured before the tests. The three-point bending test was made in the mode of controlled displacement with velocity of 0.01 mm/s.

3. Results

Evaluation of results was done as described in the Methodology section. The evaluation was considering the elastic part of the stress-strain diagram and individual load curves are shown on Fig. 4. The thinnest plates (Fig. 4 a)) showed a great signal scatter and the resulting values are influenced by this effect. All graphs show two different curves determined as mean values of all measurements. The difference in curve slope and ultimate load force is influenced by additional surface acid etching and removal of imperfectly melted powder grains [6]. This treatment is supposed to eliminate the release of small particles that results in aseptic loosening of the implants. The specimens showed a 2.5% decrease in weight (acid etching) in plates with 2000 μm thickness and up to 11.0% in specimens with 250 μm thickness.

Final values of global elastic modulus for individual variants are shown in Table 1. The value of theoretical modulus of elasticity E_t are based on theoretical dimensions of the specimens and the value of E_r is recalculated from measured dimensions adjusted for real dimensions (that eliminates the influence of surface unevenness that distorts dimensions measurement.

Table 1. Overview of recalculated values of global elastic moduli, where E_t is calculated from theoretical dimensions (geometry before printing) and E_r is calculated from real dimensions adjusted for 3D printing-related imperfections.

plate dim.	250 μm	500 μm	750 μm	1000 μm	1250 μm	1500 μm	1750 μm	2000 μm
modulus				without surface etching				
E_t [GPa]	55 ± 18	65 ± 7	92 ± 6	111 ± 4	116 ± 4	119 ± 3	119 ± 3	119 ± 4
E_r [GPa]	118 ± 23	118 ± 9	119 ± 7	119 ± 5	118 ± 4	119 ± 4	119 ± 4	119 ± 2
modulus				with surface etching				
E_t [GPa]	120 ± 21	119 ± 6	119 ± 6	118 ± 5	119 ± 4	119 ± 3	119 ± 2	119 ± 3
E_r [GPa]	119 ± 24	118 ± 8	118 ± 7	119 ± 6	119 ± 5	119 ± 5	119 ± 4	119 ± 3

4. Conclusions

Recalculated values of global elastic moduli confirmed the dependance of final values on the thickness of the specimen. While specimens with height of 1.5–2.00 mm approach values of moduli stated by the manufacturer, thin specimens with thickness of 0.25–0.75 mm provide distorted values when computed. These disparities attest to formation of defects and cracks inside the structure during manufacturing or the following heat treatment. The defects are exacerbated in small specimens and potential biomedical use has to account for this effect. Therefore, the results of the numerical computations have to be weighted with this material uncertainty.

Acknowledgements

This work was carried out with the financial support of an internal CTU grant n. SGS 161 1612078A132.

References

[1] Ridzwan M., Solehuddin S. et al., "Problem of Stress Shielding and Improvement to the Hip Implant Designs: A Review," *Journal of Medical Sciences*, Vol. 7, pp. 460–467 (2007). ISSN 1682-4474.

https://doi.org/10.3923/jms.2007.460.467.

[2] Sumner D.R., Turner T.M. et al. "Functional adaptation and ingrowth of bone vary as a function of hip implant stiffness," *Journal of Biomechanics*, Vol. 31, pp. 909–917 (1998). ISSN 0021-9290. https://doi.org/10.1016/s0021-9290(98)00096-7.

[3] Jíra, A., Šejnoha M. et al., "Mechanical Properties of Porous Structures for Dental Implants: Experimental Study and Computational Homogenization," *Materials*, Vol. 14 no. 16, pp. 1–25 (2021). ISSN 1996-1944. https://doi.org/10.3390/ma14164592.

[4] Taniguchi N., Fujibayashi S. et al., "Effect of pore size on bone ingrowth into porous titanium implants fabricated by additive manufacturing: an in vivo experiment," *Materials Science and Engineering: C*, Vol. 59, pp. 690–701 (2016). ISSN 0928-4931. https://doi.org/10.1016/j.msec.2015.10.069.

[5] Jae-Hoon L., Val F., Keun-Woo L. Wright R. F., "Effect of implant size and shape on implant success rates: A literature review", *The Journal of Prosthetic Dentistry*, Vol. 91, no. 4, pp. 377–381 (2005). ISSN 0022-3913. https://doi.org/10.1016/j.prosdent.2005.04.018.

[6] Fojt J., Kačenka Z., Jablonská E., Hybášek V., Průchová E, "Influence of the surface etching on the corrosion behaviour of a three-dimensional printed Ti-6Al-4V alloy", *Materials and Corrosion-Werkstoffe und Korrosion*, Vol. 71, no. 10, pp. 1691–1696 (2020). ISSN 0947-5117. https://doi.org/10.1002/maco.202011658.

Investigation of Washer Influence on Bolt Connections in Non-Ferrous and Non-Metal Constructions

Radka Jírová[1], Lubomír Pešík[2] & Anett Maria Kupka[3]

Abstract: Modern design of production and transport machines is currently characterised by a wide range of non-ferrous and non-metal materials. Typical reasons are demands on cost and weight. Further reasons are requirements on physical properties, e.g. thermal and electrical non-conductivity, the ability to dump vibrations and noise. For connecting machine parts from non-ferrous and non-metal materials, mechanical engineers usually use standard iron bolts and nuts that provide low costs and a wide range of goods. However, the functional reliability of bolt connection necessitates special attention and must reflect specific design principles. Operation difficulties often appear by the incorrect design process when diverse mechanical properties of connecting bolts and connected mechanical parts are not considered. In that case, joint diagrams of bolt connections are inappropriate and may lead to unreliable operation or machine failure. Therefore, the project aims to propose a design principle for reliable bolt connection of non-ferrous and non-metal mechanical parts based on joint diagram optimisation.

Keywords: Bolt Connection; Preload; Washer; Composite Materials

1. Introduction

Reliability of bolt connections is highly demanded in the context of modern engineering design. Connected bodies are more often produced by non-ferrous, non-metal and composite materials. Requirements for current engineering design increases considering costs, energy consumption, ecology or weight of components. Therefore, traditional engineering materials are replaced in a wide range. While bolt connections in common steal constructions are reliable enough, there are a few issues to solve in modern materials. The root cause of reliability decrease is loss of preload, which provides strength and tightness of connection [1,2]. This phenomenon occurs as a consequence of the embedding process of connected bodies by thermal and vibrational conditions. The design of bolt connection in terms of load characteristics connected and connecting bodies is of great importance for the above reasons, primarily for preload securing of bolt connection in various operating conditions [3,4]. Thus, the paper focuses on several conditions that increase reliability.

2. Methods

The load characteristics of common bolt connection (see Fig. 1) shows a steep deformation character depending on the preload of connecting bodies. The deformation of connecting bodies is relatively tiny against the deformation of connected bodies. This condition leads to significant preload loss after the embedding process of the bolt connection, as even a small deformation change generates a high decrease of load in the connection. By using a special-shaped washer under the bolt head, the degressive load characteristic of connecting bodies may be reached (see

[1] *Ing. Radka Jírová; Technická univerzita v Liberci; Studentska 2, 460 01 Liberec, Czech Republic;* radka.jirova@tul.cz
[2] *prof. Ing. Lubomír Pešík, CSc.; Technická univerzita v Liberci; Studentska 2, 460 01 Liberec, Czech Republic;* lubomir.pesik@tul.cz
[3] *Dipl.-Ing. Anett Maria Kupka; Hochschule Zittau/Görlitz; Theodor-Körner-Allee 16, 02763 Zittau, Germany;* A.Kupka@hszg.de

Fig. 2) [5]. And thus, minimalise the preload loss after the embedding process.

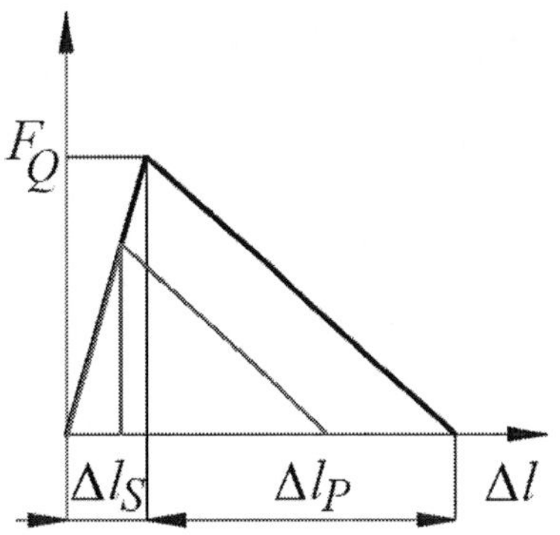

Fig. 1. Preload decrease in bolt connection without any washer.

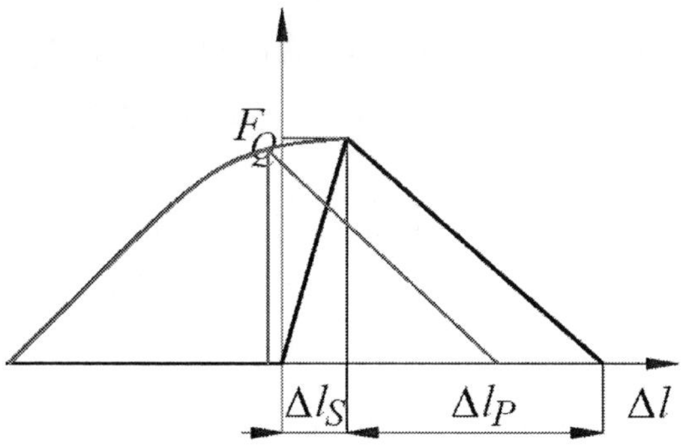

Fig. 2. Preload decrease in bolt connection with the special-shaped washer.

To investigate load characteristics of connecting bodies without using any washer and with using the special-shaped washer, the simplified 3D model was created, and deformations of connecting bodies via FEM analysis (Finite element method) were calculated.

In the FEM analysis (see Fig. 3 and Fig. 4), the bolt made of steel material was pulled in the axial direction by a remote displacement boundary condition that was placed on the bolt's rotational axis. Connecting bodies were fixed to the rigid block by frictional connection.

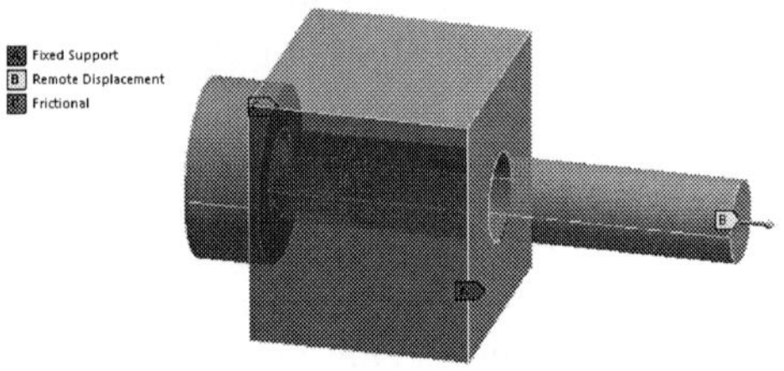

Fig. 3. Boundary conditions and contact connections – Bolt without any washer.

Fig. 4. Boundary conditions and contact connections – Bolt with the special-shaped washer.

For creating the load characteristics of both cases, the deformation of connecting bodies and reaction forces of remote displacement conditions were calculated.

3. Results and Discussion

The figures (see Fig. 5 and Fig. 6) show deformation without using any washer and with using the special-shaped washer.

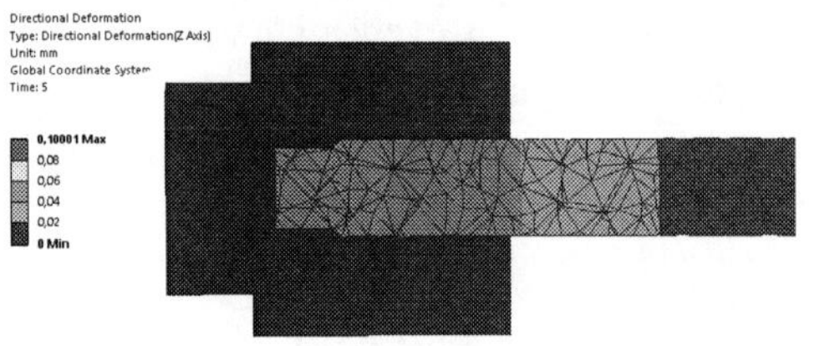

Fig. 5. Deformation in Z-axis – Without any washer.

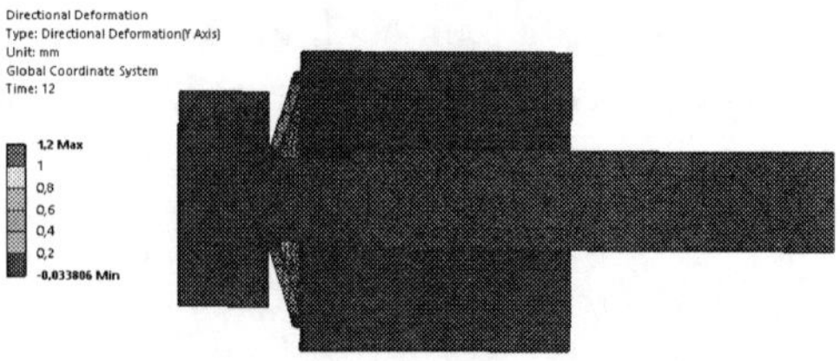

Fig. 6. Deformation in Z-axis – With the special-shaped washer.

Table 1 summarises the results of reaction forces in the case without using any washer and with using the special-shaped washer.

Table 1. Reaction forces of remote displacement boundary conditions

Bolt without any washer		Bolt with the special-shaped wahser	
Displacement [mm]	Force [N]	Displacement [mm]	Force [N]
0.00	0.00	0.00	0.00
0.02	2596.00	0.10	2010.00
0.04	5189.80	0.20	3864.50
0.06	7781.50	0.30	5551.20
0.08	10371.00	0.40	7046.10
0.10	12959.00	0.50	8387.80
		0.60	9531.30
		0.70	10524.00
		0.80	11326.00
		0.90	11983.00
		1.00	12459.00
		1.10	12774.00
		1.20	12883.00

The data in Table 1 were used for charting the load characteristics without using any washer and with using the special-shaped washer (see Fig. 7).

From the graph in Fig. 7, the great importance of stiffness decrement of connecting bodies might be observed by using the special-shaped washer placed under the bolt head. While the stiffness of the bolt without using any washer is significant, by using the washer in the shape of, e.g. a disc spring, the stiffness decreases markedly. The load characteristics of bolt connection that uses the washer of disc spring shape reports degressive characteristics of connecting bodies stiffness. In the case of the embedding process of connected bodies, the bolt connection preload drops minimally.

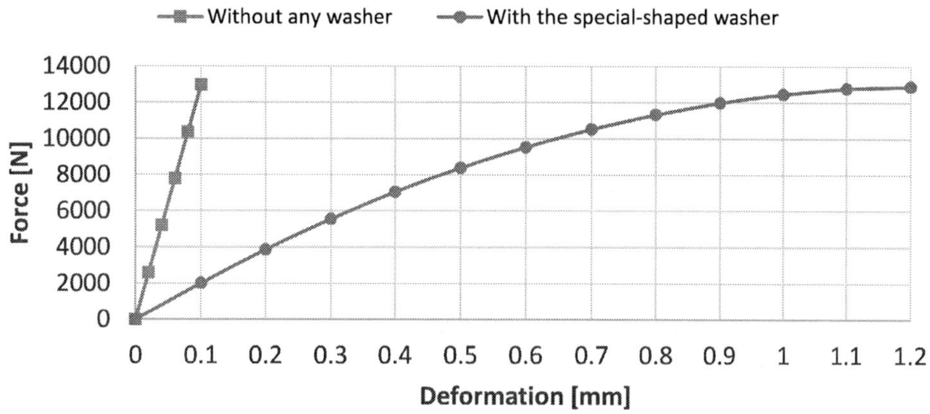

Fig. 7. Comparison of load characteristics – Without any washer and with the special-shaped washer.

4. Conclusions

The paper focused on the problematics of preload loss due to the embedding process of connected bodies in constructions of non-ferrous or non-metal material. In the solved case, the preload loss is tightly related to the steep load characteristics of connecting bodies that are usually made of standard steel materials. To minimise preload loss, the special-shaped washer under the bolt head may be used for reaching the degressive load characteristic of connecting bodies. Then, the deformation change after the embedding process generates a small decrease of preload in the connection.

Acknowledgements

This work was supported by the Student Grant Competition of the Technical University of Liberec under project No. SGS-2022-5047.

References

[1] Otter den Ch. and Maljaars J., "Preload Loss of Stainless Steel Bolts in Aluminium Plated Slip Resistant Connections," *Thin-Walled Structures*, Vol. 157, (2020). ISSN 0263-8231. https://doi.org/10.1016/j.tws.2020.106984.

[2] Jírová R. and Pešík L., "Analysis of Screw Connection in Air Conditioning Systems," *Scientific Journal of Silesian University of Technology-Series Transport*, Vol. 73, pp. 41–48 (2022). ISSN 0209-3324. http://dx.doi.org/10.20858/sjsutst.2022.114.4.

[3] Kassay P., Homišin J., Greaga R. and Krajňák J., "Comparison of Selected Pneumatic Shaft Couplings," *Scientific Journal of Silesian University of Technology-Series Transport*, Vol. 73, pp. 41–48 (2011). ISSN 0209-3324.

[4] Krajňák J., Homišin J., Greaga R. and Kassay P., "Effect of Helium on Mechanical Properties of Flexible Pneumatic Coupling," *Scientific Journal of Silesian University of Technology-Series Transport*, Vol. 73, pp. 63–69 (2011). ISSN 0209-3324.

[5] Kober M. and Kühhorn A., "Development of Optimized Washer-Shape for Damage-Free Load Application to Composites," *Composite Structures*,Vol. 92, pp. 1985–1991 (2010). ISSN 0263-8223. https://doi.org/10.1016/j.compstruct.2009.09.014.

Design and Control of Uniaxial Testing Machine

Pavel Kacálek[1] & Martin Hermann[2]

Abstract: This paper focuses on the design, control, and testing of an in situ testing machine that in combination with a commercial system Q-400 μDIC measures the macroscopic stress-strain behavior of small-scale rubber specimens. The deformation measurement of specimens is key to the direct tensile method and directly affects the accuracy of the obtained mechanical parameters.

Keywords: Digital image correlation; μDIC; In situ testing machine

1. Introduction

This paper focuses on the design, control, and testing of an in situ testing machine that in combination with a commercial system Q-400 μDIC measures the macroscopic stress-strain behavior of small-scale rubber specimens. The deformation measurement of specimens is key to the direct tensile method and directly affects the accuracy of the obtained mechanical parameters. The digital image correlation method (DIC) has the advantage of being a non-contact approach with high accuracy that can be amplified on small-scale specimens using a stereo light microscope (SLM) [1]. The DIC method is widely used to measure the shape, displacement, and full-field deformation of objects and has become the most active method in optical measurement.

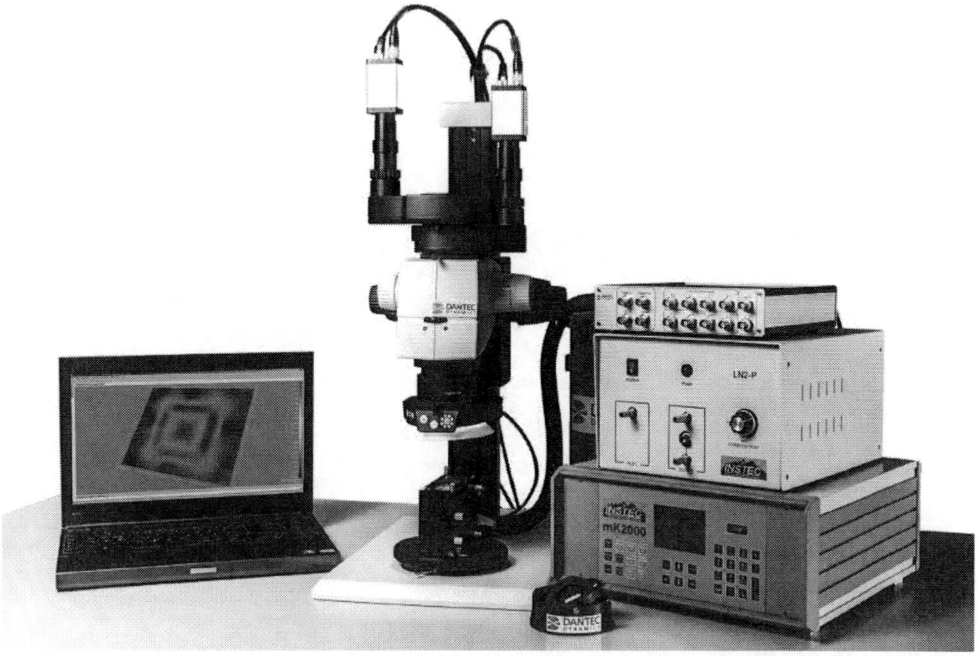

Fig. 1. Commercial system Q-400 μDIC from Dantec Dynamics.

[1] *Ing. Pavel Kacálek; Technical University of Liberec Faculty of Mechanical Engineering; Studentská 2, 461 17 Liberec, Czech republic;*

[2] *Ing. Martin Hermann; Technical University of Liberec Faculty of Mechanical Engineering; Studentská 2, 461 17 Liberec, Czech republic;*

2. System Specification

Since the Q-400 μDIC on Fig. 1 is the dominant part of the system, it leads to the strategy of making the tensile test machine be able to nest with the SLM so it can pull the specimen under the SLM field of view [2].

The center of the sample is the region of interest and must not translate so that there can be a correlation between SLM scans. It leads to the need for machine concepts that can pull each end of the sample at the same precise rate. It has been achieved by using right and left-hand lead screws driven by a stepper motor controlled by Arduino that receives commands from the PC. Stepper motor control is realized using the AccelStepper library for Arduino [3].

The requirements for the functional parameters of the equipment were determined in consultation with the laboratory staff of the Department of Mechanics, Elasticity and Strength of the Faculty of Mechanical Engineering of the Technical University of Liberec. The functional requirements for the machine were:

- maximal sample size: 20×90×4 mm
- maximal strain: 100% for a 45 mm long sample
- maximal tensile and compression load: 200 N
- loading method: prescribed displacement
- maximal loading rate: 300 mm/min
- Minimum displacement resolution: 0.01 mm

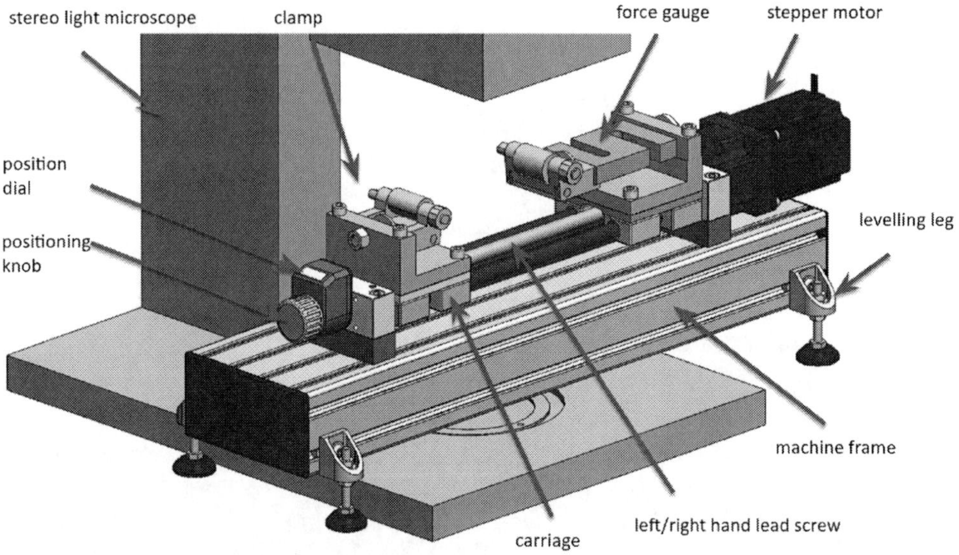

Fig. 2. Designed device under simplified SLM microscope model.

3. Connection and controll

The design of electrical components was carried out concerning the price and availability of components so as not to restrict the flow of energy from the power source to the target component.

Stepper motor MOT-AN-S-060-020-056-L-A-AAAA is powered by driver HY-DIV268N-5A and power supply WX-DC2416. Arduino executes incoming commands from PC, particularly motion of the stepper motor by pins 8, 9, 10 and signalization of measuring procedure start by pin 13. For the acquisition of force, the Timing Box is used. The force data signal obtained from force gauge Zemic H3 25 kg is conditioned using analog signal conditioner HBM AE101, powered by step-down converter LM2596. The Timing box is also connected to pin 13 for acquisition startup. This circuit was assembled into the custom 3D printed box.

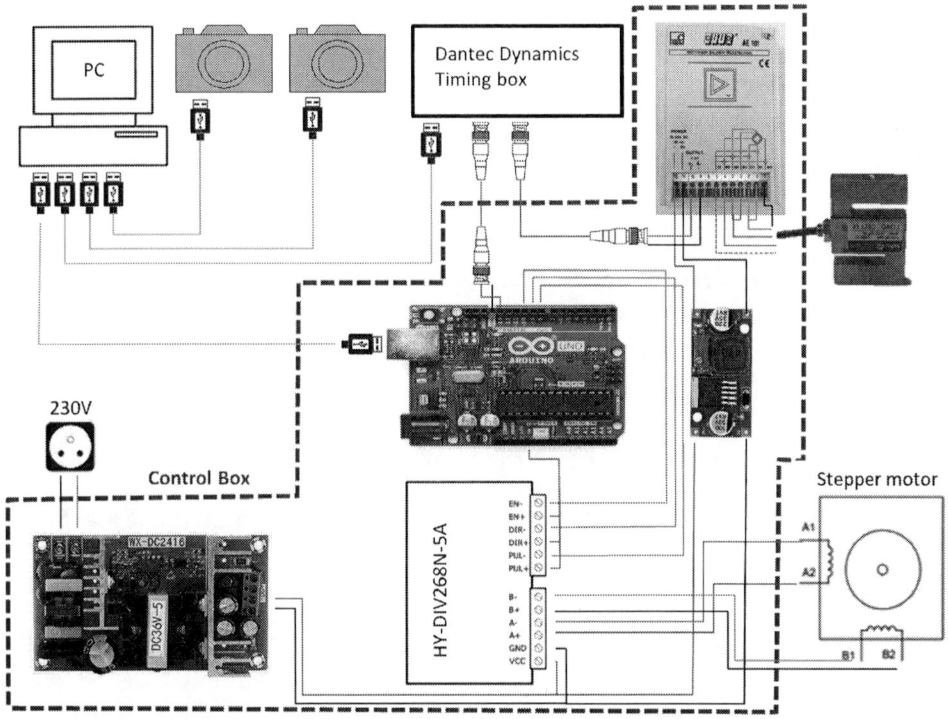

Fig. 3. Illustrative wiring diagram of power, logical and measuring circuit.

A program with a structure resembling a state machine was loaded on the Arduino (Fig. 5). It allows the execution of a parabolic cam profile via AccelStepper library and other utilities like waiting on position, powering up and powering down the stepper motor. States are changed by an incoming message via USB cable and Arduino reports status changes back to the control PC after the command is executed. Messages are identified by alphabetical character and a various number of arguments as required. For purposes of quasistatic loading of samples and control over the testing machine following states/messages were defined:

- A, Arduino has lost connection to PC.
- B, power down stepper motor.
- C, power up stepper motor.
- D x, set JOG speed to x and start motor moving in clockwise direction.
- E x, set JOG speed to x and start motor moving in counter clockwise direction.
- F, stop motion and report position.
- G x y z u, change settings, stepsize to x, maximal speed to y, maximal acceleration to z and maximal JOG speed to u.

- H x y, set speed to y, and initiate relative motion of distance x.
- I, set pin 13. to logical High of TTL.
- J, set pin 13. to logical Low of TTL.
- K x, wait for x miliseconds.
- Z, reports motor direction and remaining steps to target position.
- Y, reports readines for next task.
- X, reports stepsize, maximal speed, maximal acceleration and maximal JOG speed.
- W, reports current position.

GUI created in Matlab is used on the side of the controlling PC to send state-changing messages via buttons to the Arduino. The GUI was created so that only buttons that have defined transition state from the current state can be pressed. For example, the user cannot access the JOG mode button if the stepper motor is not powered up.

The GUI also allows users to create a loading series of arbitrary lengths consisting of four block types. The blocks serve for relative motion, waiting and setting on/off TTL signal. On Fig. 4 there is loading series consisting of setting TTL level to high, relative motion of distance 5 mm and speed 0.1 mm/sec, waiting on position for 1000 miliseconds, relative motion in the opposite direction of distance 5 mm and speed 0.1 mm/sec and turning the TTL level low.

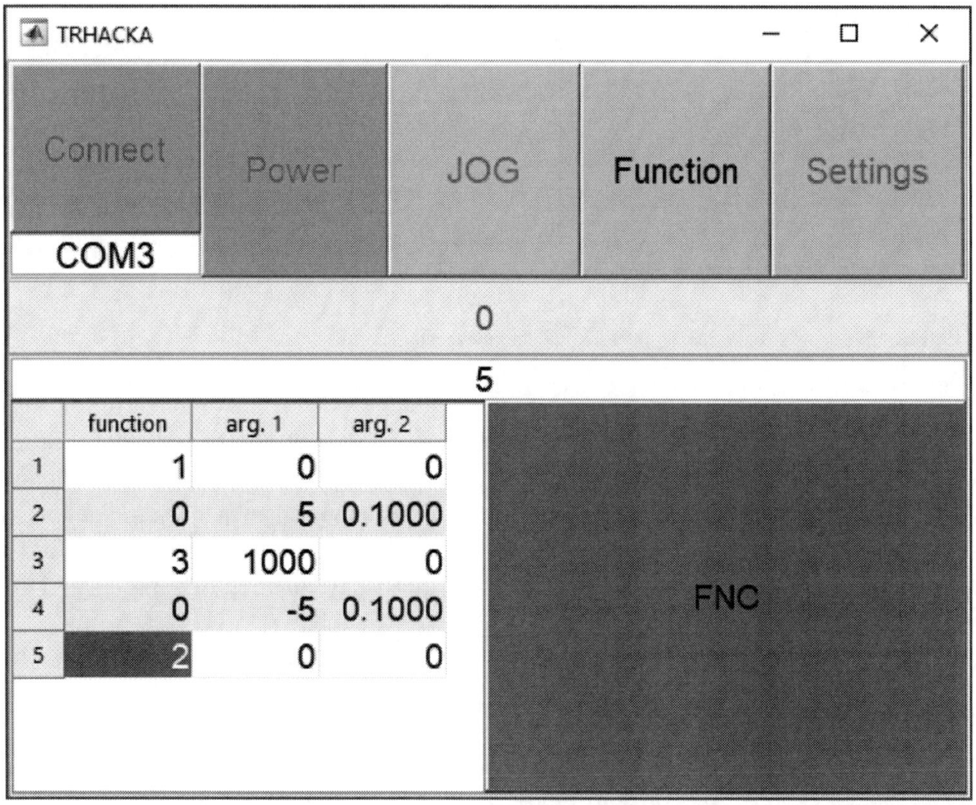

Fig. 4. Example of loading series inside created GUI.

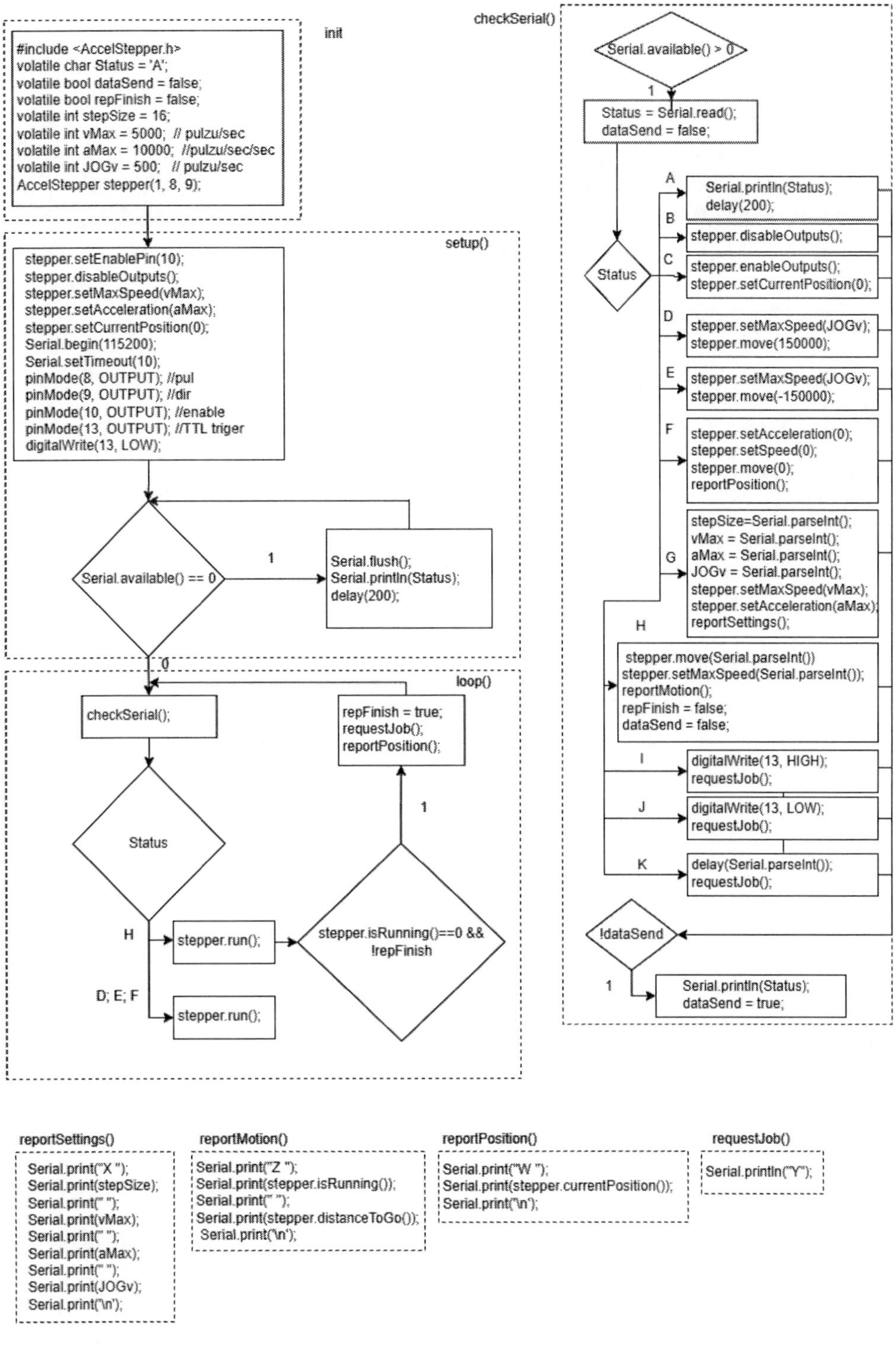

Fig. 5. Block diagram of the Arduino program.

4. Design verification

In order to validate that the prescribed movement is executed by the designed uniaxial testing device, following measurements were executed. On the device carriage a optical Non-Contact Displacement Transducer ILD1420-200 from micro-epsilon was attachad as on Fig. 6.

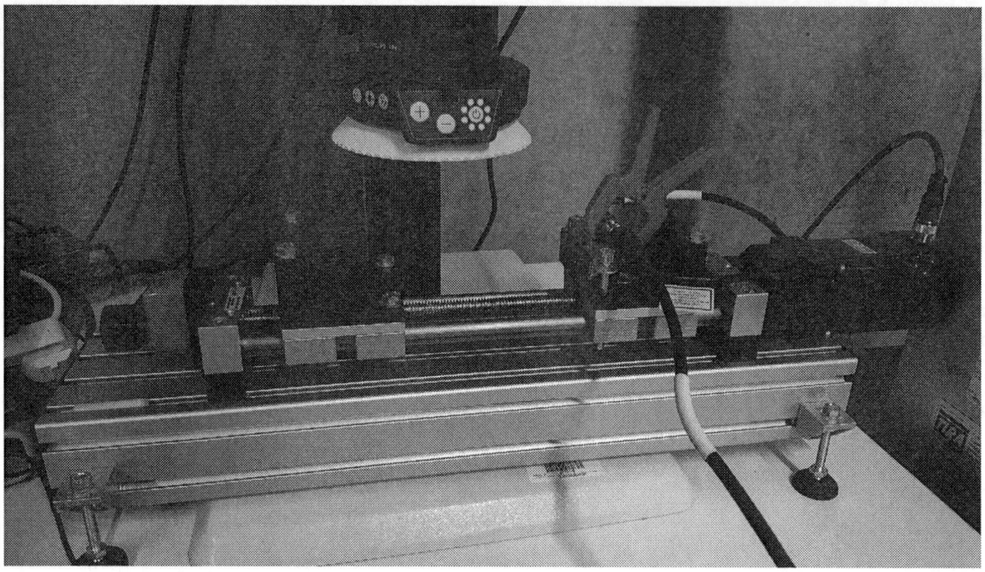

Fig. 6. Uniaxial testing machine under the Q400 Stereo Light Microscope, with ILD1420-200 attached to carriages for measuring distance between carriages.

First, a constant speed motion was carried across the entire range of the testing device. Second, the entire range of the testing device was stepped through with a 0.25 mm steps separated by small time delays. The stepping was finally performed in a series consisting of always a pair of steps forward and one step backward. Distance between carriages denoted S were ackquired at 4 kHz rate. Then convolution with chebyshev window of length 750 samples were performed. For positioning with steps, stationary sections were identified and the average position of the given step was determined from them. Velocity denoted v was obtained by numerical derivation of filtered position. Normalized velocity was defined as obtained velocity v divided by prescribed velocity denoted v0.

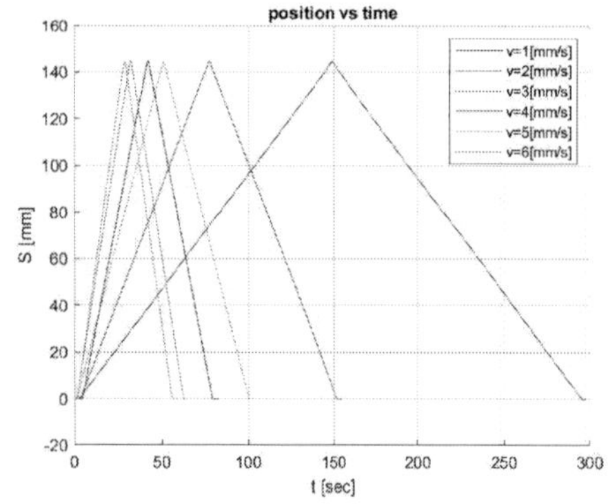

Fig. 7. Distance between carriages throughout the motion.

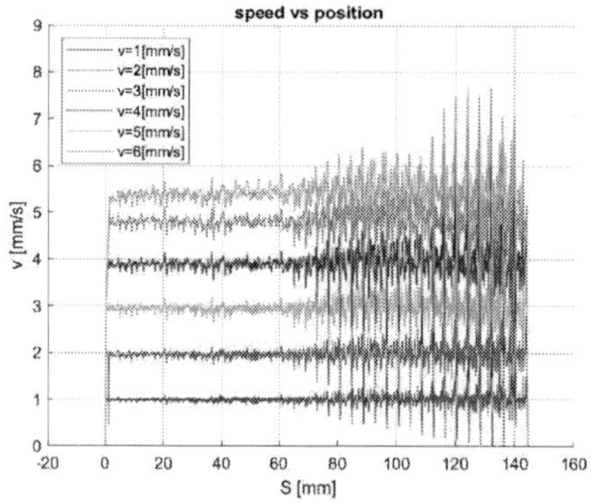

Fig. 8. Mutual speed of carriages.

In Fig. 7 there is depicted that the range of the testing machine was swept by 6 speeds, namely 1, 2, 3, 4, 5 and 6 mm/s. The movement started from position where carriages were closest to each other and went further apart. In Fig. 9 there is the testing machine range swept in reverse order but only with speeds of 1, 2 and 3 mm/s.

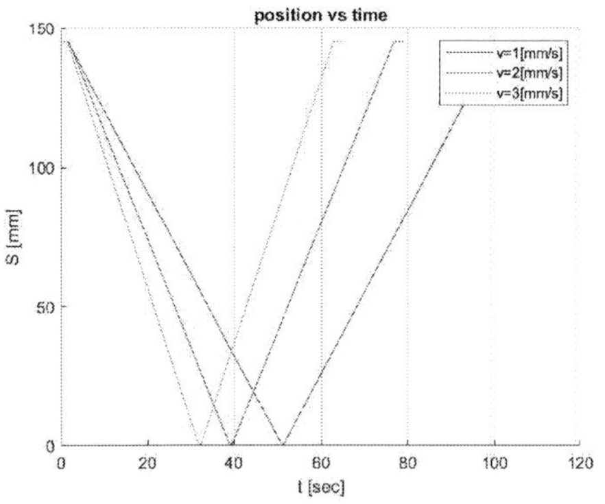

Fig. 9. Distance between carriages throughout the motion.

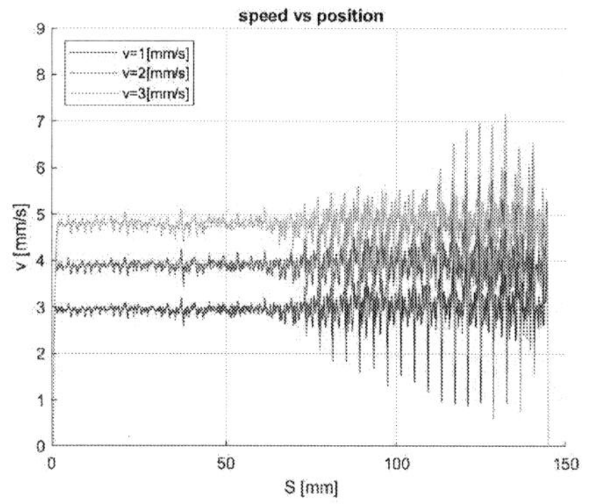

Fig. 10. Mutual speed of carriages.

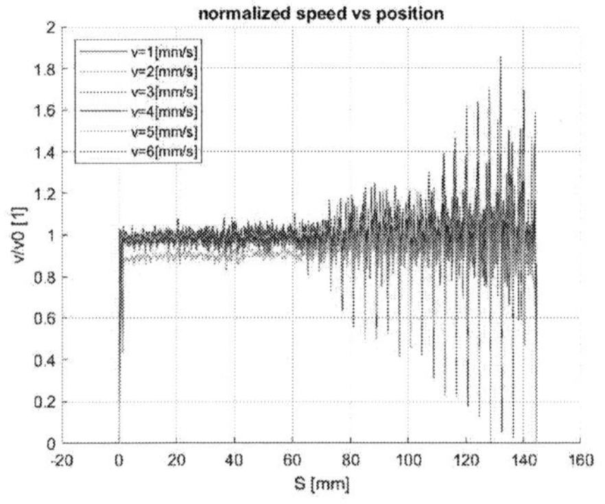

Fig. 11. Normalized speed for movement from Fig. 7.

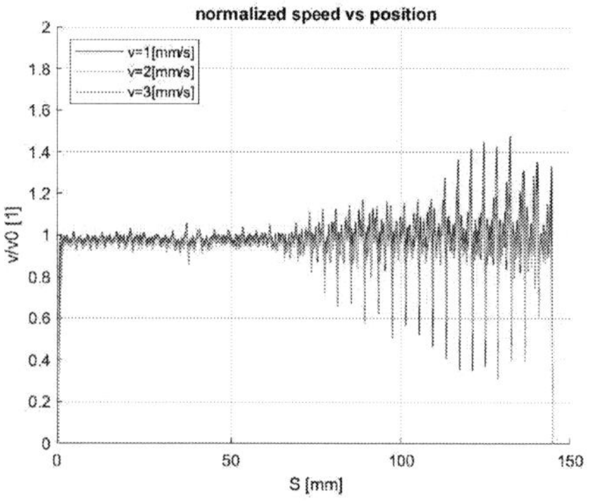

Fig. 12. Normalized speed for movement from Fig. 9.

In Fig. 8 and Fig. 10 are the speeds displayed versus distance between the carriages. It can be seen that the speed is not constant throughout the motion and the variance in speed grows rapidly after the carriages are 65 mm apart. The Fig. 11 and Fig. 12 shows normalized speed throughout the range. Average values of normalized speeds accros motion range for prescribed speeds 1, 2, 3, 4, 5 and 6 mm/s are 1.0040, 0.9916, 0.9914, 0.9809, 0.9633, 0.9073 respectively. The peaks of the speed deviations are 4 mm apart, this value corresponds to the combined pitch of the left-right hand lead screw.

In Fig. 13 there is zoomed in beggining of the first stepping procedure. In Fig. 14 there is a relative distance between consecutive steps. It can be obserwed as in velocity sweep measurement that in the first 65 mm of motion the steps have higher repeatibility and after 65 mm the deviation of step size increases. The mean step size throughout the stroke was 0.2495 mm with standard deviation of 0.0177 mm in opposite to required 0.25 mm.

In Fig. 15 there is the second stepping zoomed in. In Fig. 16 there is relative distace between consecutive steps. As during the steping the direction of step was changing there can be seen the backlash between leadscrew and nut. While the carriages were aparting the backlash was between 0.1 mm and 0.15 mm and after 65 mm the backlash has started increasing up to 0.2 mm.

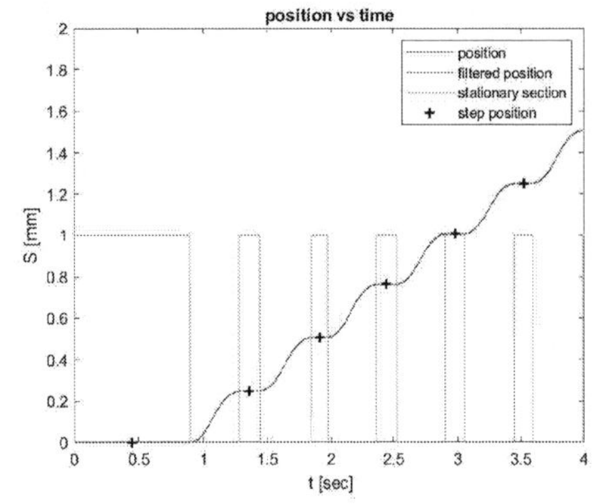

Fig. 13. Carriages moving apart in 0.25 mm steps.

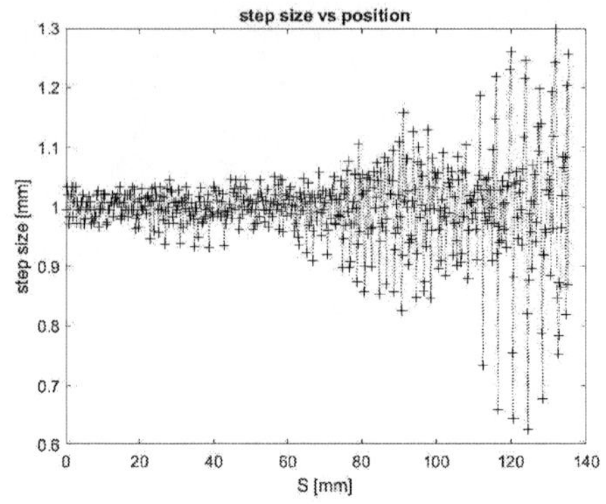

Fig. 14. Distances between individual steps.

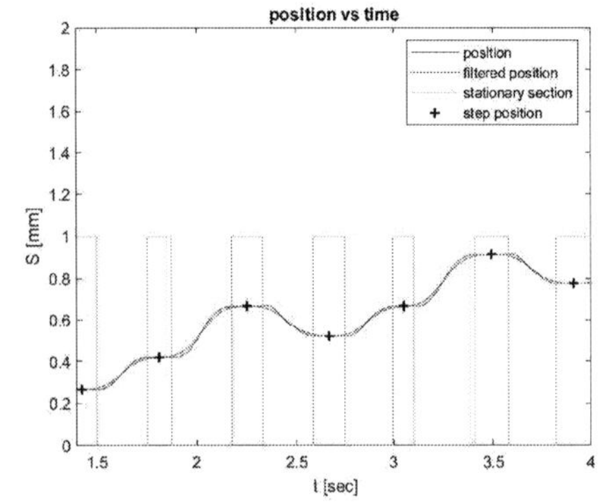

Fig. 15. Second stepping motion zoomed in.

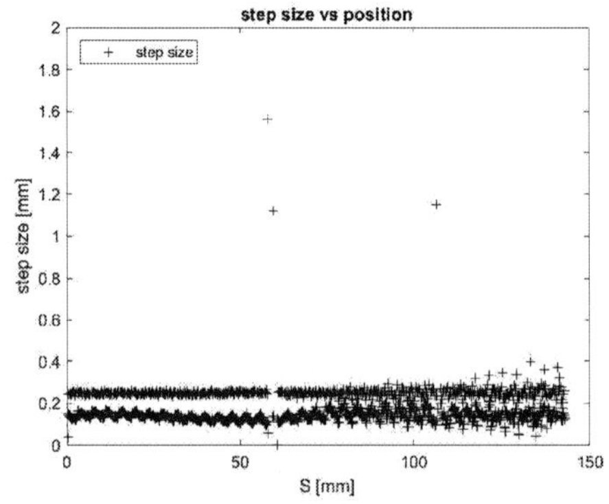

Fig. 16. Distances between individual steps.

5. Conclusions

Based on the research of similar mini tensile testing machines a testing device based on Igus SLW-1040 was made [4, 5]. Control of stepper motor for quasi-static loading was solved using Arduino and AccelStepper library. Moreover, for convenience of use, a graphical user interface (GUI) was created in Matlab. The ability of the device to prescribe position loads was investigated. it was found that the lead screw that is part of the linear unit has a larger pitch error while the carrages are 65 mm or more apart. And the speed performed by the testing device is always a bit slower than the speed required. With regard to the identified properties of the device, this device will be used for uniaxial quasi-static tests without changes of the loading direction.

Acknowledgements

This publication was written at the Technical University of Liberec as part of the project "Research of advanced materials, and application of machine learning in the area of control and modeling of mechanical systems" nr. SGS-2022-5072 with the support of the Specific University Research Grant, as provided by the Ministry of Education, Youth and Sports of the Czech Republic in the year 2022.

References

[1] Peters W.H. and Ranson W.F., "Digital Imaging Techniques In Experimental Stress Analysis," *Opt. Eng.*, Vol. 21, no. 3, Art. no. 213427 (1982). https://doi.org/10.1117/12.7972925.

[2] Dantec Dynamics: MicroDIC Available from https://www.dantecdynamics.com Accessed: 2.2.2022.

[3] MIKEM. AccelStepper. *Airspayce* Available from https://www.airspayce.com/mikem/arduino/AccelStepper Accessed: 2022-01-14

[4] Liu X. and Lu R., "Testing System for the Mechanical Properties of Small-Scale Specimens Based on 3D Microscopic Digital Image Correlation," *Sensors*, Vol. 20, no. 12, Art. no. 3530 (2020). https://doi.org/10.3390/s20123530.

[5] Bhaskar L. K. et al., "Design and development of a miniaturized multiaxial test setup for in situ x-ray diffraction experiments," *Review of Scientific Instruments*, Vol. 92, no. 1, Art. no. 015116 (2021). https://doi.org/10.1063/5.0031805.

Experimental and Numerical Analysis of a Shear Loaded Thin Composite Panel

Martin Kadlec[1], Jan Raška[2] & Adam Karkulín[3]

Abstract: Knowledge of the deformation response in relation to the load allows an optimal approach to the design already in the prototype phase. During the experiment, a square specimen of a thin composite panel was placed in a frame that introduced mainly shear stress. The deformations obtained from the experiment were evaluated by digital image correlation system Aramis. The experiment was simulated by using ABAQUS/Standard 6.14 software. The field of transverse displacements according to panel loading was compared with the simulation results. The correlated finite element model (FEM) will also be used in the damage analysis, and it will also serve as data source for structural health monitoring (SHM) systems.

Keywords: Composite; Experiment; Simulation; Structural Health Monitoring

1. Introduction

In aerospace technologies, problems with shell structures are often encountered [1]. Reinforcing effect of the skin is important in operating modes in which surface of structures transmits shear and compressive stresses. Knowledge of the deformation response in relation to the load allows an optimal approach to the design already in the prototype phase [2–4]. This paper is related to experimental and numerical analysis of a thin composite panel that transmits mainly shear stress.

2. Methods

2.1. Material and experiment

The frame shear test was used for an experiment on a carbon fibre reinforced polymer laminate with the thickness of 1.58 mm and layup $[-45/+45/0/90/0/0]_S$. The material parameters were optimized according to Ref. [5] and are listed in Table 1 and Table 2. The basic test principle described in ASTM D8067 [6] is that a square specimen is gripped along all four edges by four pair of clamps that are pin-connected at their ends to form a square frame into which the specimen is clamped (Fig. 1). Elastic modulus of the frame was 70 000 MPa. An axial load is applied at two diagonal corners to induce the shear loading. The specimen was loaded until certain level to prevent its failure. The deformations were captured by a digital image correlation (DIC) system and by a net of strain gage rosettes.

Table 1. In-Plane elastic properties of laminate

E_{11} [MPa]	E_{22}[MPa]	G_{12} [MPa]	μ_{12} [1]
129840.80	13341.71	4892.13	0.26

[1] *Ing. Martin Kadlec, Ph.D.; VZLU – Czech Aerospace Research Centre; Beranových 130; 199 05 Prague; Czech Republic;* kadlec@vzlu.cz

[2] *Ing. Jan Raška, Ph.D.; VZLU – Czech Aerospace Research Centre; Beranových 130; 199 05 Prague; Czech Republic;* raska@vzlu.cz

[3] *Ing. Adam Karkulín; VZLU – Czech Aerospace Research Centre; Beranových 130; 199 05 Prague; Czech Republic;* karkulin@vzlu.cz

Table 2. Out-of-plane elastic properties of laminate

G_{23} [MPa]	G_{13} [MPa]	μ_{13} [1]	μ_{23} [1]
4627.52	4892.13	0.26	0.44

Fig. 1. Test setup of the square composite panel loaded in a shear frame.

2.2. *Modeling*

The experiment was simulated by using ABAQUS/Standard 6.14 software [7]. The model of laminate was created following the homogenization method with help of one continuum-shell element thru the thickness, while 3 integration points per each of 12 layers are considered. In case of the frame model, the fully integrated linear brick elements were used. There were 2 elements used per every part thickness of the jig, which is the minimum to respect the present bending mode. The dimension of shell elements was 3.5 mm and 5 mm in case of the bricks. For the calculation to correspond to the experiment, the interaction of the frame and the test specimen must be correctly understood and, if necessary, the key parameters of the simulation must be adapted. The individual responses of the model are scaled relative to the nominal expected stiffness of the fixture.

Because "buckling modes" can occur immediately, this is a technological test rather than a component test. The test fixture also contains several imperfections in the form of surface irregularities and clearances at the point of contact, which are reflected in the interaction between the fixture and the test specimen. Therefore, the function of the test frame in the experiment must be properly understood and, if necessary, the key parameters of the simulation must be adapted. The individual responses of the model are scaled relative to the nominal expected stiffness of the fixture. We assume the validity of relation (1), where C is the ideal bending stiffness, E^i is the modulus of elasticity of the i-th component, J_o^i is the cross-sectional characteristic. Equation (2) expresses the relationship of the ideal stiffness S with respect to the real S_r through the interaction coefficient k. Ideally, it can take a maximum value of 1. It is shown in the paper

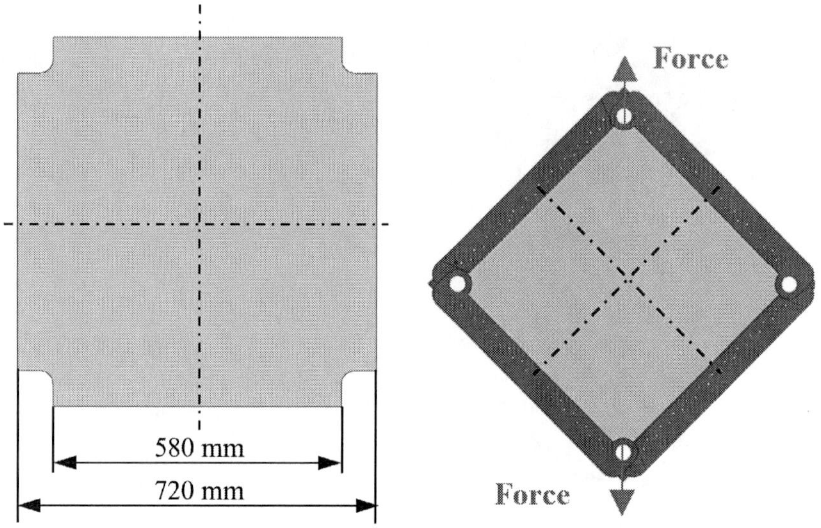

(a) Composite plate specimen.

(b) Assembled frame model with a specimen and load mean.

Fig. 2. Specimen shape and clamping in the frame.

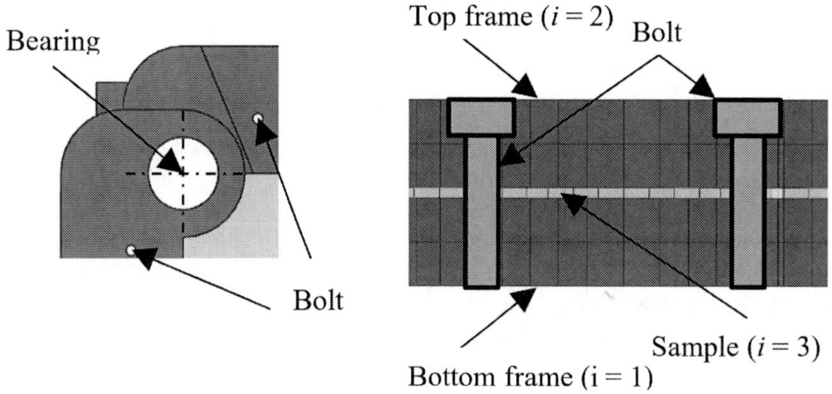

(a) Frame corner joint detail.

(b) Cross-section through the frame.

Fig. 3. Model schematic.

that varying this parameter the formation of the appropriate buckling mode and the real reduced stiffness of the structure could be controlled.

$$S = \sum_{i=1}^{2} E_x^i \cdot J_o^i \tag{1}$$

$$S_r = S \cdot k \tag{2}$$

The relative stiffness K_r is explained by equation (3). The percentage deviation of resultant S_d from the real stiffness is explained with help of equation (4). $F(u)^E$ is the force determined by experimental and $F(u,k)^M$ is the force calculated from the model, while the similar displacement (u) of control point is considered.

$$K_r = k \cdot 100\% \tag{3}$$

$$S_d = \frac{F(u)^E - F(u,k)^M}{F(u)^E} \cdot 100\% \tag{4}$$

These quantities are defined here to be further used to evaluate the results.

3. Results

Fig. 5f shows the measured out-of-plane displacement of the panel at the load level of 80 kN. The buckling shape started developing just after the loading start. It has one positive diagonal wave and two negative side waves along the diagonal wave with maximum amplitude of 7 mm.

There were several simulations performed for different coefficients k, which can control the origin of the respective buckling mode and the overall stiffness. The simulated field of displacements in Fig. 5e that used $k = 0.2$ was in good comparison with the experiment.

The dependence of the interaction coefficient with the actual simulation and experiment shapes is shown in Fig. 5a to Fig. 5f. The relationship between the relative stiffness and the control parameter k in relation to the experimentally determined stiffness is shown in Fig. 4. The evaluation was performed using the relations (1) to (4). The load characteristic of the model with the optimal interaction coefficient is shown in Fig. 6.

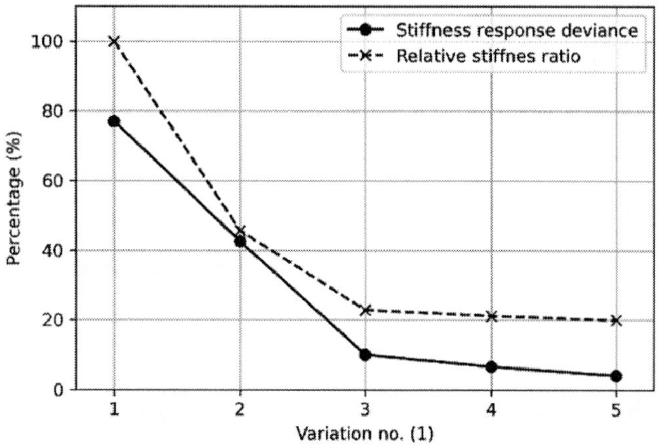

Fig. 4. Stiffness response deviance and relative stiffness ratio for various simulation variations.

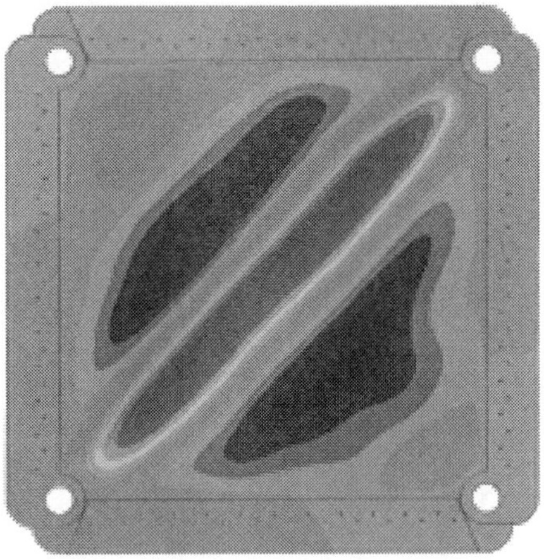

(a) Simulation 1, $k = 1$

(b) Simulation 2, $k = 0.46$

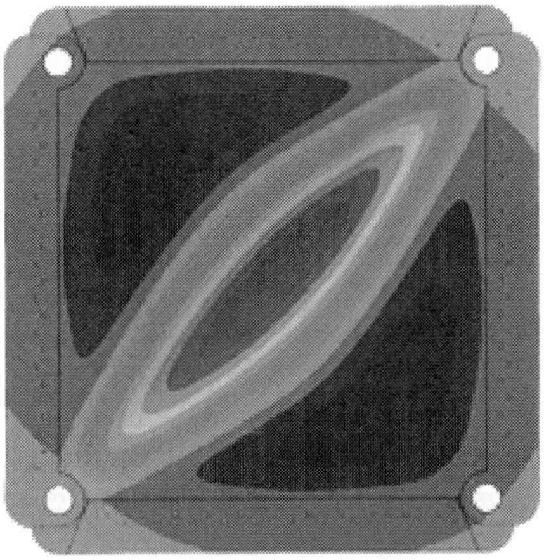

(c) Simulation 3, $k = 0.23$

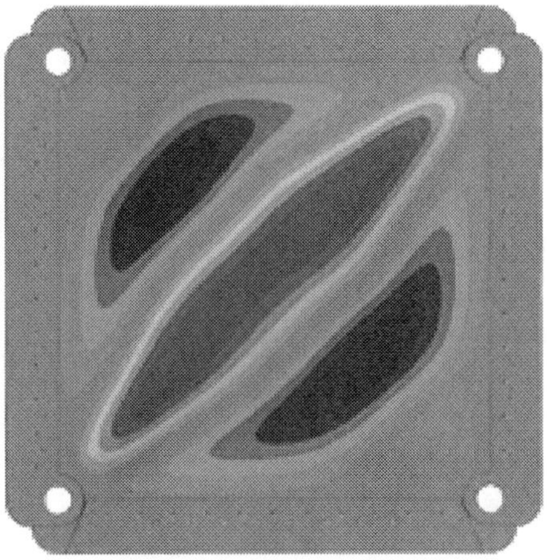

(d) Simulation 4, $k = 0.21$

(e) Simulation 5, $k = 0.20$

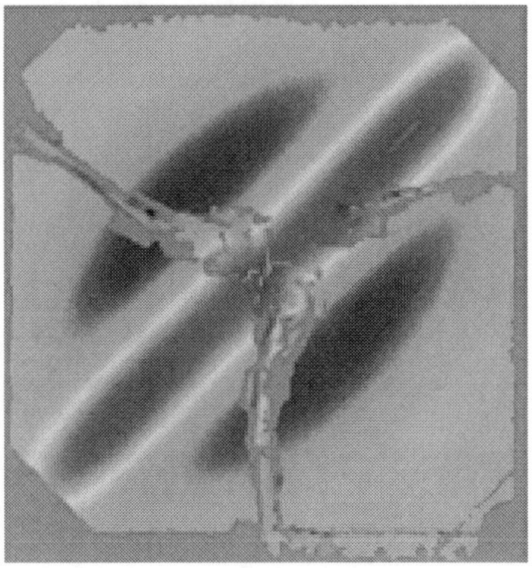

(f) Experiment

Fig. 5. Distribution of perpendicular displacement: a)–e) simulation; f) experiment.

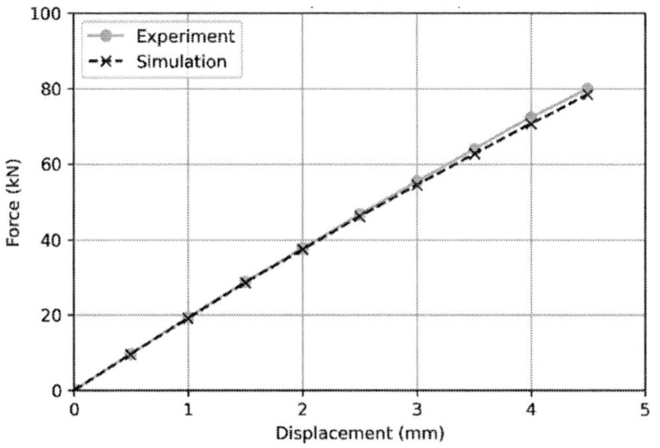

Fig. 6. Force/displacement response for simulation and experiment (simulation variation no. 5 for $k = 0.2$).

4. Conclusions

The model reaches optimal values that can be compared with the experiment if we consider 20% of the nominal stiffness of the arm. This indicates the presence of clearance in test jig. The correlated finite element model (FEM) will also be used in the damage analysis, and it will also serve as data source for structural health monitoring (SHM) systems.

Acknowledgements

Results incorporated in this paper received funding from Clean Sky 2 Joint Undertaking (JU) under grant agreement No 886703. The JU receives support from the European Union's Horizon 2020 research an innovation programme and Clean Sky 2 JU member other than the Union.

References

[1] Pagani A., et al., "Stress states in highly flexible thin-walled composite structures by unified shell model," *AIAA Journal*, Vol 59, no. 10, pp. 4243–4256 (2021). https://doi.org/10.2514/1.J060024.

[2] Wang M.N., et al., "Mechanical Respond and Failure Mode of Large Size Honeycomb Sandwiched Composites under In-Plane Shear Load," *Molecules*, Vol. 24, no. 23, Art. no. 4248 (2019). https://doi.org/10.3390/molecules24234248.

[3] da Silva D.C., "A semi-analytical model for shear buckling analysis of stiffened composite panel with debonding defect," *Thin-Walled Structures*, Vol. 171, pp. 108636 (2022). https://doi.org/10.1016/j.tws.2021.108636.

[4] Korcz-Konkol N and Iwicki P., "Corrugated Sheeting as a Member of a Shear Panel Under Repeated Load—Experimental Test," *Materials*, Vol. 13, no. 18, Art. no. 4032 (2020). https://doi.org/10.3390/ma13184032.

[5] Vlach J., Raška J., Horňas J. and Petrusová L., "Impacted area description effect on strength of laminate determined by calculation,“ *Procedia Struct. Integrity*, Vol. 35, pp. 132–140 (2022). https://doi.org/10.1016/j.prostr.2021.12.057.

[6] ASTM D8067, *Standard Test Method for In-Plane Shear Properties of Sandwich Panels Using a Picture Frame Fixture* (ASTM International, West Conshohocken, PA, 2017).

[7] ABAQUS-6.14. *Theory Manual*. Available online: http://130.149.89.49:2080/v6.14/ (accessed on 8 December 2021).

Experimental Verification of the Bolted Connection Relaxation

Jan Kanaval[1] & Eliška Cézová[2]

Abstract: The phenomenon of relaxation (reduction of assembly preload) of bolted connections has been known for a long time. At present, it is possible to take this phenomenon into account already in the design and strength calculation of prestressed bolted connections. The amount of relaxation (plastic permanent deformation, settlement) of a bolted connection in micrometers is usually estimated based on recommendations. These recommendations for estimating the amount of relaxation (plastic permanent deformation, settlement) of a bolted connection are based on performed experiments and practical experience.

Keywords: Prestressed Bolted Connection; Experimental Measurement of the UntighteningTorque; Assembly Preload Reduction

1. Introduction

This paper describes one of the ways to verify the magnitude of relaxation (plastic permanent deformation, settlement) of a bolted connection by a simple experiment. There are given guide values in [mm] for permanent squeezing of the clamped parts (including threads) in Table 1 (values in [in] are given in parenthesis) [1].

Table 1. Guide values in [mm] for permanent squeezing of the clamped parts (including threads) are given in the following tables (values in [in] are given in parenthesis).

Number of division joints	Roughness of joints	
	$Ra \geq 6.3$	$Ra \leq 3.2$
2	0.013 (0.00051)	0.010 (0.00039)
3	0.016 (0.00063)	0.012 (0.00047)
4	0.020 (0.00079)	0.014 (0.00055)
5	0.025 (0.00098)	0.016 (0.00063)
6	0.030 (0.00118)	0.018 (0.00071)

This table is valid for tensile/pressure loading of the connection.

2. Experimental measurement of the untightening torque as a method to verify the magnitude of a bolted connection relaxation

The magnitude of the relaxation of the bolted connection can be determined by the following experimental methods:

- by direct measurement of bolt elongation immediately after assembly and subsequently in time intervals from assembly up to 24 hours,
- by tensometric measurement of the axial preload of the bolt immediately after assembly and subsequently in time intervals from assembly up to 24 hours,

[1] *Ing. Jan Kanaval, Ph.D.; Department of Designing and Machine Elements, Fac. of Mech. Engineering, Czech Technical University in Prague; Technická 4, 166 07 Prague, Czech Republic;* jan.kanaval@fs.cvut.cz

[2] *Ing. Eliška Cézová, Ph.D.; Department of Designing and Machine Elements, Fac. of Mech. Engineering, Czech Technical University in Prague; Technická 4, 166 07 Prague, Czech Republic;* eliska.cezova@fs.cvut.cz

- by measurement of the untightening torque immediately after assembly and subsequently in time intervals from assembly up to 24 hours.

2.1. *Screw connection for experimental testing*

Used screws M6×16 (30 pieces of testing screw connections) are made of stainless steel A2-80. A threaded hole is made in the aluminium alloy. To perform the measurements tightening and untightening torques were delivered 5 pieces assembled products. The cover of Pulse Generator assembled according to the specified technological procedure (see Fig. 1, Fig. 2). The prescribed tightening torque 10.5 Nm corresponds to the EN 14831: 2005 standard [2].

Fig. 1. Tightening the Test Screw with a Torque Wrench.

Fig. 2. Tightening and Untightening Torque Were Measured Using a Calibrated Sensor HBM T20WN/20 Nm.

2.2. *Calculation of tightening and untightening torques of the test screw connection*

It should be emphasized here that, according to the theory of bolted joints, the untightening torque of a bolted joint is usually less than the tightening torque [3]. Tightening torque calculation (screw M6×16/A2-80), see Eq. (1,2).

$$M_{\mathrm{KKu1}} = M_{\mathrm{KZ}} + M_{\mathrm{TH}} = Q_{\mathrm{o1}} \cdot \frac{d_2}{2} \cdot \tan(\gamma + \varphi') + Q_{\mathrm{o1}} \cdot \rho_{\mathrm{TH}} \cdot f_{\mathrm{TH}} \tag{1}$$

$$M_{\mathrm{KKu1}} = 7900 \cdot \frac{5.35}{2} \cdot \tan(3.405° + 10.481°) + 7900 \cdot 4.15 \cdot 0.16 = 10470\,\mathrm{Nmm} \approx 10.5\,\mathrm{Nm} \tag{2}$$

Untightening torque calculation (screw M6x16/A2-80), see Eq. (3, 4).

$$M_{\mathrm{KKp1}} = M_{\mathrm{KZ}} + M_{\mathrm{TH}} = Q_{\mathrm{o1}} \cdot \frac{d_2}{2} \cdot \tan(\varphi' - \gamma) + Q_{\mathrm{o1}} \cdot \rho_{\mathrm{TH}} \cdot f_{\mathrm{TH}} \tag{3}$$

$$M_{\mathrm{KKp1}} = 7900 \cdot \frac{5.35}{2} \cdot \tan(10.481° - 3.405°) + 7900 \cdot 4.15 \cdot 0.16 = 7869\,\mathrm{Nmm} \approx 7.9\,\mathrm{Nm} \tag{4}$$

Used symbols: γ - thread pitch angle [°]; φ' - friction angle in thread [°]; d_2 - flank diameter of the bolt thread [mm]; ρ_{TH} - friction radius under the screw head [mm]; f_{TH} - coefficient of friction under the screw head [1]; M_{KKu1}, M_{KKp1} - tightening and untightening torques [Nmm]; M_{KZ} - torque [Nmm]; M_{TH} - friction torque under the screw head [Nmm]; Q_{o1}, Q_{o2} - pretensions forces [N].

There are given measured values of untightening torques for original assembly and reassembly in Table 2 and Table 3.

Table 2. Untightening torque measurement – original assembly.

Measured by HBM T20WN/20 Nm torque sensor (calibrated)					
Test sample 1		Test sample 2		Test sample 3	
Screw 1	6.894 Nm	Screw 1	7.590 Nm	Screw 1	7.270 Nm
Screw 2	7.800 Nm	Screw 2	7.329 Nm	Screw 2	7.009 Nm
Screw 3	6.733 Nm	Screw 3	7.179 Nm	Screw 3	6.778 Nm
Screw 4	7.120 Nm	Screw 4	7.028 Nm	Screw 4	6.591 Nm
Screw 5	7.137 Nm	Screw 5	6.917 Nm	Screw 5	7.390 Nm
Screw 6	7.266 Nm	Screw 6	7.082 Nm	Screw 6	7.562 Nm
Test sample 4		Test sample 5		Statistical evaluation of measured untightening torque	
Screw 1	6.733 Nm	Screw 1	8.685 Nm	Selection average value M_{Kpm2}:	
Screw 2	7.215 Nm	Screw 2	6.862 Nm		
Screw 3	7.046 Nm	Screw 3	7.382 Nm		7.274 Nm
Screw 4	7.442 Nm	Screw 4	7.584 Nm	Selection standard deviation:	
Screw 5	7.273 Nm	Screw 5	8.333 Nm		
Screw 6	7.452 Nm	Screw 6	7.504 Nm		0.447 Nm

Table 3. Untightening torque measurement – reassembly.

Measured by HBM T20WN/20 Nm torque sensor (calibrated)					
Test sample 1		Test sample 2		Test sample 3	
Screw 1	7.888 Nm	Screw 1	8.512 Nm	Screw 1	8.612 Nm
Screw 2	8.315 Nm	Screw 2	8.291 Nm	Screw 2	7.564 Nm
Screw 3	8.598 Nm	Screw 3	8.397 Nm	Screw 3	7.709 Nm
Screw 4	8.681 Nm	Screw 4	9.031 Nm	Screw 4	7.913 Nm
Screw 5	8.411 Nm	Screw 5	8.191 Nm	Screw 5	7.917 Nm
Screw 6	8.027 Nm	Screw 6	8.496 Nm	Screw 6	7.762 Nm
Test sample 4		Test sample 5		Statistical evaluation of measured untightening torque	
Screw 1	8.221 Nm	Screw 1	8.683 Nm	Selection average value M_{Kpm2}:	
Screw 2	8.131 Nm	Screw 2	8.048 Nm		
Screw 3	7.351 Nm	Screw 3	8.069 Nm		8.203 Nm
Screw 4	7.857 Nm	Screw 4	8.634 Nm	Selection standard deviation:	
Screw 5	8.233 Nm	Screw 5	8.484 Nm		
Screw 6	8.035 Nm	Screw 6	8.033 Nm		0.376 Nm

If we substitute the measured selection average torque value after relaxation 7.274 Nm into the equation for calculation of the untightening torque, see Eq. (5), then the axial preload of 7 302.8 N is obtained.

$$
\begin{aligned}
Q_{o2} &= \frac{M_{\mathrm{Kpm2}}}{\frac{d_2}{2} \cdot \tan(\varphi' - \gamma) + \rho_{\mathrm{TH}} \cdot f_{\mathrm{TH}}} = \\
&= \frac{7274}{\frac{5.35}{2} \cdot \tan(10.481^\circ - 3.405^\circ) + 4.15 \cdot 0.16} = 7302.8\ \mathrm{N}
\end{aligned} \tag{5}
$$

Difference between measured selection averages untightening torques for original assembly and for reassembly is 0.9 Nm, see Eq. (6).

$$\Delta M_{\mathrm{KKp}} = M_{\mathrm{Kpm1}} - M_{\mathrm{Kpm2}} \approx 8.2 - 7.3 = 0.9\ \mathrm{Nm} \tag{6}$$

Difference between calculated pretensions forces for original assembly and reassembly is 597.2 N, see Eq. (7).

$$\Delta Q_{\mathrm{o}} = Q_{\mathrm{o1}} - Q_{\mathrm{o2}} = 7900 - 7302.8 = 597.2\ \mathrm{N} \tag{7}$$

3. Experimental measurement results

Selection average value of the experimentally determined untightening torque of bolt joints for original cover assembly is 7.274 Nm and selection standard deviation of the experimentally determined untightening torque of bolt joints is 0.447 Nm. In this case, the measured selection average value of the untightening torque is approx. 0.9 Nm lower, which is a consequence of the relaxation of the screw connection.

Selection average value of the experimentally determined untightening torque of bolt joints for reassembly cover assembly is 8.203 Nm and selection standard deviation of the experimentally determined untightening torque of bolt joints is 0.376 Nm. In this case, the screw connections did not relax, as the screw connections were loosened immediately after assembly [3].

If we substitute the measured torque after relaxation 7.274 Nm (selection average) into the modified equation for calculation of the untightening torque, see Eq. (5), then the axial pretension of 7 302.8 N is obtained. So the axial pretension of the screw connection decreased by 597.2 N, which is by 7.6%.

4. Conclusion

Thus, it was verified that using the experimental measurement of the untightening torques of the screw connections, it is possible to estimate the amount of relaxation of the screw connections.

Acknowledgements

Thanks to my colleague Ing. Marek Štádler for the preparation of measuring stand.

References

[1] MITCalc – Mechanical, Industrial and Technical Calculations. [2022-04-16]. Available from https://www.mitcalc.com.

[2] Standard EN 14831: 2005 – Fasteners – Tightening performance – Torque/Angle simplified test method.

[3] Guideline VDI 2230, Systematic calculation of high duty bolted joints, Part 1, Joints with one cylindrical bolt, 2003.

[4] Kanaval J., *Special Process Qualification "Torque Tightening"*, Department of Designing and Machine Elements, Faculty of Mechanical Engineering, Czech Technical University in Prague, Prague, 2020.

Experimental Verification and Stress Analysis of Thermoplastic Aircraft Door

Adam Karkulín[1], Roman Růžek[2] & Radek Doubrava[3]

Abstract: The paper deals with test design and experimental verification of a new thermoplastic door panel manufactured from a high-performance thermoplastic material. The main goal of the work was to proof mechanical properties of the designed door panel with stop fitting. The structure was verified in compliance with Damage Tolerance philosophy. The overall design of the experiment is presented. That means static and fatigue evaluation of structure with presence of impact damages and finally to define the residual static strength of the damaged structure after fatigue loading. The structure complied with expected mechanical characteristics and ultimate structure load.

Keywords: Experimental testing; Composite material; Fatigue; Non-destructive testing; Fractography; Finite element analysis

1. Introduction

The ever-expanding use of composite material finds use in many new structures and thanks to the leaps in numerical simulations the parts can be designed in such a way to maximize the strength and minimise the weight [1]. Despite this, experimental testing is still necessary to validate structures or approve and obtain certification [2]. The paper documents a testbed part of design process which is still necessary for proof of a new structure. The new developed and optimized thermoplastic passenger door is presented. Experimental procedures, test conditions definition and stress analysis of the pressurized passenger door panel in accordance with Damage Tolerance philosophy is presented.

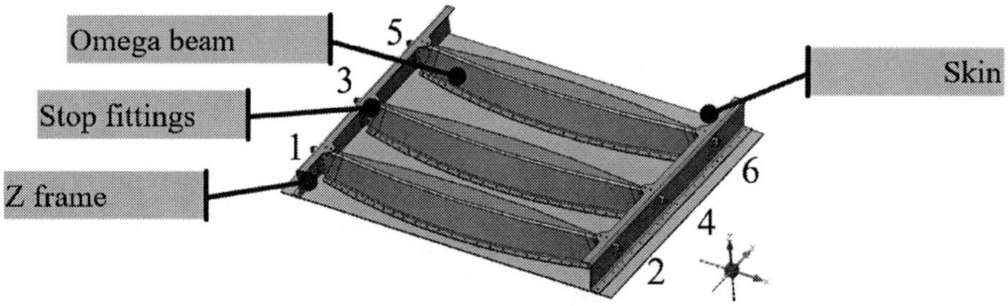

Fig. 1. Pristine test panel prior to testing with stop fitting numbering.

2. Material

Carbon materials using thermoplastic resin (PPS (PolyPhenylene Sulfide) and LMPEAK (Low-Melt PolyArylEtherKetone)) system was used. The hot forming method was applied for fabri-

[1] *Ing. Adam Karkulín; VZLU – Czech Aerospace Research Centre; Beranových 130; 199 05 Prague; Czech Republic;* karkulin@vzlu.cz

[2] *Dr. Ing. Roman Růžek; VZLU – Czech Aerospace Research Centre; Beranových 130; 199 05 Prague; Czech Republic;* ruzek@vzlu.cz

[3] *Ing. Radek Doubrava, Ph.D.; VZLU – Czech Aerospace Research Centre; Beranových 130; 199 05 Prague; Czech Republic;* doubrava@vzlu.cz

cation of individual door composite parts (skin [3], Z profiles [4] and Omega profiles [5]). Stop fittings were made of Titanium alloy [6]. Individual parts of the door were joined using steel bolts. The overallgeometry of the door panel was length of 1250 mm, width of 950 mm and high of 114 mm. The panel scheme is shown in Fig. 1.

3. Methods

3.1. Test campaign and BVID calibration

The test campaign based on damage tolerance philosophy, e.g. with assumption of flaw existing from the beginning of part operation, was designed. The experimental verification of the structure consists of following individual sections:

1. implementation of barely visible impact damages (BVID);
2. static loading up to limit load representing an overpressure value corresponding to service altitude of the airframe;
3. fatigue loading representing required service life of the structure whereas the overpressure was the main part of loading;
4. static loading up to limit load with aim the confirm ability to sustain the service loads after service life achievement;
5. static loading up to designed ultimate load representing overpressure value including the safety factors required by airworthiness requirements and;
6. continuous loading up to a failure;
7. non-destructive inspection (NDI) and the structure analysis after failure;
8. verification of numerical models and comparison with experimental data.

VZLU's SUPR drop tower (Fig. 2 (left)) using mass drop method was used for application of BVID. A calibration measurement on small panel (Fig. 2 (right)) was conducted prior to BVID implementation on test panel to obtain impact energy to dent depth dependence (Fig. 3). Considering the slight difference in door panel thickness compared to small test panels the 40 J energy was chosen as a base for door panel impacting.

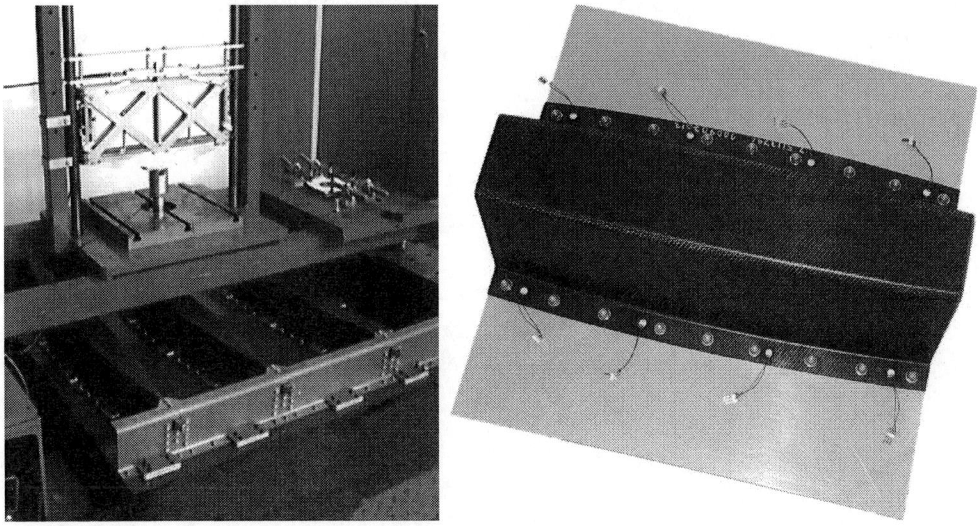

Fig. 2. VZLU SUPR drop tower for impact tests (left), Small panel for BVID calibration (right).

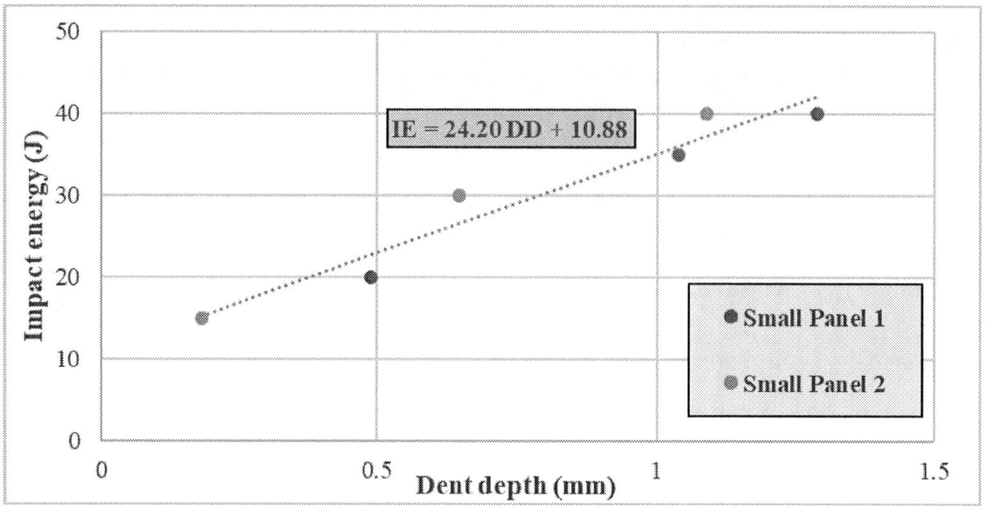

Fig. 3. Impact energy vs. dent depth dependence of small panels.

3.2. Substitution of real aerodynamic overpressure & instrumentation

The main load affecting aircraft door is directly tied to pressure difference resulting in overpressure. The real aerodynamic overpressure loading is in the experiment substituted through whiffle tree loading system (see Fig. 4 & Fig. 5) into 24 individual points the position of which was optimised. The transformed overpressure was replaced by corresponding forces and different load cases were defined:

- limit load – 115 kN corresponding to 1.3DP or 921 mbar
- ultimate load – 172 kN corresponding to 2DP or 1386 mbar
- fatigue load – stress ratio 0.1 with maximum force of 101 kN corresponding to 1.17DP or 693 mbar

The static and fatigue tests were conducted using uniaxial hydraulic test machine [7]. The panel was instrumented with strain gauges, LVDT measuring system and the deformation was measured by the means of digital image correlation measuring system. A transition between the Omega beam and stop fitting (stop fitting was previously optimized as part of the TRIO program [1]) was the area of interest.

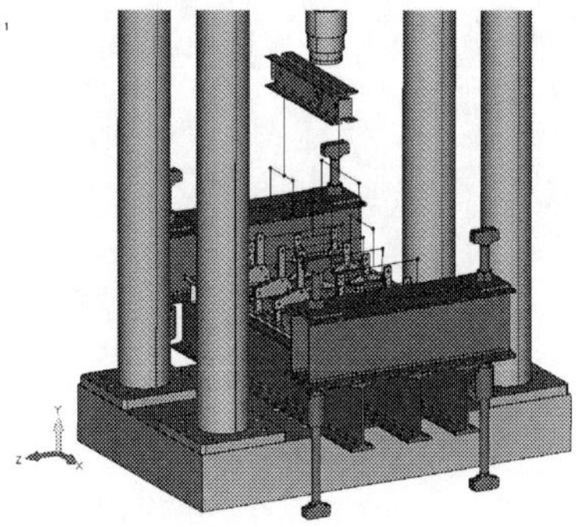

Fig. 4. Model of whiffle tree loading system.

Fig. 5. Whiffle tree loading system set-up.

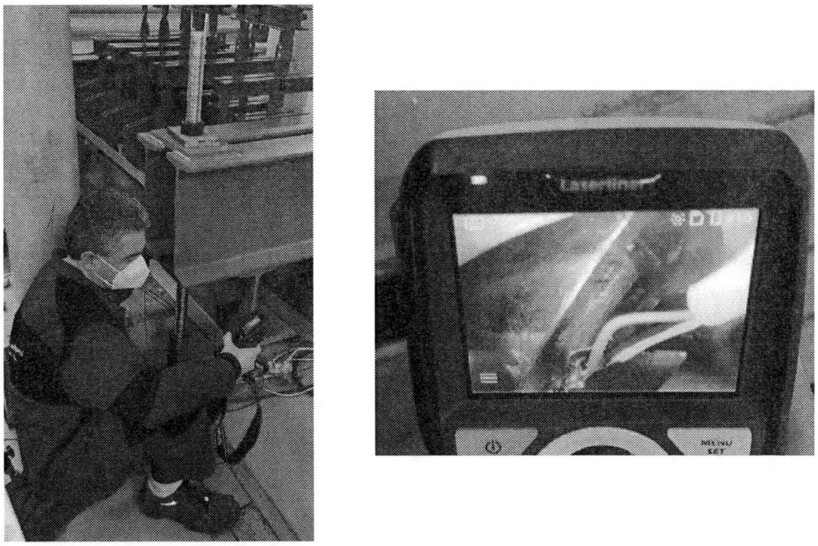

Fig. 6. Non-destructive inspection test using visual method (VT).

The experimental results were compared with data obtained through finite element analysis. The non-destructive inspections of BVID using the ultrasonic method (A-scan) were conducted between different stages of the experiment to observe a potential damage propagation. Furthermore, visual (Fig. 6) and eddy current NDI tests were performed.

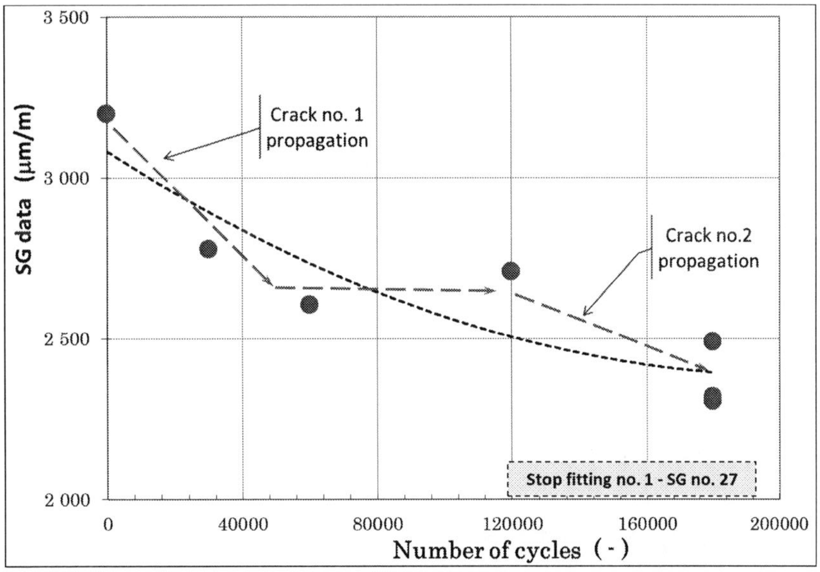

Fig. 7. Strain decrease as result of crack propagation – fatigue loading.

4. Results

The door structure sustains all required load (static and fatigue) without failure. The residual static failure of the door panel was achieved under static load of 125% of required ultimate load. The structure collapsed due to metallic stop fitting failure. The experimental and finite element analysis indicated a decrease of strain by 6.5%, a non-conservative estimate when compared with actual strain gauge measurement result of 18% (crack no. 1 – Fig. 7).

Panel displacement was observed using LVDT at two ends of panel and the measurement of LVDT no.1 was compared with DIC measurement, see Fig. 8. The strain gauge measurement of Omega profiles during loading up to failure is presented in Fig. 9. Non-destructive inspection showed failure of Omega profile, see Fig. 11. Damage of stop fitting was found after disassembly, extensive analysis was performed by means of electron microscopy. All stop fittings containing crack were analysed. The analysis exposed crack initiation sites and transition areas between fatigue crack growth and final quasi-static loading were also located and analysed (Fig. 10).

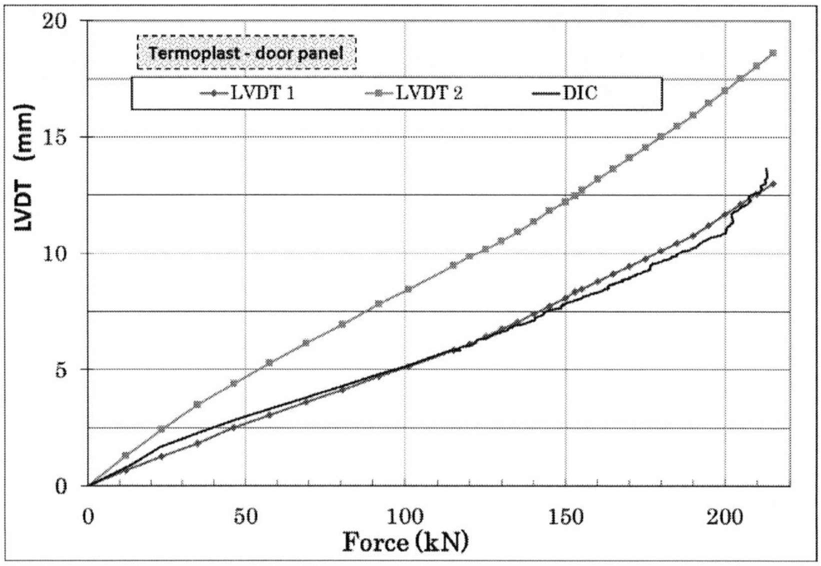

Fig. 8. LVDT and DIC displacement measurement up to door panel failure.

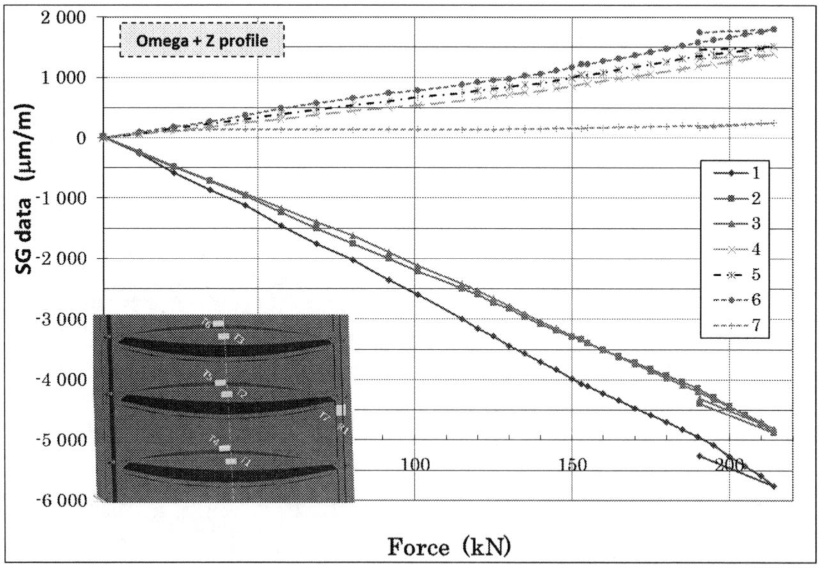

Fig. 9. SG data of Omega profile up to door panel failure.

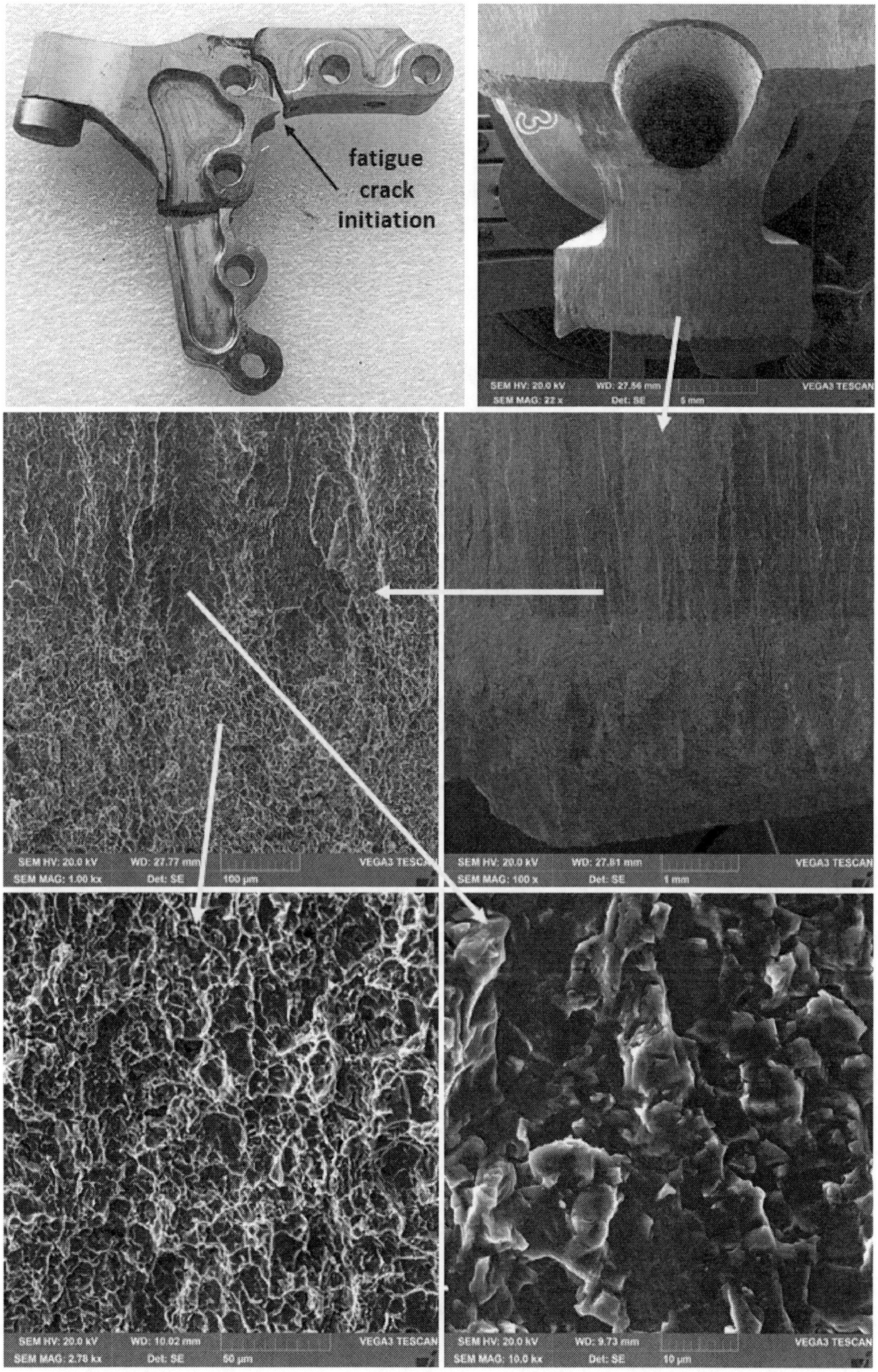

Fig. 10. Example of typical boundary area between the fatigue crack growth and final quasi-static loading.

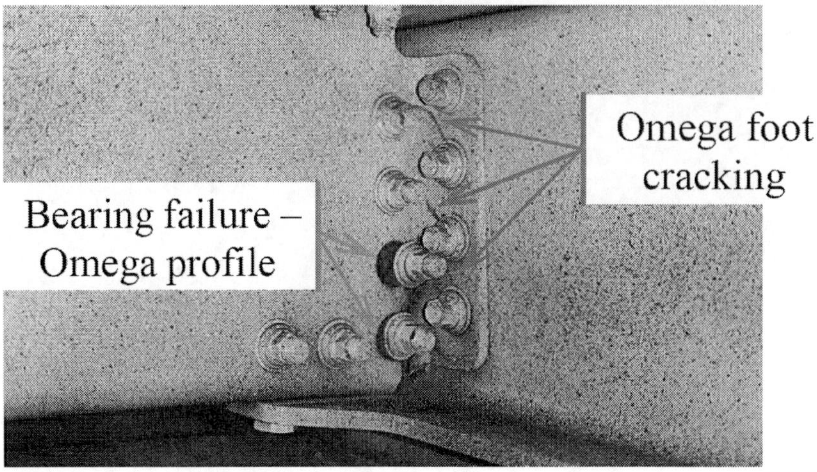

Fig. 11. Omega profile damage – loading up to failure.

Comparison of principal strain behaviour calculated using FE model and DIC measuring system at limit load before fatigue loading is depicted in Fig. 12. Critical area, where the failure occurred during experiment (see Fig. 11), is highlighted. The FE prediction of strain distribution shows good correlation to experimental verification.

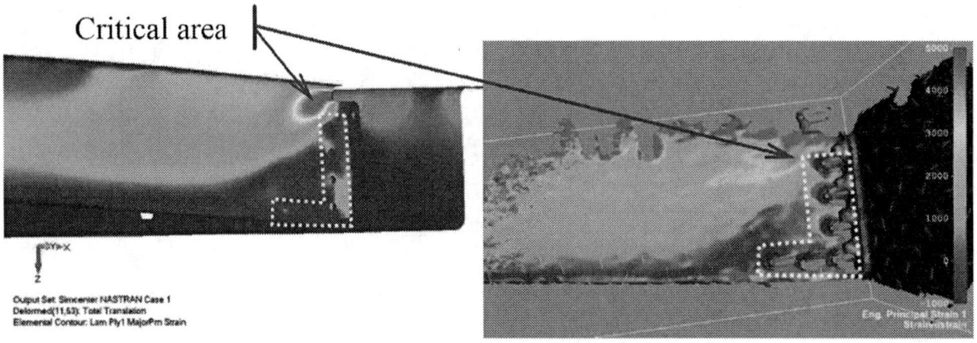

Fig. 12. Comparison of principal strain contour map before fatigue at limit load between: FE simulation (left) & DIC measurement (right).

5. Conclusion

The thermoplastic door panel was designed, manufactured, numerically analysed, and experimentally verified. Following main conclusions can be stated based on performed work:

- All main parameters and proposed requirements to mechanical properties of passenger door panel with compliance of Damage Tolerance philosophy were proved and completed.
- The door panel residual strength of 125% of designed ultimate load was achieved.
- No impact damage propagation in composite parts was observed during mechanical tests despite BVID damage that was introduced close to the stop fitting.
- No crack was detected in metallic parts during mechanical loading by non-destructive inspection.
- After disassembling of the door panel fatigue cracks in stop fittings were found out. Detailed fractographic analyses detected relative very fast crack propagation of most loaded stop fitting.

- No significant differences in total deformation of the panel measured using LVDT sensors and DIC method were observed.

All findings and conclusions will be used for improvement of a door panel design considering all the damage tolerance requirements and especially a structure inspectability and health monitoring.

Fig. 13. Core project team in COVID period.

Acknowledgements

This research was funded by the Ministry of Industry and Trade of the Czech Republic (MPO), program TRIO, grant number FV30033 "Structural and technological development of primary aircraft parts made by composite with thermoplastic matrix" and with the help of institutional funding of the MPO in the DKRV01 program dedicated to the development of research organizations.

References

[1] Doubrava R., Růžek R. and Křena J., "Design and optimisation of thermoplastic composite aircraft passenger door," in Proceedings of 58th International conference on Experimental Stress Analysis 2020, Ostrava, 2020, Czech Republic, *Available from* http://experimentalni-mechanika.cz/cs/konference/konference/2020.html?download=2890:design-and-optimisation-of-thermoplastic-composite-aircraft-passenger-door *Accessed: 2022-05-18.*

[2] Růžek R., Křena J., Doubrava R., Tkadlec J., Kadlec M. and Bělský P., "Optimal Design and Testing of a Thermoplastic Pressurized Passenger Door Manufactured Using Thermoforming," *Polymers*, Vol. 13(19), pp. 3394 (2021). https://doi.org/10.3390/polym13193394

[3] TC1225, *Cetex. Product data sheet* . s.l. : Toray Advanced Composites, G.van der Muelenweg 2, 7443 RE Nijverdal, The Netherlands, 2020.

[4] BMS8-353K. Reinforced Thermoplastic Laminate (Boeing Material Specification). s.l. : BOEING, 2007.

[5] ABS5045. "Carbon fibre fabrics reinforce Polyphenylene Sulphide resin (PPS - Laminates (consolidated plates) with or without lighting protection, with or without anti-corrosien protection," s.l. : AIRBUS S.A.S. 31707 Blangnac Cedex, France, 2006.

[6] AIMS03-18-006. "Titanium alloy (Ti-6Al-4V) Annealed plate 6,0 mm<a<100 mm Damage Tolerant," s.l. : AIRBUS S.A.S. 31707 Blangnac Cedex, France, 2010.

[7] Vlach J., Doubrava R., Růžek R., Raška J., Horňas J. and Kadlec M. "Linearization of Composite Material Damage Model Results and Its Impact on the Subsequent Stress–Strain Analysis," *Polymers*, Vol. 14, pp. 1123 (2022). https://doi.org/10.3390/polym14061123.

Influence of Hydrotest on Fatigue Growth of Artificial Crack in DN800 Pipeline

Jan Kec[1], Ivo Černý[2], Jakub Mlnařík[3] & Barbora Kyselá[4]

Abstract: The hydrotest, i.e., a static overload to 110% of the specified minimum yield strength (SMYS) of a pipeline, is a technique used to detect cracks of critical dimensions and to slow down the growth of subcritical cracks by blunting the tip. Although this is a technique known for decades for the treatment of pipelines with crack-like defects, the effect of a hydrotest is at least ambiguous. This is mainly because a certain group of cracks opens and tears due to static overload, which results in the activation of crack growth in the subsequent operation by either stress corrosion cracking (SCC) or a fatigue mechanism. For evaluating the effect of a hydrotest, a bare pipe was fatigue pre-cracked from a notch to the depth of 7 mm. The pipeline with this artificial defect, which simulates a crack from a long-term operation, was statically overloaded to 110% SMYS. Subsequently, a fatigue test was applied within the set range corresponding to general operation, a pressure fluctuation of 4–7.35 MPa and a total of 20,000 cycles. Part of this article is the evaluation of crack growth kinetics during tests and measurement of the stress-strain state on DN800 pipe.

Keywords: Pipeline; Hydrotest; X60 steel

1. Introduction

Pipeline transport has been one of the most widely used and most efficient methods of transporting oil, gas, and other petrochemical commodities. It is more efficient than truck, rail, cars, and tanker transport. For this reason, there are now around 2 million kilometres of transit pipelines worldwide [1]. The gas industry in our country has a long tradition. The Czech gas network includes tens of thousands of kilometres, and most transit pipelines are almost five decades old [2]. As a result of the long-term operation of gas pipelines, various types of defects occur, including corrosion, dents, gouges, and crack-like defects. The following can be considered to be a crack-like defect: SCC, fatigue crack, lamination, hot crack, lack of fusion, lack of penetration, etc. Although fatigue failures are rather uncommon in gas pipelines since thousands of cycles occur over the lifetime of the pipeline. Fatigue failures can occur in some cases. For instance, this was the case in the failure of the pipeline DN600 in a double submerged arc weld [3]. SCC, however, has been identified as a cause of pipeline failures worldwide [4].

A technique called a hydrotest has been used to mitigate the effect of crack-like defects. It has been known for 40 years, and it was first used in Texas [5]. This technique involves static overloading up to 110 % SMYS to detect crack-like defects of critical size or to block the growth of subcritical defects by forming a plastic zone at the crack tip. However, in the case of dormant cracks, the hydrotest may cause activation and the start of propagation [6]. Although this is a technique known for decades for the treatment of pipelines with crack-like defects, the effect

[1] *Ing. Jan Kec; Strength laboratory, SVÚM a.s.; Tovární 2053, 250 88 Čelákovice, Czech Republic;* kec@svum.cz
[2] *Ing. Ivo Černý, Ph.D.; Strength laboratory, SVÚM a.s.; Tovární 2053, 250 88 Čelákovice, Czech Republic;* ivo.cerny@seznam.cz
[3] *Ing. Jakub Mlnařík, Ph.D.; Corrosion laboratory, SVÚM a.s.; Tovární 2053, 250 88 Čelákovice, Czech Republic;* mlnarik@svum.cz
[4] *Ing. Barbora Kyselá; Strength laboratory, SVÚM a.s.; Tovární 2053, 250 88 Čelákovice, Czech Republic;* kysela@svum.cz

of the hydrotest is at least ambiguous, especially in cracks that have a depth of over half of the wall thickness.

2. Experiment

2.1. Procedure

A 3 m long defect-free DN800 pipe (820×10.6 mm) was used to investigate the effect of the hydrotest. A cut out body approximately 1 m long was used to evaluate the mechanical properties. The most important mechanical properties are summarised in Tab. 1. The pipe was made of X60 steel according to API 5L designation. It was from Mannesmann Rohre Werk with longitudinal submerged arc weld (LSAW). A detailed analysis of the mechanical properties and determination of the ductile-brittle transition temperature in relation to the microstructure for X60 steel is discussed in [7]. The LSAW was placed in the 12 o'clock position during testing. An artificial notch was manufactured exactly in the centre between the two closing heads and at the 5 o'clock position. The semi-circular notch had an initial depth of 5.5 mm. It was prepared with an ultrathin cutting wheel with a diameter of 26 mm and a thickness of 0.2 mm.

Table 1. Mechanical properties of DN800 pipe.

Specimen orientation	$R_{p0.2}$ [MPa]	$R_{t0.5}$ [MPa]	R_m [MPa]	A [%]	Z [%]	KV_2 at 0 °C [J]
Longitudinal	444	414	621	26.7	57.7	106
Circumferential	453	456	627	23.0	48.8	55

The artificial crack depth was designed so that the increment of fatigue cracks from the notches would meet the minimum crack depth requirements of the standards for fatigue crack growth rate and fracture toughness evaluation. Eq. (1), well known as the Paris law, was used to evaluate the growth rate during fatigue pre-cracking.

$$\frac{da}{dN} = C \cdot \Delta K^m \tag{1}$$

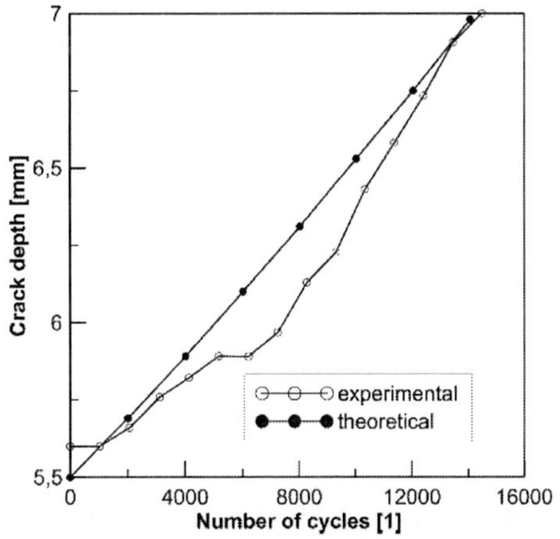

Fig. 1. Theoretically estimated crack depth based on the Paris law and experimentally measured crack depth during fatigue pre-cracking.

Crack depth was measured by the alternating current potential drop (ACPD) method; details of the innovative approach of measurement by this method are described in [8]. Strain gauges were placed around the crack mouth to measure the strain fields during the tests. Strain gauges were also placed on the bare pipe to measure longitudinal and circumferential strain. Fatigue pre-cracking was carried out in a pressure range of 2–7.35 MPa to accelerate the fatigue crack initiation and growth of the artificial notch. A comparison of the theoretically designed growth curve and the actual measured one can be observed in Fig. 1. It is possible to observe a good agreement, where the experimental curve follows the theoretically designed one.

2.2. Results

After reaching the desired 7 mm depth of the artificial crack, the hydrotest proceeded. The hydrotest consisted of three dwells. The first dwell was at 75% SMYS, the second dwell was at 110% SMYS, and the third and final dwell was at 90% SMYS, which was termed as leaking test. All three dwells took at least one hour (see Fig. 2). The loading rate was constant and was set at 5 bar/minute.

Elastoplastic behaviour can be seen in the stress-strain curves of strain gauges TEN1 and TEN 4 (see Fig. 3) which were placed in the mouth of the artificial crack. Slight strain flow (strain increment of 500 μm/m) was observed at a pressure dwell of 11.9 MPa. No yielding was observed at other levels (75% and 90% SMYS). A residual strain of about 1500 μm/m was observed after unloading. This can be considered to be substantial, and it was probably caused by the formation of a stretch zone.

The longitudinally oriented strain gauge, TEN6, showed only a linear response during the hydrotest without any manifestation of elastoplasticity. For the circumferentially oriented strain gauge, TEN5, a deviation from linearity and converging towards higher strains can be observed above 10 MPa, indicating the first manifestations of yielding (see Fig. 3). In addition to the mentioned yielding, resulting in a change in the dimensions of the pipe, a slight change in the mechanical properties of the pipe can occur. Gajdoš et al. [9] studied the effect of a hydrotest on the change in mechanical and fracture properties of X46 steel and concluded that there was an improvement in notch toughness but a decrease in fracture toughness.

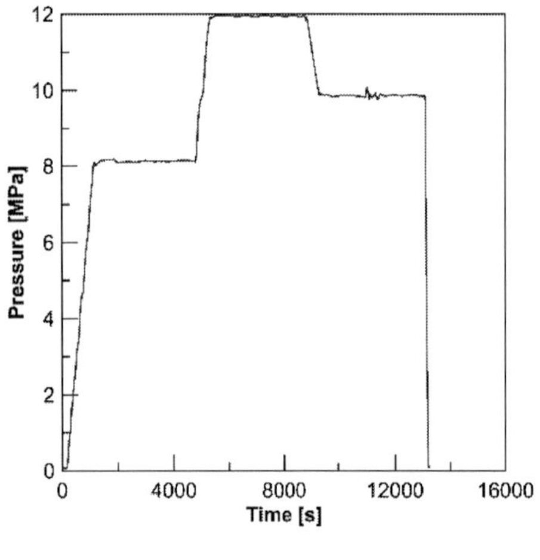

Fig. 2. Pressure response during hydrotest.

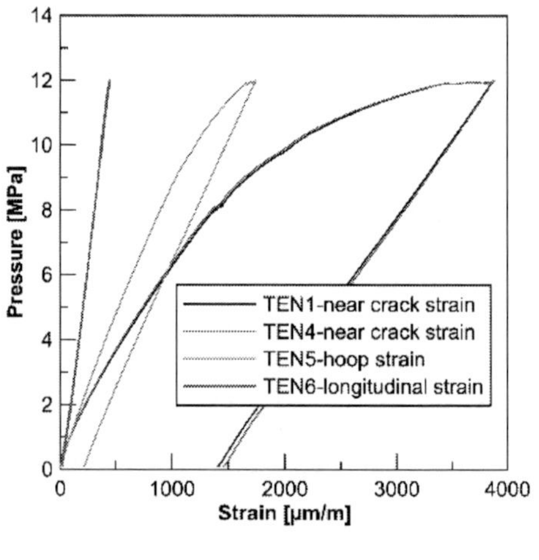

Fig. 3. Pressure-strain curves measured during the hydrotest.

After relieving the body after the hydrotest, the fatigue pressure test immediately followed. Pressurization was carried out with a pressure range from 4 to 7.35 MPa. The total number of cycles was set at 20,000 cycles. The measured artificial crack growth rate during the fatigue pressure test is shown in Fig. 4. There are apparently three growth phases – (a), (b), and (c). During phase (a), growth retardation is evident, caused by the formation of a monotonic plastic zone at the crack tip. After 400 cycles, however, phase (b) followed. Phase (b) is characterized by accelerated crack growth. The final phase (c) represents crack growth rate stabilization. The size of the plastic zone for the plain strain condition (β = 3) was calculated by Eq. (2) and Eq. (3). The plastic zone diameter under cycling loading is given by

$$r_{p\ pre-cracking} = \frac{1}{\beta.\pi}\left(\frac{\Delta K}{2.SMYS}\right)^2 = 0.07 \text{ mm} \tag{2}$$

and the monotonic plastic zone diameter is given by

$$r_{p\ hydrotest} = \frac{1}{\beta.\pi}\left(\frac{K_{max}}{SMYS}\right)^2 = 0.07 \text{ mm} \tag{3}$$

where is the cycling stress intensity factor and is the maximum stress intensity factor, corresponding to the stress level of 110% SMYS. According to the above-mentioned equation, the plastic zone would extend to a depth of 8 mm, i.e., the entire increment resulting from the fatigue pressure test would be in the plastic zone. This assumption is not in agreement with changes in crack growth rate. Thus, it appears that the plastic zone was not as large as theoretically calculated from Eq. (3). Assuming accelerated fatigue growth behind the plastic zone, the plastic zone was one order of magnitude smaller. This phenomenon can be attributed to the small residual ligament below the crack, as already commented by Linhart et al. in [10].

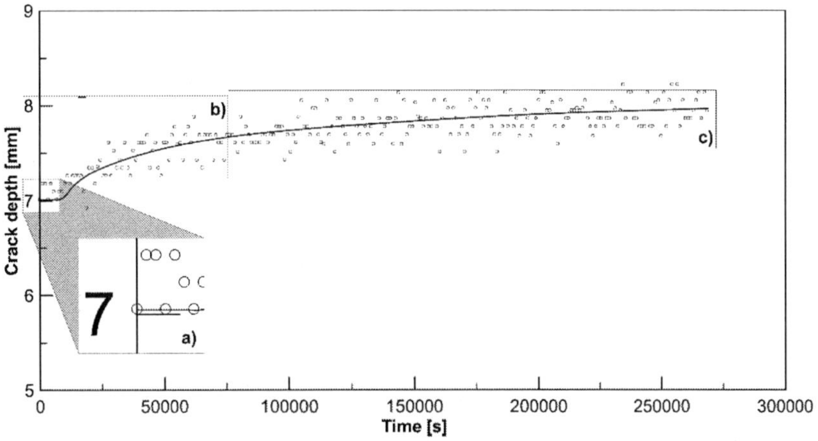

Fig. 4. Crack growth measurement during fatigue pressure test with distinguish phases.

After reaching the final number of 20,000 cycles, the pipeline was unloaded to zero pressure and the Mostro valve with an IP55 servo motor was disconnected from the pressure system. All strain gauges were balanced before the burst pressure test began.

The first significant plastic deformation around the crack mouth could be observed on strain gauges TEN1 and TEN4 above 12 MPa. A significant yielding of the pipeline occurred above a pressure of 14 MPa (σ_{hoop} = 541 MPa), which is a slightly higher value compared to the previously tested DN800 pipelines. Generally, yielding of the DN800 pipelines occurs around 12 MPa (σ_{hoop} = 462 MPa). This implies that the DN800 was strain hardened during the hydrotest. The crack depth was constant during the static loading up to the pressure of 15 MPa (σ_{hoop} = 580 MPa). Although the fracture depth is constant, high plasticity and necking occurs on the strain gauges in the crack mouth. When the pressure reached 15 MPa, the crack depth increased rapidly (see Fig. 5) due to ductile tearing of the residual ligament. Rupture of residual ligament followed briefly at a pressure of 16.08 MPa. Due to the semi-circular shape of the crack, the higher stress intensity factor was at the crack tip and instead of the surface, which resulted in the application of the Leak-Before-Break criterion. Also, for such a high pressure, there was no shear fracture propagation but only leakage.

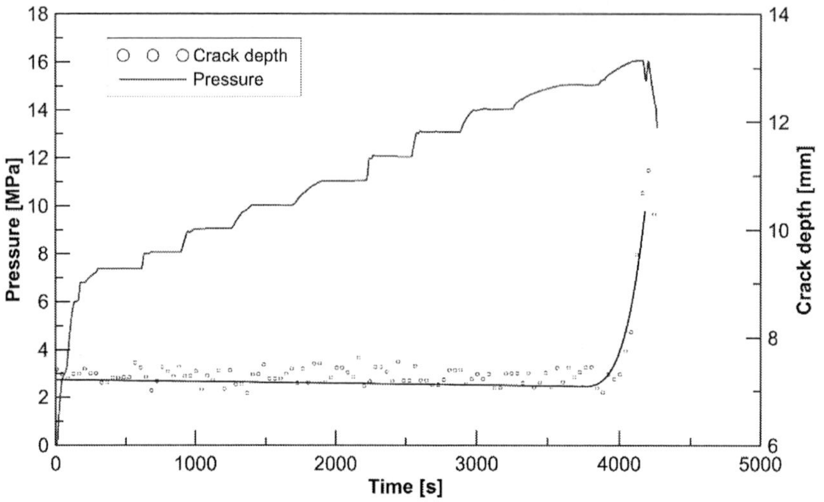

Fig. 5. Pressure-time dependence and crack growth measurement during the burst pressure test.

The area with the crack was cut out from the pipeline and then fractured in liquid nitrogen to eliminate the plastic deformation of the fracture during breakage. The macromorphology of the crack, including the most important dimensions, can be seen in Fig. 6. Five macromorphological regions can be observed. The artificial notch had a depth of 5.76 mm and a surface length of 23.05 mm. This was followed by a fatigue pre-cracked region up to 7.02 mm. A stretch zone, which is not very obvious from Fig. 6, separates the fatigue pre-cracked region from the fatigue part of the fracture which was produced during the fatigue pressure test with 20,000 cycles. Although the increment appears to be significantly smaller in comparison with the fatigue pre-cracking region, it should be noted that the stress intensity factor ΔK was reduced from 29.8 MPa.m$^{0.5}$ to 20 MPa.m$^{0.5}$. The dark-looking part of the fracture that extends to the pipe's inner surface is the static increment that occurred during the burst pressure test. The edges of the fracture surface are composed of cleavage facets that were only formed during breakage in liquid nitrogen. The micromorphology of the fatigue pre-crack region and fatigue growth region consists of ductile striations. The width of the stretch zone reaches about 5 μm, and it is uneven and non-compact along the crack tip.

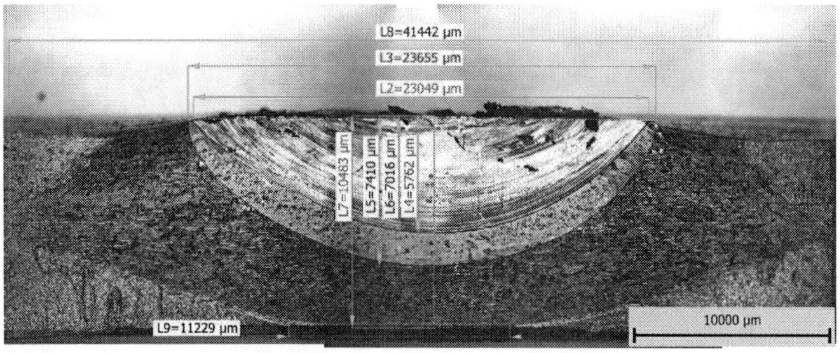

Fig. 6. Artificial crack in DN800 pipeline after tests.

3. Conclusions

This paper evaluates the effect of a hydrotest on the fatigue growth of an artificial crack in DN800 pipe. The artificial crack was fatigue pre-cracked into the pipeline from a notch to a final depth of 7 mm. The pipeline with this artificial defect was hydrotested. Subsequently, a fatigue test was applied within the set range corresponding to general operation, a pressure fluctuation of 4-7.35 MPa and a total of 20,000 cycles. Crack growth kinetics during tests and measurement of the stress-strain state on the pipe DN800 were measured.

During the hydrotest, the formation of a static plastic zone at the crack tip was observed. This should serve as a barrier to subsequent fatigue growth. The formation of the plastic zone is also associated with plastic deformation of the crack mouth, detected with help of strain gauges. The formation of the plastic zone is also associated with blunting of the crack tip and the formation of stretch zones. From the data, it is evident that stretch zones are formed only at pressures corresponding to the yield stress and higher.

Three areas of growth were observed during the fatigue test: the retardation area, the accelerated growth area, and the stable growth area. The growth retardation region was short, and after 400 cycles crack re-initiation and accelerated growth occurred. According to theoretical calculations, the diameter of the plastic zone should be about 1 mm. Taking into account the fact that the whole fatigue increment for 20,000 cycles was 0.41 mm, it can be estimated that the plastic zone did not reach the predicted dimensions and even the hydrotest was very ineffective. This phenomenon can be attributed to the small residual ligament under the artificial crack, thus the material lacked the plasticity needed to create a plastic zone corresponding to the theoretical calculations.

Acknowledgements

The support of institutional financing, grant MSM257970000, is gratefully acknowledged.

References

[1] Cheng Y.F., *Stress Corrosion Cracking of Pipelines* (Wiley, USA, 2013). ISBN 978-1-118-02267-2.

[2] Gajdoš L'., *Spolehlivost plynovodních potrubí* (ČVUT, Praha, 2000). ISBN 80-01-02143-2.

[3] Fazzini P.G., Belmonte J.C., Chapetti M.D. and Otegui J.L., "Fatigue assessment of a double submerged arc welded gas pipeline," *International Journal of Fatigue*, Vol. 29, no. 6, pp. 1115–1124 (2007). ISSN 0142-1123. https://doi.org/10.1016/j.ijfatigue.2006.09.013.

[4] Manfredi C. and Otegui J.L., "Failures by SCC in buried pipelines," *Eng. Failure Analysis*, Vol. 9, no. 5, pp. 495–509 (2002). ISSN 1350-6307. https://doi.org/10.1016/S1350-6307(01)00032-2.

[5] Kiefner J.F. and Maxey, W.A., "The benefits and limitations of hydrostatic testing," in *API's 51st Annual Pipeline Conference & Cybernetics Symposium*, (2000).

[6] Li J., Elboujdaini M., Gao M. and Revie R.W., "Investigation of plastic zones near SCC tips in a pipeline after hydrostatic testing," *Materials Science & Engineering. A*, Vol. 486, no. 1, pp. 496–502 (2008). https://doi.org/10.1016/j.msea.2007.09.046.

[7] Kec J., Černý I., Poloch A., Kyselá B., Poupa M., Pavelková R. and Janovec J. "Investigation of mechanical properties and microstructure of X60 line-pipe steel," *Procedia Structural Integrity*, Vol. 17, pp. 230–237 (2019). ISSN 2452-3216. https://doi.org/10.1016/j.prostr.2019.08.031.

[8] Mlnařík J., Kec J. and Chuchvalec P., "Innovative device for measuring crack depth by ACPD potential method" in Proceedings of Zvyšování životnosti komponent energetických zařízení v elektrárnách, Srní, 2020 (Západočeská univerzita v Plzni), pp. 165–170. ISBN 978-80-261-0959-4

[9] Gajdoš L'., Šperl M. and Pařízek P., "The effect of overloading on toughness characteristics," *Materials Today: Proceedings* Vol. 4, pp. 5803–5808 (2017). ISSN 2214-7853.

[10] Linhart V., Pavelková R., Brož P. and Čipera M., "Effect of overloading on the retardation of the growth of cracks in high-pressure pipelines," *Plyn*, Vol. 93, no. 4, pp. 76–84 (2013). ISSN 0032-1761.

Design and Optimization of Load-Bearing Steel Structure Using the Probabilistic Method SBRA

Petr Kesl[1] & Zdeňka Rendlová[2]

Abstract: This paper aims to present the use of a probabilistic approach for the design and subsequent optimization of a load-bearing steel structure.The assessment itself is performed in commercial software that uses a probabilistic approach, and the obtained solution is compared with the solution performed in software working on the principle of the finite element method. The results of both approaches are compared.

Keywords: Steel Structure; Design Structure; Limit State; Optimization; European standards; Probability of Failure; Simulation-Based Reliability Assessment (SBRA)

1. Introduction

At present, many calculations closely related to the strength analysis of structures are performed only using the finite element method (FEM). A less frequently used approach for structural analysis and optimization of structures is the probabilistic approach, namely Simulation Realibity Assessment Method (SBRA). This is an approach where, on the basis of the obtained histograms of the courses of internal forces obtained from the evaluated load cases, we obtain their maximum values and on the basis of them the probability of failure of the structure is evaluated [4].

2. Probabilistic approach principle

The SBRA method is based on probabilistic approach that uses Monte Carlo algorithms for simulations [1,2]. The main idea of methods of this type lies in the repeated solution of relatively simple equations that are of many parameters (e.g. dimensions of the body, loads, mechanical properties) defined as constant values or by histograms. Based on the performed calculations, the probability of failure is then determined. Principle of the mentioned method is shown in Fig. 1.

Probability of failure is given by so-called safety function

$$P_{f(i)} = R_{(i)} - E_{(i)}\,, \tag{1}$$

where the load effect is represented by $E_{(i)}$ and the structural resistance is given by $R_{(i)}$. Then, the probability of failure P_f can be expressed as

$$P_f = \frac{N_f}{N}\,, \tag{2}$$

where N_f is number of results that do not fulfill safety function (1) and N is total number of results.

[1] *Ing. Petr Kesl, Ph.D.; Department of Mechanics,Faculty of Applied Sciences, University of West Bohemia; Technická 8, 301 00 Pilsen, Czech Republic;* pkesl@kme.zcu.cz

[2] *Ing. Zdeňka Rendlová, Ph.D.; Department of Mechanics,Faculty of Applied Sciences, University of West Bohemia; Technická 8, 301 00 Pilsen, Czech Republic;* zrendlov@kme.zcu.cz

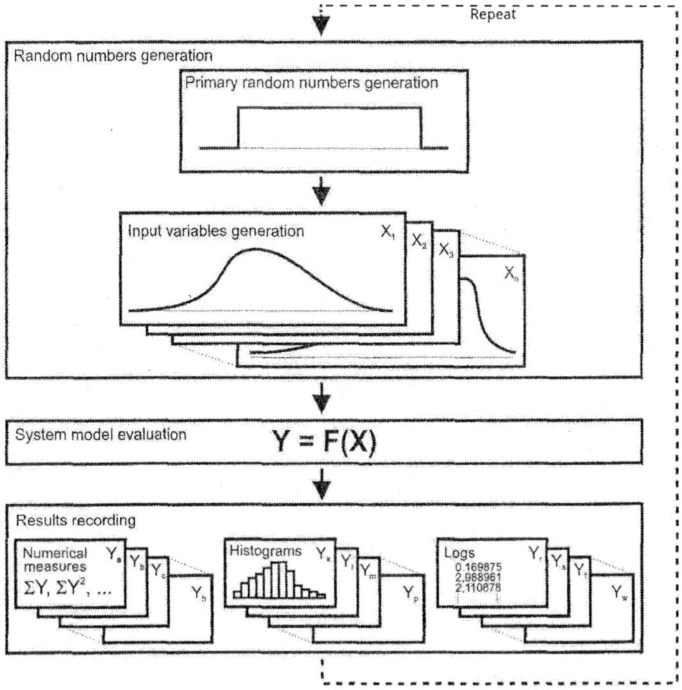

Fig. 1. SBRA principle.

3. Application

The above proposed method is used for the design and optimization of a steel bridge structure using valid European standards (ČSN EN 1991-1-1, ČSN EN 1993-1-1) [3]. After simplification, the computing model of the steel bridge construction is created using four types of structural elements. Primary beams and secondary beams are of IPE profile, closing beams are of UPE profile and stiffenning beams are of L profile. The proposed bridge structure is to be made of S 235 steel with a Young's modulus of elasticity $E = 2.1 \cdot 10^{11}$ MPa. The main construction is of symmetrical desing, see Fig. 2.

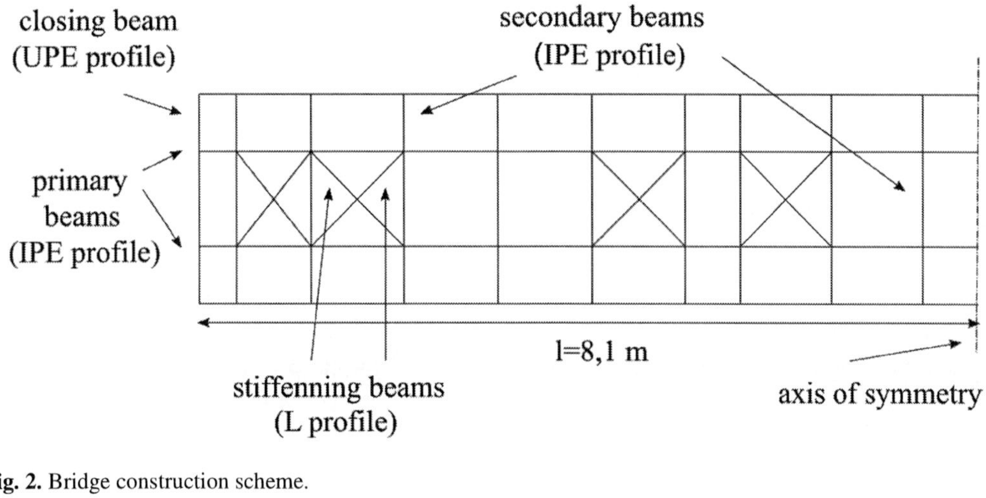

Fig. 2. Bridge construction scheme.

The bridge construction is assessed for 13 load cases including static and dynamic loads (for example weight of the structure or influence of weather conditions). To determine the response of the structure 230 load combination envelopes are asembled using probabilistic generator. For the needs of using the probabilistic approach, the load functions and the load condition coefficients are defined by means of the histograms shown in Fig. 3 and Fig. 4, respectively.

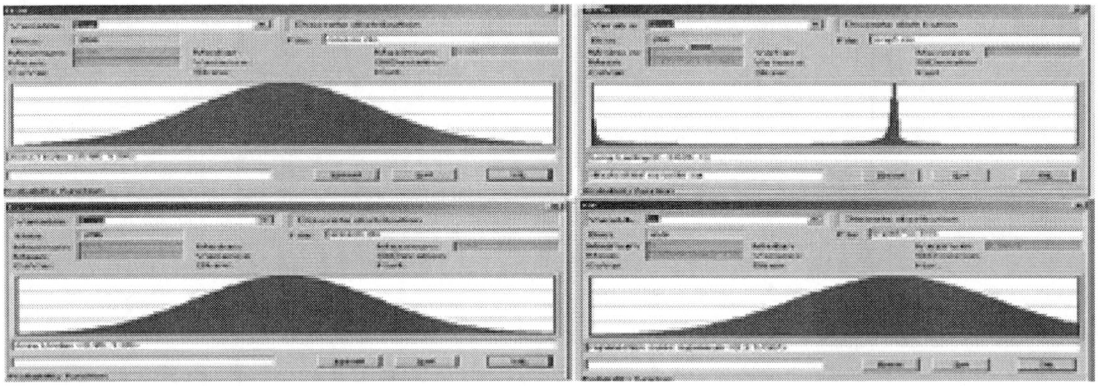

Fig. 3. Load function histograms $E_{(i)}$.

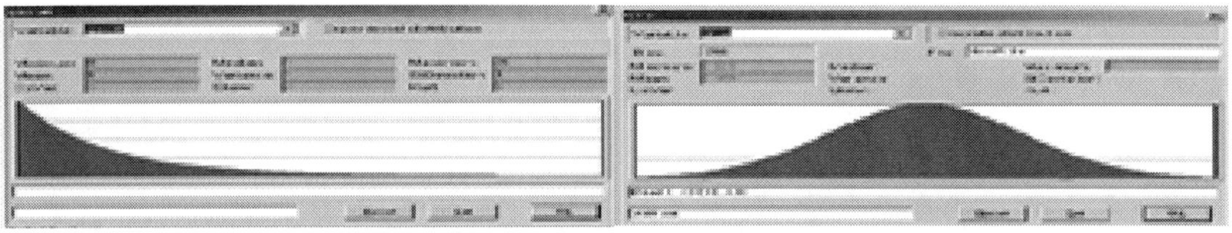

Fig. 4. Histograms of load condition coefficients $\gamma_{f(i)}$.

The design optimization was performed using the probabilistic approach in the SBRA Anthill software, where the caculations were performed for 5 million simulation cycles. Cross-sectional assessment is performed with respect to applicable standards using user-implemented relationships in the language of computational software. Fro example, for assessment of the cross section under torque M_{ed}

$$ohyb = \frac{k_{yy} \cdot M_{ed}}{M_{yrd}}, \; M_{yrd} = \frac{CapaLT \cdot Wply \cdot Fy\,var}{1,0}, \tag{3}$$

where M_{ed} is design combination of bending moment and M_{yrd} reduced bearing capacity of the cross section in bending.

4. Conclusions

The bridge steel structure was analyzed using two approaches. For the defined combinations of load cases, the values of shear forces and bending moments on the partial elements of the structure were obtained and subsequently the probability of failure. The analysis shows that the use of a probabilistic approach to optimize the steel structure can be considered a suitable approach providing very good results, taking into account the many influences to which the bridge construction is exposed. The obtained results in the Table 1 show that the use of the probabilistic approach allowed significant material savings with respect to the internal forces arising due to the load inside the most stressed segment of the bridge structure.

Table 1. Comparison of results – primary beams.

Equations	Obtained results		Difference in %
	ČSN EN 1993-1-1	SBRA Anthill	
$M_{pla,Rd}$, $M_{plc,Rd}$, plasticity	43.2–43.8 %	30.2–35.8 %	8–13
$\sigma_{a,dol}$, $\sigma_{c,dol}$, elasticity	94.6%	88.6–89.6%	6–7
$P_{f(i)}$	$P_{fd(i)} = 7.20 \cdot 10^{-5}$–$7.28 \cdot 10^{-5}$	$1.12 \cdot 10^{-5}$–$1.375 \cdot 10^{-5}$	$P_{f(i)} < Pfd\,(i)$

References

[1] Marek P., Guštar M. and Anagnost T., *Simulation-Based Reliability Assessment for Structural Engineers* (CRC-Press, Boca Raton, Florida, USA, 1996). ISBN 978-0-8493-8286-4.

[2] Pláníčka F., Šolc, M. and Marek, P., "From Partial Factors Design to the SBRA Probabilistic method," *Acta Mechanica Slovaca*, Vol. 15, no. 2, pp.22–27 (2011). ISSN 1335-2393.

[3] The EN Eurocodes https://eurocodes.jrc.ec.europa.eu/ .

[4] Marek P., Brozzetti J., Guštar M. and Tikalsky P., E. Editors, *Probabilistic Assessment of Structures using Monte Carlo Simulation. Basics,Exercises, Software*, Second edition, (ITAM CAS, Praha, 2003).

The Experimental Analysis of the Pressure Field in Breast Implants, the Effect of Surface Structure Failure on the Biomechanical Properties in the Body

Barbora Klíčová[1], Aleš Richter[2] & Aleš Lufinka[3]

Abstract: The paper presents stress analysis of breast implants. This study introduces the verification of measurement principles and methodology. The experimental results on selected types and sizes of breast implants are mentioned in conclusion for example.

Keywords: Biomechanical analysis of pressure field; Breast implants; Stress diagnostics breast implants

1. Introduction

Breast augmentation is the third most often performed aesthetic surgical procedure globally [1]. The detailed experimental analysis of pressure fields in breast implants is the principal research due to prediction of dangerous situations of humans in everyday life. For example the slight damage to the surface during implantation is unacceptable, but not completely excluded [1–3]. A risky situation can occur in contact sports activities, diving but also by activating the airbag during a car accident [4]. The correct evaluation of these risk conditions can only be made on the right experimental analysis with the right methodology. The paper includes the verification of measurement principles and methodology and presentation of experimental results on selected types and sizes of breast implants.

2. Methods

The FU250 test device was used for the tested explant breast implants loading. The device is equipped with a linear hydraulic engine, which is controlled by a computer control system. Different load variants in a wide range of parameters can thus be easily set. The test equipment was supplemented by a GTM 2.5 kN force sensor for measuring the total loading force. A special system for measuring the pressure in the tested breast implant was inserted in the lower movable table. The implant compression value was measured from the hydraulic engine piston rod position.

The course of the test was further recorded by a camera. All signals and video from the camera were connected to the Dewe 5000 measurement system. Synchronous dates and video recording are possible in this system. The start and development of possible cracks in the implant can thus be unambiguously assigned to the values of the measured quantities (Fig. 1).

[1] *Ing. Barbora Klíčová; Faculty of Health Studies, Technical University in Liberec; Studentska 140/02, 461 17 Liberec 1, Czech Republic; Barbora.Klicova@tul.cz and Faculty of Biomedical Engineering; Nam.Sitna 3105, 272 01 Kladno, Czech republic;* Barbora.Klicova@fbmi.cvut.cz

[2] *prof. Ing. Aleš Richter, CSc; Faculty of Health Studies, Technical University in Liberec; Studentska 140/02, 461 17 Liberec 1, Czech Republic;* Ales.Richter@tul.cz

[3] *Ing. Aleš Lufinka, Ph.D.; Department of the Design of Machine Elements and Mechanism, Faculty of Mechanical Engineering; Studentska 1402/2, 461 17 Liberec 1, Czech Republic;* ales.lufinka@tul.cz

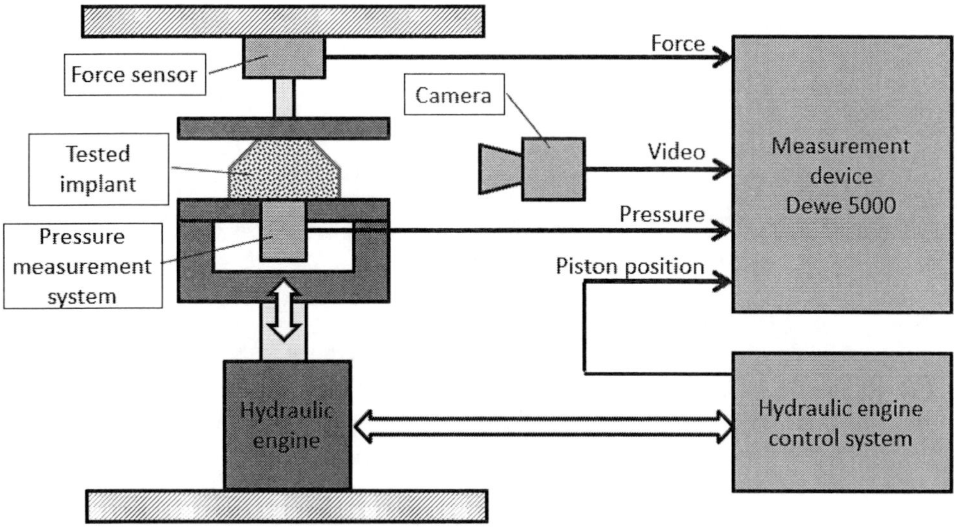

Fig. 1. The experiment equipment assembly, functional diagram.

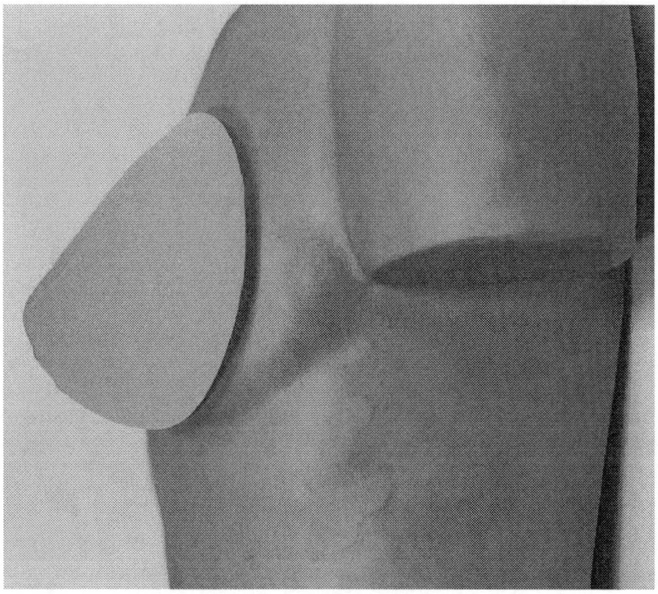

Fig. 2. The anatomical description of the problem under investigation.

2.1. The analysis of pressure balance, the theoretic background

The calculation of the pressure field inside the implant is based on the following simplification.

$$p_i(z) = p_{\text{ext}} + \rho \cdot (c - z) \tag{1}$$

$$p_{\text{atm}} = \rho \cdot (c - z) \tag{2}$$

$$p_{\text{ext}} = \frac{F_{\text{ext}}}{S};\ \rho = \frac{\Delta m}{\Delta V};\ S = \frac{\pi D S^2}{4} \tag{3}$$

$$p_{\text{ext}} \gg p_{\text{atm}} = \rho \cdot (c - z);\ c < 0.2\,m \tag{4}$$

We assume that atmospheric pressure (p_{atm}) is minimal in relation to external pressure (p_{ext}), which brings about dangerous life situations. The pressure (π) inside the implant will depend primarily on external pressure (p_{ext}) caused by undesirable circumstances.

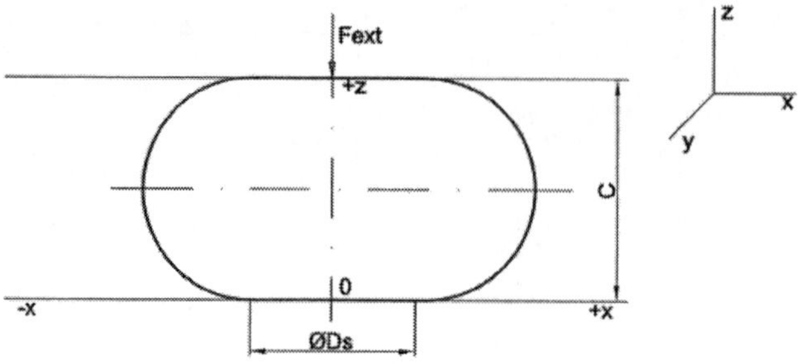

Fig. 3. The substitute geometric model of the implant during the stress testing; F_{ext} – external force, c – compres in direction of z-axis, ϕ D_{s} – diameter of pressed surface.

3. Results

The example of experimental analysis results is introduced on Fig. 4. The pressure conditions inside the implant are then derived from the course of the force.

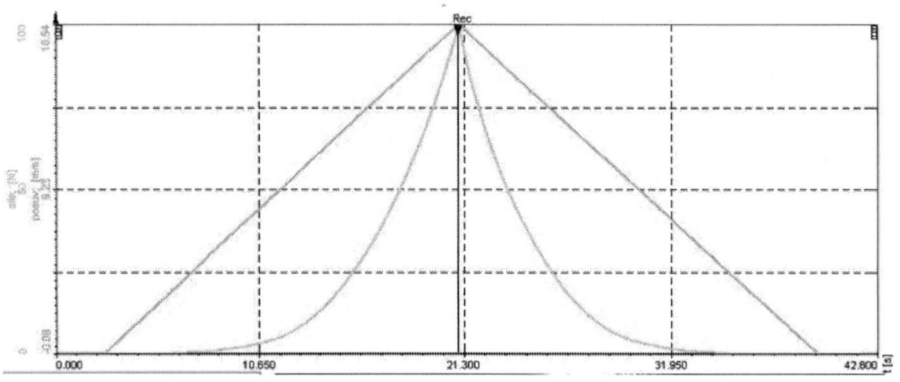

Fig. 4. The behaviour of force by gradual compression (volume of implant 250 cm^3, total compression by 18.5 mm from the starting position at zero external pressure, maximal force 100 N).

The particularised mechanical dates from experimental analysis are used for pressure field modelling in the breast implants. The principal research is to predict dangerous situations and reduce the following possible risk of damage to the implant within the body.

Table 1. The general summary of pressure experiments.

M_{imp}[g]	c_0-c	F_{ext}[N]	p_{ex}[N/m^2]	D_{s}[mm]
250	18.5	99.3	8130	125
	20.5	148	11400	130
360	18.8	99.4	9094	118
	20.2	148.5	12580	122
550	31.2	197.4	9560	162
	32.5	250.1	11040	167

The M_{imp} mass of implants, the c_0-c compress from initial position, the F_{ext} external force, p_{ext} the external pressure transferred to the implant, D_{s} diameter of contact pressure surface, the $\rho = 0.97$ g/cm^3.

This experimental setup and measured results confirm the correctness of the chosen methodology. In further experiments, it will be necessary to solve the exact determination of the contact surface of the contact force (F_{ext}). It will be necessary to supplement the test kit with image capture with an accurate coordinate system. We assume the incompressibility of gelatin inside the implant, then it is possible to focus further research on the relationship of contact force and surface in relation to the elasticity of the surface. The analysis of mechanical surface stress is the main purpose of this research [5].

4. Conclusion

This first static analysis of the pressure field in the breast implant showed that it was insufficient to estimate the risk of damage and its consequences in the human body. In the future, we also plan to focus on dynamics. We will monitor the time changes in the pressure field, dynamics that may occur during sports and other daily activities. The predominant influence also has the anatomical position, whether the implant is placed directly under the skin or under the pectoral muscle [6].

This initial study looks for basic ways to test the breast implants damage due to mechanical surface disruption. The main purpose of future research will be to find a suitable anatomical model and the method of evaluating dangerous life situations that could cause their destruction inside the human body.

Acknowledgements

This work was supported by the Grant Agency of the Czech Technical University in Prague, grant No. SGS22/132/OKH4/2T/17. This is work supported by the programme *H2020-EU.4.b. - Twinning of research institutions* from the call *H2020-WIDESPREAD-2018-03* under *Grant Agreement ID: 857061*, under the project „Networking For Research And Development Of Human Interactive And Sensitive Robotics Taking Advantage Of Additive Manufacturing (R2P2)".

References

[1] Ramião N.G., Martins P.S., Barroso M.L., Santos D.C. and Fernandes A.A., "Biomechanical analysis of intact versus ruptured Poly Implant Prothèse breast implants," *Interface Focus*, Vol. 9, Art. no. 20180086 (2019). doi: 10.1098/rsfs.2018.0086.

[2] Klíčová B. and Kneppo P., *Identification of breast implant damage during diagnostics from the point of view of biomechanics*, Dissertation thesis: Unpublished, (Czech Technical University in Prague, 2022).

[3] Hillard C., Fowler J.D., Barta R. and Cunningham B., "Silicone breast implant rupture: a review,"*Gland Surg*, Vol. 6, no. 2 (2016). ISSN 2227-8575. doi:10.21037/gs.2016.09.12.

[4] Krajcová A., Hurt K., Kufa R. and Molitor M. "Breast implant rupture: A sports trauma report," *Česká Gynekolie*, Vol. 85, no. 2, pp.116–119 (2020). English. PMID: 32527105.

[5] Kedlesová E., Bittner V., Klíčová B., Molitor M. and Čapek L., "Indentation test of explant breast implant," in *Proceedings of Biomedical Engineering Trends*, Liberec, 2021 (Technical University in Liberec, Liberec, 2021), pp. 8485. 9788074945854.

[6] Jurat D. and Wenzel D. "Breast Augmentation in Athletic Women: A Retrospective Survey Assessing Pectoral Muscle Function and Implant Aesthetics Post-Augmentation," *Aesthetic Plast Surg.*, Vol. 46, pp. 1075–1081 (2022). doi: 10.1007/s00266-021-02692-8. Epub ahead of print. PMID: 35075503.

Initial Yield Locus Tracking for Stainless Steel Produced by SLM Technology

Michal Kořínek[1], Radim Halama[1], Marek Pagáč[2] & Dušan Stříkla[3]

Abstract: The study of initial anisotropy is done for 316L stainless steel produced by 3D printing. A non-destructive tracing method is used to determine the initial yield locus. Experiments were performed in two printing directions and three loading paths were tested. The specimens were loaded proportionally and a combination of axial and torsional loading was used. The yield stress is defined as the deviation from the linearity. The resulting dispersion between the probing paths is negligible. The direction of printing affects the shape of the initial yield locus. For the horizontally printed specimens, the points can be approximated according to the von-Mises criterion. For vertically printed specimens, the shear stress component was dominant, and the initial yield surface appears as an ellipse.

Keywords: Yield locus; Tracking; Stainless steel; SS316L; Initial anisotropy; Proportional loading

1. Introduction

Additive technologies are progressive and modern manufacturing processes that are increasingly being used today. These technologies can be used to produce components not only from plastics but also from iron alloys. Additive technology can replace conventional manufacturing for complex shapes obtained, for example, by topological optimization [1], even in the field of mechanical engineering. However, especially in mechanical engineering, where functional elements are used, it is necessary to describe the mechanical properties of the material well, and additive technologies have a significant effect on these properties.

This paper deals with experimental determination of the initial anisotropy for stainless steel 316L (SS316L) produced by 3D printing, specifically by Selective Laser Melting technology (SLM). The direction of the printing has an influence on the mechanical properties of the material. The study continues our previous research that has dealt with the response of the material in monotonic tests with notched specimens of the same material [2]. Now, we focus on the determination of the initial yield locus for the specimen printed in different directions - horizontal and vertical, and its subsequent comparison. The tracking method is used to determine the yield locus [3, 4]. This is a non-destructive method, measurement is evaluated online, and can be repeated using one specimen.

2. Specimens and measurement

For the first experiments, a tubular specimen for low cycle fatigue tests [5] was used, see Fig. 1, and build direction was horizontal. This specimen was made from a virgin powder. The specimen for the next experiment was printed vertically and was in the "as printed" state. It was

[1] *Ing. Michal Kořínek, prof. Ing. Radim Halama, Ph.D.; Department of Applied Mechanics, Faculty of Mechanical Engineering, VSB - Technical University of Ostrava; 17. listopadu 2172/15, 708 00 Ostrava, Czech Republic;* michal.korinek@vsb.cz, radim.halama@vsb.cz

[2] *doc. Ing. Marek Pagáč, Ph.D.; Department of Machining, Assembly and Engineering Metrology, Faculty of Mechanical Engineering, VSB - Technical University of Ostrava; 17. listopadu 2172/15, 708 00 Ostrava, Czech Republic;* marek.pagac@vsb.cz

[3] *Dušan Stříkla.; LabControl s.r.o.,; Přerovecká 65/36, 747 95 Opava, Czech Republic;* strilka@labcontrol.cz

manufactured from a recycled powder. The geometry of the specimens was different. For the second experiment, the outer diameter was 14 mm and the inner 11 mm. The geometry is not as accurate because we cannot achieve the exact dimensions when printing as in machining. However, according to [6], the machining does not affect the yield stress values. This specimen was tested only twice according to Fig. 2b and Fig. 2c. Both specimens were made from SS316L produced by 3D printing, specifically by SLM technology. Process parameters and details regarding 3D printing are reported elsewhere [7]. The tracking method was applied to determine the yield stress using LabControl 100kN/1000Nm hydraulic testing machine. The Epsilon 3550 extensometer was used with a gauge length of 25 mm for data collection, which allows the actual measurement of axial and torsional strains. This method allows one to determine the yield stress for any proportional combination of axial stress and shear stress. Therefore, it is a stress-controlled test with a stress rate of 1 MPa per second. To determine the specific yield stress, the tracing method with center point was used, as is the case with [8]. Specific yield stress values were obtained from a total of 16 probing paths in this experiment. Each specific yield stress is characterized by the angle ψ_i between the axis of the axial stress and the loading path, see Fig. 2, where the sequence of the specific yield stresses is defined. Yield stress is determined as a divergence from linear behavior. This deviation is defined as the increase in plastic strain in the axial or shear strain component.

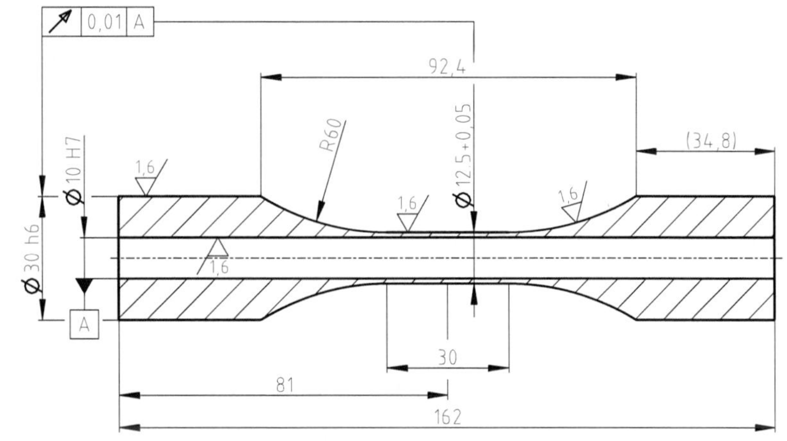

Fig. 1. Specimen dimension for determination of the yield locus by tracking method.

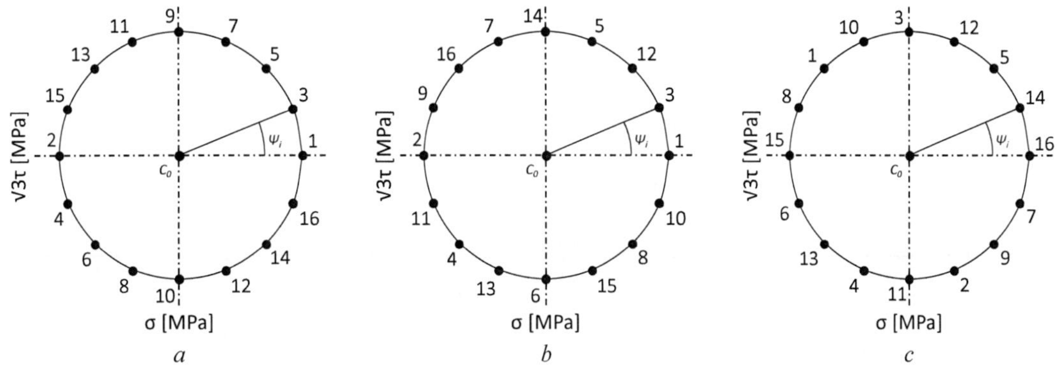

Fig. 2. Tracking for determine the initial yield surface for test 1(a), test 2(b) and test 3(c). The values of angle ψ_i are regularly graded, after 22.5 deg.

We used the 0.01% deviation from linearity, which corresponds to the value of 100 με. Equations (1) were used to solve this value as two criteria that ended the probing path, and the probing path was terminated as soon as at least one of these conditions was met.

$$\varepsilon_{pl} = \varepsilon - \varepsilon_{el} - \varepsilon_R, \qquad \gamma_{pl} = \gamma - \gamma_{el} - \gamma_R, \tag{1}$$

where ε and γ are total strain measured directly by the extensometer, ε_{el} and γ_{el} are elastic strain components and ε_R a γ_R are residual stresses that are caused by scattering of the measured data [9]. Ideally, these values are zero. The shear component of the plastic deformation γ_{pl} must be divided by the value $\sqrt{3}$ to correspond to the values in the stress space. The components of the elastic strains are evaluated during the experiment, and linear regression of the least squares method is used. This process is fully automatic, and the calculation takes place during the measurement. After fulfilling this condition, the unload to the center point C_0 follows, and another probing path in the sequence follows. During the experiment, a total of 16 specific yield stresses are determined by proportional loading on a thin-walled hollow specimen. Next, a total of 3 different probing paths were used to check the results in the experiment, see Fig. 2. Each measurement was done only once.

3. Conclusions

The initial yield locus is shown in the space of axial stress and shear stress ($\sigma - \sqrt{3}\tau$). The dispersion between specific yield stress for different loading paths is small and the results indicate that the order of the individual probing paths does not have a significant effect on the yield locus, as is shown in the Fig. 3. From the resulting initial yield locus obtained on a horizontally printed specimen (Fig. 3a), it can be stated that its shape can be described by the von-Mises yield function, even if it is produced by 3D printing, which introduces some degree of anisotropy into the material.

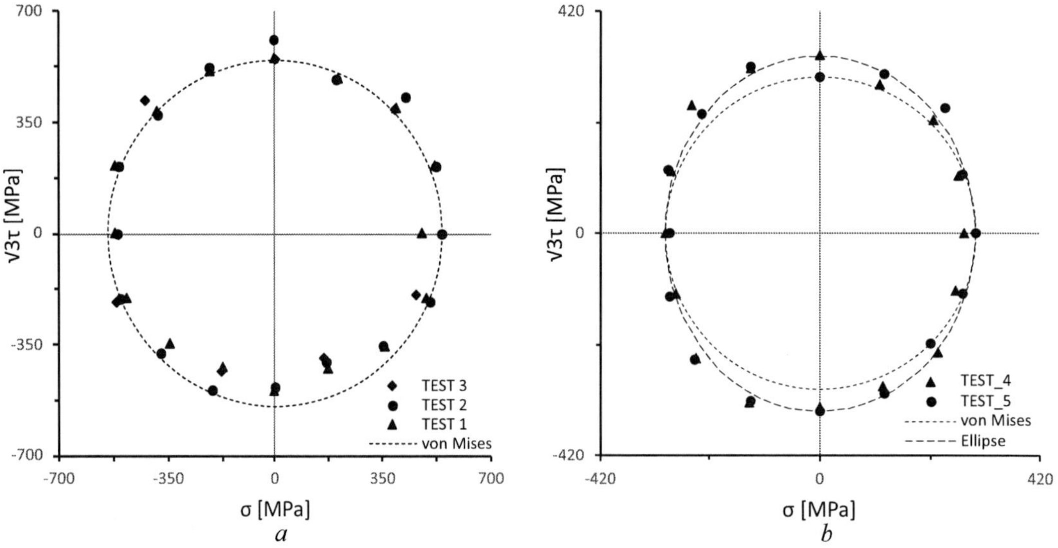

Fig. 3. The initial yield locus for SS316L produced by 3D printed for different probing paths. Horizontally printed specimen (a) and vertically printed specimen (b).

To plot the von-Mises criterion, it is necessary to determine the radius of the circle because this condition appears precisely as a circle in the space of stress. The required radius is thus determined as the average of all the partial values of the specific yield stress. Therefore, this value is set at 527 MPa. In the vertically printed specimen, the shear components are dominant, and so the initial yield locus corresponds to the ellipse shape. To render an ellipse, we need to know the size of the main and minor semiaxis. The values were determined in a manner similar to that of the first specimen. The size of the main semiaxis is the average of the pure shear measurement and has a value of 324 MPa. The minor half-axis is evaluated from the axial load and takes value of 287 MPa. Fig. 3b shows the initial yield locus for the horizontally printed specimen in the "as printed" state. For comparison, the same figure also shows the condition according to von-Mises. The radius value corresponds to the size of the minor semiaxis according to the shape of the ellipse, although the average yield stress value is 313 MPa. Thus, from the experiments, it can be concluded that the direction of printing affects the mechanical properties in different directions of loading. The yield stress observed by tensile tests is less for vertically printed specimens than for horizontally printed one [6]. It is worth mentioning that the usage of a recycled powder also reduces the yield stress.

Acknowledgements

This work was supported by The Ministry of Education, Youth and Sports from the Specific Research Project (SP2022/26), and has been done is connection with the DMS project reg. no. CZ.02.1.01/0.0/17_049/0008407 financed by Structural Funds of Europe Union.

References

[1] Šotola M., Maršálek P., Rybanský D., Fusek M. and Gabriel, D., "Sensitivity Analysis of Key Formulations of Topology Optimization on an Example of Cantilever Bending Beam," *Symmetry*, Vol. 13, Art. no. 712 (2021). https://doi.org/10.3390/sym13040712.

[2] Kořínek M., Halama R., Fojtík F., Pagáč M. Krček, J. and Krzikalla, D., "Monotonic Testing of 3D Printed SS316L under Multiaxial Loading", in Proceedings of Experimental Stress Analysis - 58th International Scientific Conference, EAN 2020, (Ostrava, 2020), pp. 213–224.

[3] Štefan J., Parma S., Marek R., Plešek J., Ciocanel C. and Feigenbaum, H., "Overview of an Experimental Program for Development of Yield Surfaces Tracing Method," *Applied Sciences*, Vol. 11, no. 16, Art. no. 7606 (2021). https://doi.org/10.3390/app11167606.

[4] Iftikhar Ch M. A. and Khan A. S., "The evolution of yield loci with finite plastic deformation along proportional and non-proportional loading paths in an annealed extruded AZ31 magnesium alloy," *International Journal of Plasticity*, Vol. 143, (2021). ISSN 0749-6419. https://doi.org/10.1016/j.ijplas.2021.103007.

[5] ASTM E2207 - 15 Standard Practice for Strain-Controlled Axial-Torsional Fatigue Testing with Thin-Walled Tubular Specimens. West Conshoshocken, USA: ASTM International, 2015.

[6] Kořínek M., Halama R., Fojtík F., Pagáč M., Krček J., Krzikalla D., Kocich R. and Kunčická, L., "Monotonic Tension-Torsion Experiments and FE Modeling on Notched Specimens Produced by SLM Technology from SS316L," *Materials*, Vol. 14, no. 1, Art. no 33 (2021). https://doi.org/10.3390/ma14010033.

[7] Halama R., Pagáč M., Paška Z., Pavlíček P. and Chen X., "Ratcheting Behaviour of 3D Printed and Conventionally Produced SS316L Material," in Proceedings of the ASME 2019 Pressure Vessels & Piping Conference PVP2019 July 14-19, 2019, San Antonio, TX, USA, paper number PVP2019-93384.

[8] Phillips A. and Tang J. L., "The effect of loading path on the yield surface at elevated temperatures, " *Int. J. Solids Struct.*, Vol. 8 no. 4, pp. 463–474 (1984). https://doi.org/10.1016/0020-7683(72)90017-0.

[9] Wu H. C. and Yeh,W. C., "On the experimental determination of yield surface and some results of annealed 304 stainless steel," *International Journal of Plasticity*, Vol. 7, pp. 803–826 (1993). https://doi.org/10.1016/0749-6419(91)90019-U.

Probabilistic Reliability Assessment of Energetic Devices

Klára Kotassková[1], Jana Marková[2], Miroslav Sýkora[2] & Radovan Šťastný[3]

Abstract: A key issue for the operators of power plants is to extend the working life of energetic devices and their components beyond the design working life. Currently used procedures are presently focused on the application of common deterministic methods. Thus, it is usually sufficient to use the conservative values of the basic variables for actions and resistance for evaluation of energetic components. On the other hand, it is widely recognized that the input parameters for deterministic assessment may be considered as parameters having stochastic (random) character which can be replaced by statistical distributions in calculations. Deterministic methods should be optimized and potential resources of currently applied procedures identified. Due to a number of influencing factors, the actual state of the energetic device might be difficult to be identified. Probabilistic and statistical approaches can be effectively used to analyze the basic problems in data collection from diagnostic measurements and their subsequent application in estimating the residual life of energetic devices and to optimize processes for their effective maintenance, diagnostics, renovation and also replacement. Data collection and quality assessment can considerably reduce number of failures and can bring major financial benefits for operators. However, this fact is often overlooked, so the main goal of the proposed paper is to present the advantages of the probabilistic approach for assessment the condition of selected devices in the energetic sector in comparison with deterministic methods.

Keywords: Service life; Probability of failure; Steam lines; Degradation mechanism; Risk analysis; Predictive maintenance; Asset management

1. Introduction

The tool for optimizing the maintenance, repair or replacement of the device(s) with regard to the financial capabilities of the company (or state) is important when extending the life of the device beyond its planned (design) working life. The purpose is to ensure a safe and reliable end to the operation of the device. The main goal of this approach is to propose a unified system for evaluating devices using interrelationships between collected data, based on historical data development, uncertainties between data, information on failures, outages and material degradation. The proposed new approach to reliability and residual working life assessment should generally be applicable to various energetic devices and their components in conventional and nuclear power plants. The approach is focused on the analysis and optimization of the service life of the components of the boiler pressure system called steam pipelines.

2. Evaluation procedure

The process of assessing the condition of the facility consists of following key activities:

- description of the device (system or sub-system),
- preliminary risk analysis,

[1] *Ing. Klára Kotassková, Institute of Applied Mechanics Brno, Ltd, Resslova 972, Brno, Czech Republic, kotasskovak@uam.cz and Czech Technical University in Prague Klokner Institute*

[2] *doc. Ing. Jana Marková, Ph.D., doc. Ing. Miroslav Sýkora, Ph.D., Czech Technical University in Prague, Klokner Institute, Šolínova 1903/7, 166 08, Czech Republic, Jana.Markova@cvut.cz, Miroslav.Sykora@cvut.cz*

[3] *Ing. Radovan Šťastný, Inspections and diagnostics RaD, Dolní Domaslavice 378,73938, Czech Republic*

- data preparation and control,
- reliability analysis.

2.1. Description of device

To evaluate the condition of the device, the information given in the operator's control documents needs to be made available. These documents facilitate to identify individual parts (components) of the device, the type of degradation mechanisms, used diagnostic methods, warning levels of monitored diagnostic parameters (limiting values which should not be exceeded), description of data collection and measurement frequency (or number of collected data from one diagnostic measurement), development of monitored degradation parameter, measurement uncertainty, description of deterministic evaluation of the device. To compare devices, general information about the device should be provided. The procedure for collecting information is shown in Fig. 1.

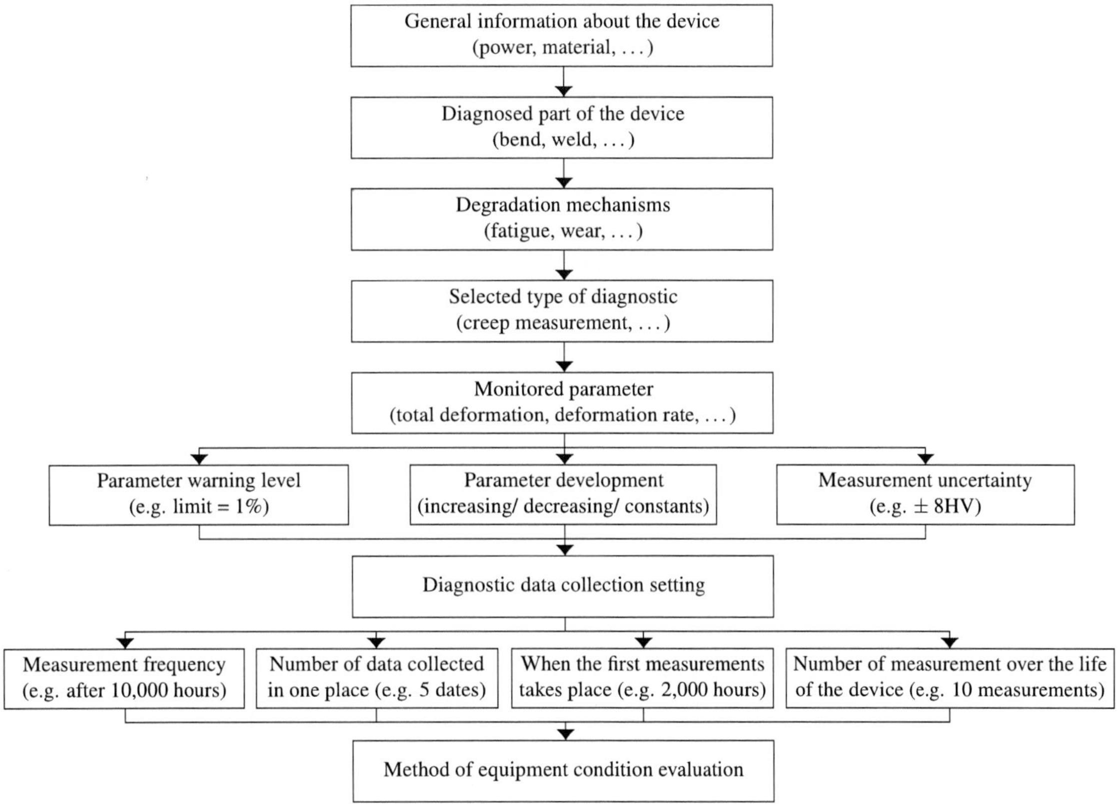

Fig. 1. Flowchart of assessment.

The output of the device condition evaluation is a report, an example of which is given in Table 1. The report describes the data important for evaluating the reliability of the device. For operators, places where the information in the management documentation is not available are marked in red (NA - not available, see Table 1) and should be acquired later.

Table 1. Description of the analysed system (device).

Diagnostic parameter	Number of data per 1 measurement	Measurement uncertainty	First check [hours]	Control frequency	Development	Warning level
1	2	NA	20000	1× in 1 year	Growing	4%
2	7	8%	12000	1× in 5 years	NA	NA

2.2. Preliminary risk analysis

Preliminary risk analysis is the initial step to identify the most critical parts of the facility. This process is divided into several activities – Fig. 2 shows the flow-chart of preliminary risk analysis. The first step describes significant parts of the evaluated device. The second step identifies the substantial types of degradation mechanisms for relevant facility and their consequences. This is followed by the risk analysis and evaluation of risk acceptance. Based on these results, the critical parts of the device are identified and a suitable diagnostic measurement is assigned to them.

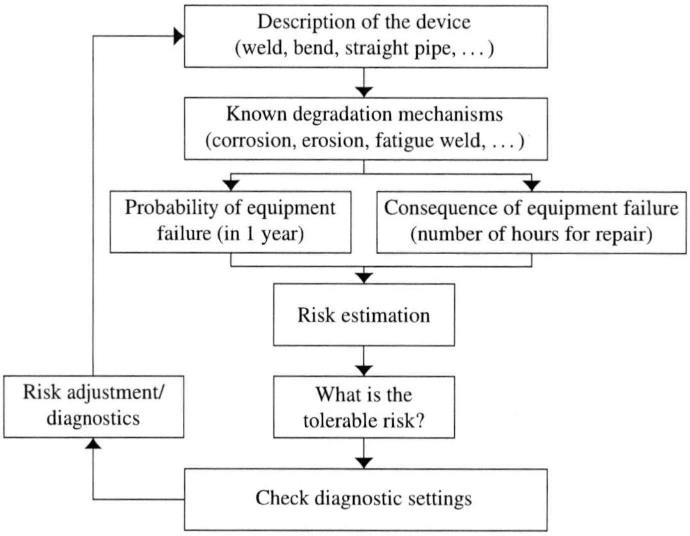

Fig. 2. Preliminary risk analysis of device.

Table 2. Preliminary risk analysis and evaluation of the potential need for diagnostic change.

Failure location	Degradation mechanism / Occurrence	Estimated preliminary risk	Conclusions
Part 1	1 (corrosion)	0.81	High need for diagnostic changes
	2 (erosion)	1.65	High need for diagnostic changes
	3 (fatigue)	0	No diagnostic changes
Part 2	3 (fatigue)	0.01	Change to basic diagnostics
	2 (erosion)	0.09	Change to basic diagnostics

The preliminary risk analysis needs to be checked with respect to the actual diagnostic settings. The preliminary analysis could lead to results that more or less locations of the relevant device or its parts should be surveyed or monitored. Table 2 illustrates the level of estimated preliminary risk malfunctioning of device due to the degradation mechanism and potential need for diagnostic change. This should indicate the correctness of diagnostic settings.

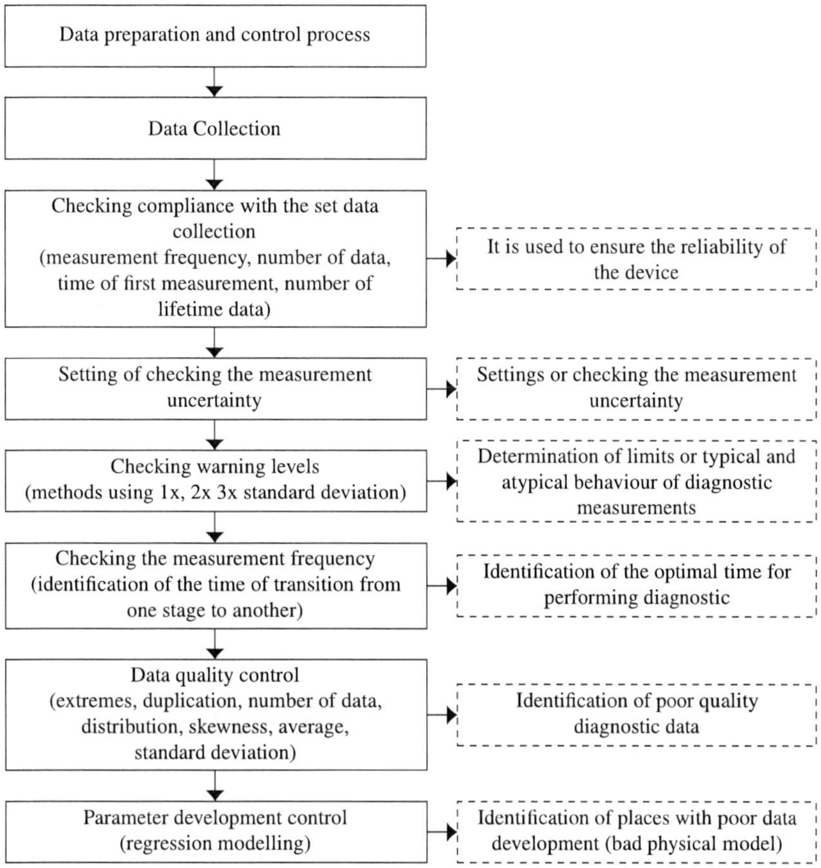

Fig. 3. Data preparation and control process.

2.3. Data preparation and control

The operator of the facility encounters a number of significant uncertainties during its operation. Measurement uncertainty affects the quality (reliability) of the evaluation. The data is often inconsistent and scattered, the specialist or expert then doubts their authenticity and he/ she may decide to exclude the data from the evaluation. The main sources of uncertainty in the evaluation include:

- natural randomness of loads, material properties and geometric data,
- statistical uncertainties due to the limited range of data available,
- uncertainties of computational models due to simplification of real conditions,
- uncertainties caused by inaccurate definitions of functional requirements,
- lack of knowledge about the behavior of new materials in real conditions,
- gross errors in the design, manufacture, installation and operation of the device.

The procedure for data preparation and control considering common sources of uncertainty in assessing the condition of the device is illustrated in Fig. 3. It should be noted that described uncertainties cannot be completely removed, but they can be reduced.

The output of these activities leads to six reports informing about the status of the operator's data. Places where diagnostics are not performed according to the specified rules or where there is a need for change the diagnostics settings should be marked up and further analysed in detail.

2.4. *Reliability analysis*

For the final evaluation of reliability of the device, the data from the previous subsection are processed to provide the basis for analysis of probability of device failure. The basic goal in determining the reliability of the device is whether the measured diagnostic value exceeds the specified critical value which indicates a non-acceptable state of the device. For example, the value of the total deformation did not exceed the critical value (1%) of the total deformation. Fig. 4 shows the procedure for evaluating the reliability of the device.

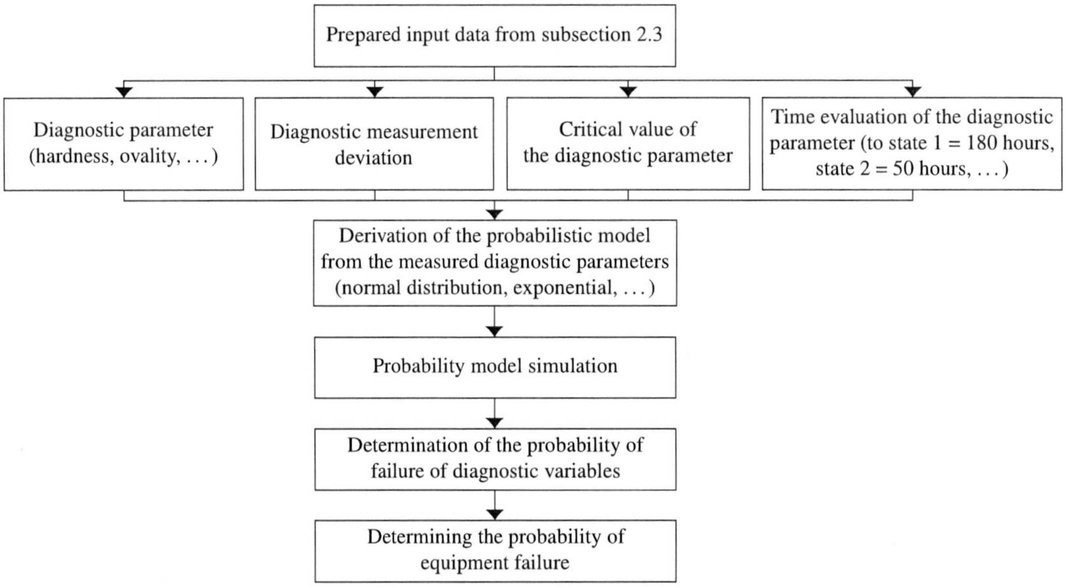

Fig. 4. Reliability analysis.

3. Conclusions

The contribution is focused on the evaluation of data collection taking into account uncertainties from different diagnostics to support efficient decision-making in the long-term management of devices in power plants. The procedure should increase the operability and reliability of devices.

Presently, the differences in data collection, their control, interpretation and evaluation lead to the urgent need to develop a comprehensive reliability and service life evaluation of devices systematically based on the use of probabilistic methods and risk assessment, applicable to all key facilities in energy operations.

Acknowledgements

This study has been conducted at the CTU in Prague, Klokner Institute within the framework of the Grant LTT18003 supported by the Ministry of Education, Youth and Sports of the Czech Republic.

References

[1] Kotassková K., *Statistical Analysis of Hardness Test Results According to NTD A. S. I.*, in Czech (ÚAM Brno, report ZP5886, 2016).

[2] Kotassková K., *Analysis of Results of Tests of Replicas*, in Czech (ÚAM Brno, report ZP6037, 2016).

[3] ASI, *Analysis of Microstructure using Replicas*, in Czech (A.S.I., Praha, NDT_ASI_III_3_2017, 2018).

[4] Kotassková K., *Development of Methodology for Estimating Service Life of Paper Isolation of Transformers to Reduce Operational Risks, Phase 5* in Czech, (ÚAM Brno, report ZP5089, 2012).

[5] Kotassková K., *Assessment of Technical Conditions of Transformers*, in Czech (ÚAM Brno, report ZP5012, 2012).

[6] Kotassková K., *Specification of Criteria for Measurements of Base Material Creep*, in Czech (ÚAM Brno, report ZP6719, 2018).

[7] Kotassková K., *Data Audit*, in Czech (ÚAM Brno, report ZP7566, 2020).

[8] Markova J., Jung K. and Stastna-Kotasskova K., "Probabilistic analyses of existing power producing facilities," In: ESREL 2018, Trondheim, Norway, (2018).

Advances in Experimental Mechanics Related to Novel Materials Tested under Monotonic and Cyclic Loading Conditions

Zbigniew L. Kowalewski[1], Mateusz Kopec[1] & Aneta Ustrzycka[1]

Abstract: This paper presents several attempts to use novel optical techniques for damage evaluation of materials subjected to either monotonic or cyclic loading and its monitoring on specimens made of the nickel super-alloy and power engineering steel. Effectiveness in damage analysis of Digital Image Correlation (DIC) and Electronic Speckle Pattern Interferometry (ESPI) is compared.

Keywords: Fatigue; Damage; Optical full-field methods

1. Introduction

The most conventional method for material testing includes extensometer recordings during static, monotonic and cyclic loading. Such methodology enables continuous recording of strain changes in a particular direction defined at the beginning of mechanical test. It can only monitor a displacement on the limited strain gauge, and more importantly, give only an average value of it. This is a serious limitation of the technique, particularly in the case of fatigue investigations. Although fatigue phenomenon has been investigated by many research centres for more than 200 years, there are still a lot of difficulties in prediction of crack initiation under cyclic loading. It is well known that the process of the fatigue damage development and structural degradation is of local nature, and as a consequence, an application of the above mentioned conventional extensometers for strain measurements cannot reflect strain distribution along the gauge length of specimen tested. Indication of a point of crack initiation within the gauge length is practically impossible using the conventional extensometers. Such a problem may be effectively solved by application of either ESPI (Electronic Speckle Pattern Interferometry) [1] or DIC (Digital Image Correlation) full-field optical methods [2, 3].

ESPI is the holographic interferometry type technique based on the laser beam analysis distracted from the optically rough surface. The specimen surface is captured as an image by the CCD (charge-coupled device) sensor and transferred to a working station. As the first beam illuminates the specimen surface, the reflected beam interferes with the reference beam, and thus, the subtraction process of the speckle interferograms (before and after loading up to the defined levels) is performed and correlation fringes are obtained. The fringes enable to generate a phase map with distribution of displacement components in each direction, separately [4]. Subsequent mathematical operations under the fixed boundary conditions (measurement area dimensions) and the material parameters (Young's modulus and Poisson's ratio) led to the final full-field stress and strain phase maps. ESPI, although different in terms of operation principles, it is quite similar to DIC optical technique for which also the displacement maps during loading can be obtained. It should be mentioned however, that it has significantly higher than DIC resolution. More importantly, it is mainly laboratory scale dedicated due to a high vibration sensitivity, that disturbs seriously the results captured.

[1] *Professor Zbigniew L. Kowalewski; Institute of Fundamental Technological Research; Pawinskiego 5B, 02-106 Warsaw, Poland;* zkowalew@ippt.pan.pl

DIC is a stereoscopic technique in which two CCD cameras, light sources and computational software are used. A mathematical theory of DIC was presented by Chu et al. [5]. In this method, a specimen needs to be covered with a special pattern (black dots on a bright background) [5, 6]. Such pattern defines the x and z coordinates which are further used to run the test under the strain control. Origins of rectangular or square shaped pattern are directly applied to calculate the displacement/strain. The results obtained are presented in the form of full-field, strain distribution maps [6]. DIC method is mainly used for static measurements in which tensile and compressive behaviour [7], fracture toughness [8] and the geometrical imperfection effects on mechanical response [6] can be investigated in detail.

Both ESPI and DIC have many applications. DIC was used to study different kinds of materials, such as soft and hard biological tissues [9–12], dental materials [13] or materials for high temperature applications [14]. It was also applied for structural dynamics assessment [15], residual stress and its error estimation [16], in situ monitoring of additive manufacturing [17], and many others. On the other hand, ESPI has been recently employed for assessing the integrity of structural materials, evaluating residual stress fields in pipes, measuring surface displacements during inflation testing of ocular tissues or to assessing strain in bone-implant interfaces [18].

This paper presents several attempts to use such techniques for damage evaluation of materials subjected to different types of loading.

2. Materials and Methods

Two types of plane specimens were tested. The first one, just the standard, and the second with a hole located in the central part of the gauge length. Engineering drawings of specimens designed for fatigue tests are shown in Fig. 1.

Two kinds of materials were tested. The specimens of the first material were manufactured using X10CrMoVNb9-1 (P91) polycrystalline steel. This is a low carbon, creep-resistant steel, typically used for tubes, plates and structural components in the power plant industry. The content of alloying elements in P91 includes C (0.2–0.5)%; Mn (0.3–0.6)%; Cr (8–9.5)%; Mo (0.85–1.1)%; V (0.18–0.25)%; Ni (< 0.4)%; Cu (< 0.3)%; Si (0.08–0.12)%; S (< 0.01)%. The specimens of the second material in question were produced applying nickel super-alloy of the following chemical composition: C – 0.09%, Cr – 8.8%, Mn – 0.1%, Si – 0.25%, W – 9.7%, Co – 9.5%, Al – 5.7%, Ta+Ti+HF – 5.5%.

The fatigue tests for both tested materials were force controlled with zero mean level and a constant stress amplitude with a frequency of 10 Hz or 20 Hz using the MTS-810 hydraulic testing machine. The tests for P91 steel were carried out in the range of stress amplitude from $\pm$380 MPa to $\pm$640 MPa. In the case of nickel super-alloy in each test the maximum cyclic stress range and stress amplitude were equal to 600 MPa and 300 MPa, respectively. Both these parameters were lower than the yield point of the material in question.

The fatigue development was monitored by Dantec Dynamics Q100 ESPI system and DIC Aramis 12M equipped with lenses of total focal length of 75 mm and calibration settings appropriate to the measuring area equal to 170$\times$156 mm.

In order to eliminate vibrations of the testing machine during ESPI optical measurements, the loading process of the specimen was executed manually using a special device designed originally by the IPPT PAN workers. The loading program is presented in Fig. 2.

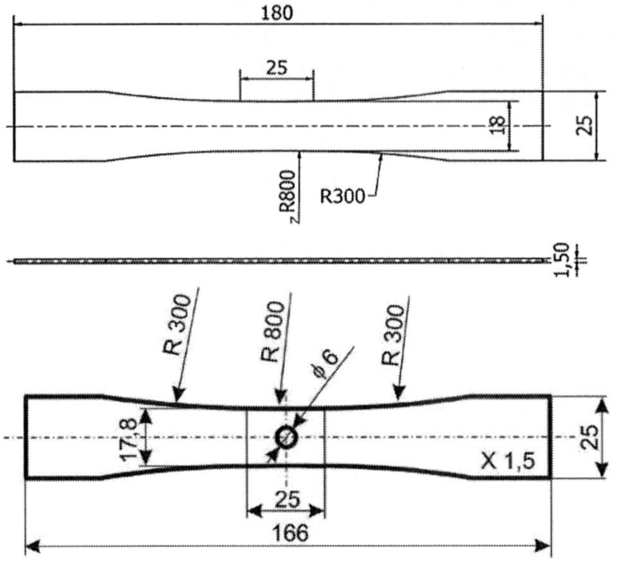

Fig. 1. The geometry of the specimens: top – typical plane; bottom – with central hole in the middle of gauge length (units are in mm).

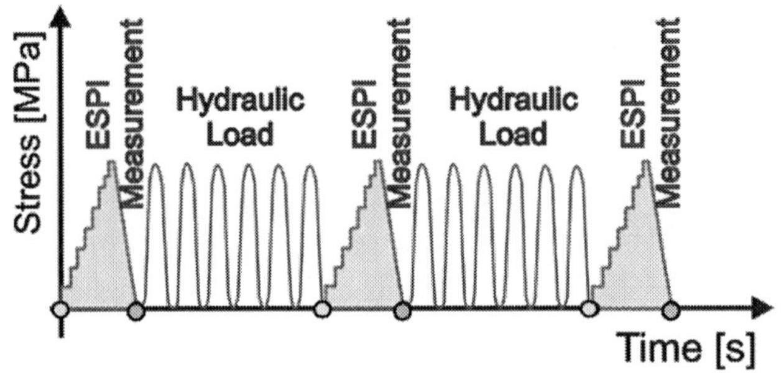

Fig. 2. Loading program for cyclic loading tests supported by ESPI system.

3. Results

3.1. The results for the P91 steel

The purpose of tests on P91 steel was to provide experimental data on the evolution of micro-strain regime during high cycle fatigue crack initiation. The whole fatigue process was divided on the blocks of cycles and carried out using the hydraulic servo-controlled testing machine. The process of cyclic loading was interrupted several times after selected numbers of cycles in order to perform displacement measurements by means of the ESPI camera. Strain distribution maps at the maximum load applied on the plane specimen surface using ESPI are presented for increasing number of the loading cycles in Figs. 3, 4, 5, 6 and 7. Three types of specimens were tested: specimen with relatively rough surface, electropolished specimen and specimen with the hole. The different stages of the fatigue process, from the beginning up to the moment of crack initiation and subsequent propagation are presented. During the process of cyclic loading the material deforms heterogeneously and numerous strain concentration spots are visible in the case of rough specimen, Fig. 3. The lateral profiles of the maximal cross-sectional strain distribution for increasing cycles number for rough specimen on the surface are presented in Fig. 4. It is expected that the microstructure of the material plays an important role in the strain localization. In order to reduce roughness and remove some modification of the material properties resulting from electro-machining (local heating-cooling), the surface of specimen was electro-polished in the vicinity of cracks generated in the fatigue process. The strain distribution maps are presented in Fig. 5. The lateral profiles of the maximal cross-sectional strain distribution for increasing cycles number for electropolished specimen are presented in Fig. 6. As it is seen from the profiles, Fig. 4 and Fig. 6, the places of damage initiation are easy for identification. They are well represented by the significant increase in strain.

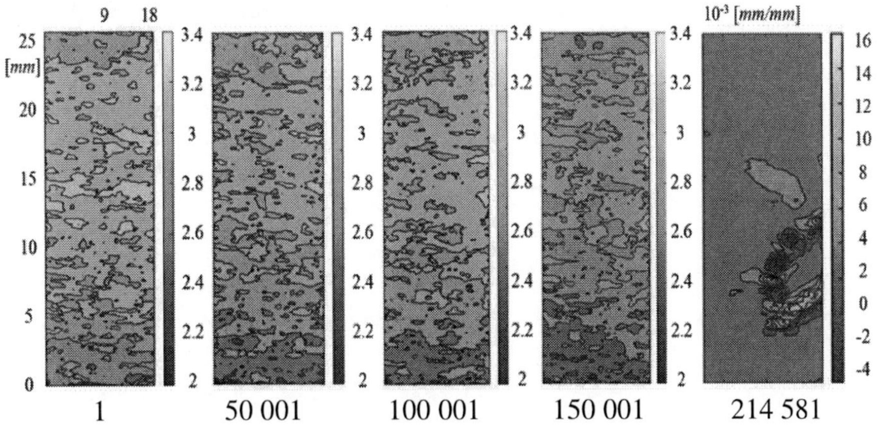

Fig. 3. Strain distribution maps – rough specimen on the surface.

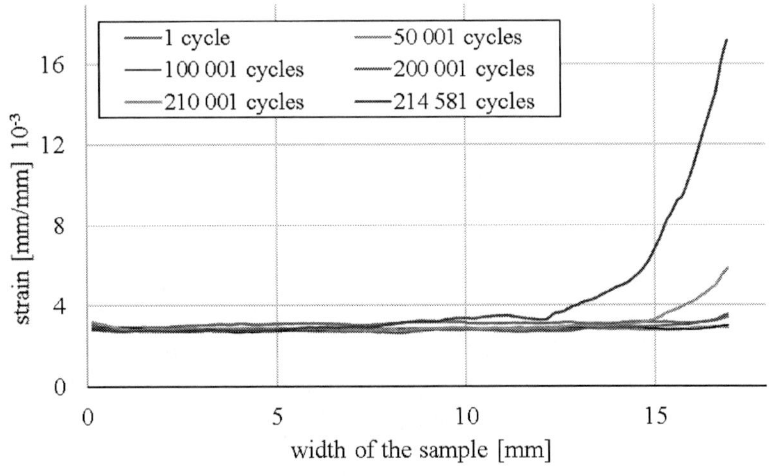

Fig. 4. Lateral profiles of the maximal cross-sectional strain evolution.

In the case of specimen with the hole, in the first cycle the strain accumulation spot is visible near the hole, (Fig. 7, 1 cycle). In the next cycles the material is strengthened (Fig. 7, 2 cycles), but the local zones of strain accumulation in the vicinity of the hole can be still observed, however, they have a slightly lower level in comparison to the first cycle. The crack localisation on the left side of the hole is presented in Fig. 7 after 78 057 cycles.

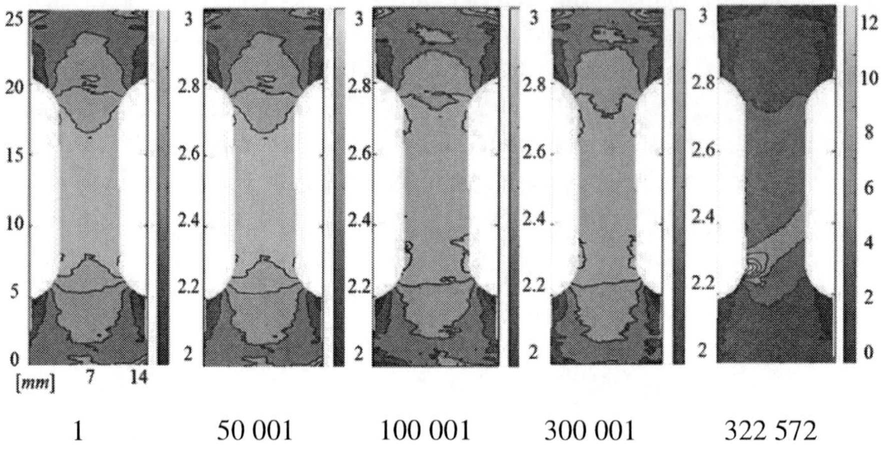

Fig. 5. Strain distribution maps – electropolished specimen.

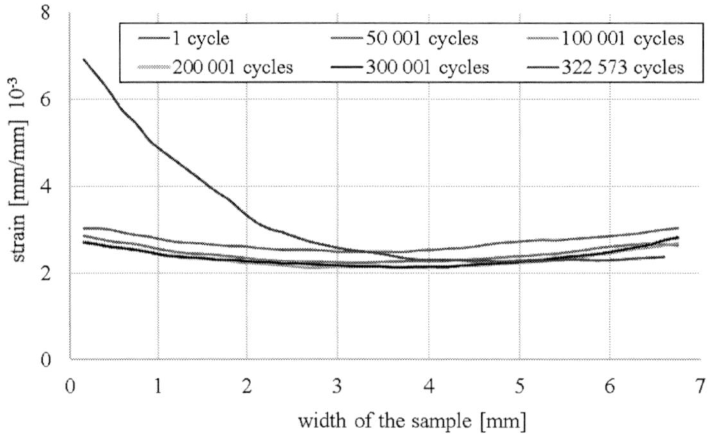

Fig. 6. Lateral profiles of the maximal cross-sectional strain evolution.

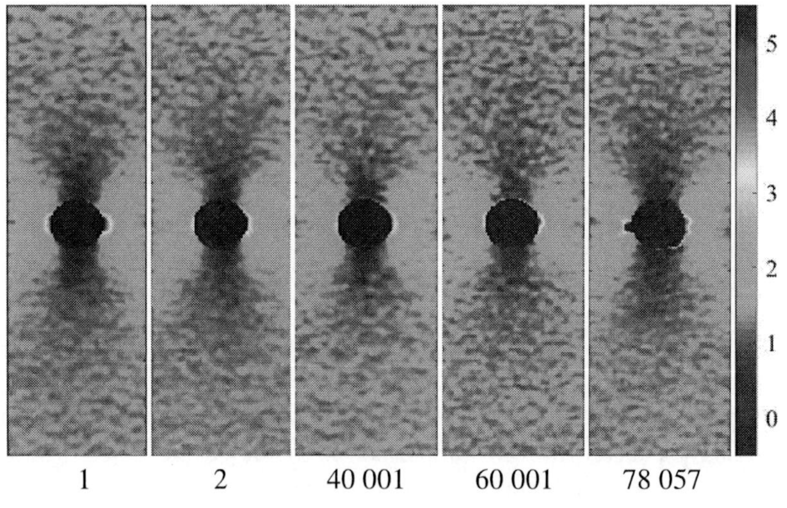

Fig. 7. Strain distribution maps – specimen with the hole.

Introduction of a geometric discontinuity in the form of the hole in the central part of specimen introduces advantageous condition (stress and strain concentrations) for crack initiation. On the basis of strain fields specified for different forces a local hysteresis curve can be generated for selected points of the specimen, (Fig. 8 (1–8)). The local stress cannot be measured directly in the experiment, however, the hysteresis curve shows the nominal stress versus local strain.

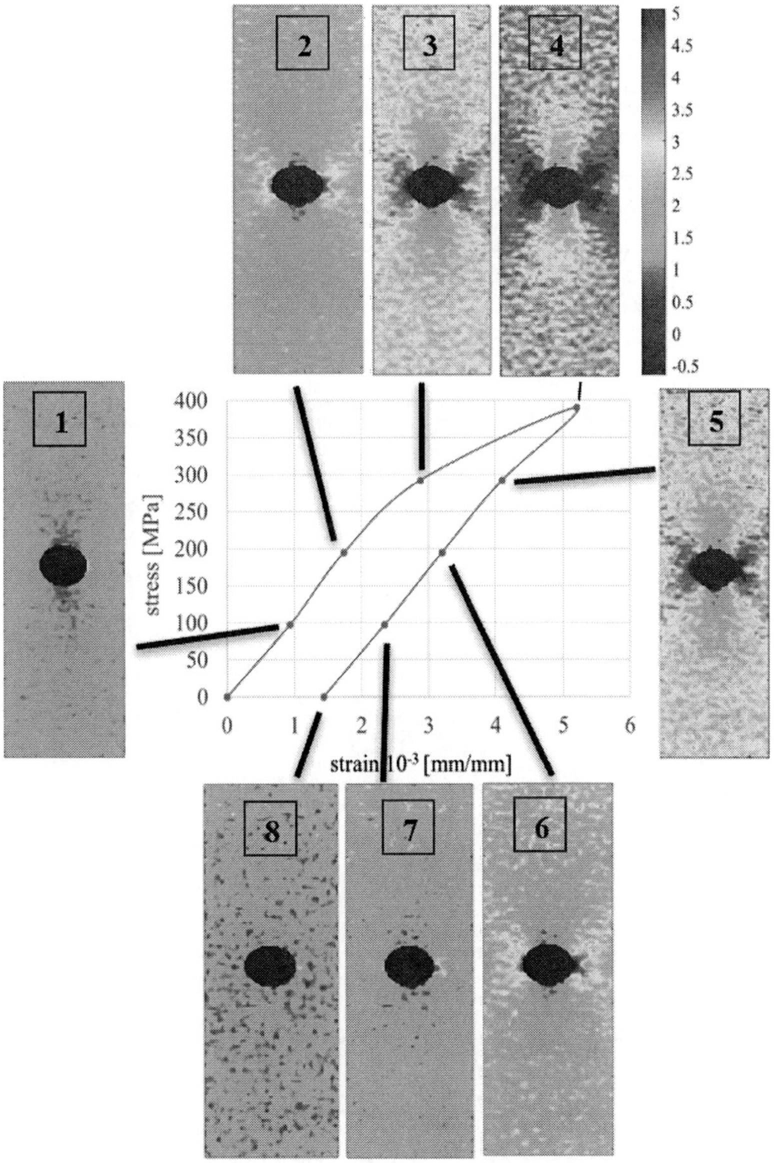

Fig. 8. Hysteresis curve showing the single loading cycle paused at selected stress levels during the strain distribution maps capturing.

The latter corresponds to the maximum stress/strain concentration point. The aim of the experiment was to follow the distribution of local strain fluctuations during a loading/unloading cycle. The strain distribution maps corresponding to the selected points of the curve are also presented in Fig. 8. Near the hole, the local zone of strain concentration was created. As shown in Fig. 8 (3, 4, 5), during specimen loading, the strain concentration occurs in shape of "V". Two plasticity bands inclined at certain angle are created. The initial state of fatigue crack nucleation mechanism begins by a formation of shear bands. This is a result of deformation mechanism that localizes large shear strains. As a consequence of the inhomogeneous plastic deformation at the root of the hole, the residual strain is left on the surface, Fig. 8 (8).

The ESPI results in the form of local strain values are in good agreement with the numerical results that can be found in [1].

In order to check the effectiveness of the ESPI method, additional tests were performed under the same conditions on the plane specimens without holes using also Digital Image Correlation (DIC). Each specimen fractured within 1000 cycles from the crack initiation. It should be highlighted, that the ESPI measurements enabled to observe damage almost close to the specimen fracture only. In order to better recognition of this problem, tests were also performed for the stress amplitudes equal to 550 MPa, 600 MPa and 630 MPa, however, the same difficulties for optical crack detection were met. Therefore, strain maps for 500 MPa were presented only, Fig. 9.

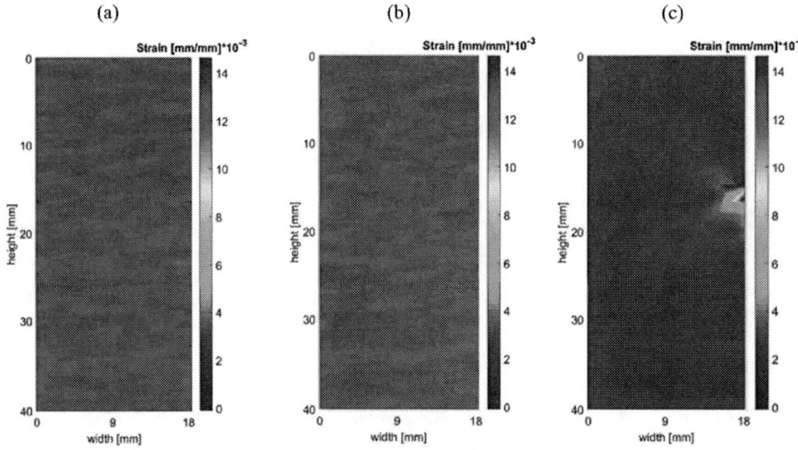

Fig. 9. Comparison of ESPI measurements performed for the stress amplitude equal to 500 MPa with the unified scale after: 100 000 cycles (a); 250 000 cycles (b); 299 207 cycles (c).

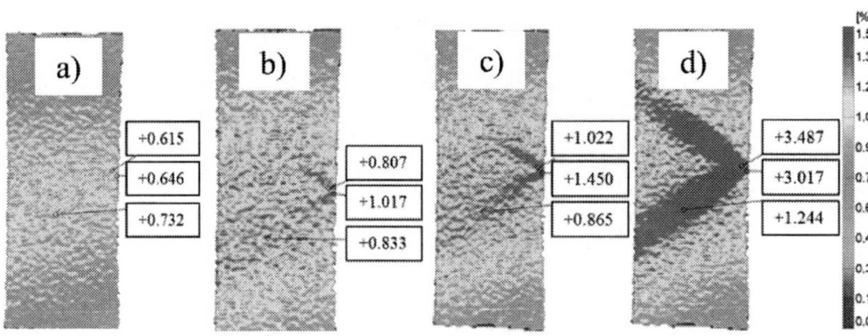

Fig. 10. DIC measurements performed for the stress amplitude equal to 500 MPa with unified scale after: 1 cycle a); 100 000 cycles b); 250 000 cycles c); 301 251 cycles d).

In order to check applicability of the DIC methodology, measurements were performed for the same stress amplitude as that for the ESPI used, i.e. 500 MPa, after selected number of cycles to the specimen fracture.

Unlike the ESPI measurements, DIC technique captured a strain localisation area after just one cycle (Fig. 10a). The subsequent evolution of fatigue damage to 100 000 cycles enabled to clearly indicate the area of potential crack initiation (Fig. 10b) and its development (Fig. 10c-d) up to specimen fracture.The effectiveness of DIC method was further confirmed by performing an additional measurements for the stress amplitude equal to 600 MPa, 630 MPa and 640 MPa. For the each measurement, the area of potential crack initiation was precisely captured after initial number of fatigue cycles and specimens fractured exactly in this specified region. It should be mentioned however, that the application of highest values of stress amplitude required to modify the scale in order to clearly present the strain distribution.

3.2. The results for the nickel super-alloy

Fatigue investigations on nickel super-alloy were also carried out under force control using the MTS-810 hydraulic testing machine. In each test the maximum cyclic stress range and stress amplitude were equal to 600 MPa and 300MPa, respectively.

The first cycle was conducted manually, and subsequently, a block of cycles under the frequency of 10 Hz was carried [19]. The process of cyclic loading was interrupted several times in order to perform displacement measurements by means of the ESPI camera. The experimental programme provided displacement measurements at the beginning of test and after 20 000, 40 000 and 50 000 cycles. The number of cycles to failure was N_f = 54 315.

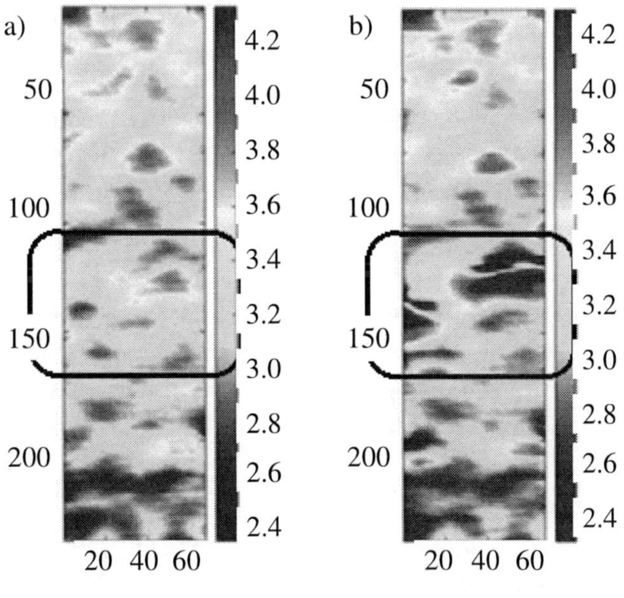

Fig. 11. The field strain distribution along Y axis corresponding to the acting stress direction. Measurements performed: a) in the first loading cycle; b) after 50 000 cycles. Scale is matched to the extreme strain values in the first measurement.

ESPI observations carrying out at various stages of the fatigue degradation represent a status and dynamics of the damage development. They enable a determination of the areas of the greatest stress concentrations and reflect a local character of the fatigue damage initiation.

The field strain distributions along the vertical axis corresponding to the acting stress direction in the specimen are presented in Fig. 11. The figure shows strain distributions for different stages of the fatigue process, i.e. after: (a) first cycle; and (b) 50 000 cycles. It has to be mentioned, that all these maps were obtained for the same scale in order to enable a direct comparison of the results achieved. As it is seen, the method enables identification of places where damage initiates. Fig. 12 demonstrates a location of decohesion on the specimen gauge length, that well agrees with the largest displacement concentration occurring on the phase maps captured by means of the ESPI measurements.

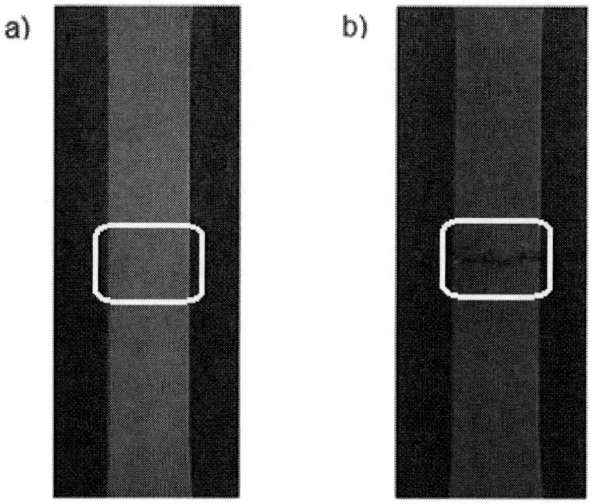

Fig. 12. Fatigue specimen photo: a) before fatigue test; b) after fatigue test (N_f = 54 315 cycles).

In order to find which optical technique is more suitable for damage identification for this material, similar fatigue investigations were carried out in tests supported by the DIC system. These tests were performed at stress amplitude equal to 600 MPa. The loading frequency was 20 Hz, and the image was recorded every 5 seconds, i.e. each map was recorded every 100 cycles. Fig. 13 presents images from selected fatigue cycles. The specimen broke after 46 364 cycles.

For this specimen, the first signs of localization were visible after 45 thousands loading cycles. The increase in deformation at this stage was connected with the formation of a crack in the material, which propagates until the cohesion is entirely lost. On this basis, the moment of crack formation can be estimated, although the detection threshold is lower than that in the technically comparable ESPI method obtained.

4. Conclusions

The paper presents the results of fatigue tests carried out on the P91 power engineering steel and nickel super-alloy with special emphasis on damage development identification. Contemporary novel optical techniques were used for damage initiation detection and further crack propagation monitoring. The results of damage monitoring during fatigue tests supported by Digital Image Correlation or Electronic Speckle Pattern Interferometry proved their great suitability for effective identification of places of damage initiation.

Both optical techniques were assessed in terms of their effectiveness to fatigue damage development monitoring. It was found, that DIC enables to monitor the fatigue behaviour and accurately indicates the area of potential failure within the early stage of fatigue damage development.

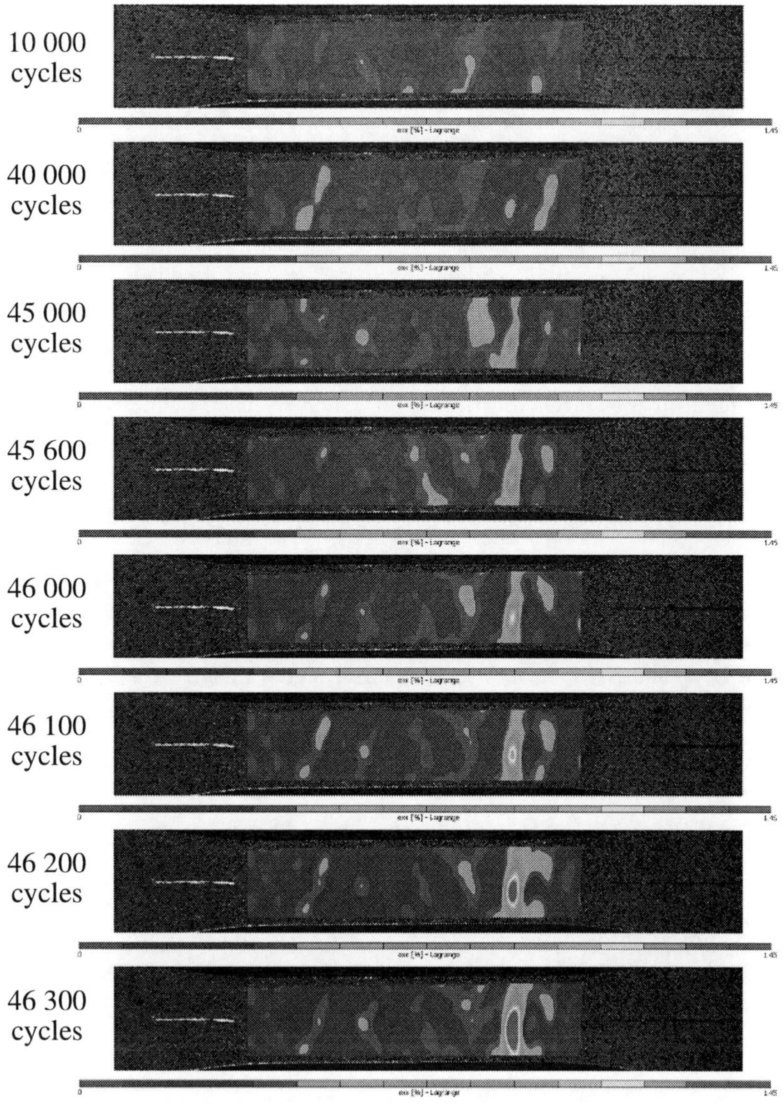

Fig. 13. Images showing strain distribution maps captured by DIC.

Contrary to this, the application of ESPI method was not so successful. It also enabled to indicate a location of potential damage area, however, significantly later than DIC. The main limitation of ESPI technique usage results from its high sensitivity which procures many difficulties during working with the servo-controlled hydraulic testing machines. Such machines generate high frequency vibrations during experiments due to oil flow supplying the machine loading systems. The vibrations disturb significantly a work of ESPI cameras and narrows greatly its measuring capabilities.

Acknowledgements

This work has been supported by the National Science Centre through the Grant No 2019/35/B/ST8/03151.

References

[1] Ustrzycka A., Mróz Z., Kowalewski Z.L. and Kucharski S., "Analysis of fatigue crack initiation in cyclic microplasticity regime," *International Journal of Fatigue*, Vol. 131, pp. 105342 (2020). https://doi.org/10.1016/j.ijfatigue.2019.105342.

[2] Kopec M., Brodecki A., Kukla D. and Kowalewski Z.L., "Suitability of DIC and ESPI optical methods for monitoring fatigue damage development in X10CrMoVNb9-1 power engineering steel," *Arch. Civ. Mech. Eng.*, Vol. 21, pp.167-1-13 (2021). https://doi.org/10.1007/s43452-021-00316-1.

[3] Altenbach H., Brünig M. and Kowalewski Z.L., *Plasticity, Damage and Fracture in Advanced Materials*, (Springer, 2020). ISBN 978-0-030-34850-2. https://doi.org/10.1007/978-3-030-34851-9.

[4] Andersson M., "The influence of notches on fatigue of heat treated sintered steel," in: *International Powder Metallurgy Congress and Exhibition* Euro PM2013 in Gothenburg on September 17, (2013).

[5] Chu T.C., Ranson W.F., Sutton M.A. and Peters W.H., "Application of digital-image-correlation techniques to experimental mechanics," *Exp. Mech.*, Vol. 25, pp. 232–244 (1985). DOI 10.1007/BF02325092.

[6] Lord J.D., Digital image correlation (DIC). In: Eaton-Evans J., Dulie-Barton J.M., Burguete RL (eds), *Modern Stress and Strain Analysis. A State of the Art Guide to Measurement Techniques*, British Society for Strain Measurement, pp. 14–15, 2009.

[7] Forster J., Theobald A., Engel S. and Pasmann R., "Using optical measuring system for identification of material parameters for finite element analysis," in: *11. LS-DYNA, DYNAmore GmbH*, Ulm, pp. 1–9 (2012).

[8] Durif E., Réthoré J., Combescure A., Fregonese M. and Chaudet P., "Controlling Stress intensity factors during a fatigue crack propagation using digital imagecorrelation and a load shedding procedure," *Exp. Mech.*, Vol. 52, pp. 1021–1031 (2012). doi 10.1007/s11340-011-9552-6.

[9] Han Y., Kim D.W. and Kwon H.J., "Application of digital image cross-correlation and smoothing function to the diagnosis of breast cancer," *J Mech Behav Biomed Mater*, Vol. 14, pp. 7–18 (2012). doi:10.1016/j.jmbbm.2012.05.007.

[10] Hokka M., Mirow N., Nagel H., Irqsusi M., Vogt S. and Kuokkala V.T., "In-vivo deformation measurements of the human heart by 3D digital image correlation," *J Biomech.*, Vol. 48, pp. 2217–2220 (2015). doi:10.1016/j.jbiomech.2015.03.015.

[11] Juszczyk M.M., Cristofolini L. and Viceconti M., "The human proximal femur behaves linearly elastic up to failure under physiological loading conditions," *J Biomech*, Vol. 44, pp. 2259–2266 (2011). https://doi.org/10.1016/j.jbiomech.2011.05.038.

[12] Zani L., Erani P., Grassi L., Taddei F. and Cristofolini L., "Strain distribution in the proximal Human femur during in vitro simulated sideways fall," *J Biomech.*, Vol. 48, pp. 2130–2143 (2015). doi:10.1016/j.jbiomech.2015.02.022.

[13] Yoon S., Jung H.-J., Knowles J.C. and Lee H.-H., "Digital image correlation in dental materials and related research: A review," *Dental Materials*, Vol. 37, pp. 758-771 (2021). https://doi.org/10.1016/j.dental.2021.02.024.

[14] Berke R. and Lambros J., "Ultraviolet digital image correlation (UV-DIC) for high temperature applications," *Rev. Sci. Instrum.*, Vol. 85, no. 4, (2014). https://doi.org/10.1063/1.4871991.

[15] Niezrecki C., Avitabile P., Warren C., Pingle P. and Helfrick M., "A review of digital image correlation applied to structural dynamics," *AIP Conference Proceedings*, American Institute of Physics, Vol. 1253, pp. 219–232 (2010). http://dx.doi.org/10.1063/1.3455461.

[16] Lunt A.J.G. and Korsunsky A.M., "A review of micro-scale focused ion beam milling and digital image correlation analysis for residual stress evaluation and error estimation," *Surf and Coati Tech*, Vol. 283, pp. 373–388 (2015). https://doi.org/10.1016/j.surfcoat.2015.10.049.

[17] Cunha F.G., Santos T.G. and Xavier J., "In Situ Monitoring of Additive Manufacturing Using Digital Image Correlation: A Review," *Materials*, Vol. 14, pp. 1511 (2021). https://doi.org/10.3390/ma14061511.

[18] Pagliarulo V. and Ferraro P., "New applications of electronic speckle pattern interferometry in novel materials and structures," *Proc. SPIE 11059, Multimodal Sensing: Technologies and Applications*, 1105910, 21 June 2019; https://doi.org/10.1117/12.2527997.

[19] Kopec M., Grzywna P., Kukla D. and Kowalewski Z., "Evaluation of the fatigue damage development using ESPI method," *Inżynieria Materiałowa*, Vol. 4, no. 212, pp. 1–4, (2016). https://doi.org/10.15199/28.2016.4.9.

EAN 2022 60th annual conference on experimental stress analysis

Flexural and Torsional Loading of Integrated Loop Technology Joint

Bohumil Kropík[1], Tomáš Ponížil[2], Anna Malá[3], Tereza Zámečníková[4], Nikola Schmidová[5], Karel Doubrava[6], Milan Dvořák[7] & Tomáš Mareš[8]

Abstract: This contribution deals with the experimental part of the development of an integrated loop technology (ILT) joint. The first development of the ILT was based on tensile and compressive loading scenarios, where the best design was chosen from more different designs. The final design was then loaded with flexural and torsional loading scenarios to identify the behaviour under these conditions and to verify that this ILT design is safe even under these conditions. Both loading conditions were also analysed using the finite element method analysis. All approaches identified similar critical places of design.

Keywords: CFRP; Integrated joint; Filament winding; ILT

1. Introduction

The motive of the work is to improve the development of an easily automatable integrated joint, thus increasing the share of automation in the joining of composite materials. Most spread methods, such as, for example, manual lamination or bonding, are highly dependent on human work, which also affects the quality and repeatability of the mechanical properties of these joints. Some automated techniques for connection have already been used (e.g., wound pin joints of metal inserts as shown in figure (Fig. 1) or as presented in [1] and many others), but, for example, the connection of a composite to the composite is done through an additional metal part, which causes an increase in weight of the construction. Thus, a better solution is the integrated loop technology (ILT) joint, which allows the connection of the composite tube to any other tube with the possibility of automated production without necessarily added parts.

The first generation of specimens was discussed in [2]. This development was supported by the finite element method (FEM) analysis, which was presented in [3]. Based on the results of the first-generation specimens, some improvements were made and the FEM analysis of the second-generation specimens was discussed in [4]. The experimental part of the work was supported by a wide range of monitoring methods, such as acoustic emission, which was discussed for the first-generation of specimens in [5] and together with digital image correlation, strain

[1] *Ing. Bohumil Kropík; Department of Mechanics, Biomechanics and Mechatronics, Fac. of Mech. Engineering, Czech Technical University in Prague; Technická 4, 166 07 Prague, Czech Republic;* bohumil.kropik@fs.cvut.cz

[2] *Ing. Tomáš Ponížil; Compo Tech PLUS, spol. s r. o., Nová 1316, 342 01 Sušice, Czech Republic;* tponizil@gmail.com

[3] *Ing. Anna Malá; Department of Mechanics, Biomechanics and Mechatronics, Fac. of Mech. Engineering, Czech Technical University in Prague; Technická 4, 166 07 Prague, Czech Republic;* anna.mala@fs.cvut.cz

[4] *Ing. Tereza Zámečníková; Department of Mechanics, Biomechanics and Mechatronics, Fac. of Mech. Engineering, Czech Technical University in Prague; Technická 4, 166 07 Prague, Czech Republic;* tereza.zamecnikova@fs.cvut.cz

[5] *Ing. Nikola Schmidová; Department of Mechanics, Biomechanics and Mechatronics, Fac. of Mech. Engineering, Czech Technical University in Prague; Technická 4, 166 07 Prague, Czech Republic;* nikola.schmidova@fs.cvut.cz

[6] *Ing. Karel Doubrava, Ph.D.; Department of Mechanics, Biomechanics and Mechatronics, Fac. of Mech. Engineering, Czech Technical University in Prague; Technická 4, 166 07 Prague, Czech Republic;* karel.doubrava@fs.cvut.cz

[7] *Ing. Milan Dvořák, Ph.D.; Department of Mechanics, Biomechanics and Mechatronics, Fac. of Mech. Engineering, Czech Technical University in Prague; Technická 4, 166 07 Prague, Czech Republic;* milan.dvorak@fs.cvut.cz

[8] *doc. Ing. Tomáš Mareš, Ph.D.; Department of Mechanics, Biomechanics and Mechatronics, Fac. of Mech. Engineering, Czech Technical University in Prague; Technická 4, 166 07 Prague, Czech Republic;* tomas.mares@fs.cvut.cz

measurement, and video analysis in [6] where the development of the ILT was presented and second-generation specimens were discussed. The ILT joint was used in the construction of a bicycle frame for enduro-bike races, which was presented in [7].

Fig. 1. Integrated joint with wound pins T-IGEL (JEC Europe composites show & conferences Paris, May 2015, exhibition stand Teufelberger).

2. Specimen description

Based on the development of the ILT joint, the most suitable design for the tube with the ILT joints was identified, which is described in [6]. Since the development was based on tensile and compressive loading scenarios, experimental verification under torsional and flexural loading conditions was the next logical step in identifying any potentially critical behaviour due to the possible complex loadings of the real construction. This led to the creation of two different types of specimens. The layout of the tube layers is given in Table 1. Since the production process of the tube with joint is the filament winding, the layers are numbered from the mandrel to the outside surface of the tube.

Table 1. Specimen layout [6].

Layer position	Layer type
First layer	Hoop
Second layer	A1
Third layer	A1
Fourth layer	SN(45)
Fifth layer	SN(45)

The specimens consist of three different types of layers, which are indicated in the figures (Fig. 2 – Fig. 4). These are the layers A1, Hoop and SN(45), which are described in more detail in [6]. The specimen of the first type corresponds to the final design developed previously, its length is about 350 mm and it is shown in the figure (Fig. 5). The second type was the same design cut in half as shown in the figure (Fig. 6). The cross section of the specimens is a tube shown in figure (Fig. 7).

Fig. 2. Area A1 of the specimen [6].

Fig. 3. Area Hoop of the specimen [6].

Fig. 4. Area SN(45) of the specimen [6].

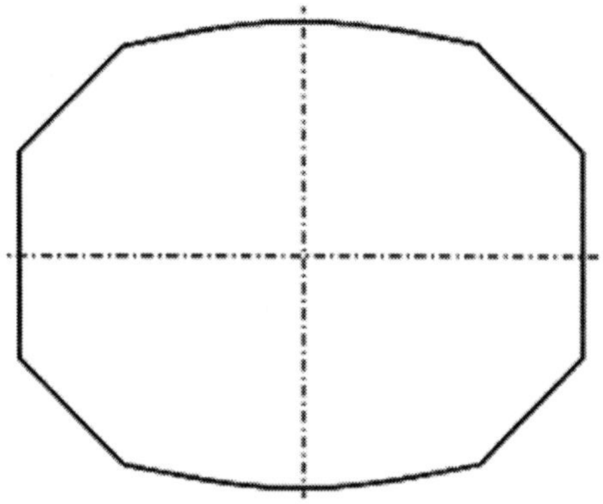

Fig. 5. Cross section of the tube.

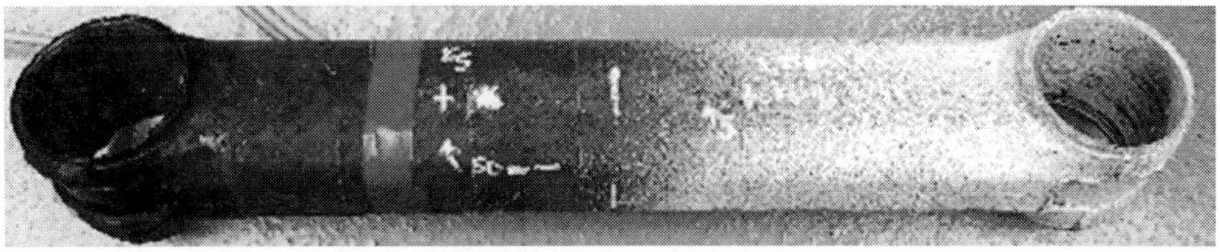

Fig. 6. Typical geometry of specimens of the first type (torsional loading).

Fig. 7. Typical geometry of specimens of the second type (flexural loading).

3. Flexural loading

This section deals with the flexural loading of specimens and describes experimental measurement and numerical analysis. In this contribution, we use Normalized Strain, Normalized Bending Moment and Normalized Torque. All of these quantities are related to their maximal measured values during the experimental part of the relevant loading scenario.

3.1. Loading case

The specimen was placed on a rigid steel shaft and the loading force (F) was applied using a textile strap. Part of the specimen under the strap was supported by a 3D printed plastic insert. The experimental setup is shown in the figures (Fig. 8 – Fig. 9), where dimension A shows the width of the strap, which was 26 mm, and dimension B shows the position of the strain gauges (SGs), which was usually 40 mm. The identification of strain was carried out using two linear 1-axis strain gauges (SG1 and SG2) installed on the narrower surface of a tube. The length of the specimen is marked as $L/2$.

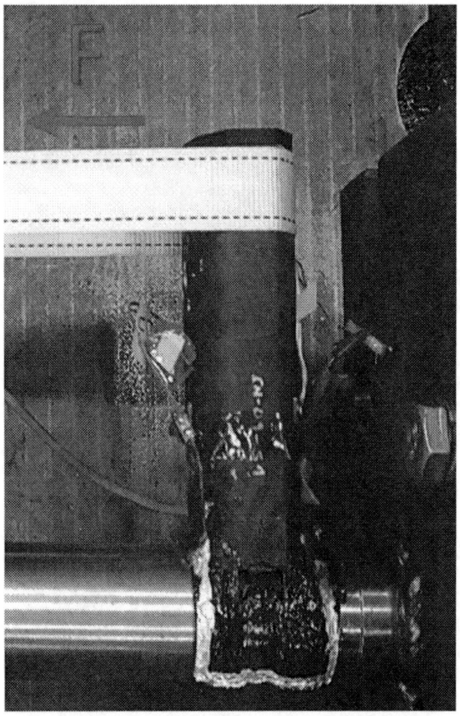

Fig. 8. Picture of the loading setup.

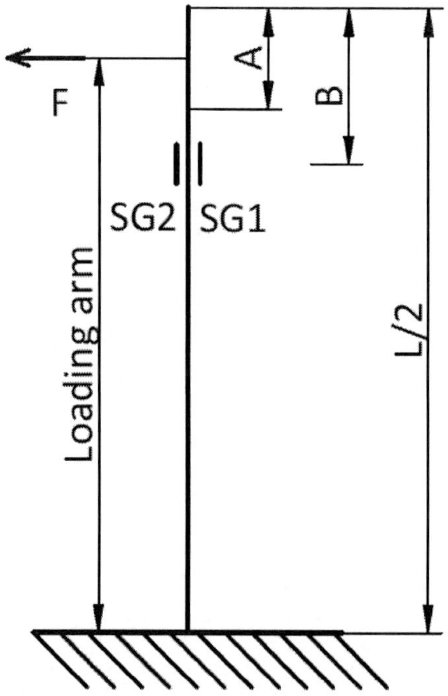

Fig. 9. Simplified loading setup scheme.

3.2. FE analysis

A detailed explanation of the FE analysis will be published separately. In this contribution, we will only explain the basic ideas to compare the results with the experimental part. The contact between the strap and the specimen was simplified in the FE model. This part was modelled as rigid, and the insert was not included in the model. The loading was applied as a force through this rigid part, and the inner tube connected by the ILT was restrained in all movements and rotations. The SGs were modelled at the position B = 40 mm and an inverse reverse factor (IRF) was used to identify the critical places of the specimen.

3.3. Experimental measurement

The experimental measurement was carried out in displacement-controlled mode with a loading speed of 8 mm/min. In total, 14 specimens were subjected to this loading scenario. For some specimens, the position of the SGs was slightly changed, which is described together with the length of each specimen in Table 2. The marker "x" is used for specimens where the SGs were not installed. With knowledge of the measured force (F) and length of the specimen ($L/2$), respectively, the length of the loading arm (L_{LA}), the bending moment (M_o) was calculated from (1). From the SGs, the bending strain was calculated using (2).

$$M_o = F \cdot L_{LA} \tag{1}$$

$$\varepsilon_o = \frac{\varepsilon_{SG_1} - \varepsilon_{SG_2}}{2} \tag{2}$$

Table 2. Specimens lengths and SGs positions.

Specimen number	SGs position "B" [mm]	Specimen $\frac{L}{2}$ [mm]
SN_09_8	40	174
SN_09_9	40	173
SN_09_10	38	171
SN_09_11	40	171
SN_09_12	34	181
SN_09_13	40	175
SN_09_14	34	176
SN_09_15	40	174
SN_09_30	x	176
SN_09_31	x	173
SN_09_32	x	174
SN_09_33	x	172
SN_09_34	33	174
SN_09_35	32	173

3.4. Results and discussion

The results of the presented FE analysis are at 62% of the maximal force measured during the experimental part. The change in shape of the tube is shown in the figure (Fig. 10). In the figures (Fig. 11 and Fig. 12), there are IRF values on the tension and compression sides of the tube, respectively. The red colour in the figures (Fig. 11 and Fig. 12) means that the element is considered by the IRF criteria as failed.

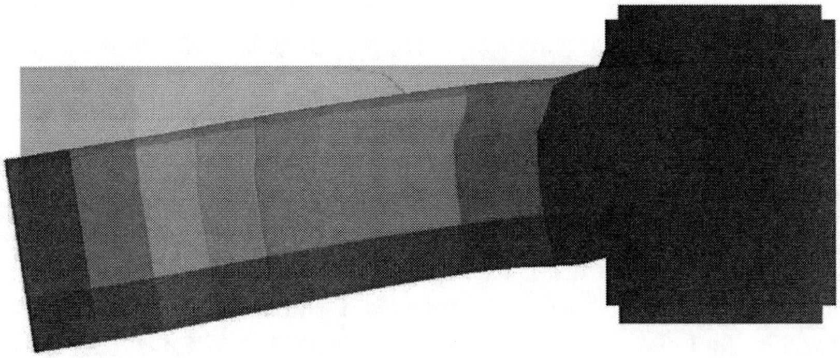

Fig. 10. Bending of the specimen after applied load

In the experimental part, all specimens had the final failure on the compressive side of the ILT joint, as shown in the figure (Fig. 13). The buckling problem was also monitored during this experiment on the compressive side of the specimen. The average Normalized Bending Moment strength identified experimentally was $M_o = 72.9 \pm 11.6\%$.

Fig. 11. Tension side of the specimen IRF.

Fig. 12. Compression side of the specimen IRF.

Fig. 13. Typical specimen failure during the flexural test.

The behaviour of the specimens with the ILT joints was identified under the flexural loading scenario. FE analysis showed critical places with the use of the IRF factor, and the experimental part monitored the final failures of the specimens and the strength of the specimens. All specimens had a final failure on the compressive side with loss of stability of the specimen tube. The moment-strain behaviour was identified by both FE analysis and experiment. The comparison is shown in Fig. 14. The FE analysis result is more rigid than the experiment, which is the expected result, since all clearances were not modelled. The Exp average curve presents the linearization of the average value of the first 250 measured points of all measured specimens. It was picked to include only linear part of the experimental curves to extrapolate linear behavior. The difference between the specimens is most likely due to the location of the SGs close to the loading area and the experimental setup with the use of a 3D printed insert. The composite tube after applying the load probably deformed over the insert, causing differences between the measured strains. In addition, the length of specimens varied and the SGs position was measured from the cut edge. This means that the position of the SGs was always a bit different with respect to the crossing of the tows on the specimen surface. The manufacturing process and the repeatability of it taking into account manual cutting of the specimens and the installation of the SGs could also affect it.

We must also consider the influence of the shape of the cross section.

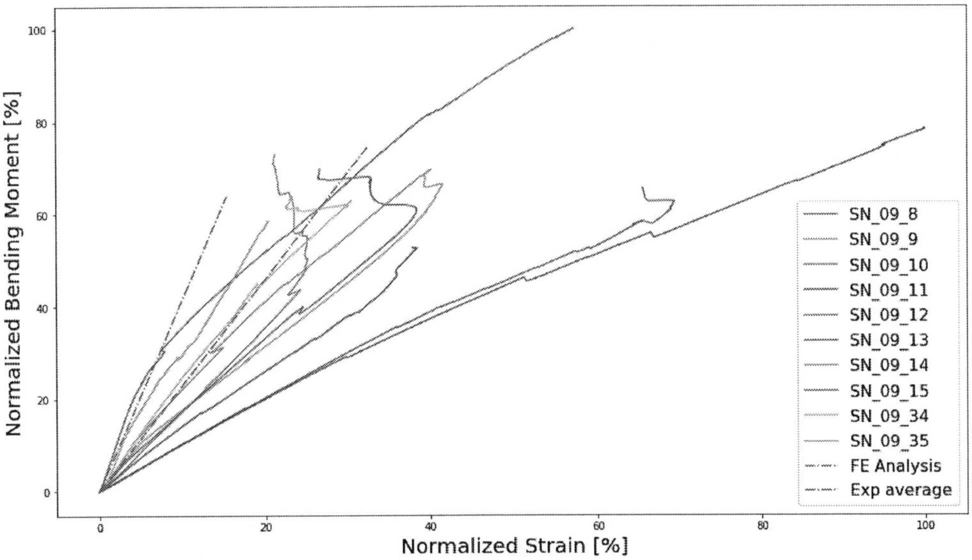

Fig. 14. Comparison between specimens and FE analysis.

4. Torsional loading

This section deals with the torsional loading of specimens and describes experimental measurement and numerical analysis.

4.1. Loading case

The experimental setup for the torsional loading scenario consists of two separate parts. One side of the experimental machine is the rigid part shown in the figure (Fig. 15) on the left side. The other side is the hydraulic load actuator itself, which applies the load as shown in the same figure. The specimen is attached to the machine through pins P1 and P2 between these two parts. A simplified scheme of the setup is shown in the figure (Fig. 16). Usually, one shear SG was used, located on the wider side of the tube in position marked C. The basic position was in the middle of the tube in a distance $L/2 = 175$ mm.

Fig. 15. Setup of the torsional experimental measurement.

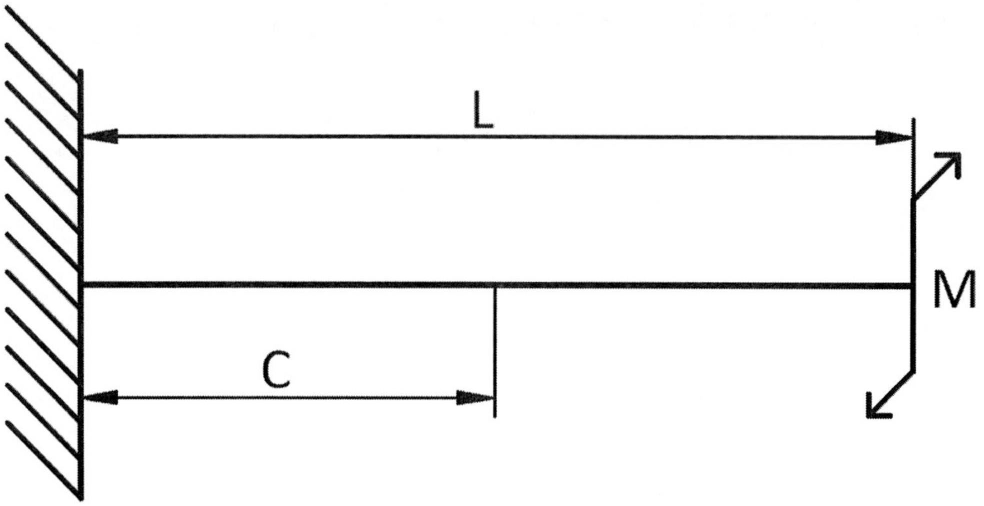

Fig. 16. Simplified scheme of the experimental setup.

4.2. FE analysis

The FE model was simplified in comparison to the described loading case, since most of the connections were not modelled. The specimen was analysed with steel pins inside the ILT connections. On one side of the specimen, the pin was restrained from all movements and rotations, and the other was used to apply the axial torque to the specimen. Two shear SGs were modelled on opposite sides of the wider surface of the tube, as is also shown in the figure (Fig. 17). The IRF factor was used again to identify critical places in this design.

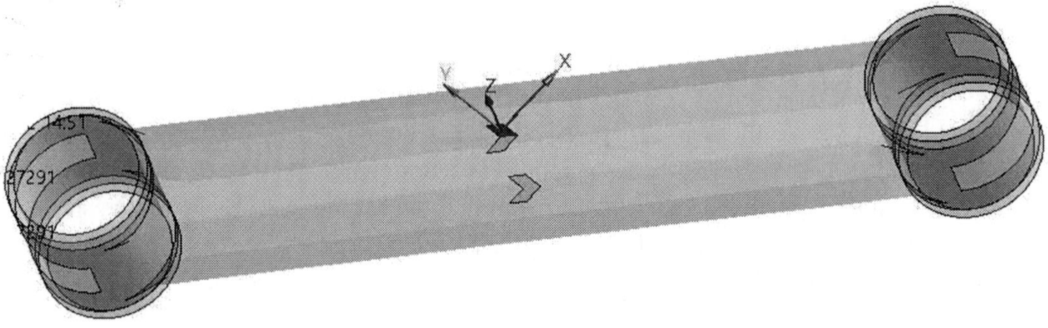

Fig. 17. Positions of the SGs in the FE analysis.

4.3. Experimental measurement

During the experimental procedure, the loading was carried out in displacement-controlled mode as a rotation of 2°/min. In total, 10 specimens were subjected to this loading scenario. All specimens had carbon fibre sensors (CFSs) integrated into their structure. Results from these sensors will be published separately. Because of these CFSs it was necessary to change the position of some SGs. In general, this should not be a problem, as the torsional loading is constant for the entire tube. In the case of CFRP composite material loading, local strain behaviour might affect the result. Thus, the position was changed on the same surface first and some specimens had the SGs on a narrower and wider surface to compare if there is a major difference in strain on these surfaces of the specimen. The positions of the SGs are shown in Table 3. For the 90° rotational location from the wider surface to the narrower surface, there is an "R" marker in Table 3 and the doubled positions (2x) have a full bridge connection of two SGs, one on each side of the tube. All specimens had an installed rectangular strain gauge rosette, except SN_09_40 and SN_09_41, which had the shear SGs. Torsional part of the strain (ε_T) was calculated from the rectangular strain rosette as in (3) because the loading was perpendicular to the position of the SGs.

$$\varepsilon_T = \frac{\varepsilon_c - \varepsilon_a}{2} \quad (3)$$

Table 3. Specimens SGs positions.

Specimen number	SG position "C" [mm]
SN_09_18	165
SN_09_19	175
SN_09_20	175
SN_09_21	175
SN_09_40	2x 130 (R) + 1x 175
SN_09_41	2x 125 (R) + 1x 170
SN_09_42	120
SN_09_43	120
SN_09_44	120
SN_09_45	120

4.4. Results and discussion

The result of FE analysis is shown at a level of 100% of the Normalized Torque in the figure (Fig. 18). The red colour in the figure (Fig. 18) means that the element is considered by the IRF criteria as failed.

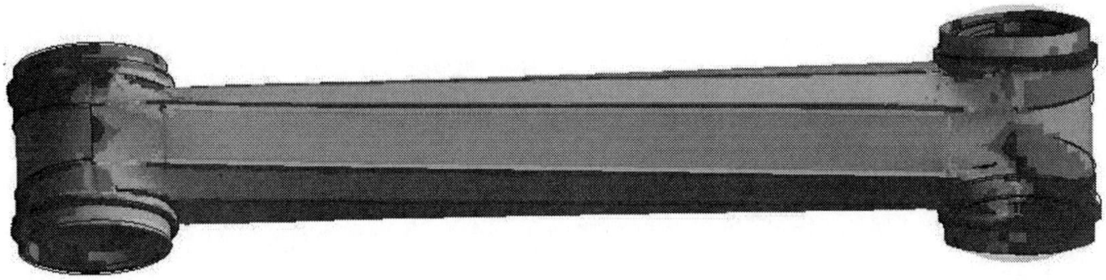

Fig. 18. Torsional specimen IRF.

Two different types of final failue were observed during experimental part of torsional loading, which are shown in the figures (Fig. 19 and Fig. 20). These are the final failures of the tube in the direction of the SN(45) layer, which was probably initiated by the failure of the ILT loop tows (as indicated by the FE analysis) and the final failure of the tube close to the ILT in the cross-direction. The experimentally identified average Normalized Torque strength τ_{max} was 75.2±11.4%.

The behaviour of the specimens with the ILT joints was identified under the torsional loading scenario. FE analysis result showed critical places with the use of the IRF factor, and the experimental part monitored the final failures of the specimens. Two different types of final failures were observed. The torque-strain response was identified by both the FE analysis and the experiment. The comparison is shown in the figure (Fig. 21). Torque-strain behaviour obtained from FE analysis is more rigid than the one obtained from experiment, which is an expected result, since all clearances were not modelled. From experimental data, it could be seen that there appears to be some relation with the position of the SGs, but apparently also some other influences can be seen. These could be, for example, sensors integrated in different places inside the specimens but also many others from the ones discussed in subsection 3.4. Most of the measured strain responses are quite similar regardless of position. A separate comparison of rotated SGs is given in the figure (Fig. 22), where we can see that the position of SG, texture and cross-section of the specimen can affect the strain response as both specimens act differently.

Fig. 19. Typical final failure of the specimen in the direction of the SN(45) layer.

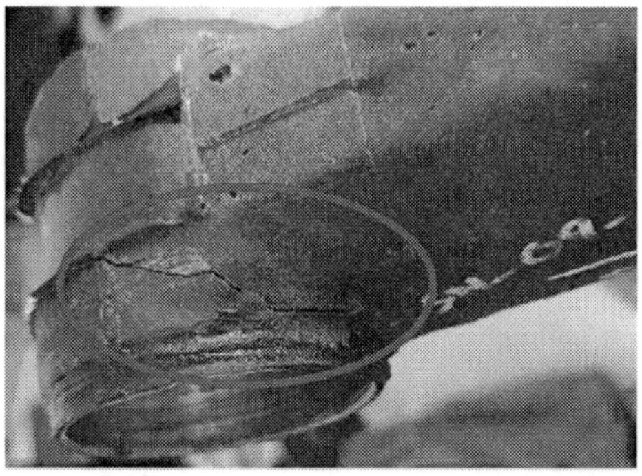

Fig. 20. Typical final failure of the specimen close to the ILT.

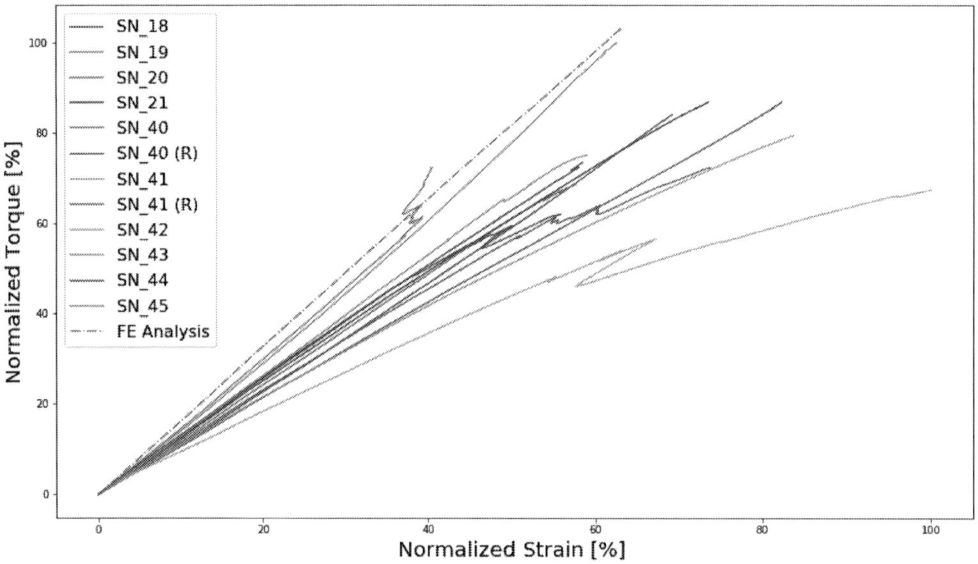

Fig. 21. Measured Normalized Strain compared to FEM analysis.

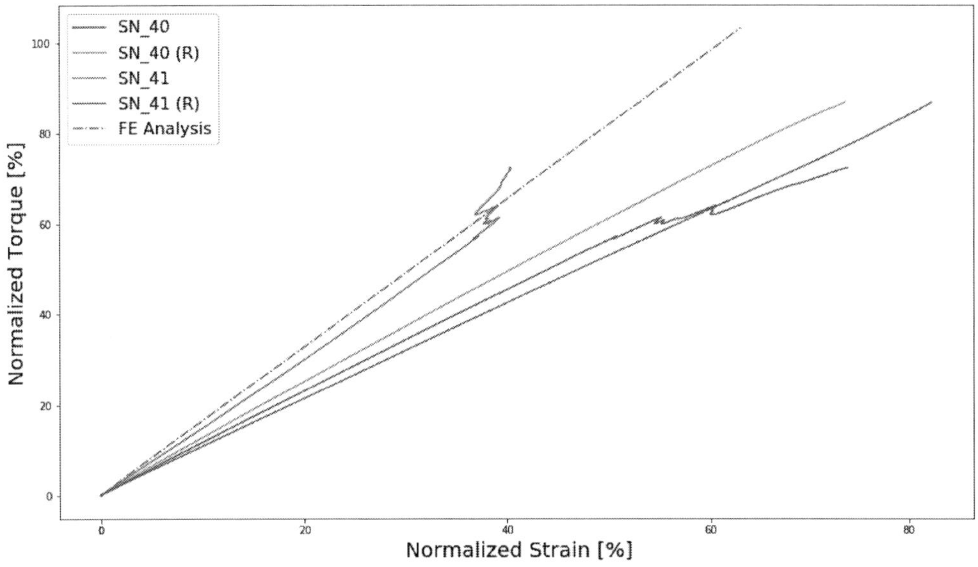

Fig. 22. Measured Normalized Strain of rotated and regular SGs compared to FEM analysis.

5. Conclusion

The critical places of the tubes with the ILT joints were identified by experimental measurements and FEM analysis. Both approaches were in agreement with the identified critical places, which were usually in the tube area close to the ILT joint or rarely in connection with a probable ILT tow failure. Critical places in the area of the specimen where the tube design changes into the ILT joint were previously identified during development.

Unlike the torsional loading, we had an idea about the damaging of the specimens under the flexural loading from the tensional and compressional loading scenarios done before. In principle, the bended beam has one side tensioned and the other side compressed. Since we had all previous failures near the joints, we did not expect changes in behaviour. Actually, failure was expected on the compression side, as the maximal compressive strength was lower than the maximal tensional strength of the specimens during the development of an optimal design under tension and compression. Buckling was also expected, as it was previously observed during the compressive loading scenario. All specimens loaded with flexural loading had a final failure on the compressive side of the tube. The average Normalized Bending moment strength identified experimentally was $M_o = 72.9{\pm}11.6\%$. The difference in measured signals may have been affected by many influences. Since the length of the specimens varied and the SGs position was measured from the cut edge, the position of the SGs was always a bit different considering the cross-shape sections of the composite. Also, the manufacturing process and repeatability of it, the accuracy of manual division of specimens, and the installation of SGs could affect the results. Considering the FE analysis, with different lengths of the specimen, the shape of the bent specimen was probably slightly different. Another unknown that may have affected the measurement was the 3D printed plastic insert. The fit of the insert was designed as a clearance fit. This allowed the tube to slightly deform around the insert, which probably also affected the measured results. As the insert was not included in the FE analysis and should simply support the tube under the strap, its influence could be considered for future investigation. We must also consider the influence of the shape of the cross section and the buckling, which changes the stress distribution.

Torsional loading was a completely new type of experimental loading case for these specimens with the ILT. From previous measurements on the bicycle frame, we identified that this type of load is not as dominant as the bending load, so it was not the primary loading scenario for the development of tube specimens with the ILT. However, its behaviour must be identified for safe use of the ILT in general cases. All specimens loaded with torsional loading had a final failure on random side of the tube near the ILT joint with a rare failure of the ILT tows and the tube layer SN(45). The experimentally identified average Normalized Torque strength was τ_{max} = 75.2±11.4%. Differences in measured behaviour may have been affected by many factors that are mostly similar to those of flexural specimens.

The results of the experimental work performed do not indicate another potential critical place of the given type of specimen design. The significant benefit of composite materials is that the final failure usually does not mean that the material is unable to transfer the loading. The final failure means that the final strength of the specimen was achieved and a significant drop in transferred loading was identified along with a visible failure. Unless the usual behaviour of metal designs, which usually fail completely, all specimens could still transfer the load up to some level which could be considered as a safety factor of this design. Identification of critical places of design and its behaviour is a necessary step for design of safe structures, which means that all described observations should be investigated more in future.

Acknowledgements

This work has been supported by project No. SGS21/151/OHK2/3T/12 of the Grant Agency of the Czech Technical University in Prague and by project No. TJ02000252 of the Technological agency of the Czech Republic.

References

[1] Kulíšek V., "Strength and durability of carbon fibre loops," Available from https://www.researchgate.net/publication/265028651_Strength_and_durability_of_carbon_fibre_loops Accessed: 2022-06-20.

[2] Malá A. et al., "Analýza tuhosti a pevnosti vinutých trubek s integrovanými spoji," in Proceedings of Mechanika kompozitních materiálů a konstrukcí 2020 – sborník, Kamenice, 2020 (ČVUT v Praze, Praha, 2020), pp. 15 – 19. ISBN 978-80-01-06767-3.

[3] Malá A. et al., "Finite Element Analysis of Composite Tubes with Integrated Loop Connections," in Proceedings of computational mechanics 2019, Srní, 2019 (ZČU v Plzni, Plzeň, 2019), pp. 114 – 116. ISBN 978-80-01-06767-3.

[4] Malá A. et al., "Finite element analysis of composite tubes with integrated loop technology joints," in PROCEEDINGS OF COMPUTATIONAL MECHANICS 2021, Srní, 2021 (ZČU v Plzni, Plzeň, 2021), pp. 146 – 148. ISBN 978-80-261-1059-0.

[5] Kropík B. et al., "Detection and Monitoring of Failures in CFRP Specimens with Integrated Joints by Acoustic Emission Method," in 27th Workshop of Applied Mechanics – Proceedings, Praha, 2019 (ČVUT v Praze, Praha, 2019), pp. 20 – 23. ISBN 978-80-01-06680-5.

[6] Kropík B. et al., "Tension and Compression Optimization of Integrated Loop Technology Joint" in EAN 2021 59th conference on experimental stress analysis, Litomyšl, 2022 (ČVUT v Praze, Praha, 2022), pp. 99 – 106. ISBN 978-80-01-06885-4.

[7] Kropík B. et al., "Identification of MTB Enduro Bicycle Frame Behaviour," in 58th Conference on Experimental Stress Analysis - Book of Full Papers, Sobotín, 2020 (ČSM, Brno, 2020), pp. 252 – 262. ISBN 978-80-248-4451-0.

Relating Digital Image Correlation and Infrared Thermography using Image to Image translation

Lovre Krstulović-Opara[1], Alen Grebo[1], Branislav Panić[2] & Marko Nagode[2]

Abstract: The purpose of this paper is to show how an estimation of effective strain (through digital image correlation) can be obtained given thermography data using artificial intelligence. This estimation can potentially be used on the data the network hasn't observed during the training process. In turn this provides the possibility of mapping new thermography acquired image samples to effective strain.

Keywords: Digital Image Correlation; Infrared Thermography; Image to Image Translation; GAN; Pix2Pix

1. Introduction

Knowledge about material inner behavior during loading has always been of particular interest in mechanical sciences. Several methods such as strain gauge measuring, photo elasticity, digital image correlation, etc. have been develop through the years with the goal of determining stress and strain distribution of loaded body. Although strain gauges are appropriate approach when acquiring strain of dynamically loaded structure, they are providing just point wise information about the strain distribution. Digital image correlation (DIC) is providing full field information about dynamically loaded structures up to certain velocities, depending on photographic performances, optics, and sensitivity of digital cameras (CMOS or CCD sensor based). In our previous work [1] we have shown similarity of infrared (IR) image patterns and effective strain distribution obtained by DIC. It was shown that infrared thermography based on the InSb focal point array detector operating on cryogenic temperatures can be used for higher loading velocities where DIC is not providing results. Also, infrared thermography cannot be used for lower loading velocities as thermal effect is dissolved due to cooling of unloaded material.

Here we present an approach that can be used to relate these individual measurements without involving both methods for cases when results of one method is not available.

2. Relating DIC to IR

In [1] the comparison between DIC and IR showed that IR temperature distribution is similar to effective stress obtained by DIC. Fig. 1 depicts two datasets of aluminium alloy specimens during tensile tests with predefined displacement rate of 284 mm/s. These images were obtained by Chronos 1.4 digital camera and Flir SC 5000 MW infrared camera, triggered and capturing simultaneously at a frequency of 500 Hz. The DIC analysis was performed by open-source software NCorr [2]. When comparing here presented results with ones reported in [1], the differences are in the development of parallel acquisition trigger system for DIC and IR images and development of a Matlab/Python code for image processing. This enabled the integration of these two methods.

[1] Lovre Krstulović-Opara, Alen Grebo; University of Split, Faculty of Electrical Engineering, Mechanical Engineering and Naval Architectur, Ruđera Boškovića 32, 21000 Split, Croatia, Lovre.Krstulovic-Opara@fesb.hr & Alen.Grebo@fesb.hr

[2] Marko Nagode, Branislav Panic; University of Ljubljana, Faculty of Mechanical Engineering, Aškerčeva ulica 6, 1000 Ljubljana, Slovenia, marko.nagode@fs.uni-lj.si & branislav.panic@fs.uni-lj.si

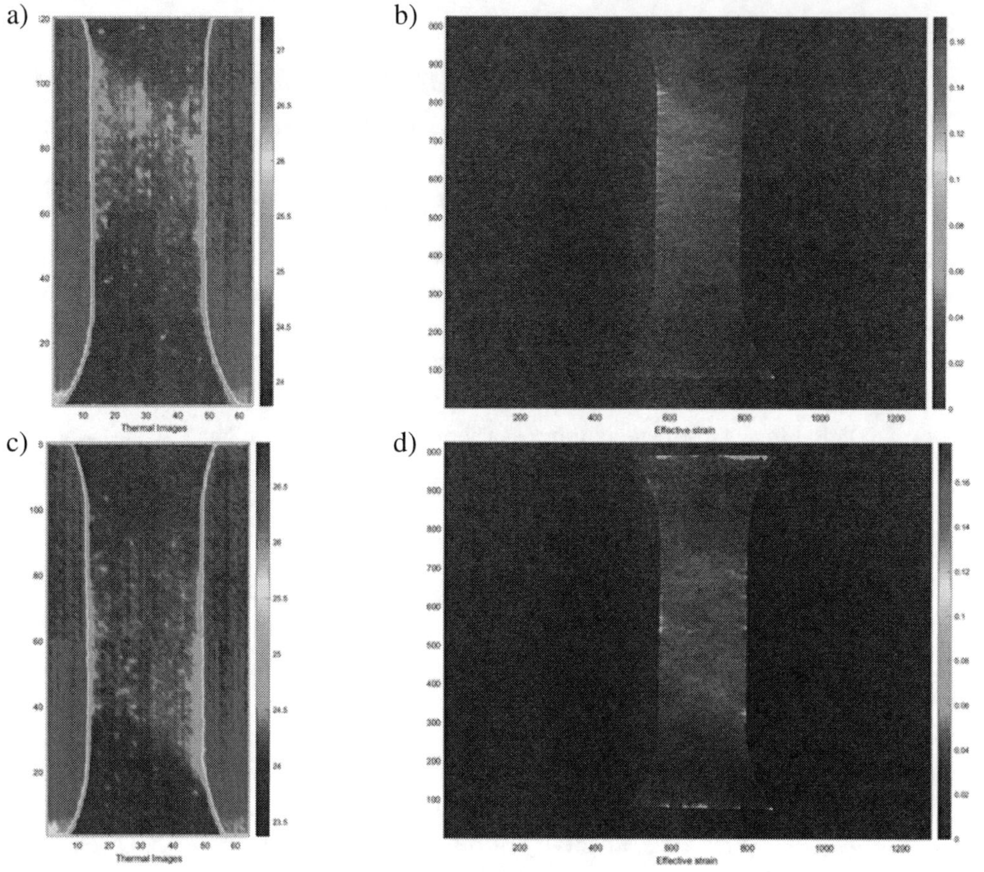

Fig. 1. Temperature and strain distributions for two datasets a) Dataset 1 – thermal image, b) Dataset 1 – DIC, c) Dataset 2 – thermal image and d) Dataset 2 – DIC.

As the relation between DIC and IR results is not well established there is room for a novel technique based on neural networks. Image to image translation techniques were developed for this idea of mapping one image to another while Pix2pix is an AI technique based around Generative adversarial models [3]. Architecture such as Pix2Pix are used to translate images from one space to another with sufficient accuracy. Considering Pix2Pix approach, dataset used in the training consisted of 233 paired images from two datasets.

In Fig. 1 one random moment can be observed per dataset. As a pre-processing technique a feature scaling (min-max normalization) was implemented on both datasets, this enabled comparison and unification of both datasets as colour distributions now matched through both datasets.

As image input size for Pix2Pix is 600×600 pixels smaller subsets of images were chosen for training as shown in Fig. 2. Also, for the case of thermography images upscaling had to be performed. Generative adversarial networks are a type of network that estimates the generative part of the model through an adversarial process, in which the generator and discriminators are trained simultaneously [3].

Pix2pix architecture adds a patch-GAN discriminator which works by classifying an $n \times n$ patch into real, or fake. This is used to provide more constraint for decision making process of the discriminator [4]. An illustration of the model architecture is shown in figure below.

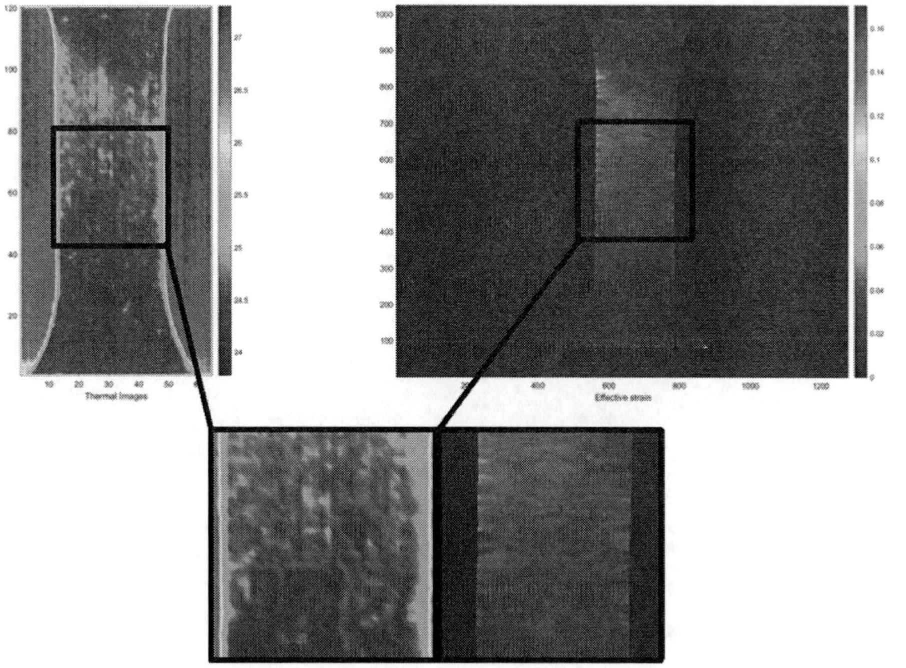

Fig. 2. Subsets for training, two subsets 600×600 pixels merged into 1200×600 pixels.

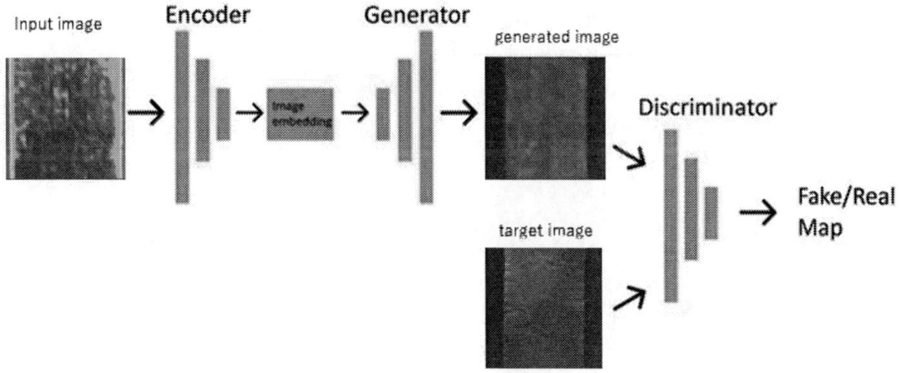

Fig. 3. Ilustration of Pix2pix architecture with addition of training data [3].

In simple terms during training one thermal image (input image) is passed through the encoder, after which the generator tries to generate an image as closely to the target image. The Discriminator tries to determine if the image is fake or real. This process loops for a number of epochs (one pass through all the data in the training set), until the training is complete.

The model presented herein was trained with the following hyperparameters:

- Learning rate = $2 \cdot 10^{-4}$

- Batch size of 8
- Epochs = 15000
- $\lambda_{gp} = 10$
- $L_1\lambda = 400$

The training was performed on a Nvidia Quadro RTX4000 graphics card that took two days and three hours. The results of the training are shown on the figure below.

Fig. 4. Training results.

3. Conclusion

The training results look promising as the overall distribution of color was matched. The error between all the images in a datasets per pixel values wes 3.9% which is satisfactory. It was shown how advanced AI techniques and methods can be applied to experimental mechanics offering new perspectives and courses of research. Current state of the research is in its final stages as the proof-of-concept network is undergoing extended training on larger datasets mainly three-point bending and compression tests. Also, the research will be expanded with reduced order modelling using REBMIX finite mixture modelling package [4].

References

[1] Krstulović-Opara L., Surijak, M., Vesenjak, M., Tonković, Z., Frančeski, J., Kodvanj, J. and Domazet, Ž., "Determination of material properties using dynamic tests supported by thermography, digital image correlation and numerical simulation," *8th International Congress of Croatian Society of Mechanics*, eds.: I. Kožar, N. Bićanić, G. Jelinić, M. Čanađija, 29.9.-2.10., Opatija, Croatia, 53, (2015).

[2] Blaber J., Adair B. and Antoniou A., "Ncorr: Open-Source 2D Digital Image Correlation Matlab Software ," *Experimental Mechanics*, Vol. 55, pp. 1105–1122 (2015). https://doi.org/10.1007/s11340-015-0009-1.

[3] Goodfellow I.J., Pouget-Abadie J., Mirza M., Xu B., Warde-Farley D., Ozair S., Courville A., and Bengio Y., "Generative adversarial nets," *Adv. Neural Inf. Process. Syst.*, pp. 2672–2680 (2014). https://doi.org/10.48550/arXiv.1406.2661.

[4] Isola P., Zhu J.Y., Zhou T. and Efros A.A., "Image-to-Image Translation with Conditional Adversarial Networks," in 2017 IEEE Conference on Computer Vision and Pattern Recognition (CVPR) Honolulu, HI, USA 2017. https://doi.org/10.48550/arXiv.1611.07004.

[5] Nagode M. and Klemenc J., "Modelling of Load Spectra Containing Clusters of Less Probable Load Cycles," *Int. J. Fatigue*, Vol. 143, Art. no. 106006 (2021). doi: 10.1016/j.ijfatigue.2020.106006.

Replacement of a Metal Fan Blade with a Composite One

Jan Krystek[1], Tomáš Mandys[2], Roman Kroft[3], Vladislav Laš[4],
Lukáš Horák[5] & Kateřina Patková[6]

Abstract: Currently, there are a lot of fan applications where they have been part of a composite blade composed of long fibers and epoxy resin. The usage of epoxy resin is limiting for high temperature applications. Therefore a composite material with carbon fibers and a geopolymer matrix, that has a very high temperature resistance in comparison with epoxy resins, was chosen in this work. An experimental modal analysis of the existing metal alloy blade and new composites one was performed. The obtained values were used to verify the computational models. Furthermore, bending tests of the mentioned blades were performed. Based on the comparison of experiments and numerical analyses of FEM, some previously unknown mechanical parameters of the used composite structure were specified.

Keywords: Bending test; Composite fan blade; Geopolymer; Modal analysis

1. Introduction

Fans are used in a variety of industries. We can find them, for example, in coal-fired power stations, traffic tunnels but also in households. A typical material for the production of a fan blades is a metal or metal alloys. Its disadvantage is its high density and low resistance in aggressive environments. Currently, there are many fan applications where composite blades have been used [1]. These composite blades usually consist of long fibres (carbon or glass) and epoxy resin. These materials appear, for example, on blades in aerodynamic tunnels or wind turbines [2]. Applications in high operating temperature environments make the limit for epoxy resins. This disadvantage of the epoxy resins is overcomes by the good temperature resistance of the geopolymers matrices demonstrated in works [3].

In this work, the application of blades of the ventilators in traffic tunnels was chosen. The temperature resistance is required while the fan must be able to run for 120 minutes at 400 °C. Aluminium alloy blades are currently used for this application. A weight of the aluminium blade in selected tunnel fan application is in current time about 4.8 kg. Due to energy savings, the aim is to replace aluminium blades with lighter composites one. Specifically, it is a sandwich construction of the blade. The core is made of heat-resistant lightweight material and the cover is made of carbon fabric and geopolymer matrix. To preserve the aerodynamic properties, the composite blade had the same external shape of the blade body as the aluminium one. The weight of the new composite blade is only 1.7 kg.

[1] *Ing. Jan Krystek, Ph.D.; Department of Mechanincs, Faculty of Applied Sciences, University of West Bohemia; Univerzitní 8, 301 00 Plzeň, Czech Republic;* krystek@kme.zcu.cz

[2] *Ing. Tomáš Mandys, Ph.D.; NTIS – New Technologies for the Information Society, Faculty of Applied Sciences, University of West Bohemia; Univerzitní 8, 301 00 Plzeň, Czech Republic;* tmandys@ntis.zcu.cz

[3] *Ing. Roman Kroft; NTIS – New Technologies for the Information Society, Faculty of Applied Sciences, University of West Bohemia; Univerzitní 8, 301 00 Plzeň, Czech Republic;* rkroft@ntis.zcu.cz

[4] *prof. Ing. Vladislav Laš, CSc.; Department of Mechanics, Faculty of Applied Sciences, University of West Bohemia; Univerzitní 8, 301 00 Plzeň, Czech Republic;* las@kme.zcu.cz

[5] *Bc. Lukáš Horák; Department of Mechanics, Faculty of Applied Sciences, University of West Bohemia; Univerzitní 8, 301 00 Plzeň, Czech Republic;* horakl@students.zcu.cz

[6] *Kateřina Patková; Department of Mechanics, Faculty of Applied Sciences, University of West Bohemia; Univerzitní 8, 301 00 Plzeň, Czech Republic;* patkka@students.zcu.cz

There are generally high demands on the fan blades. To verify them and ensure safe operation, it is necessary to perform a variety of tests and verification of resistance, such as centrifugal loading. The lifetime of the blade is affected by its bending stiffness. To ensure safe operation of rotating parts, it is necessary to know the modal properties of the blades and the entire fan.

2. Experiments

Experimental modal analysis was performed for aluminium and composite blades. For both types of the blades were identified and compared eigenfrequencies and corresponding eigenmodes. Two types of fixing were tested (freely stored blade and clamped blade). The results in Table 1 show that the first eigenfrequencies of both blades are above the excitation frequency caused by the operating speed of the fan (25 Hz). The composite blades are designated by the number that indicates the produced prototype number.

Table 1. Eigenfrequencies [Hz].

Freely stored blade						Clamped blade					
Mode:	1	2	3	4	5	Mode:	1	2	3	4	5
Aluminium	477	881	1223	1742	2312						
Blade 13	413	581	885	1113	1559	Blade 13	>100	351	510	854	1057
Blade 14	418	590	892	1133	1577	Blade 14	>100	364	533	877	1111
Blade 15	418	586	921	1133	>1600						
Blade 16	416	586	947	1142	>1600						

Fig. 1. Setup of bending test.

Fig. 2. Failure of blade 3 after bending test.

Bending tests of the clamped aluminium and composite blades were performed (Fig. 1). Composite blades were clamped in a turning chuck using specially shaped jaws. These blades were placed so that the end of the blade was in horizontal position to the ground and perpendicular to the load steel bar attached to the top crosshead of the test machine (Zwick/Roell Z050). The outer edge of the round bar 20 mm diameter was placed in alignment with the top edge of the blade. The composite blade no. 3 was loaded to the failure (Fig. 2), the blade no. 13 was repeatedly loaded to the non-destructive limit.

The deflections of the blades at the point of the loading were measured with the extensometer, the deflections in other selected points placed on the face in the axis of the tested blades were measured by laser sensors. The results achieved on the aluminium and composite blades for load $F = 200$ N are graphically compared in Fig. 3. The designed composite blade has less bending stiffness. This fact can negatively affect the lifetime of the composite blade.

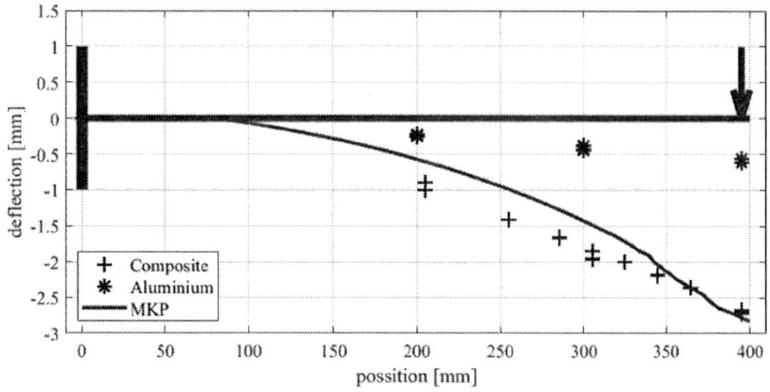

Fig. 3. Deflection of blades in selected points.

3. Numerical simulations of experiments

Numerical model of composite blade was created in finite element software Abaqus. Computational mesh of composite face was modelled by shell elements S4 (with 4 nodes) and core and steel ring by solid elements C3D8 (with 8 nodes).

Selected mechanical parameters of the composite (e.g. modulus of elasticity) were obtained from performed tensile and compressive static tests of the material specimens. A very good agreement between the experiment and numerical simulation with identified static parameters were found in the case of the bending test (Fig. 3). While a large difference was found with the same parameters in the case of numerical modal analysis. Therefore, based on optimization in the Isight system, material parameters were identified so that the results of numerical (Fig. 4) and experimental modal analysis (Fig. 5) had the best agreement. The results of the modal analysis for both sets of material parameters for the freely clamped blade are given in Table 2.

Table 2. Eigenfrequencies of freely stored blade [Hz].

	Mode 1	Mode 2	Mode 3	Mode 4	Mode 5
Experiment	416	586	911	1130	1584
Material parameters for static test	254	326	568	620	1030
Material parameters for modal analysis	430	603	909	1102	1581

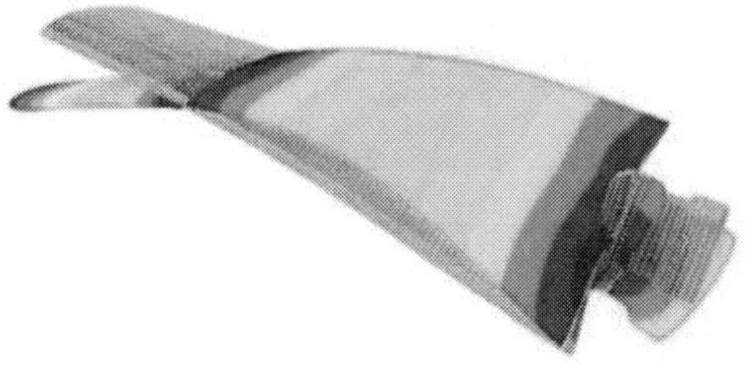

Fig. 4. Numerical modal analysis – first bending mode of composite blade.

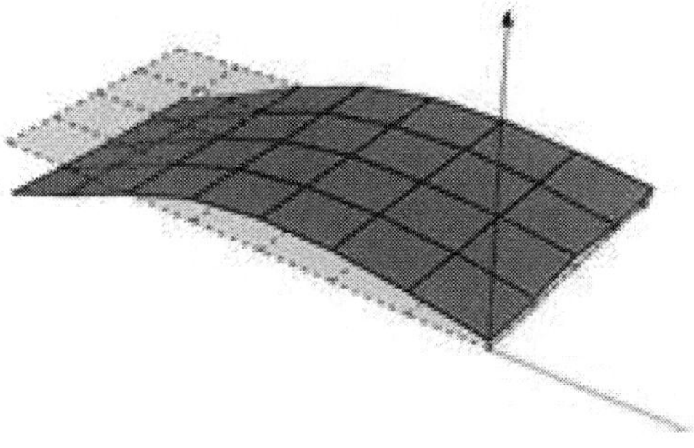

Fig. 5. Experimental modal analysis – first bending mode of composite blade.

4. Conclusion

The aim of the work was to examine the possibilities of replacing the aluminium blade of the axial fan in the traffic tunnels with a composite blade. This replacement would lead to energy savings in the operation due to a significant reduction of the weight of rotating bodies.

To verify the suitability of the selected application, some of the necessary tests were performed - modal analysis and bending test of the blades. Modal analysis proved the possibility of using a composite blade. A new set of material parameters was found for numerical simulations of modal analyses. In the case of bending tests, a good agreement between experimental and numerical data was found for the original material parameters. The bending test showed less stiffness of the composite blade in comparison with the aluminium alloy one. For possible application of the composite blade, it would be necessary, for example, to increase the bending stiffness by a shape modification of blades body or by a change in the design of the composite outer skins, which would ensure the longer lifetime.

Acknowledgements

The results published in presented paper were obtained during the project of Technology Agency of the Czech Republic and Ministry of Industry and Trade no. FW01010153. This work was supported by project SGS-2022-008.

References

[1] Rao J.S., Sheelavant K. and Bombale B., "Concept optimal design of composite fan blades," in Proceedings of the ASME Turbo Expo 2011, Vol. 6, Vancouver, Canada, ASME pp. 811–820 (2012). ISBN 978-0-7918-5466-2. https://doi.org/10.1115/GT2011-45994.

[2] Chen X., Zhao W., Zhao X.L. and Xu J.Z., "Failure test and finite element simulation of a large wind turbine composite blade under static loading," *Energies*, Vol. 7, no. 4, pp. 2274 – 2297 (2014). ISSN 19961073. DOI: 10.3390/en7042274.

[3] Haincová E., Hájková P. and Kohout J., "Prepregs for temperature resistant composites," *Material*, Vol. 12, no. 23, Art. no. 4012 (2019). ISSN 19961944. https://doi.org/10.3390/ma12234012.

The Effect of a PNNI on Mechanical Properties of Rat Skins

Kristýna Kubášová[1], Radek Sedláček[1] & Lucia Machová Urdzíková[2]

Abstract: This paper deals with the measurement of the mechanical properties of rat skin depending on the administered medication. The question is whether the treatment performed after spinal cord injury does not affect the properties of the surrounding tissues and is therefore safe. The performed analyses have shown that a change in the mechanical properties of rat skins after the medication wasn't demonstrated.

Keywords: Tensile test; Mechanical properties; Rat skins; Perineuronal nets

1. Introduction

Spinal cord injury is often incurable. Modern medicine focuses primarily on the treatment of acute conditions, but the treatment of chronic diseases isn't so much studied. One of the few ways to cure spinal cord injury is to block the creation of perineuronal nets (PNNs), which are enwrapping certain types of neurons in the central nervous system. The investigated PNN inhibitor is the drug 4-methylumbelliferone (4MU), which increases post-traumatic regeneration. For the drug to be safe to use, its effect on all tissues needs to be verified. The subject of this work is its influence on the mechanical properties of skins [1–3].

2. Methods and materials

The methodology of the experiment was based on the determination of the maximum measured stress, longitudinal deformation, and modulus of elasticity at the tensile test. The samples were tested on the testing system MTS Mini Bionix 858.02 (MTS, Minnesota, USA) with a load cell (range of 0–500 N) and a pneumatic jaw. The tensile test was performed at a constant loading speed 20 mm/min. Displacement, time and force were recorded with a sampling frequency 20 Hz. A video extensometer X-sight ONE (X-sight, Brno, Czech Republic) with image recording was used to evaluate the longitudinal deformation. The flow of the experiment is shown in the figure below (see Fig. 1).

A total of 41 explanted rat skins were tested (see Fig. 2). The samples were stored at -18 °C after extraction. Prior to the experiment, they were thawed and soaked in isotonic saline at 8 °C for 24 hours. Before mechanical testing, all samples were removed from the saline. The samples were divided into four groups depending on the concentration of administered medication (Table 1). All samples were trimmed into a rectangular shape, the width of all samples was from 10 to 15 mm, and the thickness ranges were from 1.0 to 1.3 mm. Black markers (in the shape of a cross) were drawn on the specimens for longitudinal strain measurement, sensing using the video extensometer.

[1] *Ing. Kristýna Kubášová, Ing. Radek Sedláček, Ph.D. Department of Mechanics, Biomechanics and Mechatronics, Faculty of Mechanical Engineering, Czech Technical University in Prague. Technická 4, 166 07 Prague 6, Czech Republic; kristyna.kubasova@fs.cvut.cz; radek.sedlacek@fs.cvut.cz*

[2] *MUDr. Lucia Machová Urdzíková Ph.D., Institute of Experimental Medicine CAS, v.v.i. Vídeňská 1083, 142 20 Prague 4, Czech Republic; lucia.machova@iem.cas.cz*

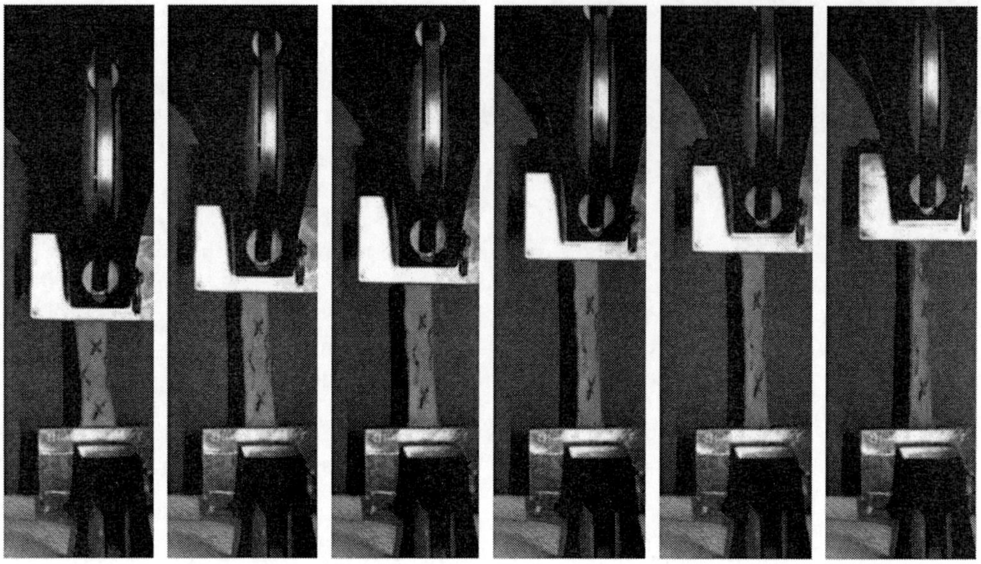

Fig. 1. The flow of the experiment and the pneumatic jaws.

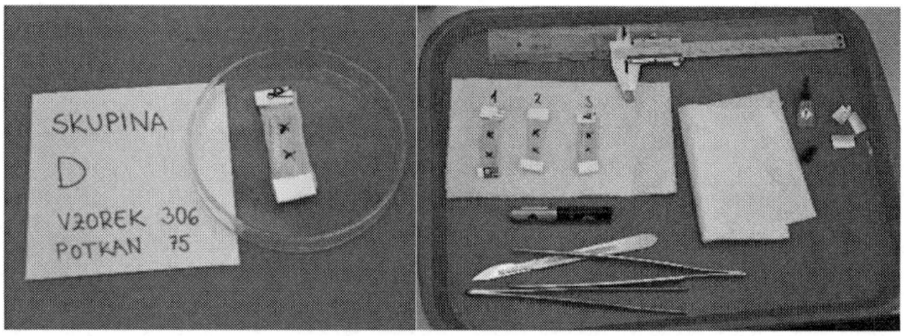

Fig. 2. Tested samples (Sample 306, rat 75, group D).

Table 1. Description of groups.

Group	4MU concentration	Description
A	0	Control group (for group B)
B	2.5%	Group of rats with 2.5% 4MU application
C	0	Control group (for group D)
D	5%	Group of rats with 5% 4MU application

The force F [N] and longitudinal deformation (relative strain) ε [%] were recorded directly from the test system and the video extensometer. Just the maximum measured stress was evaluated as the ratio of the maximum force and the initial cross-section. The evaluated data were further processed into the graphical stress-strain relationship.

The maximum strain $\varepsilon_{\max}$ [%] corresponds to the maximum measured stress value in the graphical dependence of stress-strain. The modulus of elasticity E [MPa] was determined from the direction of the tangent in the linear part of the dependence of the stress on the longitudinal deformation.

3. Results and statistical analysis

The table below (see Table 2) shows the mean values and their variances for the measured and evaluated quantities for all groups A–D. From this table, it follows that the mean values across the groups fluctuate. The outlier removal methodology did not remove any outliers. The coefficients of variation of the evaluated parameters range from 22–77%, group C values have the highest variance.

Table 2. Statistical parameters for all groups (arithmetic mean ±SD (coefficient of variation)).

	A	B	C	D
σ_{max} [MPa]	8.2±2.2 (27%)	8.0±4.0 (50%)	3.6±2.8 (78%)	6.7±4.5 (67%)
ε_{max} [%]	33.5±7.4 (22%)	36.7±11.7 (32%)	22.2±10.9 (49%)	24.4±10.6 (43%)
E [MPa]	50.9±15.8 (31%)	41.9±15.7 (37%)	23.2±14.9 (64%)	37.7±29.1 (77%)

The normal distribution of the data was verified by the Shapiro-Wilk test and also visually according to the Q-Q plots. Data that come from normal distribution were further evaluated by a parametric T-test, and data that didn't meet the conditions for normal distribution were evaluated by a non-parametric Mann-Whitney U-test (both tests with a 95% confidence interval).

According to the requirements, two groups were always compared by pair test – A vs. B and C vs. D. Next, the effect size was assessed using Cohen's delta. The results of statistical significance and effect size for all evaluated groups and parameters are presented in the tables below (see Table 3 and Table 4).

Table 3. Statistical evaluation A vs. B (2.5% 4MU).

	Statistical significance				Effect size		
Parameters	Test	p-value	Significance	Test force	The ratio of arithmetic means	Cohen's d	
σ_{max}	T-test	0.897	not proven	6.5%	-2%	0.059	Very small
ε_{max}	T-test	0.475	not proven	17.4%	+10%	0.327	Small
E	T-test	0.220	not proven	33.6%	-18%	0.568	Medium

Table 4. Statistical evaluation C vs. D (5% 4MU).

	Statistical significance				Effect size		
Parameters	Test	p-value	Significance	Test force	The ratio of arithmetic means	Cohen's d	
σ_{max}	U-test	0.113	not proven	57.6%	+84%	0.832	High
ε_{max}	T-test	0.651	not proven	11.5%	+10%	0.200	Small
E	T-test	0.245	not proven	40.9%	+63%	0.640	Medium

4. Conclusion

No statistically significant differences were found in either comparison. The high effect size was demonstrated only in the comparison of groups C and D (5% of 4MU and its control) in the case of the parameter of maximum measured stress, the values of the medicated group increased by 84% compared to the control.

A change in the observed parameters between the evaluated groups wasn't demonstrated. The analyses carried out showed, among other things, a considerable dispersion of the measured values. The tables above show, among other things, that there are also differences between groups A and C, which are controls for the medicated samples, and that the variance of the values is also not negligible.

All of this leads to the conclusion that there is a need to create a single and larger sample group, which would be composed according to the same methodology and thus formed the baseline for comparison with other groups.

Acknowledgements

This research work was supported by the Czech Technical University in Prague, grant number SGS22/149/OHK2/3T/12.

References

[1] Urdzíková L. et al., "Transplantation of bone marrow stem cells as well as mobilization by granulocyte-colony stimulating factor promotes recovery after spinal cord injury in rats," *J Neurotrauma*, Vol. 23, no. 9, pp. 1379–91 (2006). https://doi.org/10.1089/neu.2006.23.1379.

[2] Urdzíková L. et al., "Flt3 ligand synergizes with granulocyte-colony-stimulating factor in bone marrow mobilization to improve functional outcome after spinal cord injury in the rat," *Cytotherapy*, Vol. 13, no. 9, pp. 1090–104 (2011). https://doi.org/10.3109/14653249.2011.575355.

[3] Štěpánková K. et al., "4-Methylumbeliferone Treatment at a Dose of 1.2 g/kg/Day Is Safe for Long-Term Usage in Rats," *International Journal of Molecular Sciences*, Vol. 24, no. 4, Art. no. 3799 (2023). https://doi.org/10.3390/ijms24043799.

Analysis of Spinopelvic Fixation Techniques for Mechanical Stabilisation of Sacral Bone Injuries

Libor Lobovský[1], Michaela Marešová[1], Tomáš Mandys[1], Drahomíra Weisová[2], Martin Salášek[2], Tomáš Pavelka[2] & Jiří Křen[1]

Abstract: Several variations of spinopelvic fixation techniques are investigated and compared. The computational prediction of displacements of fractured pelvic bone segments is performed using an experimentally verified numerical model. The model focuses on mechanical response of the pelvic ring and the last two lumbar vertebrae to physiological loading. The effect of soft tissue structures such as muscles and ligaments supporting lumbar spine is not reflected. The results indicate benefits of application of spinopelvic fixation with a cross-link transverse connector and/or with an additional iliosacral screw fixation.

Keywords: Osteosynthesis; Mini-invasive techniques; Bone injury; Sacrum; Iliosacral screw; Lumbo-sacral fixation

1. Introduction

The spinopelvic (lumbo-sacral) fixation [1, 2] is usually indicated for injuries of pelvic ring lateral segment which display significant instability, mainly for transverse sacral fractures (e.g. U-shaped or H-shaped fractures) which cause a spinopelvic dissociation [3,4], i.e. a detachment of lumbo-sacral spine from the pelvic ring. In general, the spinopelvic fixation not only enables reposition of the fractured bone parts but allows transmission of loads between lumbar spine and iliac bones. The load is transmitted along a connection rod which is mounted between the pedicle screws (which are inserted in lumbar vertebrae) and the iliac screw (which is inserted into iliac bone near the posterior superior iliac spine), see Fig. 1 and Fig. 2. The stability of bilateral spinopelvic fixation can be further enhanced by a cross-link (CL) transverse connector and possibly by a combination with another fixator, e.g. the iliosacral screw (ISS) [5].

The above mentioned traumatic pelvic ring fractures (e.g. caused by high-energy impacts during sports or car accidents) are serious injuries with a high risk of morbidity and mortality. The treatment of the injury, including selection of an appropriate fixation, depends on the type of fracture and the level of its instability. In the following, selected spinopelvic fixation techniques are examined from the mechanical point of view and the stability of each fixation is assessed for unilateral and bilateral comminuted sacral fractures under static loading by means of computational modelling.

2. Methods

In order to thoroughly compare the studied spinopelvic fixations, a computational finite element model is developed based on an experimental campaign which employed orthopaedic models of human pelvis [6]. Although the material properties of the utilised orthopaedic models (made

[1] *Ing. Libor Lobovský, Ph.D.; Bc. Michaela Marešová; Ing. Tomáš Mandys, Ph.D.; Prof. Ing. Jiří Křen, CSc.; NTIS - New Technologies for the Information Society, Faculty of Applied Sciences, University of West Bohemia; Univerzitní 2732/8, 30100 Plzeň, Czech Republic;* lobo@kme.zcu.cz

[2] *MUDr. et Mgr. Drahomíra Weisová, Ph.D.; MUDr. Martin Salášek, Ph.D.; Doc. MUDr. Tomáš Pavelka, Ph.D.; Clinic for Orthopaedics and Traumatology of Locomotive Organs, University Hospital, alej Svobody 80, 30460 Plzeň, Czech Republic;* weisovad@fnplzen.cz

of homogenous solid foam with linear isotropic mechanical response) differ from the properties of real pelvic bones, these models (unlike cadaveric samples) assure a high repeatability of the performed experiments.

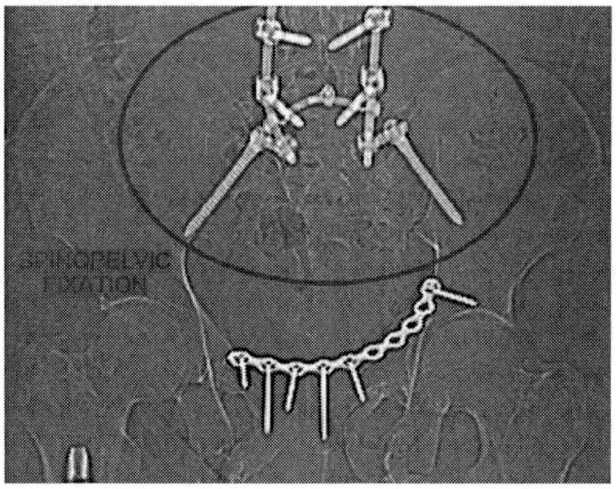

Fig. 1. CT topogram of spinopelvic fixation in male patient.

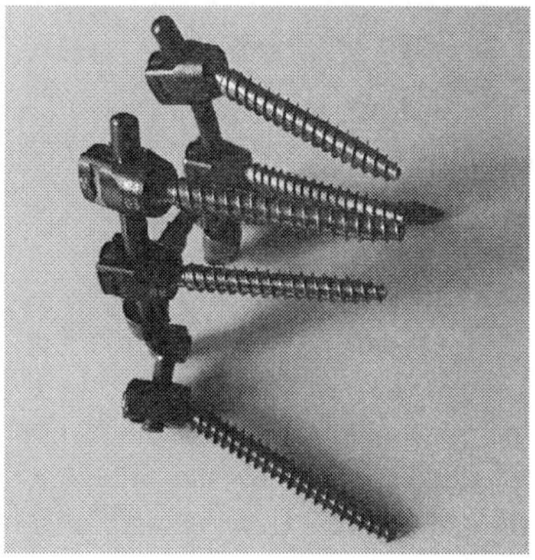

Fig. 2. Image of spinopelvic fixation screws ex situ.

The experiments are performed for pelvic models with absence of lumbar spine and for fixation techniques using ISS [7] and transiliac internal fixator (TIFI) [8]. Similarly to spinopelvic fixation, TIFI employs polyaxial screws which are inserted into iliac crests at spina iliaca posterior superior.

The studied case features a unilateral linear transforaminal fracture without a comminuted zone. The resulting set of experimental data is used for validation of the computational model of fractured pelvic ring with a selected fixation applied. The quality of the computational model and its agreement with the experiment is assessed based on the ratio R between stiffness of the pelvis with fracture treated by a given fixation technique and stiffness of the intact pelvis. Results for the examined fixation techniques indicate that the computatinal model fully reflects the trends observed during the experiments. The deviation of the computationally predicted stiffness ratio from the experimentally measured data is in a range of percents, Table 1.

Table 1. Comparison of stiffness ratio for given fixation technique as recorded during experimental measurements R_{exp} and as a result of numerical computations R_{comp}.

	R_{exp}	R_{comp}	$(R_{\mathrm{comp}} - R_{\mathrm{exp}})/R_{\mathrm{exp}}$
classic TIFI	0.639	0.660	3.3%
dual TIFI	0.720	0.785	9.0%
supraacetabular TIFI	0.722	0.777	7.6%
TIFI + ISS	0.842	0.849	0.8%

In order to model the spinopelvic fixation, the computational model is extended by the 4th and the 5th lumbar vertebrae (referred to as L4 and L5 in the following) and intervertebral discs based on computed tomography (CT) scans of a selected male patient with the lumbosacral fixation applied into both L4 and L5 (Fig. 3). Thus positioning of the fixator screws (as depicted in Fig. 4) reflects this specific case. All structures – pelvic bones, vertebrae, fixator screws and also intervertebral discs are modelled as linear homogeneous isotropic materials represented with solid elements. The material properties of the solid foam (bone parts) are identified experimentally, the properties of the homogenised intervertebral discs are based on data from literature [9, 10] and the fixator screws are made of titanium, Table 2.

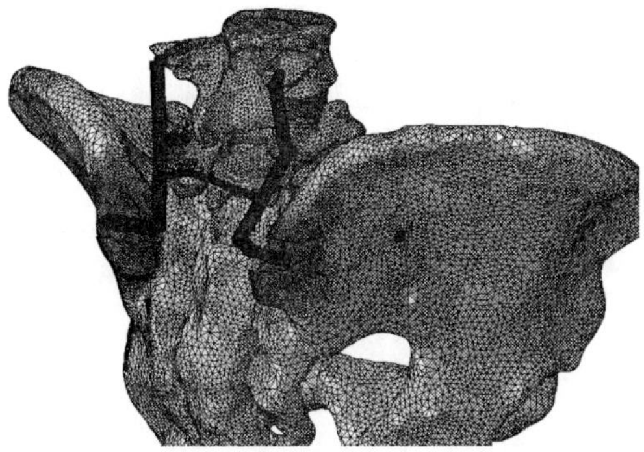

Fig. 3. Computational model of pelvis with spinopelvic fixation (pedicle screws are mounted into both L4 and L5) combined with iliosacral screw fixation.

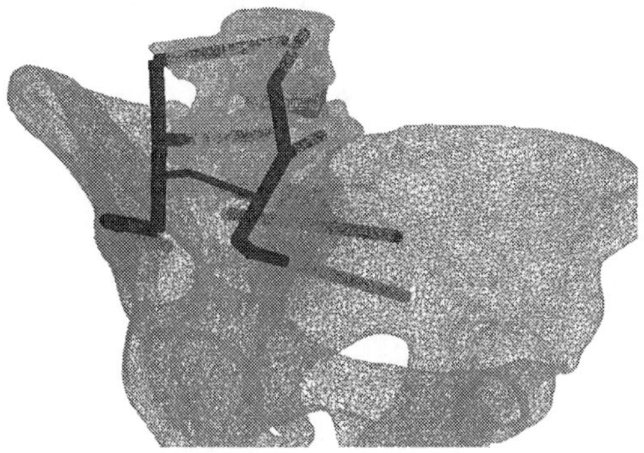

Fig. 4. Orientation of fixating screws within the model.

Table 2. Parameters of the applied materials.

Material	Young's modulus	Poisson's ratio
solid foam	194 MPa	0.2
cartilage	30 MPa	0.4
titanium	120 GPa	0.33

In the model, both acetabula are fixed in space and the load of 500 N is applied at L4 vertebra in cranio-caudal direction. In order to assess the spinopelvic fixation quality, magnitude of the sacral base displacement for the given load is recorded and compared in between the studied techniques. The stiffness of the treated pelvic structure is enumerated.

3. Results

All selected variations of spinopelvic fixation were examined for two types of sacral bone injury - unilateral vertical comminuted fracture and bilateral vertical comminuted fracture. A matrix of examined spinopelvic fixation types focuses on a positioning the lumbar pedicle screws (into L4 vertebra only, into L5 vertebra only or into both L4 and L5 vertebrae), an application of the CL connector and a combination with the ISS. The stability of each fixation is assessed based on the displacement magnitude at the base of sacrum. Results for both types of injury are summarised in Fig. 5 and Fig. 6, where the x-axis reflects the type of the pedicle screws positioning and the y-axis displays the sacral base displacement magnitude for load of 500 N (which is exerted at the superior surface of L4 vertebral body).

In general, the positioning of lumbar pedicle screws influences the stiffness of the treated pelvis as long as there is no ISS applied. In that case, L4+L5 fixations are superior regardless of application of the CL connector. Nevertheless, the CL connector enhances the stability of the applied fixation. For the L4+L5 fixation, the sacral base displacement is about 19% lower for the unilateral injury and about 22% lower for the bilateral fracture when the CL connector is applied.

A significant enhancement in stability of the applied fixation comes with application of the spinopelvic fixator together with the ISS fixation. This is pronounced especially for the bilateral type of injury (which causes spinopelvic dissociation). For the bilateral fracture, the L4+L5 fixation without CL connector combined with ISS display about 50% lower displacement than the L4+L5 fixation with CL connector but without ISS. This reduction is further improved to 59% when both CL connector and ISS are applied. When the spinopelvic fixation is combined with ISS, the differences between L4, L5 and L4+L5 variants are about negligible, in the range of several percents.

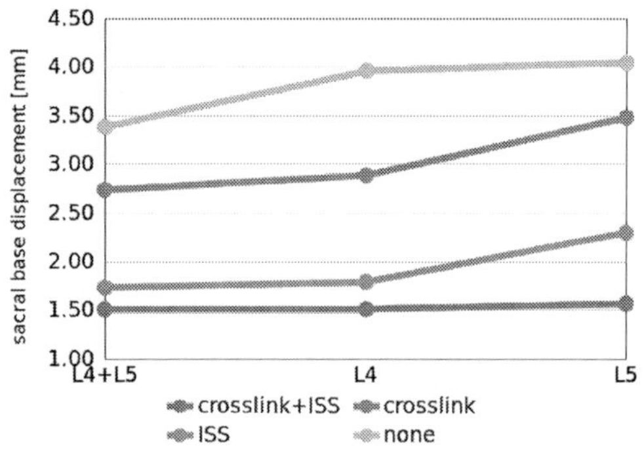

Fig. 5. Unilateral comminuted fracture: displacement magnitude at sacral base for all types of spinopelvic fixations and load 500 N.

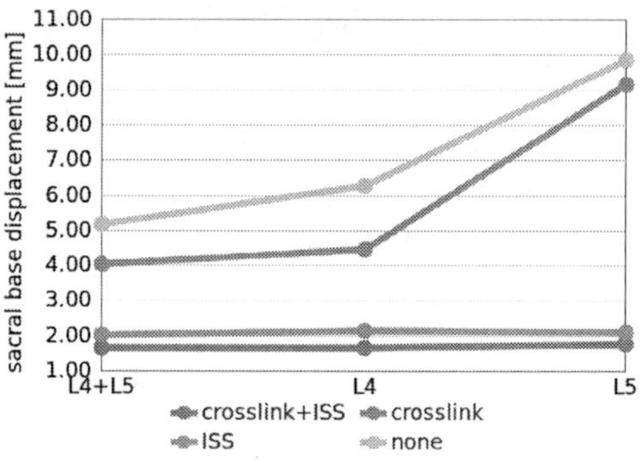

Fig. 6. Bilateral comminuted fracture: displacement magnitude at sacral base for all types of spinopelvic fixations and load 500 N.

4. Conclusions

The applied computational model is validated using the experiments on orthopaedic models of male pelvis with the unilateral linear transforaminal fracture without comminuted zone. The computational results prove that combining spinopelvic fixation with ISS is benefitial in terms of stability of the treated bone structure for all tested variants. If application of ISS is possible, i.e. if the size and state of sacral vertebrae allows the ISS treatment, the combination of the spinopelvic fixation with the ISS fixation allows for a real minimally invasive treatment of sacral fracture. When the CL connector is not applied, the incision in the lumbo-sacral region can be minimised and localised to insertions of the pedicle screws, the iliac screws and the ISS.

Acknowledgements

This publication is supported by the European Regional Development Fund-Project "Application of Modern Technologies in Medicine and Industry" No. CZ.02.1.01/0.0/0.0/17_048/0007280.

References

[1] Lindahl J., Mäkinen T.J., Koskinen S.K. and Söderlund T., "Factors associated with outcome of spinopelvic dissociation treated with lumbopelvic fixation," *Injury*, Vol. 45, pp. 1914–1920 (2014). https://doi.org/10.1016/j.injury.2014.09.003.

[2] Burns C.B., Dua K., Trasolini N.A., Komatsu D.E. and Barsi J.M., "Biomechanical Comparison of Spinopelvic Fixation Constructs: Iliac Screw Versus S2-Alar-Iliac Screw," *Spine Deformity*, Vol. 4, no. 1, pp. 10–15 (2016). https://doi.org/10.1016/j.jspd.2015.07.008.

[3] Vilela M.D., Jermani C. and Braga B.P., "Lumbopelvic fixation and sacral decompression for a U-shaped sacral fracture: case report," *Arquivos de Neuro-Psiquiatria*, Vol. 65(3-B), pp. 865–868 (2007). https://doi.org/10.1590/s0004-282x2007000500028.

[4] Weisová D., *Treatment of pelvic ring fracture using spinopelvic fixation (types of fixation, indications, functional results, mechanical characteristics)*. Doctoral Thesis, Faculty of Medicine in Plzeň, Charles University, Plzeň 2021.

[5] Schildhauer T.A., McCulloch P., Chapman J.R. and Mann F.A., "Anatomic and radiographic considerations for placement of transiliac screws in lumbopelvic fixations," *Journal of Spinal Disorders and Techniques*, Vol.15(3), pp. 199–205 (2002). https://doi.org/10.1097/00024720-200206000-00005.

[6] https://www.sawbones.com/full-male-pelvis-large-solid-foam-w-sacrum-1301.html

[7] Giráldez-Sánchez M.A., Lázaro-Gonzálvez Á, Martínez-Reina J., Serrano-Toledano D., Navarro-Robles A., Cano-Luis P., Fragkakis E.M. and Giannoudis P.V., "Percutaneous iliosacral fixation in external rotational pelvic fractures. A biomechanical analysis," *Injury*, Vol. 46(2), pp. 327–332 (2015). https://doi.org/10.1016/j.injury.2014.10.058.

[8] Dienstknecht T., Berner A., Lenich A., Zellner J., Mueller M., Nerlich M. and Fuechtmeier B., "Biomechanical analysis of a transiliac internal fixator," *International Orthopaedics*, Vol. 35(12), pp. 1863–1868 (2011). https://doi.org/10.1007/s00264-011-1251-5.

[9] Yang H., Jekir M.G., Davis M.W. and Keaveny T.M., "Effective modulus of the human intervertebral disc and its effect on vertebral bone stress," *Journal of Biomechanics*, Vol. 49, pp. 1134–1140 (2016). https://doi.org/10.1016/j.jbiomech.2016.02.045.

[10] Newell N., Carpanen D., Grigoriadis G., Little J.P. and Masouros S.D., "Material properties of human lumbar intervertebral discs across strain rates," *The Spine Journal*, Vol. 19, pp. 2013–2024 (2019). https://doi.org/10.1016/j.spinee.2019.07.012.

Contact Pressure Measurement System

Aleš Lufinka[1], Soňa Jandová[2] & Iva Petríková[3]

Abstract: When measuring human interactions with the environment, it is often necessary to show the distribution of contact pressures between the body part and the environment. The article describes the technical solution of the measuring system, which was primarily designed for measuring the load of the lower limbs when walking. The intended area of use was the monitoring of the limb's uniformity loading during rehabilitation walking after various surgeries. However, the proposed system for measuring contact pressures can be used for other applications, such as various measuring pads for seats, with a different arrangement of sensors and an increase in their numbers.

Keywords: Human-environment interaction; Pressure distributions; Load; Sensors

1. Introduction

Walking is one of the basic forms of human locomotion. After various lower limb surgeries, it is often necessary to monitor the amount of load on both limbs for rehabilitation to proceed properly [1]. For such measurements (and similar measurements, for example in the field of sports), there are various measuring systems with pressure sensors built into the insoles or different measuring pads. Commercially available systems are almost very expensive, which is why only some universities or sports laboratories or teaching hospitals use them. They are therefore only used during a patient's visit, but are not available for normal use outside the specialist's office. The main goal of this work was to develop an affordable flexible measuring device that would provide information about the load of individual parts of the human body directly to the mobile phone of the monitored person. Thanks to continuous and immediate feedback, the person could adapt the walking stereotype to current needs and it can make the healing process more effective. Similarly, for example, employers could use measuring pads to optimize the ergonomics of a person's posture or sitting position at a work machine during work process. This could eliminate chronic pain in the axial system, muscles or joints, which then causes employees to become incapacitated for work, which reduces the production and economic potential of employers.

2. Measurement system principle

The minimization of the cost of the proposed solution is achieved by two fundamental steps: reducing the number of measuring points and choosing the appropriate type of sensor. Both criteria were chosen so that the resulting system would be functional and sufficiently accurate for the purpose of the proposed application, the aim never been to achieve the parameters of high-end measurement systems. Another essential factor in the design of the measurement system was to achieve a minimum thickness to be able to be placed as an insole in a shoe. The criterion of the maximum thickness of the measuring insole was chosen to be 4 mm. The following solution was developed by respecting these criteria.

[1] *Ing. Aleš Lufinka, Ph.D.; Department of the Design of Machine Elements and Mechanism, Faculty of Mechanical Engineering; Studentska 1402/2, 461 17 Liberec 1, Czech Republic;* ales.lufinka@tul.cz

[2] *Prof. PhDr. Soňa Jandová, PhD.; Department of Applied Mechanics, Faculty of Mechanical Engineering; Studentska 1402/2, 461 17 Liberec 1, Czech Republic;* sona.jandova@tul.cz

[3] *Doc. Ing. Iva Petríková, PhD.; Department of Applied Mechanics, Faculty of Mechanical Engineering; Studentska 1402/2, 461 17 Liberec 1, Czech Republic;* iva.petrikova@tul.cz

A capacitive sensor was designed to measure the contact pressure. The elastic mass is compressed by the measured pressure. There are capacitor plates on both sides of the mass, the distance of which changes and thus the capacity of the capacitor changes. By changing the capacitance, the frequency of the connected oscillator is retuned. The pressure value is therefore converted to a frequency, which is then evaluated by the connected microcomputer. The dependence of the capacitance on the geometric dimensions of the capacitor can be seen in Eq. (1).

$$C = \varepsilon_0 \cdot \varepsilon_r \cdot \frac{S}{d} \tag{1}$$

Since the distance d is changed by the action of the contact pressure, it is clear that the dependence of the frequency on the contact pressure is nonlinear. The linearization of the dependence is then performed by software in a microcomputer.

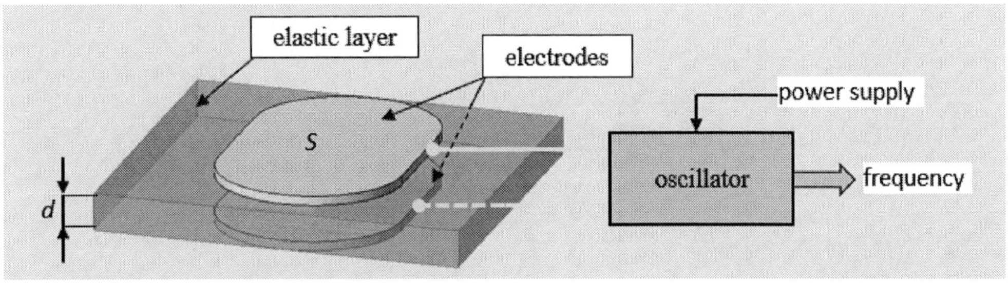

Fig. 1. The contact pressure sensor principle.

The thickness of the flexible layer, its properties and the size of the electrodes had to be optimized so that the resulting frequency was in the range measurable by a conventional microcomputer. The resulting solution uses a commercially available elastic material with a thickness of 2 mm with closed pores (similar to neoprene). Closed pores ensure minimal moisture absorption, so the layer permittivity ε_r is not affected by moisture and further minimize the elastic layer hysteresis. The 2 mm thickness also meets the required maximum thickness of the entire 4 mm insole. With an electrode area of approximately 150 mm^2, the frequency of the oscillator during compression is in the range of 250 - 70 kHz, which are values measurable by a conventional microcomputer for several channels simultaneously [1].

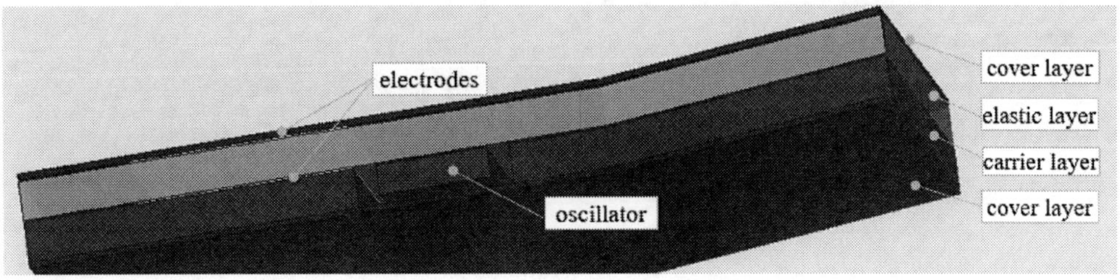

Fig. 2. The final composition of the insole layers.

The properties of the capacitive sensor are negatively affected by the parasitic capacitance of the wires between the capacitor and the oscillator. These must be as short as possible and therefore it was obvious that the oscillator must be placed directly in the insole at a minimum distance from the electrodes. The flexible printed circuit is located in the insole under the elastic layer. Electrodes are formed on it and at the same time there is an integrated oscillator circuit on its underside. The flexible printed circuit is approximately 0.1 mm thick and the height of the integrated circuit in SMD design is less than 1.5 mm. At the bottom of the insole is therefore a carrier layer of this thickness with holes for integrated circuits. The total thickness of these insole layers is therefore 3.7 mm. The insole is further finished on each side with a covering layer of fabric again about 0.1 mm thick. The set criterion of the total insole thickness of 4 mm was therefore complied with. The final composition of the insole layers is shown in Fig. 2.

The number of sensors is limited mainly by the required size of the electrodes. The typical pressure distribution under the foot was measured and used to design the distribution of the sensors and optimize their number. The foot pressure distribution shows three significant load areas [3]. If sensors are placed in these areas the entire measurement area will be covered. As a good compromise between the number of sensors and the spatial possibilities, the placement of eight pressure sensors in the area of the insole was proposed. The design of their location is shown in Fig. 3.

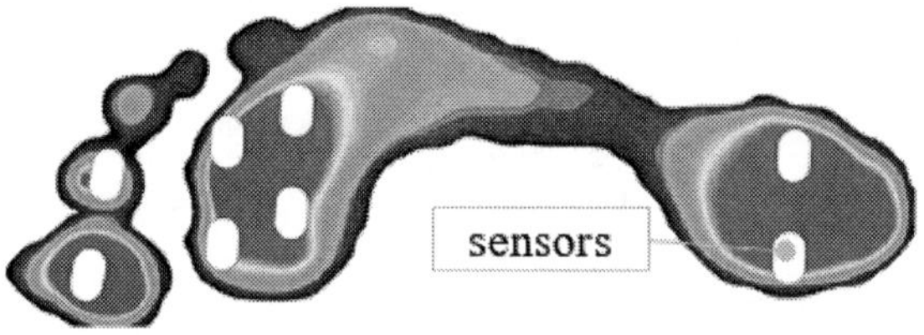

Fig. 3. Sensors location in the insole.

3. Practical implementation of the measurement sensor system

The use of a full-area flexible printed circuit was the first idea of the measurement insole design. After the production of a functional prototype, problems with the service life of the printed circuit board appeared, and therefore a second variation was created, where separated sensors are used.

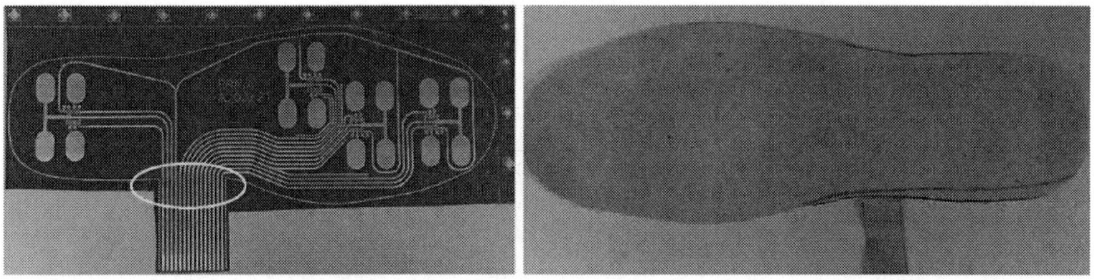

Fig. 4. Flexible printed circuit for insole on the left, insole with cable connection on the right.

3.1. The insole with full-area flexible printed circuit

In this variation, all parts of the electronic insert system are located on one common printed circuit board, which is the simplest way for series production. The disadvantage of this solution is the need to produce different variations for different shoe sizes. However, the wiring diagram would be practically the same for all sizes, it would differ only in the different position of the sensors and the length of the connecting lines. The flexible printed circuit board is commonly used, for example, to connect moving parts of printers, scanners, etc. The manufacturer sets limit parameters for the bending radius and which, according to the assumptions, should not be exceeded even when used in an insole. In practice, it turned out that the first critical point is the outlet from the insole (marked in yellow in Fig. 4 left), which is bent practically by 90° after being inserted into the shoe. While walking, the limit radius of bending was repeatedly exceeded and the copper layers were interrupted. The pattern of the printed circuits was therefore modified, and the connection was replaced by a conventional flat cable (see Fig. 4 right).

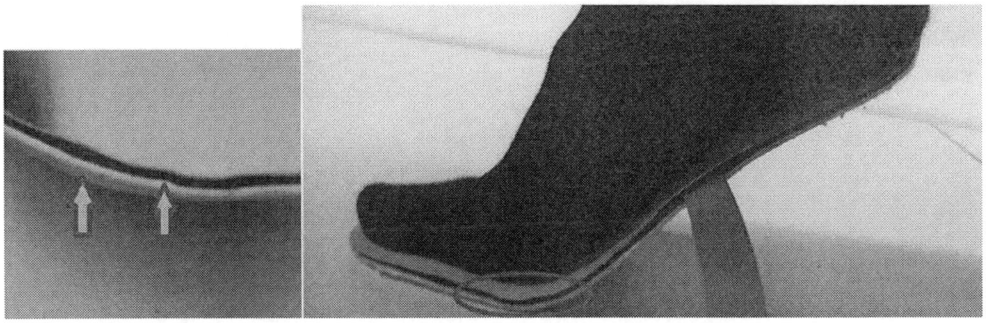

Fig. 5. Place of the maximum insole bend and the detail of the layers ungluing on the left.

Further testing showed problems with the durability and bond strength of the insole layers. Several different types of adhesives have been used, hot pressing gluing has been tried, but no method has yet yielded the expected reliability. After a certain period of use in the shoe, the layers at the critical place with maximum bend of the insole are unglued, the printed circuit is corrugated and its copper layer is interrupted.

The shortcomings described above were the main reason why the second variant of the measuring insole was designed. However, the variant with a full-area printed circuit was not rejected, the plan is to solve the problem of gluing layers with experts in the footwear industry.

Fig. 6. The insole with separate sensors.

3.2. The insole with separated sensors

The construction of the insole is similar to the previous prototype, but instead of a full-area flexible printed circuit, individual sensors are fitted in the insole. Power supply and signal connection is realized by flat cable. The insole thus does not contain a full-area layer of flexible printed circuit, the material of which is difficult to glue. The second advantage is that different printed circuits do not have to be created according to the insole size, the separate sensor is the same for all sizes. However, these advantages are paid for by the more difficult production of the insole, where the conductors of the flat cable must be placed in the insole. In addition to the small openings for the integrated circuits, channels for conductors must also be formed in the carrier layer. The pattern of the carrier layer is automatically cut on the plotter, but the subsequent deposition of conductors and alignment of the channels by the filling material complicates the production process. Fig. 6 on the left shows the sensors circuits and the cut-out shape of the carrier layer, on the right there is a completely connected insole with conductors in the carrier layer channels.

The insole is then finished with covering layers on both sides. During subsequent testing, the layers have not yet unglued, but the tests have not been performed long enough. However, the service life is already longer than in the first variation.

4. Signal processing

The basic function of the whole solution is to provide the user with up-to-date information on limbs loading while walking. The application on the smartphone was therefore chosen as the optimal user interface.

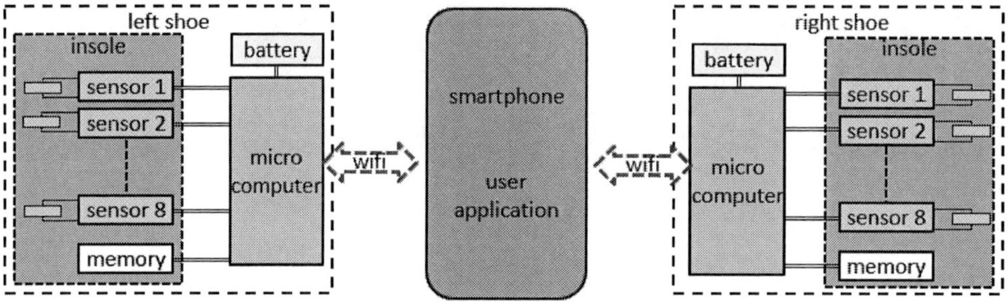

Fig. 7. Signal processing block scheme.

The insole contains eight sensors, basically eight oscillators, whose frequencies change according to the measured pressure. Because the dependence is nonlinear, there is also a memory chip in the insole, where the calibration constants of the individual sensors are stored. Each insole is connected to an autonomous microprocessor system. It measures the current frequency from eight oscillators in the insole with a sampling frequency of 40 Hz and sends data to the parent application. It also sends sensor calibration constants to the parent application before starting the measurement. The whole system works at the 3.3 V level and is powered by a battery. In the real use, therefore, two independent systems are used, one on each leg. Both systems communicate wirelessly with a user application that receives data from both insoles, converts frequencies to pressures using calibration constants, and displays measured values to the user. The final version of the application is currently under development. A computer application was also created to verify the functionality and calibration of the sensors. In this case, the insole microcomputer is connected with a USB cable. Example of measured data – the three steps of the right foot are shown in the Fig. 8.

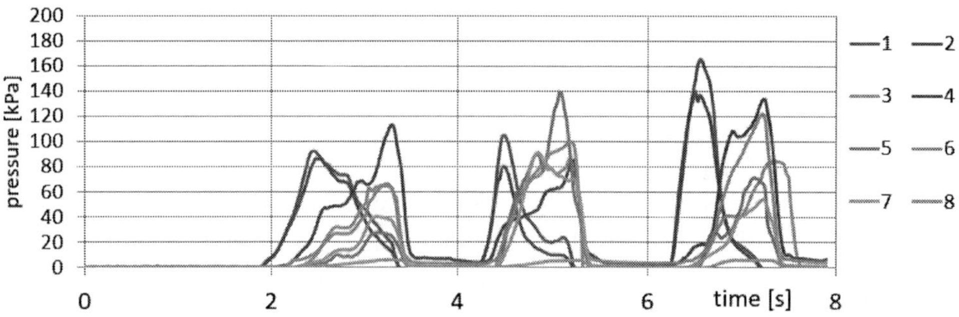

Fig. 8. Example of measured data – the three steps of the right foot.

5. Conclusion

The prototype of the measuring system of contact pressures in the insole of the shoe was built at the Technical University in Liberec. Its primary use is to monitor lower limb loading during postoperative recovery. The whole development was therefore aimed at minimizing the price of the equipment. Currently, the service life of the measuring insole is being verified and the first steps have been taken to verify the possibilities of serial production.

Acknowledgements

The project TP01010031/PROSYKO - Pro-Active System of Commercialization at TU Liberec 2 is co-financed from the state budget by the Technology Agency of the Czech Republic within the GAMA 2 Programme.

References

[1] Jandova S., Pazour J. and Janura M., "Comparison of Foot Liad in Walking after two different Surgival Treatment od Calcaneal Fracutre," *The Journal of Foot & Ankle Surgery,* Vol. 58, pp. 260–265 (2019). DOI: 10.1053/j.jfas.2018.08.051.

[2] Lufinka A., Jandová S. and Petríková I., "Development of Smart Insoles for Gait Monitoring of Patients after Low Extremity Orthopedic Treatment," in *Proceedings of 58th Experimental Stress Analysis*, Ostrava, 2020, pp. 160–164. ISBN: 978-80-248-4451-0.

[3] Jandova S., "Foot strike pattern in sport shoes with different construction," *Journal of Mechanical Engineering*, Vol. 15, no. 2, pp. 29–39 (2018).

Demonstration of the Resolution Properties of the 3-axis Neutron Diffraction Setting on the Samples of Rotary Swaged WNiCo Pseudoalloy

Pavol Mikula[1], Pavel Strunz[1], Charles Hervoches[1] & Vasyl Ryukhtin[1]

Abstract: Our contribution demonstrates high-resolution properties of the unconventional 3-axix neutron diffraction setting permitting to finer studies of individual diffraction lines. For demonstration we used 3 samples of rotary swaged WNiCo pseudoalloy. As can be seen from the obtained results, the high-resolution setting could be used in some special studies exploiting diffraction profile analysis as e.g. micro-strain/micro-stress analysis of plastically deformed samples, determination of phases having very close values of the lattice spacings.

Keywords: Neutron diffraction; High-resolution setting; Tungsten heavy alloy; Rotary swaging

1. Introduction

It is well known that tungsten heavy alloys are often used in static or dynamically loaded components due to their excellent mechanical properties and high specific mass. One of the challenging properties is a relatively easy deformation processing thanks to their structure, consisting of hard tungsten particles embedded in a soft NiCo2W matrix. The structure of the WNiCo tungsten alloy can routinely be determined by neutron or X-ray diffraction and naturally then also strain/stress fields in deformed samples [1, 2]. When a resolution of the conventional diffractometers is sufficient, microstructure development i.e. dislocation density and a mean grain size, can be studied on the basis of diffraction line profile analysis [3, 4]. The purpose of the investigation presented in this contribution is a search for possible finer structure of the NiCo2W matrix by the high-resolution 3-axis diffractometer setting [5, 6] (see Fig. 1) providing the angular resolution determined by *FWHM* of diffraction profiles by about one order of magnitude higher. As can be illustrated in the Fig. 1, the 3-axis setting employing bent perfect crystal monochromator and analyzer exploits both focusing in real and momentum space. The high resolution is achieved when minimizing the dispersion of the whole system. It is achieved by setting optimum parameters (radii of curvatures R_M, R_A, distances L_{MS} and L_{SA} from the sample for given Bragg angles θ_M, θ_A) which means that the orientation of the momentum Δk

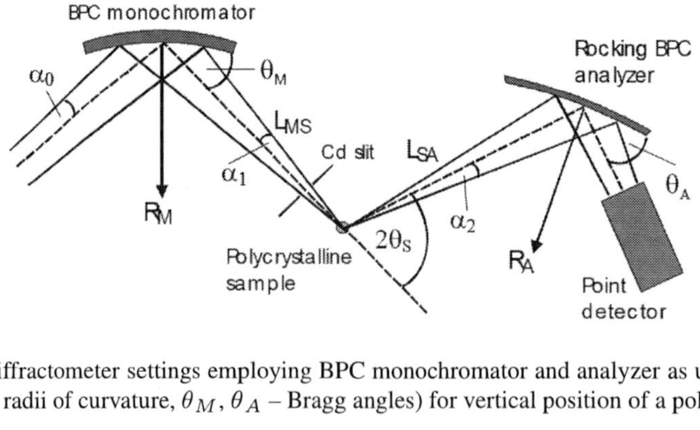

Fig. 1. Three axis diffractometer settings employing BPC monochromator and analyzer as used in the feasibility studies (R_M, R_A – radii of curvature, θ_M, θ_A – Bragg angles) for vertical position of a polycrystalline sample.

[1] *RNDr. Pavol Mikula, DrSc. Nuclear Physics Institute ASCR, 250 68 Řež, Czech Republic,* mikula@ujf.cas.cz

neutron domains related to the monochromator and analyzer are matched to that of the sample [5].

Present high-resolution neutron diffraction measurements can be considered as an addition to the extensive studies performed recently with medium resolution powder diffractometer [7] (in this paper other related references can be found).

2. Experimental

The experimental feasibility studies were carried out on the 3-axis neutron diffractometer installed at the Řež research reactor LVR-15. The diffractometer employs the Si(111) - monochromator and Si(311)-analyzer single crystals which have the dimensions of 200×40×4 mm^3 and 20×40×1.3 mm^3 (length ×width ×thickness), respectively. The monochromator Si(111) provides the neutron wavelength of 0.162 nm and has a fixed curvature with a radius R_M of about 12 m. The radius of curvature of the analyzer was changeable in the range from 3.6 m to 36 m. Other parameters of the setting were fixed. The changeable curvature of the analyzer permitted us to determine an optimum curvature minimizing the dispersion and thus optimizing the resolution of the setting. As to the samples, the W-Ni-Co (93-6-1) wt.% (80.9-16.4-2.7 at. %) pseudo-alloy was produced by powder metallurgy. The particle size of initial W, Ni and Co powders was in the range of 2–4 μm. The weighted mixture of the powders was homogeneously mixed and then sintered at 1500 °C under H_2 protective atmosphere, and subsequently quenched in water. The as-sintered material, i.e., bars with approximately 12 ×18 mm^2 elliptical cross- sections, is denoted as W0 throughout the following text. The sintered bars were further processed by rotary swaging (RS) into circular swaged bars with a diameter of 10 mm. RS was performed in two different ways: At room temperature (sample WA) and at 900 °C (sample WB). For the sake of comparison, we have reprinted and adapted the following Fig. 2 from the paper [7]. Fig. 2a shows the whole neutron diffraction spectrum related to the sample WB. This spectrum together with the related WA were used for phase determination as well as for microstrain analysis when compared their widened lines with the diffraction lines of the sample WO (before the swage deformation). In our case we focused our attention to the peak N.2 and analyzed it by the high-resolution 3-axis setting. The obtained results are shown in Fig. 3. The detailed inspection of the analyzer rocking curves reveals that in the case of WA and WB samples (after the swage deformation) and the NiCo2W phase (peak N.2), a fine substructure was identified which was earlier not detected by using the conventional medium-resolution powder diffractometer and which potentially could have some influence on the mechanical properties [7].

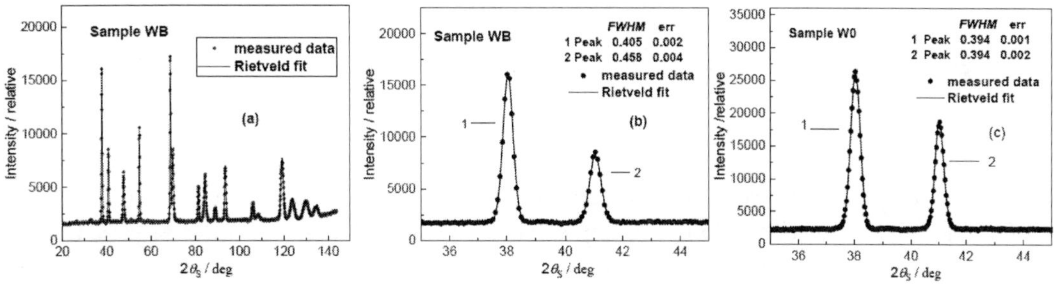

Fig. 2. Powder diffraction spectrum for the sample WB – (a), the cut part of it for smaller scattering angles – (b) and the cut part related to the reference sample W0; 1 – W matrix (BCC structure, 2 – NiCo2W phase (FCC structure).

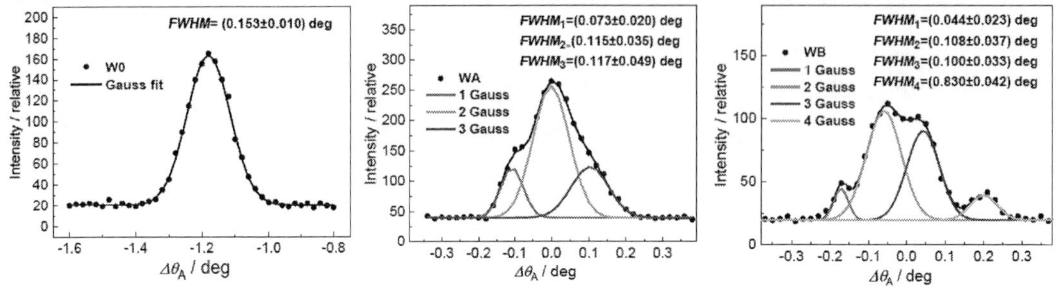

Fig. 3. Analyzer rocking curves analyzing the peak N.2 from the Fig. 2 for all three samples of the NiCo2W phase (FCC structure).

3. Conclusion

The feasibility of using the high-resolution three-axis diffractometer performance in some special studies exploiting diffraction profile analysis (e.g. strain/stress analysis of plastically deformed samples, determination of phases having very close values of the lattice spacings) is presented on three samples of the rotary swaged WNiCo pseudoalloy. Investigation was, namely, focused on the NiCo2W phase (see peak N.2 in Fig. 2). Contrary to the conventional powder diffractometer, fine substructure of the NiCo2W phase was identified. As there can be a certain relationship with the mechanical properties, further complementary experiments, as e.g. electron backscattering diffraction or high-resolution X-ray diffraction, could be very helpful for determination of the substructure phases. For the sake of comparison with the resolution of the conventional powder diffractometer, next Fig. 4 shows the analyzer rocking curves for the optimum analyzer curvature as obtained on three vertically situated α-Fe(110) standard samples of different diameters. Naturally, the resolution depends on the diameter (generally on the width and the thickness) of the sample.

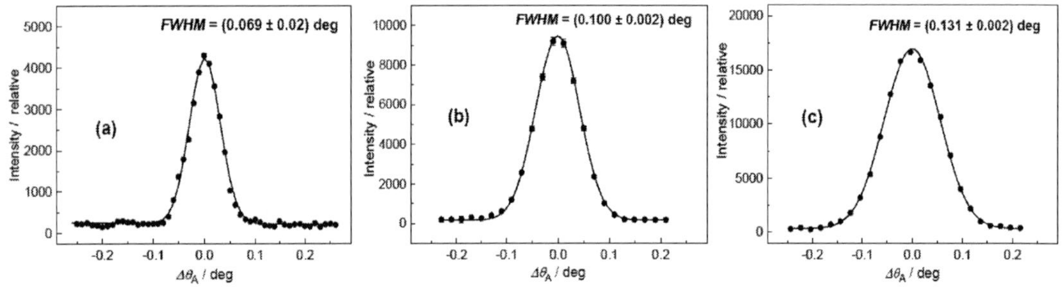

Fig. 4. Analyzer rocking curves for the optimum analyzer curvature as obtained on three vertically situated α-Fe(110) standard samples of different diameters; (a) – $\phi = 2$ mm, (b) – $\phi = 5$ mm and (c) – $\phi = 8$ mm.

The presented high-resolution 3-axis setting can also be used for macro-strain scanning; however, the measurement would be impractical because the step-by-step analysis (with the analyzer), would be time consuming. We hope that the presented neutron diffraction setting can offer an additional support to complement the information achieved by using the other conventional characterization methodologies.

Acknowledgements

Measurements were carried out at the CANAM infrastructure of the NPI CAS Řež supported through MŠMT project No. LM2015056. The presented results were also supported in the frame of the following MŠMT projects: LM2015048 and the infrastructural one – "Experimental nuclear reactors LVR-15 and LR-0." We thank Ms. B. Michalcová for a significant help in the measurements and basic elaboration of the data.

References

[1] Noyan I.C. and Cohen J.B., *Residual stress: Measurement by diffraction and interpretation*, ed. n. 1, (Springer-Verlag, New York Inc., 1987).

[2] Hutchings M.T. and Krawitz A.D. (Eds.), *Measurement of residual and applied stress using neutron diffraction*, (Springer Dordrecht, 1992).

[3] Delhez R., de Keijser T.H. and Mittemeijer E.J.,"Determination of crystallite size and lattice distortions through X-ray diffraction line profile analysis," *Z. Anal. Chem.* Vol. 312, pp. 1–16 (1982). https://doi.org/10.1007/BF00482725.

[4] Davydov V., Lukáš P., Strunz P. and Kužel R., "Single-line diffraction profile analysis method used for evaluation of microstructural parameters in the plain ferritic steel upon tensile straining", *Materials Science Forum*, Vol. 571-572, pp. 181–188 (2008). https://doi.org/10.4028/www.scientific.net/MSF.571-572.181.

[5] Vrána M., Lukáš P., Mikula P. and Kulda J., "Bragg diffraction optics in high-resolution strain measurements, " *Nucl. Instrum. Methods in Phys. Research A*, Vol. 338, pp. 125–131 (1994). https://doi.org/10.1016/0168-9002(94)90172-4.

[6] Mikula P., Šaroun J., Ryukhtin V. and Strunz P., "High-resolution strain/stress measurements by three-axis neutron diffractometer," *Materials - MDPI*, Vol. 13, pp. 5449–5456 (2020). DOI: 10.3390/ma13235449.

[7] Strunz P., Kunčická L., Beran P., Kocich R. and Hervoches C., "Correlating microstrain and activated slip systems with mechanical properties within rotary swaged WNiCo pseudoalloy," *Materials – MDPI*, Vol. 13, pp. 208–224 (2020). DOI:10.3390/ma13010208.

Knowledge Base for Evaluating the Strength and Fatigue Life of Weldments from Thin-walled Profiles Used in the Construction of Bus Bodyworks

Radovan Minich[1] & Miloslav Kepka[2]

Abstract: Systematic collection and analysis of data from calculations and experiments in laboratory and operational conditions is essential for knowledge base for evaluating the strength and fatigue life of weldments from thin-walled profiles used in the construction of bus bodyworks.

Keywords: Bus bodywork; Weldment from thin-walled profiles; Strain gauge measurement; Laboratory fatigue tests; Estimation of fatigue life

1. Introduction

The topic of the paper is the subject of doctoral studies. The aim of research is to classify systematically existing and new knowledge about the loading and fatigue strength of welded joints of bodyworks of buses, battery buses and trolleybuses so that the service life and reliability of the vehicle bodyworks can be predicted already at their design stage.

2. Methods

Data from older projects that have been solved in the past, especially in cooperation with former companies ŠKODA, KAROSA and IKARUS, have already been collected [1]. The sets of S–N lines are shown in Fig. 1.

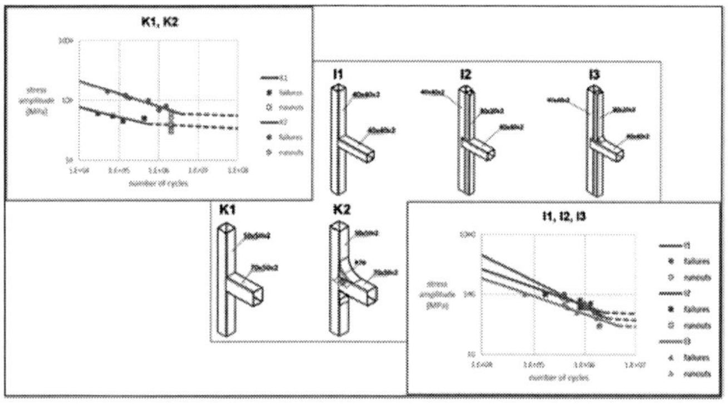

Fig. 1. The sets of S–N lines from past projects.

The data generated during the current cooperation with the bus manufacturers SOR and SOLARIS are now being evaluated. And the results of own research activities are supplemented.

The developed knowledge base contains experimentally determined S–N curves for various configurations of welded thin-walled profiles, static calculations of stress fields by the finite

[1] *Ing. Radovan Minich; Regional Technological Institute, research center of Faculty of Mechanical Engineering at University of West Bohemia; Univerzitni 8/2732, 30100 Pilsen, Czech Republic;* minichr@rti.zcu.cz

[2] *doc. Ing. Miloslav Kepka, CSc.; Regional Technological Institute, research center of Faculty of Mechanical Engineering at University of West Bohemia; Univerzitni 8/2732, 30100 Pilsen, Czech Republic;* kepkam@rti.zcu.cz

element method and their experimental verification on the basis of strain gauge measurements during loading in the laboratory. Laboratory fatigue tests are then performed to determine the fatigue strength of the evaluated structural nodes [2,3]. This is followed by the phase of measuring the service stresses at the equivalent locations of the investigated structural node during the real operation of the vehicle prototype. In Fig. 2 is an example of one such complex procedure, which was implemented in cooperation with the Research and Testing Institute Pilsen.

Fig. 2. The procedure: FEM calculation – laboratory fatigue tests including experimental stress analysis - strain gauge measurement during the real operation of the vehicle prototype.

The fatigue effects of vehicle operation can be predicted by measuring the stress at the vehicle body nodes as it runs on the test tracks. Data from measurements of various load cases are collected in the knowledge base. The measurement is usually performed with an empty bus and with a load that represents a fully occupied vehicle. For both of these load cases, the stresses in the critical bodywork nodes can be determined at various excitation types, see Fig. 3.

Maximum service stress amplitudes (or stress ranges) are estimated by strain gauge measurements on model testing track. The model testing track is usually made from artificial obstacles. The cylinder segments are laid on an even asphalt road in such composition, that overruns of them occurred gradually by right-hand wheels, by both wheels simultaneously and by left-hand wheels.

The most important are the measurement while the vehicle prototype is driving in real traffic on various unevenness's of real roads. The measured random stress-time histories are converted to fatigue damage using cumulative damage rules which have been proposed by various authors, for example Palmgren [4] and Miner [5], Haibach [5] and others.

Some customers also require measuring of the bodywork during tests of the vehicle on the proving ground that contains special tracks with defined parameters.

3. Example of knowledge base data

A detail of interest was the significantly stressed welded joint of thin-walled profiles in the upper corner of the door opening of the bus body shown in Fig. 4. S–N curve was determined in the laboratory for the critical cross-section and the same critical cross-section was measured during vehicle operation with strain gauges.

Fig. 3. Types of forced body vibration: crossing artificial standardized obstacles; driving on real roads with different surface quality; driving on special tracks on the test circuit.

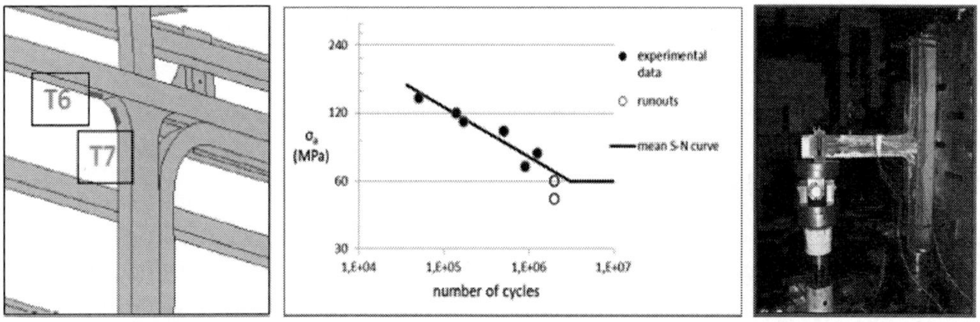

Fig. 4. Detail of interest (T6): schematic illustration; S-N line; test stand for laboratory fatigue tests.

The following Fig. 5, Fig. 6 and Fig. 7 illustrate examples of measured stresses for all the above-described load cases (the strain gauge T6).

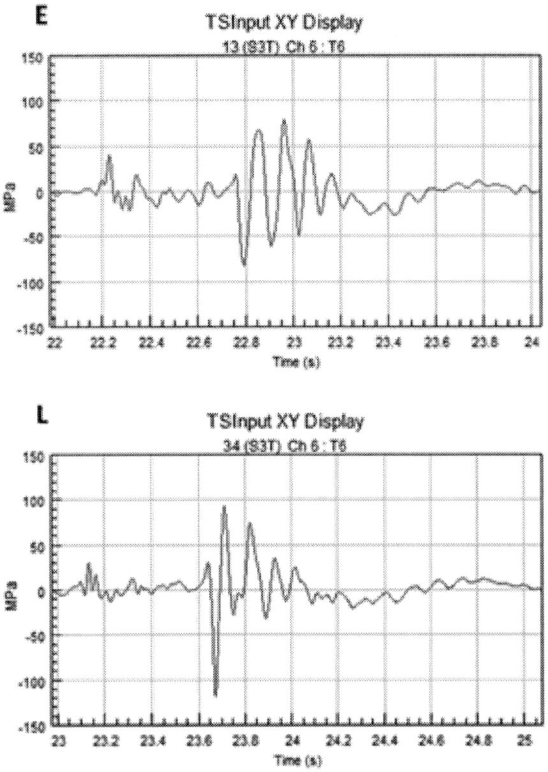

Fig. 5. Crossing of artificial obstacles by both wheels at the same time (E – empty bus, L – loaded bus).

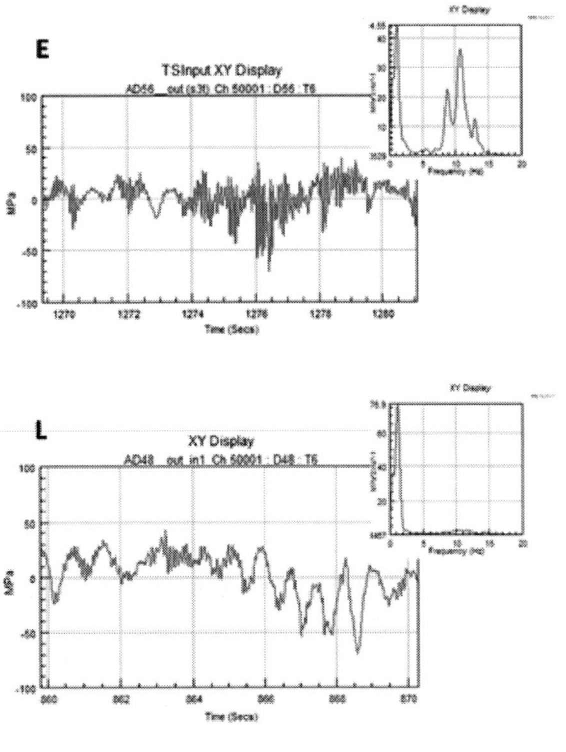

Fig. 6. Driving on real roads – short cuts from the measured signals (E – empty bus, L – loaded bus).

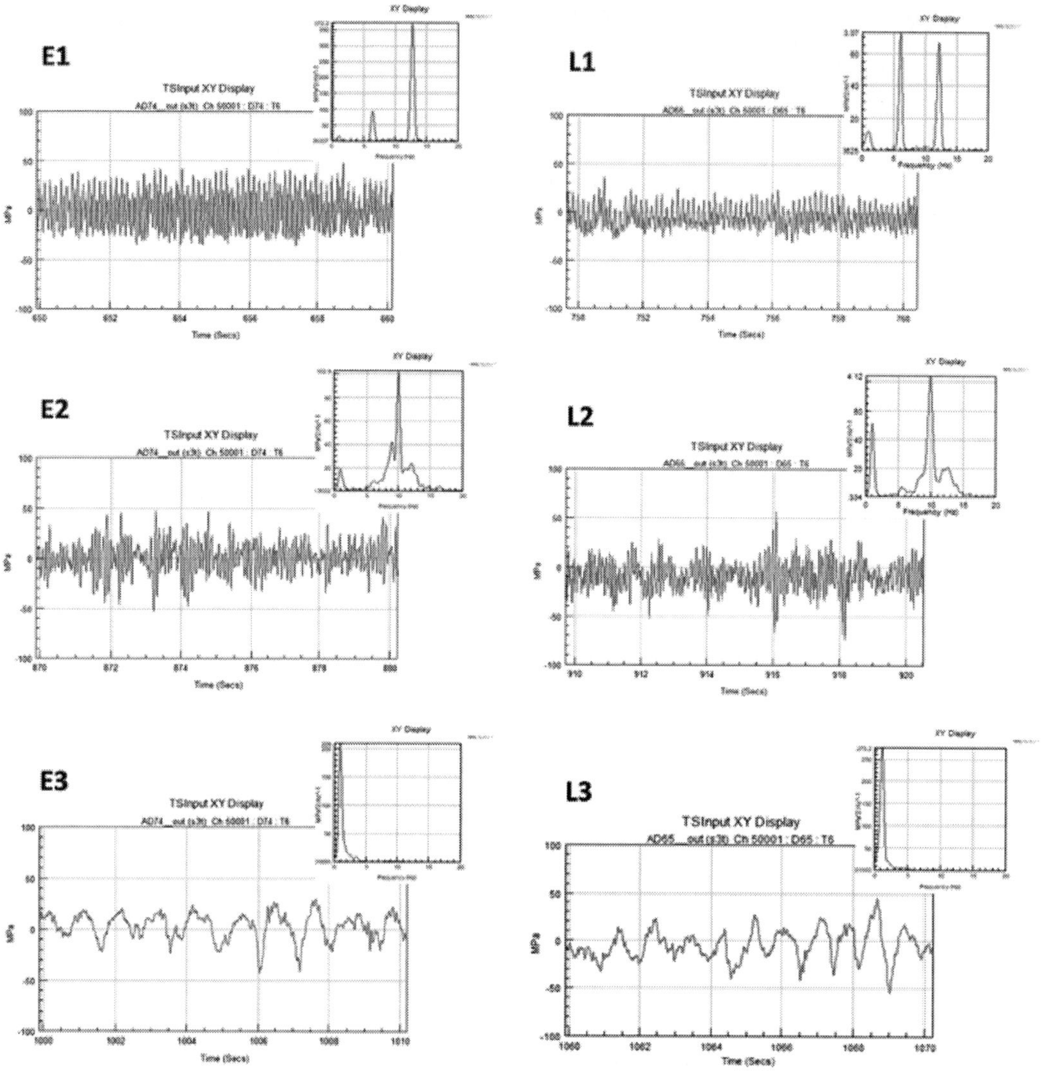

Fig. 7. Driving on special tracks on the proving ground - short cuts from the measured signals (E – empty bus, L – loaded bus, 1 – sinus resonance road, 2 – paved road, 3 – Belgian paving).

The paper [7] demonstrates on several case studies how the results of measurements (calculations) obtained under various load cases can be used.

Based on the maximum stress amplitudes measured or calculated by computer simulations of the vehicle crossing of artificial obstacles, it is possible to estimate the so-called design stress spectra usable in the initial stage of new vehicle development for dimensioning critical body nodes.

The stress-time courses, measured when the prototype vehicle is driven on uneven road surfaces, are converted into representative service stress spectra using the rain-flow counting method. These then serves as one of the inputs to the fatigue damage calculations.

Based on fatigue damage calculations, it is possible to estimate the potential shortening of the vehicle's service life test on the proving ground compared to the test in normal operation.

Fatigue damage and fatigue life prediction are usually estimated using the linear fatigue damage accumulation hypothesis. Various authors apply different boundary conditions to the fatigue damage calculation. A schematic representation of these boundary conditions is shown in Fig. 8.

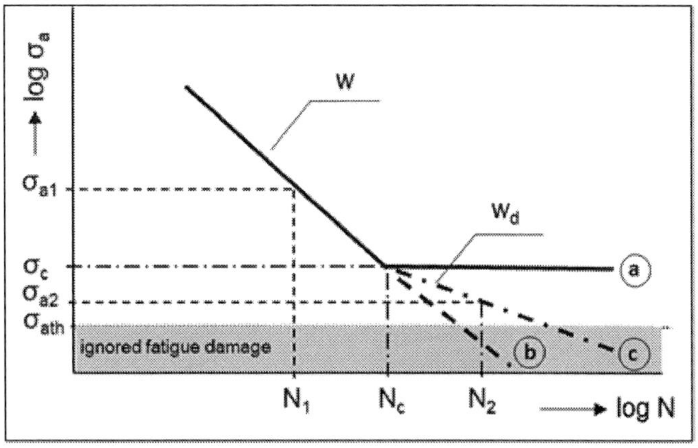

Fig. 8. Boundary conditions for calculating cumulative fatigue damage: (a – Palmgren-Miner original, b – Palmgren- -Miner elementar, c – Haibach).

4. Conclusions

The response of bus bodyworks to loading at real operation has the character of random processes. Their statistical characteristics and time courses can be precisely analysed only on the basis of long-term measurements. Fatigue analysis is fundamentally different from dynamic analysis. Dynamic analyses are performed in the frequency domain; fatigue analyses must be performed in the time domain. An important phenomenon can be the occurrence of extraordinary overloads, which occur due to extraordinary operating situations as for example driving over high road roughness or driving over a pothole.

The authors collect and analyse data from strain gauge measurements and laboratory fatigue tests and compare them with the results of computer simulations, performed mainly by the finite element method. Their goal is to contribute to the dimensioning of welded nodes of thin-walled body profiles for the required fatigue life to be possible and reliable already at the stage of digital design of the new vehicle structure.

Acknowledgements

The authors prepared this paper with the support of the Ministry of Industry and Trade of the Czech Republic, the project FV40260 "On-line measurement and analysis of the operational loading of structures with adaptive virtual models". The authors also thank the Research and Testing Institute Pilsen, which provided a number of experimental results obtained in the past.

References

[1] Kepka Jr. M., *Degradation of mechanical properties of cyclically loaded materials and structural nodes* (University of West Bohemia, Pilsen, 2021), Dissertation.

[2] Minich R. and Kepka Jr. M., *Laboratory fatigue tests of bodywork welded joints of SOR articulated electric bus: preliminary FEM analysis and design of test stands* (University of West Bohemia, Pilsen, 2020), Research report.

[3] Minich R., Kepka Jr. M. and Tittel J., *Laboratory fatigue tests of bodywork welded joints of SOR articulated electric bus* (University of West Bohemia, Pilsen, 2021), Research report.

[4] Palmgren A.Z., "Die Lebensdauer von Kugellagern," *Z. Ver. Deutsch. Ing.*, 68:339 (1924).

[5] Miner M.A., "Cumulative damage in fatigue," *Transactions of the ASME. Series E. J. Appl. Mech.*,12:159–64 (1945).

[6] Haibach E., *Modified Linear Damage Accumulation Hypothesis Accounting for a Decreasing Fatigue Strength During the Increasing Fatigue Damage* (LBF, Darmstadt, Germany, 1970), TM Nr. 50.

[7] Kepka M. and Kepka Jr. M., "Consideration of random loading processes and scatter of fatigue properties for assessing the service life of welded bus bodyworks," *International Journal of Fatigue*, 151 (2021) 106324, https://doi.org/10.1016/j.ijfatigue.2021.106324.

An Alternative Method for Residual Stress Measurement in Railway Axles

Radka Musilová[1], Jaroslav Václavík[2] & Rostislav Fajkoš[3]

Abstract: Residual stresses in railway axles should be measured according specified procedures, given in standard EN 13261. The goal of this article is to document, that groove grinding method is very suitable and sensitive for such type of measurement. The stress measurement in part of the axle was performed using groove grinding and saw cut methods and obtained results are compared and discussed.

Keywords: Residual stress; Groove grinding; Saw cut; Railway axle

1. Introduction

The companies producing railway axles are obliged to demonstrate the level of residual stresses in their products based on the specification of standard EN 13261 [1] in which the applicable methods are suggested and the allowable values of residual stresses are specified. The standard requires that surface tension residual stresses measured in the depth of 0.1 mm below the axle surface are lower than 100 MPa an in addition that the mutual deviation of the residual stresses measured at 6 points separated by the angle of 60° in the central section of the axle (a difference of the maximal and minimum values) must be within the interval up to 40 MPa. The hole drilling method is standardly used for this task. However in some cases, especially in the case of hardened or roll-burnished railway axles this method gives the uncertainties comparable with the demanded stress values [2]. Another saw cut method [3] is also recommended in [1], appendix E (Fig. 5). In that case, six strain gauges should be attached to the axle in axial direction, separated by the angle 60°. After cutting the plate of 25 mm thick with strain gauges in its middle, the residual stresses can be calculated directly from the released strains. We propose to use for this task groove grinding method [4] (Fig. 1).

2. Theory

The principle of the groove grinding method consists in grinding an annular grove at the depth increments perpendicularly to the surface. The stress is calculated from relieved strains at the strain gauge glued perpendicularly at a given distance from the groove edge with the help of the calibration constants (Fig. 1, Fig. 4) [4].

Standardly the groove is grinded immediately at end of the strain gauge HBM 0.6LY11 foil and the maximum depth of the groove is 2 mm. The groove grinding method is nearly three times more sensitive than hole-drilling method as it can be seen from Fig. 2 and we recommend it to this type of measurement.

[1] *Ing. Radka Musilová, Ph.D.; Dynamic Testing Laboratory, Research and Testing Institute Pilsen; Tylova 1581/46, 301 00 Pilsen, Czech Republic;* musilova@vzuplzen.cz

[2] *Ing. Jaroslav Václavík; Dynamic Testing Laboratory, Research and Testing Institute Pilsen; Tylova 1581/46, 301 00 Pilsen, Czech Republic;* vaclavik@vzuplzen.cz

[3] *Ing. Rostislav Fajkoš, Ph.D.; GHH-BONATRANS GROUP a.s., Revoluční 1234, 735 94 Bohumín, Czech Republic;* rostislav.fajkos@ghh-bonatrans.com

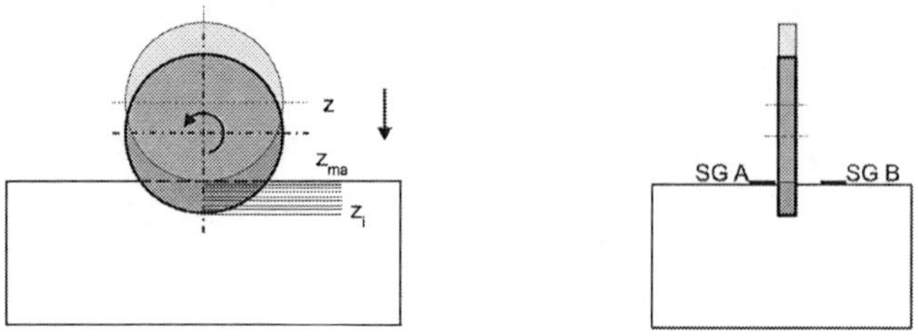

Fig. 1. Principle of groove grinding method. One or two strain gauges can be used for released strain measurement.

On the other hand, the saw cut method is suitable only for cases of homogenous stress field not significantly changed with the depth. The high stress surface peaks cannot be detected with this method. The obtained error is roughly demonstrated in Fig. 3. The curve represents the percentage of evaluated stress, if the stress occurs from surface to the z-depth and then no stress is presented in greater depths. E.g. when the stress is presented only in 1 mm surface layer, then less than 20 % of really presented residual stresses are evaluated.

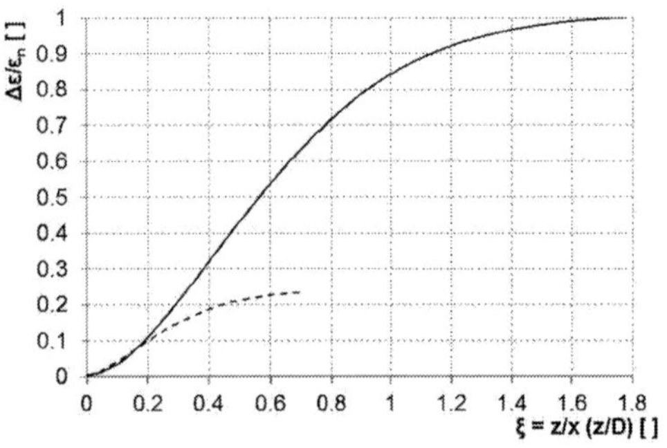

Fig. 2. Comparison of the sensitivity of groove grinding and hole drilling methods.

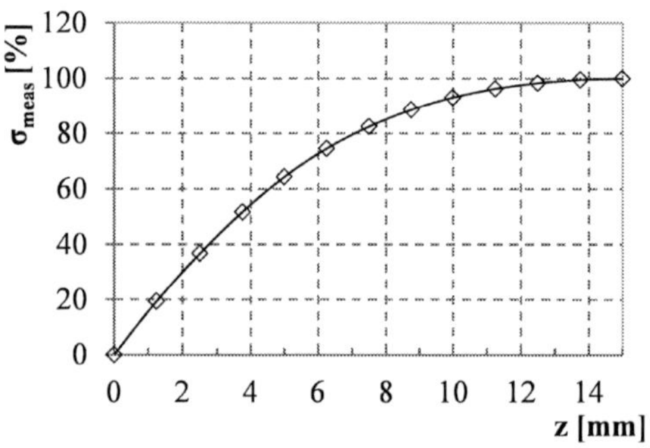

Fig. 3. Evaluated percentage stress with saw cut method for surface stress up z-depth.

Fig. 4. Measurement of residual stresses using groove grinding method on a part of the axle.

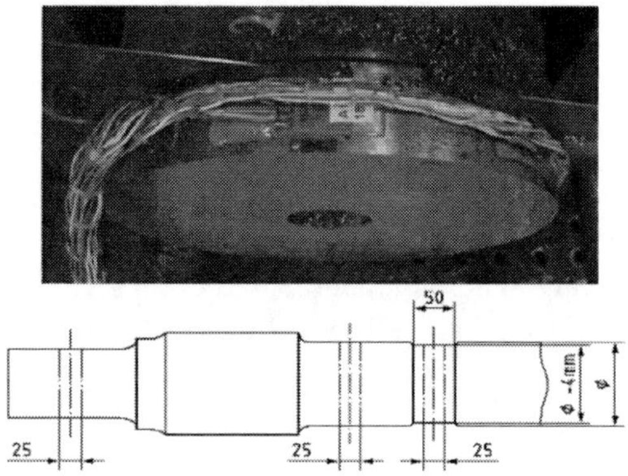

Fig. 5. Principle of saw-cut method and cut plate with strain gauges (Figure used from [3]).

3. Performed test

3.1. Groove grinding method

The investigation was made on surface-hardened middle part of railway axle (Fig. 4) with groove grinding method. The diamond cutting wheel ⌀ 38 mm was used with 15000 rpm. The relieved strains at 2 mm distance between the cutting wheel edge strain gauge middle point and evaluated axial stresses using integral method are given in Fig. 6 and Fig. 7. The measurement was performed in one axle section at six 60° separated measured points. There is high stress dispersion up to 0.5 mm depth but in higher depths is lower (Fig. 10). The dispersion of residual stresses in the 2 mm depth between measured points is given in Fig. 10. It can be seen from this figure that the stress dispersion is lower than 40 MPa just fulfilling the standard [1] requirements.

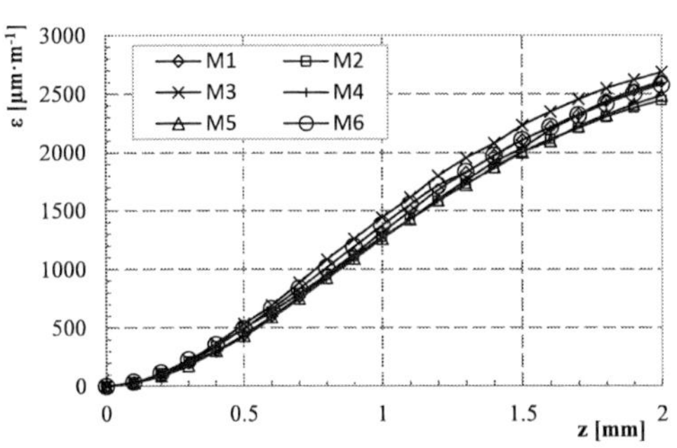

Fig. 6. Released strain by groove grinding method at six measured points in one axle section, angularly of 60° from each other separated

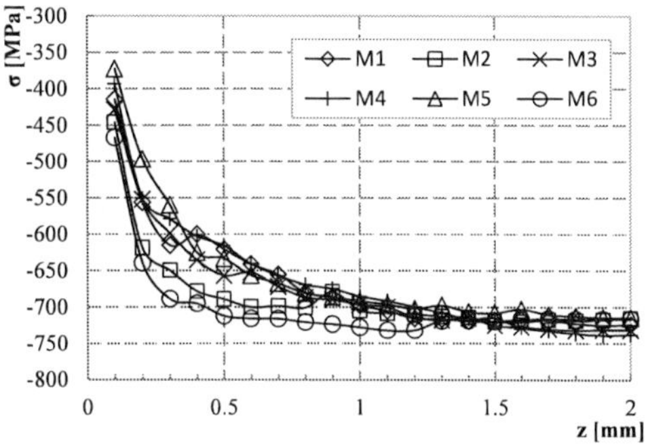

Fig. 7. Residual stresses evaluated from groove grinding and for released strains, given in Fig. 6

3.2. *Saw-cut method*

Six strain gauge rosettes HBM RY91-6/120 were glued with the adhesive HBM Z70 at six measured points in one axle section, 50 mm from the axle part end in axial direction angularly of 60° from each other separated. Only two grids from each rosette were used for measurement. Each rosette was covered with HBM SG250. The initial reading was made before the cutting the disk. Then the disc of 25 mm thickness was cut from the axle on the saw machine performing two cuts just leaving the strain gauge rosettes in the middle of the disc. The second strain gauge reading was made after cutting. Supposing the principal residual stresses are lying in axial end tangential directions the residual stresses were calculated using common relations for bi-axial stress distribution.

$$\sigma_1 = \frac{E}{1-\mu^2}(\varepsilon_1 + \mu\varepsilon_2); \qquad \sigma_2 = \frac{E}{1-\mu^2}(\varepsilon_2 + \mu\varepsilon_1) \tag{1}$$

The results of evaluated residual stresses near the surface compared with saw cut method are given in Fig. 11. The stress dispersion lower than 20 MPa was measured.

Fig. 8. Saw-cutting one face disk from the axle.

Fig. 9. Cut disc width 25 mm from the axle with installed strain gauges.

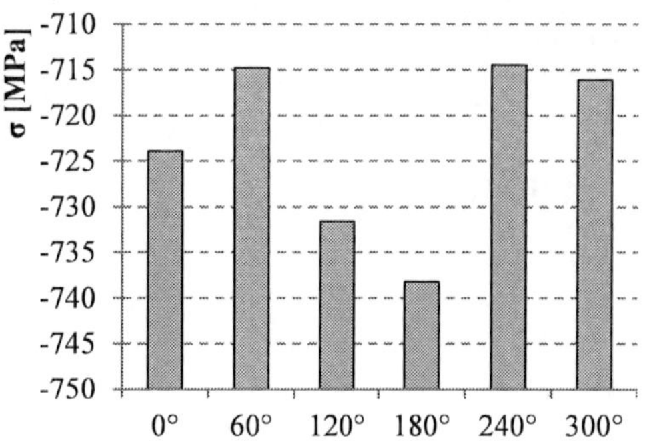

Fig. 10. Dispersion of residual stresses in the 2 mm depth measured with groove grinding method.

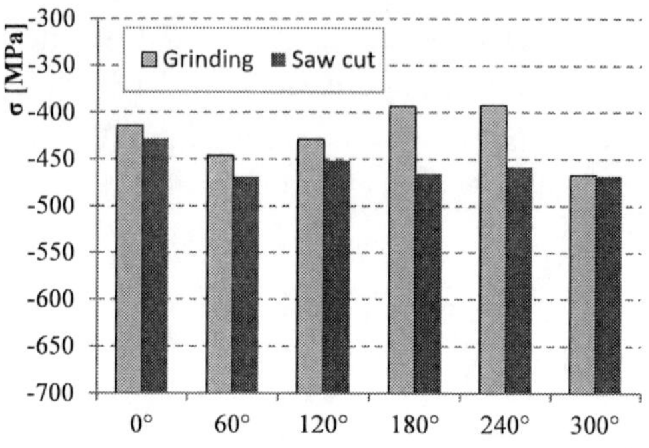

Fig. 11. Comparison of residual stress measured with groove grinding and saw cut method at the surface.

It was also verified the standard recommendation, that the tangential stress component obtained after cutting the disk is low compared with the axial component and it is not necessary to be measured.

4. Conclusion

It was found that the groove grinding method can be used for residual stress evaluation in railway axles according the standard EN 13261 with sufficient sensitivity even on hardened surfaces, where the hole-drilling method is sometimes difficult to use. The results were experimentally verified with comparison between grove-grinding, X-Ray diffraction and saw cut methods. Even if the saw cut method cannot catch the high stress peaks at the surface, both methods give surprisingly comparable results.

Acknowledgements

The article has originated in the framework of the institutional support for the long-term of conceptual development of the research organization.

References

[1] EN 13 261 Railway applications – Wheelsets and bogies – Axles – Product requirements (2021).

[2] Řeha B., Václavík J., Návrat T. and Halabuk D., "Residual stresses in railway axles," in Proceedings of 10th European Conference on Residual Stresses ECRS10, Leuven, Belgium, September 2018. Edited by Marc Seefeldt KU Leuven, department of Materials Engineering. Part of Material Research Proceedings, Vol.6, pp. 95–100, (2018). ISBN 978-1-94529188-3 (Print). ISBN 978-1-94529189-0 (eBook). http://dx.doi.org/10.21741/9781945291890-16.

[3] Berger A.W. at al, "Eigenspannungen in Radsatzwellen –Vergleichende zerstörende und zerstörungsfreie Untersuchungen," in 11. Fachtagung ZfP im Eisenbahnwesen, Erfurt, March 12,2020, NDT.net issue,(2020-08). https://www.ndt.net/article/dgzfp-rail-2020/papers/3.pdf.

[4] Václavík J., Hejman M. and Weinberg O., "Groove grinding method for near the surface residual stress measurement," in Proceedings of 57th International Conference Experimental Stress Analysis 2019, Brno University of Technology, Luhačovice, CD ROM, June 2019. ISBN 9781510889699.

The Influence of Microstructures on Rolling Resistance

Jiří Ondrášek[1]

Abstract: The subject of the paper is the presentation of the experimental work to determine the influence of microstructures on the rolling resistance. The tribological properties in the rolling contact of two bodies could be improved by a functional formation of the microstructure on the contact surface of one of the bodies that co-form the rolling or general kinematic pair. The microstructure is created by a set of micro-holes that are made by using laser technology. The accumulation of the larger volume of a lubricant between the contact surfaces of the bodies and the reduction of the contact area of the mutually rolling bodies are the possible advantages of the microstructures on the contact surfaces, which result in the rolling resistance reduction and better heat dissipation during the relative movement. The aim of the experiments is to determine the power consumption of the test device depending on the microstructure created on the specimen contact area and its density. The test principle consists of the rolling of a cylindrical specimen between three loading discs, while one of them is driven by an electric motor and at the same time a loading force is acting on it.

Keywords: Rolling resistance; Microstructure; Cam mechanism; Rolling kinematic pair

1. Introduction

Besides other activities, VÚTS, a.s. deals with the design, development and production of cams and cam mechanisms, see [1]. For this purpose, computational programs and procedures for the analysis and synthesis of combined cam mechanisms have been developed here. Conventional cam mechanisms are characterized by the ability to transfer high powers at high speeds with high positional accuracy; their applicability is mainly connected with the so-called hard automation that is characterized by unchangeable or difficult to change actions of given technical equipment.

A cam mechanism is a three-link mechanical system with single degree of freedom that contains at least one cam connected to the other links by at least one general kinematic pair (GKP), see [2–4]. The general kinematic pair is formed by the contact of the cam and the follower. The follower performs translating, oscillating or general movement. To reduce passive resistances, the follower is in many cases provided with a roller, while the contact surface of the roller is cylindrical or crowned. Another possible way to reduce the passive resistances in a cam mechanism is to apply the microstructure on the contact surface of the roller by using laser technology. The microstructure is created by a set of micro-holes. The purpose of the intentionally formed microstructure on the contact surface of the roller or the roller pin is to capture a larger volume of lubricant between the contact surfaces of the bodies and thus achieve a reduction in the passive resistances and better heat dissipation during their relative movement. Another intention of the microstructure creation is the reduction of the contact area of the mutually rolling bodies with possible impact on the reduction of the rolling resistance.

2. The tests principle

We tried to determine the microstructure influence on the rolling resistance based on the power consumption of the test rig on which the experiments were performed. Furthermore, the influence of the microstructures was evaluated using the rolling resistance coefficient during the pure

[1] *Ing. Jiří Ondrášek, Ph.D.; VÚTS, a.s.; Svárovská 619, 460 01 Liberec, Czech Republic;* jiri.ondrasek@vuts.cz

rolling between the specimen and the discs. We determined the rolling resistance coefficient in an indirect way, which consisted in the application of Newton's laws of motion in solving the dynamics of the mechanical system of the test device.

2.1. The test rig

The aim of the experiments is the determination of the size of the power consumption of the test rig in dependence on the microstructure applied to the contact surface of the loaded body 5, microstructure density, magnitude of the loading force F and magnitude of the angular velocities of the rotating bodies. In the case of the test rig, the relation applies between the constant angular velocities Ω_2 and ω of the discs and the specimen respectively:

$$\omega = 4 \cdot \Omega_2 \tag{1}$$

The kinematic scheme of the conceptual design of the test device for specimens is shown in Fig. 1.

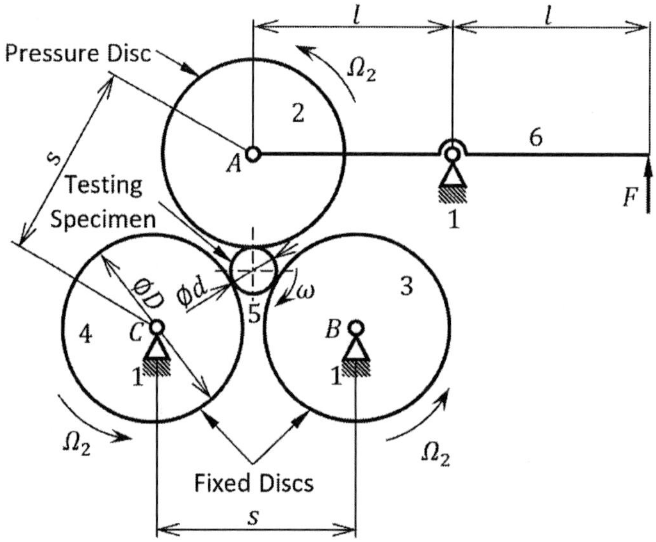

Fig. 1. Kinematic scheme of test device.

The tests principle consists of the rolling of the cylindrical specimen 5 among three loading discs 2, 3, 4, while one of the discs is driven by an electric motor and a loading force F acting on it at the same time. In this case, the stress on the contact surface of the specimen has a character of pulses. It is therefore a way of stressing by transient loading. The loading body represents a cylindrical roller and the loaded body, in analogy with the cam mechanism with a radial cam and roller follower, which corresponds to the cam. The loaded body is stressed by periodic force F by means of the loading body, the amplitude takes on the size depending on the material of the loaded body and on the method of processing its contact surfaces.

Specimen 5 of the cylindrical shape with an exact defined width is mounted on the shaft among three rotating discs 2, 3 and 4. The discs are arranged in a common plane at the apices of an equilateral triangle with the side s and their rotation axes are parallel. Disc 2 is the pressure disc and the other discs 3 and 4 are fixed. The pressure disc 2 is rotatably supported on one of the arms of the pivotally mounted two-arm lever 6 with a length of arms l. The two-arm lever is coupled through its second arm with a loading force F of the defined size. In terms of the geometrical arrangement of the loading discs and the loaded specimen, the specimen is loaded with three equally sized reactions F, from each disc once. Due to the rotation of test body 5 among the three discs rotated with the same constant angular velocity Ω_2, loading force F has the character of pulses with a period of $2\pi/3$. In the case of this type of the test device, the main aim of the experiments is to achieve fatigue damage of the specimen contact area after a certain number of cycles. The three-point contact reduces the testing time to 1/3, it makes testing much easier as each specimen can be tested for up to 10^8 cycles. The real implementation of this device is shown in Fig. 2 and more information about the test rig is in [4].

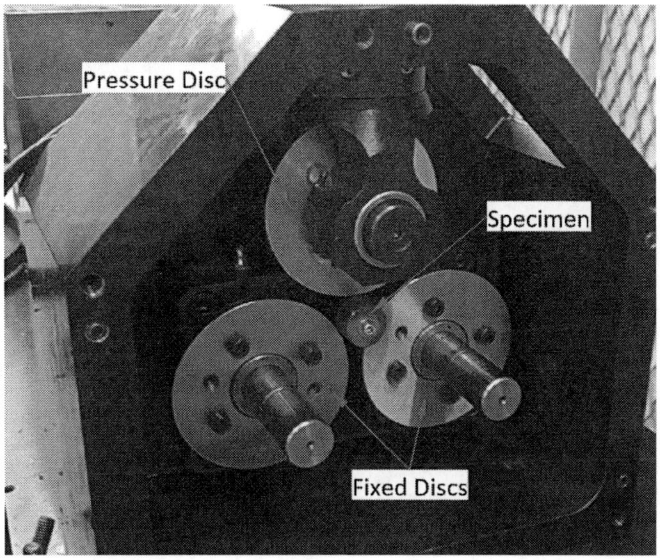

Fig. 2. Test device.

2.2. Rolling resistance coefficient

The influence of the microstructures in the contact surfaces of discs 2, 3 and 4 and specimen 5 during pure rolling is expressed by the part of power input P_ξ needed to overcome the rolling resistance of the discs and the specimen and the estimated value of the rolling resistance arm ξ. The total mechanical input P of the test device at the given uniform speed Ω_2 and a given magnitude of loading force F is determined by means of the sum of partial mechanical power inputs:

$$P = P_{Motor} + (P_A + P_B + P_C) + P_\xi \quad (2)$$

The variable P_{Motor} expresses the mechanical power input of the electric motor itself and the variables P_A, P_B, P_C are the power inputs to overcome the passive resistances in the rotary bearings of the loading discs 2, 3, 4 in locations *A*, *B*, *C*. We estimate the value of the rolling resistance arm ξ on the basis of the application of Newton's laws of motion in solving the problem of the dynamics of said mechanical system with uniform rotational movement of its members. If we divide Eq. 2 by the angular velocity Ω_2 of the rotating discs, we obtain the corresponding moment effects:

$$M = M_{Motor} + (M_A + M_B + M_C) + M_\xi \quad (3)$$

Quantity M represents the total driving torque of the device. M_{Motor} expresses the moment to drive the electric motor itself. The quantities M_A, M_B, M_C denote the rolling resistance moments of rolling bearings in the rotating bearings of the load discs. M_ξ is the moment to overcome the rolling resistance caused by contacts of the discs 2, 3, 4 with the specimen 5. The driving body of the mechanical system of the test device is pressure disc 2, which is driven by an electric motor. We determine the driving torque M_d needed to drive the mechanical system to overcome the passive resistances in the system by the calculation from Eq. 3:

$$M_d = M - M_{Motor} = (M_A + M_B + M_C) + M_\xi \tag{4}$$

We consider the problem of solving dynamics as a plane case. We compile two equations of force equilibrium and one moment equation for each body. The following relations are valid to release pressure disc 2 according to Fig. 3:

$$\begin{aligned} x: &\quad -R_{Ax} + T_{52} = 0 \\ y: &\quad -R_{Ay} + N_{52} = 0 \\ M^A: &\quad R \cdot T_{52} + M_{52} + M_A - M_d = 0 \end{aligned} \tag{5}$$

To release fixed disc 3 according to Fig. 4, the following relations apply:

$$\begin{aligned} x: &\quad -R_{Bx} + N_{53}\cos 30^\circ + T_{53}\sin 30^\circ = 0 \\ y: &\quad -R_{By} - N_{53}\sin 30^\circ + T_{53}\cos 30^\circ = 0 \\ M^A: &\quad -R \cdot T_{53} + M_{53} + M_B = 0 \end{aligned} \tag{6}$$

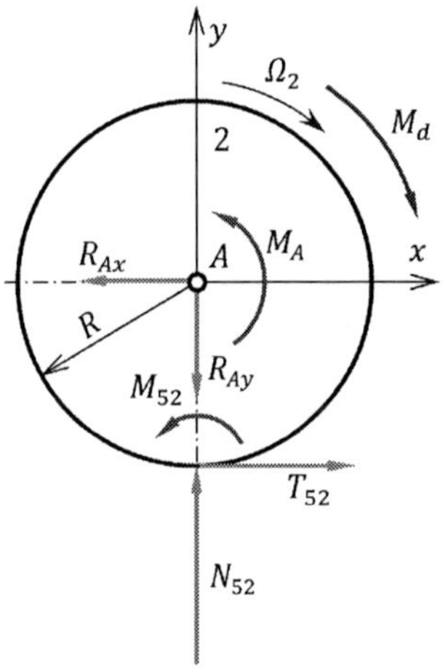

Fig. 3. Releasing the pressure disc-body 2.

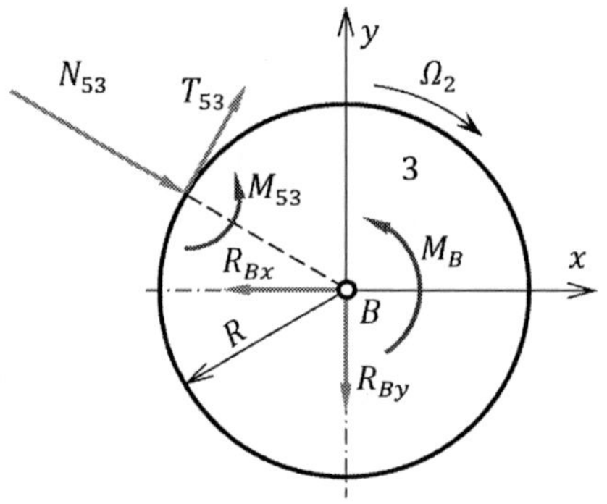

Fig. 4. Releasing the fixed disc - body 3.

To release fixed disc 4 according to Fig. 5, the following relations apply:

$$\begin{aligned} x: &\quad -R_{Cx} - N_{54}\cos 30^\circ + T_{54}\sin 30^\circ = 0 \\ y: &\quad -R_{Cy} - N_{54}\sin 30^\circ - T_{54}\cos 30^\circ = 0 \\ M^A: &\quad -R \cdot T_{54} + M_{54} + M_C = 0 \end{aligned} \tag{7}$$

To release specimen 5 according to Fig. 6, the following relations apply:

$$\begin{aligned} x: &\quad -T_{25} - (T_{35} + T_{45})\sin 30^\circ - (N_{35} - N_{45})\cos 30^\circ = 0 \\ y: &\quad -N_{25} + (N_{35} + N_{45})\sin 30^\circ - (T_{35} - T_{45})\cos 30^\circ = 0 \\ M^{S_5}: &\quad M_{25} + M_{35} + M_{45} + r \cdot (-T_{25} + T_{35} + T_{45}) = 0 \end{aligned} \tag{8}$$

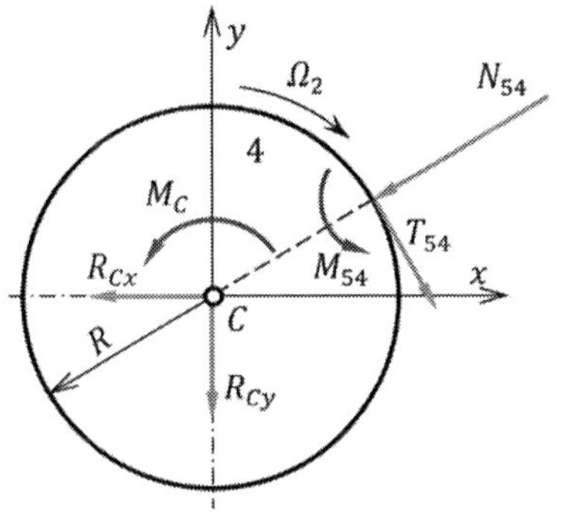

Fig. 5. Releasing the fixed disc - body 4.

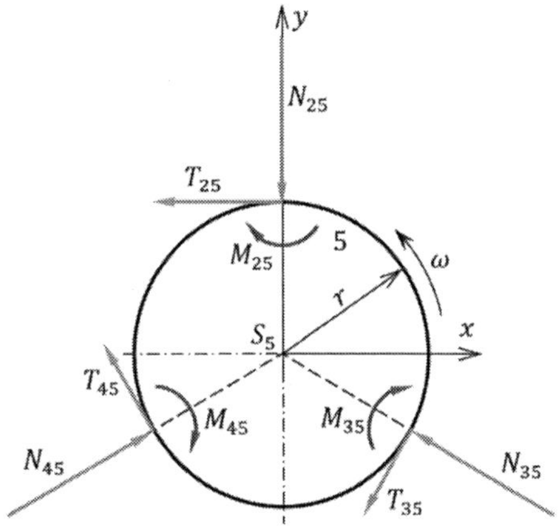

Fig. 6. Releasing the specimen - body 5.

To release two-arm lever 6 according to Fig. 7, the following relations apply:

$$\begin{aligned} x: &\quad R_{Ax} - R_{Dx} = 0 \\ y: &\quad F + R_{Ay} - R_{Dy} = 0 \\ M^A: &\quad l \cdot F - l \cdot R_{Ay} + M_A = 0 \end{aligned} \tag{9}$$

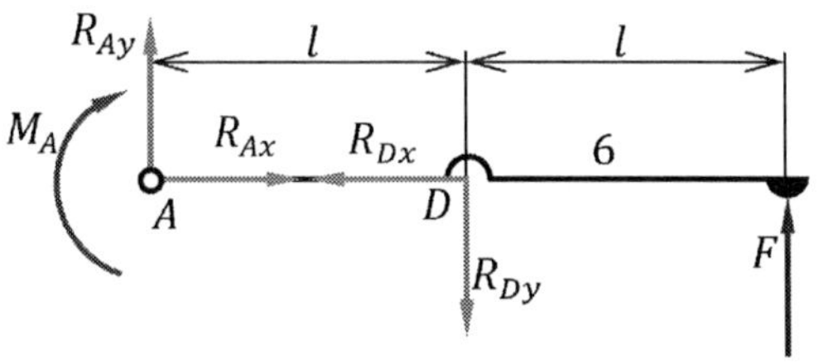

Fig. 7. Releasing the two-arm lever - body 6.

Based on Newton's third law and the expression of rolling resistance moments in general kinematic pairs using the rolling resistance arm ξ and normal forces, the following relations are valid:

$$\begin{aligned} N_{25} = &\quad N_{52}, & N_{35} = &\quad N_{53}, & N_{45} = &\quad N_{54} \\ T_{25} = &\quad T_{52}, & T_{35} = &\quad T_{53}, & T_{45} = &\quad T_{54} \\ M_{25} = M_{52} = &\quad \xi \cdot N_{52}, & M_{35} = M_{53} = &\quad \xi \cdot N_{53}, & M_{45} = M_{54} = &\quad \xi \cdot N_{54} \end{aligned} \tag{10}$$

We assume that the rolling resistance arm ξ is the same in each general kinematic pair of loading discs 2, 3, 4 and specimen 5, because the material properties of the contact areas and the structure of the contact surfaces are identical.

Based on the equilibrium equations Eq. 5 to Eq. 9 together with Eq. 10, we compile a set of equations **f**(**z**) containing a vector of unknown variables **z** to determine the size of the rolling resistance arm ξ:

$$\mathbf{f(z)=0}, \qquad \mathbf{z} = [N_{52} N_{53} N_{54} T_{52} T_{53} T_{54} \xi]^T$$

$$\begin{bmatrix} N_{52} - f + \frac{M_A}{l} \\ \xi \cdot N_{52} + R \cdot T_{52} + M_A - M_d \\ \xi \cdot N_{53} - R \cdot T_{53} + M_B \\ \xi \cdot N_{54} - R \cdot T_{54} + M_C \\ (-N_{53} + N_{54}) \cdot \cos 30^\circ - T_{52} - (T_{53} + T_{54}) \cdot \sin 30^\circ \\ -N_{52} + (N_{53} + N_{54}) \cdot \sin 30^\circ - (T_{53} - T_{54}) \cdot \cos 30^\circ \\ \xi \cdot (N_{52} + N_{53} + N_{54}) + r \cdot (-T_{25} + T_{35} + T_{45}) \end{bmatrix} = \begin{bmatrix} 0 \\ 0 \\ 0 \\ 0 \\ 0 \\ 0 \\ 0 \end{bmatrix} \qquad (11)$$

In the case of Eq. 11, it is a system of nonlinear equations, which we will solve numerically using Newton's method, see e.g. [5]. This method is based on the expansion of the system functions Eq. 11 into a Taylor series around the kth iteration $\mathbf{z}^{(k)}$:

$$\begin{gathered} \mathbf{f}(\mathbf{z}) = \mathbf{f}(\mathbf{z}^{(k)}) + \frac{\partial \mathbf{f}}{\partial \mathbf{z}^T}\bigg|_{\mathbf{z}^{(k)}} \cdot \Delta\, \mathbf{z}^{(k)} = \mathbf{f}(\mathbf{z}^{(k)}) + \mathbf{J} \cdot \Delta \mathbf{z}^{(k)} \\ \mathbf{z}^{(k)} = [N_{52}^{(k)}\ N_{53}^{(k)}\ N_{54}^{(k)}\ T_{52}^{(k)}\ T_{53}^{(k)}\ T_{54}^{(k)}\ \xi^{(k)}]^T, \quad \mathbf{J} = \frac{\partial \mathbf{f}}{\partial \mathbf{z}^T}\bigg|_{\mathbf{z}^{(k)}} \end{gathered} \qquad (12)$$

This gives the set of the linear algebraic equations for the correction of the dependent coordinates in the kth step $\Delta \mathbf{z}^{(k)}$. In the system of the linear equations Eq. 12, the matrix $\mathbf{J}$ represents the Jacobian matrix of the system of the nonlinear equations Eq. 11 corresponding to the vector of the unknown variables $\mathbf{z}$:

$$\mathbf{J} = \begin{bmatrix} 1 & 0 & 0 & 0 & 0 & 0 & 0 \\ \xi^{(k)} & 0 & 0 & R & 0 & 0 & N_{52}^{(k)} \\ 0 & \xi^{(k)} & 0 & 0 & -R & 0 & N_{53}^{(k)} \\ 0 & 0 & \xi^{(k)} & 0 & 0 & -R & N_{54}^{(k)} \\ 0 & \cos 30^\circ & -\cos 30^\circ & 1 & \sin 30^\circ & \sin 30^\circ & 0 \\ -1 & \sin 30^\circ & \sin 30^\circ & 0 & -\cos 30^\circ & \cos 30^\circ & 0 \\ \xi^{(k)} & \xi^{(k)} & \xi^{(k)} & -r & r & r & N_{52}^{(k)} + N_{53}^{(k)} + N_{54}^{(k)} \end{bmatrix} \qquad (13)$$

We express next iteration $(k + 1)$ of classical Newton's method by the equation:

$$\mathbf{z}^{(k+1)} = \mathbf{z}^{(k)} + \Delta \mathbf{z}^{(k)} \qquad (14)$$

We continue the calculation until the following condition is reached:

$$\left|\Delta \mathbf{z}^{(k)}\right| \leq \varepsilon, \qquad \left|\Delta \mathbf{z}^{(k)}\right| = \sqrt{\sum_{i=1}^{7} \left(\Delta z^{(k)}\right)^2} \qquad (15)$$

The rate of the convergence of Newton's method depends on the choice of the initial approximation $\mathbf{z}^{(0)}$. In condition 15 $\left|\Delta \mathbf{z}^{(k)}\right|$ represents the so-called Euclidean norm of corrections of the unknown quantities $\mathbf{z}$ and the symbol ϵ expresses the accuracy of the solution of the given iteration, i.e. until the approximation of the solution stabilizes at the required number of decimal places.

3. Tests results

The tests of the influence of microstructures on the rolling resistance in the general kinematic pair formed by the contact of loading discs 2, 3, 4 with specimen 5 were carried out at the load level: $F = 500$ N.

The discs revolutions: $n_2 = 800$ rpm, $\Omega_2 = \frac{2\pi \cdot n_2}{60} = 88.7758$ rad·s^{-1}.

The specimen radius is $r = 0.01$ m, the disc radius is $R = 0.04$ m and the arm length of the two-arm lever (body 6) is $l = 0.227$ m.

Four variants of the specimen design were tested: the specimen with the smooth contact surface indicated *Specimen_H* and three specimens with the contact surface provided with the microstructure are indicated as *Specimen_#3*, *Specimen_#10* and *Specimen_#11* where they differ in the geometric element density of the applied microstructure. The density is given by the spacing of the micro-holes, $s_{\#3}$ = 250 µm, $s_{\#10}$ = 500 µm, $s_{\#11}$ = 750 µm. The shape parameter of the micro-hole is given by its diameter d_{hole} = 200 µm. The discs and all the specimens with the hardened surface layer (58 + 2) HRc are made from the steel indicated as *X153CrMoV12*. A specimen example is shown in Fig. 8 and Fig. 9.

Fig. 8. Specimen.

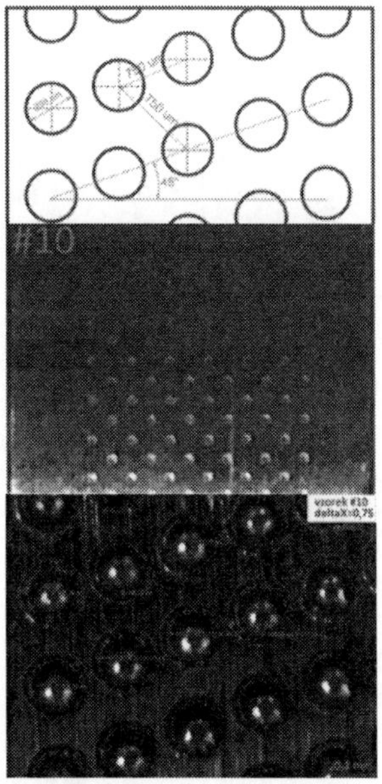

Fig. 9. Microstructure detail.

In order to achieve the most reliable results of the experimental analyses of the microstructures influence on the rolling resistance, the average values of the servomotor power consumption P_i for individual specimens i were determined from several multi-hour records, $i = H$, #3, #10, #11.

Based on the data (i.e. angular velocity Ω_2 and torque M_{Motor}) scanned from the electric motor (OMRON R88M-1M1K020C-S2) for driving pressure disc 2, its average mechanical power input at the constant angular velocity Ω_2 was determined with the following size:

$$\bar{P}_{Motor} = 10.12 \text{ W}, \qquad \bar{M}_{Motor} = \frac{\bar{P}_{Motor}}{\Omega_2} = 0.1208 \text{ Nm}$$

The mechanical power input $\bar{P}_{Motor}$ of the motor was determined as an average value from the 24 hour recording of the time course of its power consumption, as it is shown in Fig. 10.

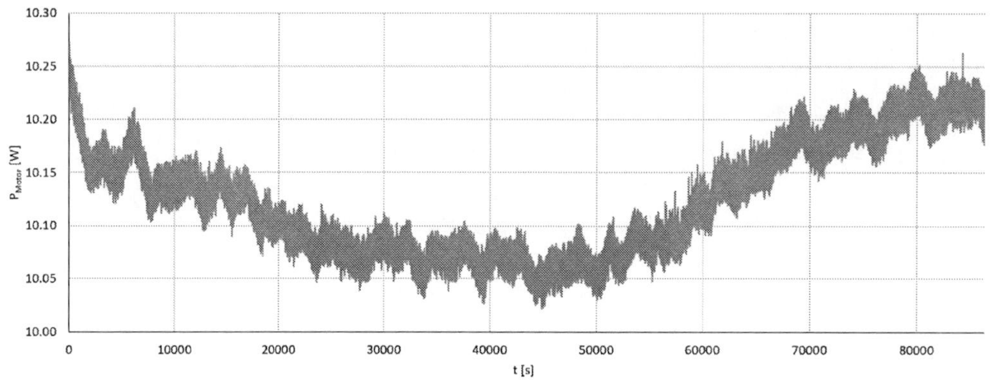

Fig. 10. The time course recording of the power input of the electric motor.

In the next step we tried to determine the magnitudes of mechanical powers P_A, P_B, P_C in order to exceed the rolling resistances of the bearings in the rotary bearings *A*, *B*, *C* of loading discs 2, 3, 4 at a given constant angular velocity Ω_2 and a given loading force F. Due to the disc mounting character, we assumed the same power losses for all the bearings:

$$P_A = P_B = P_c,$$

$$M_A = M_B = M_c$$

SKF bearings were mainly used to the discs mounting in accordance with Fig. 11. Main information on SKF bearings is in [6].

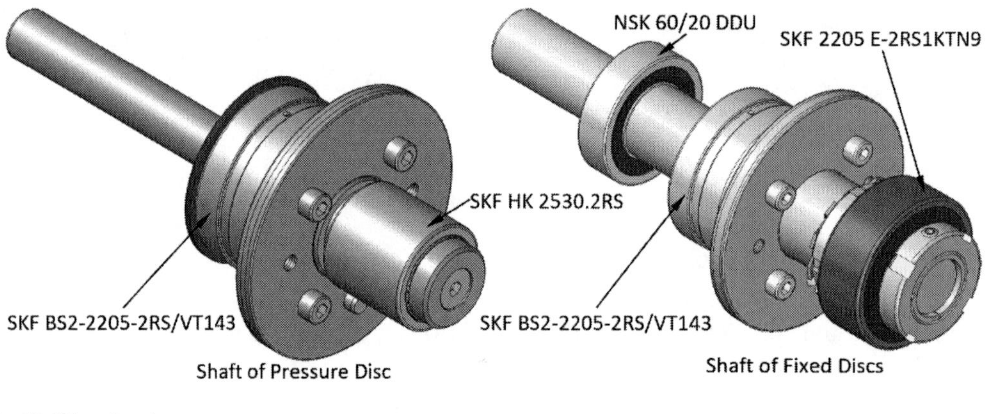

Fig. 11. Discs bearings.

We first identified the power consumption P_{A0} on the drive of pressure disc 2 without the force loading at a given angular velocity Ω_2, whereby:

$$\bar{P}_{Motor} + P_{A0} = 18.80\ \text{W}, \qquad \bar{M}_{Motor} + M_{A0} = 0.2244\ \text{Nm}$$

Therefore:

$$P_{A0} = (18.80 - 10.12)\ \text{W} = 8.68\ \text{W}, \qquad M_{A0} = (0.2244 - 0.1208)\ \text{Nm} = 0.1036\ \text{Nm}$$

The resulting value of the power input P_{A0} of pressure disc 2 was determined as the average value determined on the basis of all performed records of the time courses of the power consumption ($\bar{P}_{Motor} + PA0$) of the electric motor. Fig. 12 shows an example of a 24-hour record of the time course of the power input of the electric motor to drive pressure disc 2 without a force load.

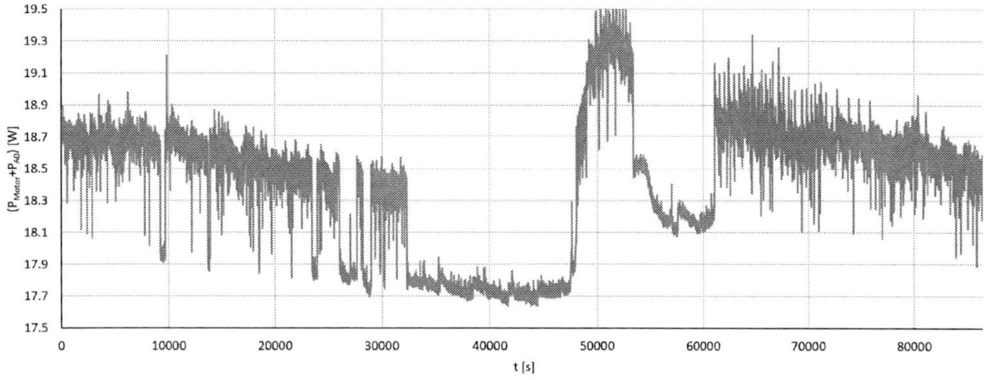

Fig. 12. The time course of the power input of the test rig to drive of the pressure disc.

Through the SKF technical support, see [7], we assessed the difference in the power input Δ_A of the loaded and unloaded bearings of the pressure disc, whereby the difference in the power input is:

$$\Delta P_A \approx 2\,\text{W}, \qquad \Delta M_A \approx 0.0239\,\text{Nm}$$

The power dissipation P_A in the bearing of pressure disc 2 at a given constant angular velocity Ω_2 and a given loading force F is determined by means of the relation:

$$P_A = P_{A0} + \Delta P_A, \qquad M_A = M_{A0} + \Delta M_A \tag{16}$$

We obtain the results by calculating Eq. 16:

$$P_A = P_B = P_C = (8.68 + 2.00)\ \text{W} = 10.68\text{W}$$

$$M_A = M_B = M_C = (0.1036 + 0.0239)\ \text{Nm} = 0.1275\text{Nm}$$

Using Eq. 2, we determine the power dissipation of the rolling resistance of the discs 2, 3 and 4 with the specimen 5 using the expression:

$$\bar{P}_{\xi i} = \bar{P}_i - \bar{P}_{Motor} + (P_A + P_B + P_C), \qquad i = H, \#3, \#10, \#11 \tag{17}$$

The values of the mechanical power inputs P_i during the testing of the individual specimens i were determined as average values on the basis of several hourly records of the data of instantaneous power consumption of the electric motor, $i = H, \#3, \#10, \#11$. The time course examples of the power input to drive of the individual specimens are shown in Fig. 13Fig. 13 to Fig. 16 Fig. 16 and appropriate average values of power consumption $\bar{P}_i$ are summarized in Table 1. Substituting appropriate values of power inputs into Eq. 17, we calculate the numerical values of the power losses $\bar{P}_{\xi i}$ for the individual specimens:

$$\bar{P}_{\xi H} = (51.10125 - 10.12 - 3 \times 10.68)\ \text{W} = 8.94125\ \text{W}$$

$$\bar{P}_{\xi \#3} = (51.05000 - 10.12 - 3 \times 10.68)\ \text{W} = 8.88750\ \text{W}$$

$$\bar{P}_{\xi \#10} = (50.23600 - 10.12 - 3 \times 10.68)\ \text{W} = 8.07600\ \text{W}$$

$$\bar{P}_{\xi \#11} = (50.60000 - 10.12 - 3 \times 10.68)\ \text{W} = 8.44000\ \text{W}$$

These are listed in Table 1.

Using the system of equations Eq. 11, we determine the values of the rolling resistance arm ξ_i for the individual specimens i $(i = H, \#3, \#10, \#11)$, which are summarized in Table 1. Next, we introduced a quantity called rolling resistance factor $f_{\xi i}$ of the specimen by the expression:

$$f_{\xi i} = \frac{\xi_i}{r}, \qquad i = H, \#3, \#10, \#11 \tag{18}$$

It is listed numerically in Table 1 for the individual specimens i.

The effect of the microstructures on the specimens was expressed as a percentage ratio of the power dissipation and rolling resistance coefficient $f_{\xi j}$ in comparison with the smooth specimen. These effects are expressed by the equations:

$$e_{Pj} = \left(1 - \frac{\bar{P}_{\xi j}}{\bar{P}_{\xi H}}\right) \cdot 100\ [\%], \quad e_{\xi j} = \left(1 - \frac{f_{\xi j}}{f_{\xi H}}\right) \cdot 100\ [\%], \quad j = \#3, \#10, \#11 \tag{19}$$

Both Eq. 19 are numerically expressed for the individual specimen types in Tab. 1. Based on the results, we can conclude that $e_{Pj} = e_{\xi j} = e_j$.

Table 1. The effect of the microstructures.

Specimen indication	Power input $\bar{P}_i$ [W]	Recording time t_{rec} [hour]	Power dissipation $P_{\xi i}$ [W]	Rolling resistance coefficient $f_{\xi i}$ [–]	Percentage ratio of the dissipation e_j [%]
Specimen_H	51.10	40	8.94125	0.00143	–
Specimen_3	50.05	21	8.88750	0.00142	0.6
Specimen_10	50.24	66	8.07600	0.00129	9.7
Specimen_11	50.60	34	8.44000	0.00135	5.6

If the values of e_P and e_ξ are positive, then the effect of the microstructures on rolling resistance is positive, otherwise negative. By comparing the respective values, we can state that the effect of all investigated microstructures on the rolling resistance at a given constant angular velocity Ω_2 and a given loading force F is positive.

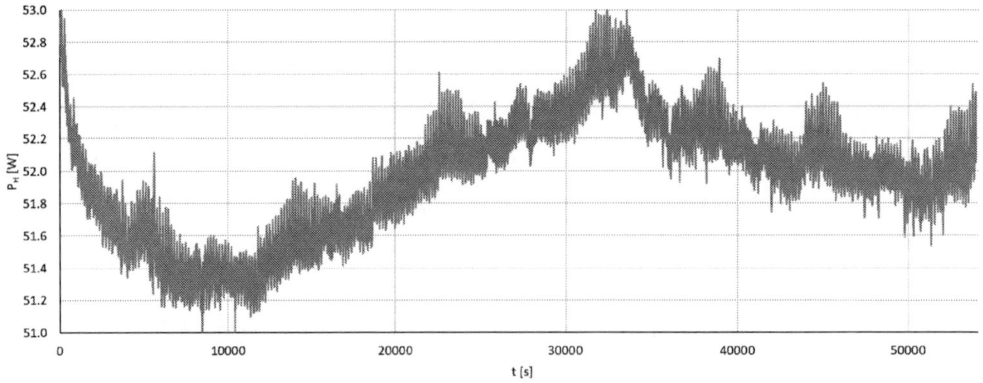

Fig. 13. The time course of the power input of the test rig with the smooth specimen.

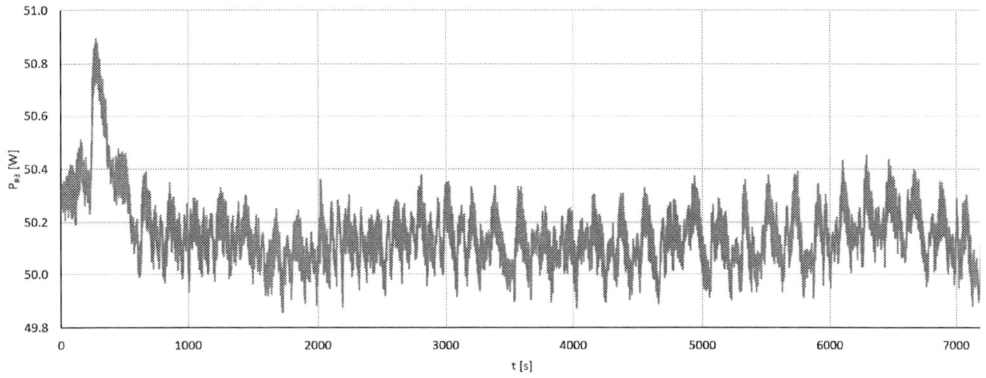

Fig. 14. The time course of the power input of the test rig with the specimen #3.

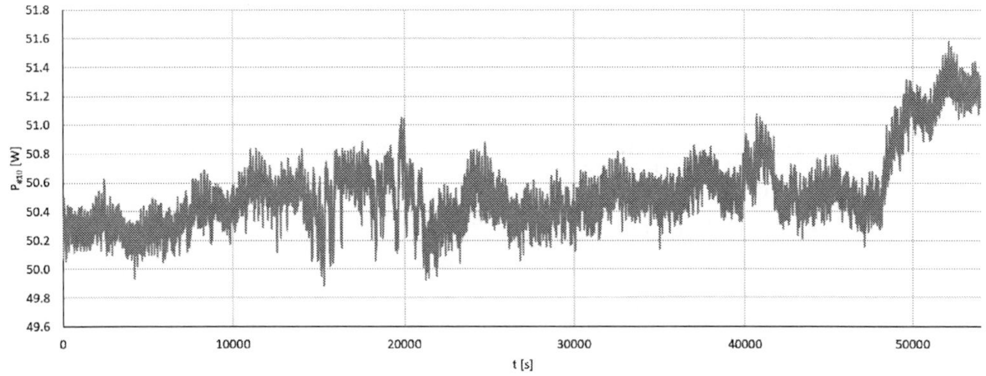

Fig. 15. The time course of the power input of the test rig with the specimen #10.

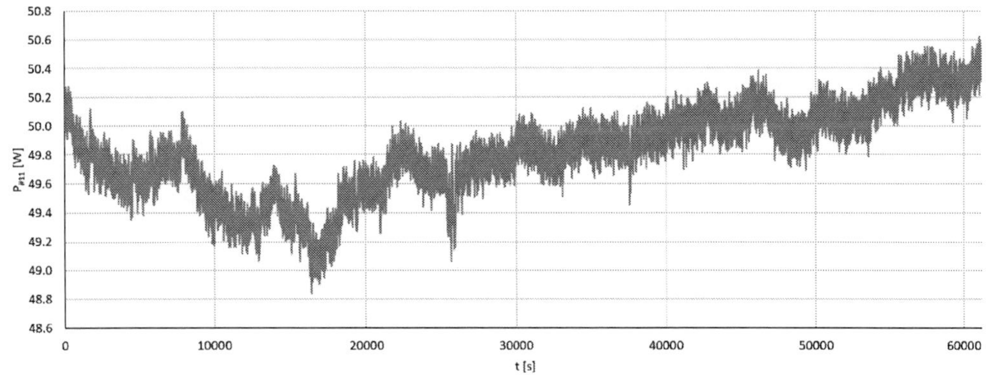

Fig. 16. The time course of the power input of the test rig with the specimen #11.

4. Conclusions

By comparing the relevant values, we can state that the influence of all the investigated microstructures on the rolling resistance at the given constant angular velocity of the discs Ω_2 ($n_2 = 800$ rpm), the specimen ω ($n = 3200$ rpm) and the given loading force $F = 500$ N is positive. The most positive effect of the microstructure on the rolling resistance has been proved in the case of *Specimen_#10*.

The contact stress expressed by the value of the reduced stress of 550 MPa and the Hertzian pressure of 1000 MPa correspond to the load for the specimen with the smooth contact surface. For the specimens with the applied microstructure in the contact surface, the mentioned values of the contact stress and pressure will differ and will be slightly higher.

In order to achieve the most reliable results of the experiments, the average values of the servomotor power consumption at the given constant angular velocity Ω_2 and the given loading force F were always determined from the several multi-hour records. However, the quantification of the effect of the microstructures on rolling resistance on an existing test device is problematic due to the mechanical passive resistances of the test device itself.

References

[1] VÚTS, a.s., *Available from* https://www.vuts.cz/

[2] Koloc Z. and Václavík M., *Cam Mechanisms* (Amsterdam, Elsevier, 1993, 424p.). ISBN 0-444-98664-2.

[3] Ondrášek J., "The General Kinematic Pair of a Cam Mechanism", in *Kinematics - Analysis and Applications*, Joseph Mizrahi, IntechOpen, DOI: 10.5772/intechopen.86682. *Available from* https://www.intechopen.com/books/kinematics-analysis-and-applications/the-general-kinematic-pair-of-a-cam-mechanism; Published: September 4th 2019; DOI: 10.5772/intechopen.77610; ISBN: 978-1-78984-491-7; Print ISBN: 978-1-78984-490-0; eBook (PDF) ISBN: 978-1-78985-700-9; Copyright year: 2019

[4] Ondrášek J., *Obecné kinematické dvojice vačkových mechanismů* (Liberec: VÚTS, a.s., 2018, 90 p.). ISBN 978-80-87184-77-6.

[5] Rektorys K. a spolupracovníci, *Přehled užité matematiky II* (Nakladatelství Prometheus, Praha, 1995) ISBN 80-85849-62-3

[6] SKF, *Available from* https://www.skf.com/group/products/rolling-bearings

[7] SKF Bearing Select, *Available from* https://skfbearingselect.com/#/bearing-selection-start

Measurement of Poisson's Ratio on the Rammed Unfired Earth

Pavel Padevět[1], Barbora Mužíková[2] & Petr Bittnar[3]

Abstract: The article is devoted to experimental measurement of the Poisson's ratio using a compression test on rammed unfired earths. The measurement was performed on 5 types of rammed earths made using illitic clay. Two strain gauges were used for contact deformation measurement. By evaluating the experimentally obtained data, the dependence of axial and transverse deformation was determined and the behavior of the building material in the linear elastic area and also in the plastic area was determined.

Keywords: Rammed earth; Compression test; Poisson's ratio; Strain; Extensometer

1. Introduction

The rammed unfired earth is one of the oldest buinding material, which is used for building on the whole world [1]. Although rammed earth is one of the oldest building materials, designing unburned clay structures is accompanied by several pitfalls. One of the problematic areas is not defining the standardized material properties of unburned rammed earth [2]. Rammed unfired earth is an orthotropic material [3]. This material will have other material properties in the direction of compaction compared to the properties defined perpendicular to the direction of compaction.

We would like to address this contribution between the deformation in the axial and transverse directions defined in the production of a building structure from rammed clay. The direction of compaction is considered to be the axial direction. The transverse direction is defined as the direction perpendicular to the direction of compaction, or is formed by the plane of the bed layer of the rammed earth. The stamped structure is created by placing the material in a mold in a layer that can be compacted. Subsequently, this layer is compacted and other compacted layers are deposited on it. In this way, structures of the required height are created by laying and compaction.

The relationship between deformations in different directions is defined by the Poisson's ratio as expressed in the equation (1), where ε_x is strain in direction of loading and ε_y is strain in lateral direction.

$$\nu = \frac{\varepsilon_x}{\varepsilon_y}. \quad (1)$$

The method of loading samples is described in the following part of the article.

[1]*Doc. Ing. Pavel Padevět, Ph.D.; Faculty of Civile Engineering, Department of Mechanics, Czech Technical University in Prague; Thákurova 7, 166 29 Prague, Czech Republic;* pavel.padevet@fsv.cvut.cz

[2]*Ing. Barbora Mužíková; Faculty of Civile Engineering, Department of Mechanics, Czech Technical University in Prague; Thákurova 7, 166 29 Prague, Czech Republic;* barbora.muzikova@fsv.cvut.cz

[3]*Ing. Petr Bittnar, Ph.D.; Faculty of Civile Engineering, Department of Mechanics, Czech Technical University in Prague; Thákurova 7, 166 29 Prague, Czech Republic;* petr.bittnar@fsv.cvut.cz

2. Tested material and system of measurement

The ability of the rammed earth material to achieve a certain compressive strength is mainly influenced by the composition and also by the moisture parameters of the material at the time of testing [4]. The type of clay, its amount in the mixture, the amount of mixing water, and also the amount of filler in relation to the clay have a fundamental influence on the strength. Illitic clay, which is marked AGL in our work, was used in this study. This type of clay consists mainly of the type of mica layers 2:1.

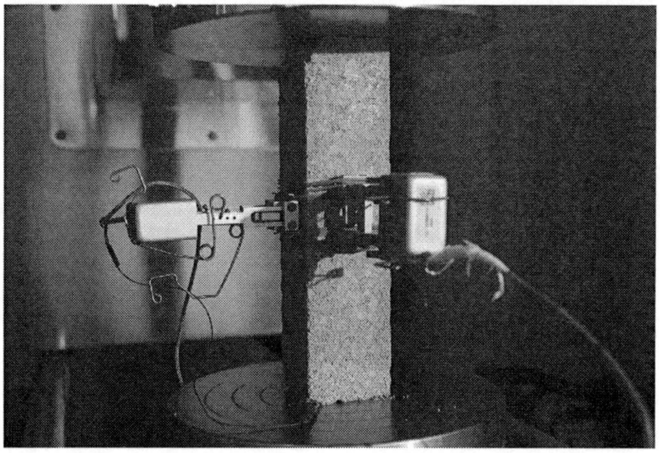

Fig. 1. The instrumentation of the specimen AGL XII for measurement of Poisson's ratio.

A detailed description of the tested mixtures is given in the Table 1. The water factor expresses the weight ratio of the amount of water to the amount of clay. The weight ratio also expresses the ratio of the amount of sand and clay. The dimension of the specimens was 40×40×160 mm. The compaction area was 40×160 mm.

Table 1. Material parameters of mixtures.

Mixture designation	Water ratio	Ratio sand/clay
AGL V	0.335	85/15
AGL VI	0.400	85/15
AGL X	0.295	75/25
AGL XI	0.400	75/25
AGL XI	0.295	70/30

3. Instrumentation of the measurement

Compressive loading was performed in the direction of the longest dimension of the specimens [5]. The measurement was performed using two attached extensometers. Deformation in the axial direction was measured in this direction. The second deformation was measured perpendicular to the first deformation, which corresponded to the direction of compaction. The basic measuring length of both extensomers was 25 mm. Fig. 1 shows that the ratio of relative strains must be considered from extensometer 1 to extensometer 2 in order to satisfy the Poisson's ratio given in equation (1).

The recording of time, force, deformation of the whole body and deformation of both extensometers was synchronized. The aim of the measurement was to obtain, in addition to recording the deformations, also the values of the force development and the total deformation of the body in the axial direction of testing.

4. Method of processing of measured data

The basic goal of determining the Poisson's ratio was to determine a specific value for the tested material. The determination of the value of ν is based on the evolution of relative deformations in the axial and transverse directions. For common building materials, the Poisson's ratio is between 0.2 and 0.25. The magnitude of the factor results from the linearly elastic behavior of the material under loading.

Firstly, a working diagram of the tested sample was determined, on which the behavior in the load part of the diagram was verified. The working diagrams show that the load part of the diagram is not elastic in the longer area, as is possible see on Fig. 2. On the contrary, the plastic behavior of the material is evident. The next step is to express the dependence of the working diagram on time and at the same time the Poisson's ratio on time. From these two curves, the area of the working diagram in which the behavior of the Poisson's factor is stable is evident. An example of displaying both curves in the graph is shown in the Fig. 3.

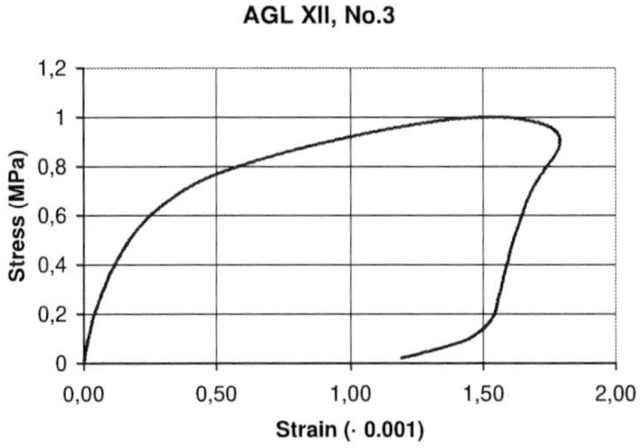

Fig. 2. Stress-strain working diagram for mixture AGL XII and specimen No. 3.

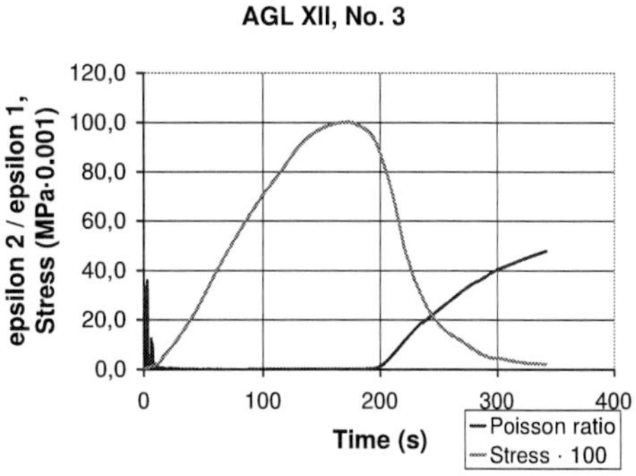

Fig. 3. Relation between stress and Poisson's ratio to time.

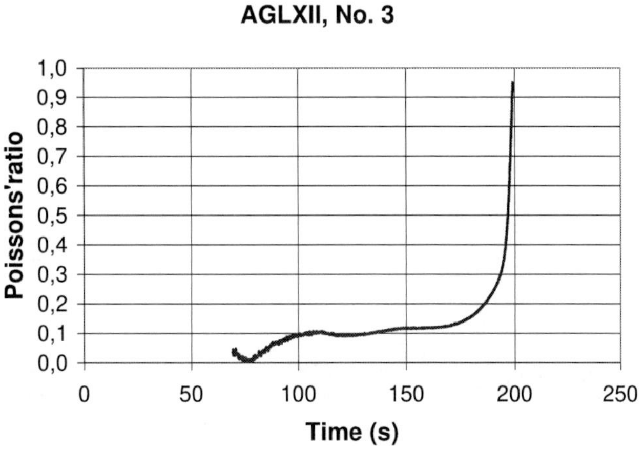

Fig. 4. Enlarged area Poisson's ratio.

The enlargement of the area described above can be obtained in the range in which the Poisson's ratio behaves stably. In none of the tests performed was it possible to find a constant value of ny for the area defined by the ascending branch of the working diagram, the area of the yield strength and the area close to the yield strength. The magnification of the stable Poisson's ratio area for body No. 3 of the AGL XII set is shown in the Fig. 4. This area corresponds to a time interval between 75 and 200 seconds in Fig. 3.

By "stable" area is meant the part of the working diagram on the ascending and descending branch, for which the values of the Poisson's ratio take values between 0 and 1. The force curve and the Poisson's ratio are related to time on the horizontal axis, see Fig. 3. The course of loading and unloading was controlled in all tests by a constant increment of the displacement of the cross member of the testing machine. Therefore, it is possible to compare the curve of force and Poisson's ratio.

5. Results

The graphs in Fig. 5 – Fig. 9 show selected parts of the Poisson diagram curves of specimens made from the mixtures shown in the Tab. 1. The AGL V set reached strength for specimens 1 and 2 at 145, respectively 135 seconds for specimen no.3. The AGL VI set reached strength in times of 100; 90 and 162 seconds. Set X had a time of strength of 179; 133 to 130 seconds. Specimens 1 to 3 reached strength in AGL XI at times 156; 161 and 163 seconds. The last AGL XII set tested had these times 167; 150 and 172 seconds. Fig. 8 contains a record of specimen No. 1, in which the extensometer remained locked in the axial direction. Its course of the Poisson's ratio is invalid.

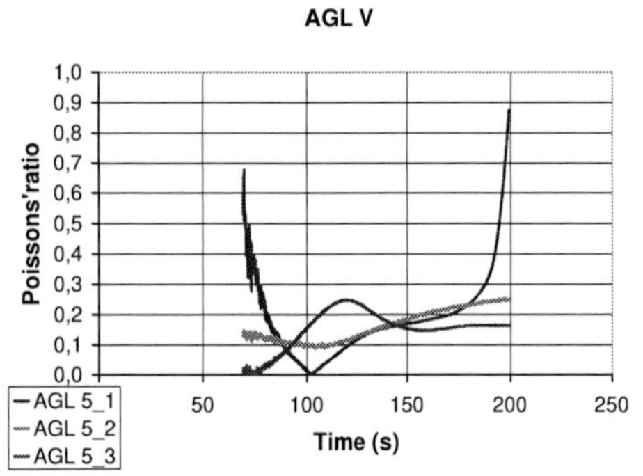

Fig. 5. Selected areas of the Poisson coefficient of the AGL V set (3 specimens).

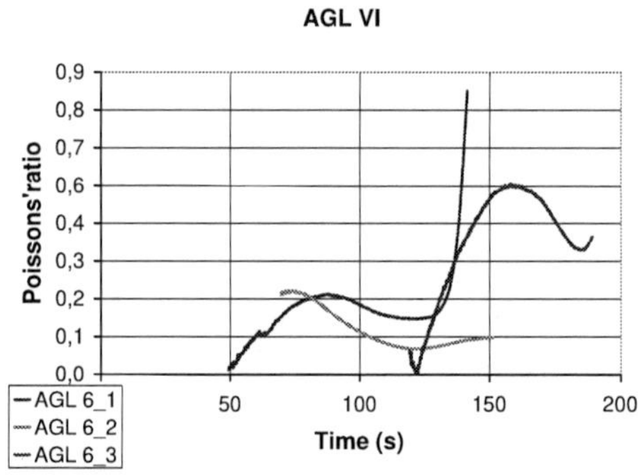

Fig. 6. Selected areas of the Poisson coefficient of the AGL VI set (3 specimens).

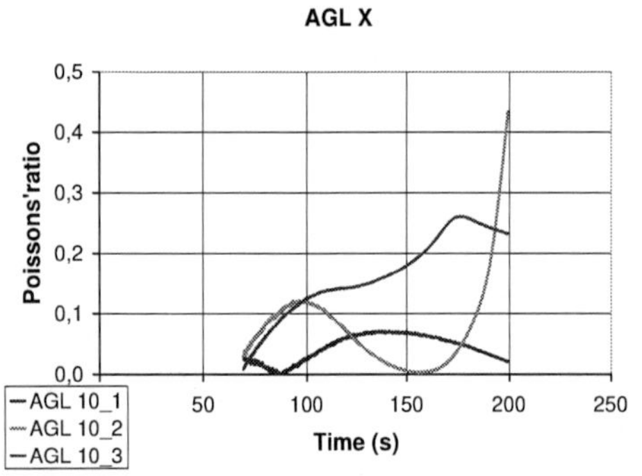

Fig. 7. Selected areas of the Poisson coefficient of the AGL X set (3 specimens).

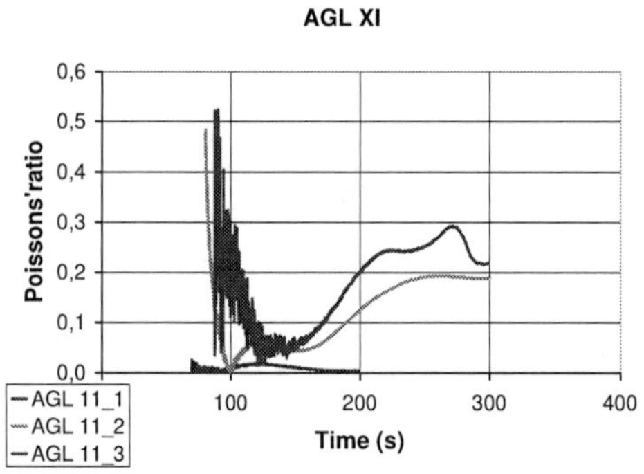

Fig. 8. Selected areas of the Poisson coefficient of the AGL XI set (3 specimens).

The results of the AGL V and AGL VI sets had a Poisson's ratio close to 0.2 for the ascending part of the working diagram. These two sets contained the lowest amount of binder – clay. At the same time, similar values can be observed for the AGL XII set with the highest proportion of illitic clay. The AGL X and AGL XI sets showed very low Poisson's ratio values approaching 0.1.

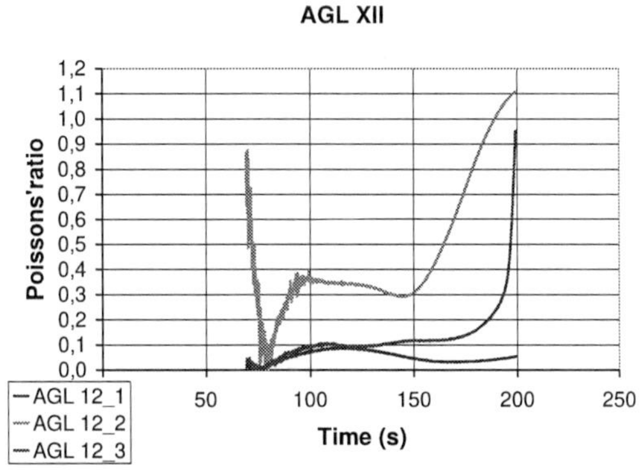

Fig. 9. Selected areas of the Poisson coefficient of the AGL XII set (3 specimens).

6. Conclusions

As can be seen from the graphs in the Fig. 5–Fig. 9, the Poisson's ratio is not determinable for rammed unfired earth by a single fixed value that would characterize this building material. On the contrary, its value is highly variable. Like building materials, its value is close to 0.2. But the AGL X and AGL XI sets showed lower coefficient values. The direct relationship between the composition of the rammed unfired earth and the magnitude of the Poisson's ratio is difficult to determine. For the Illitic clay used for the production of rammed unfired earth, values of the Poisson's ratio between 0.15 and 0.2 can be taken into account.

The method of measuring the Poisson's ratio in a contact manner using attached extensometers is very sensitive to the instrumentation of gauges on specimens. Unfortunately, measuring with adhesive strain gauges is ruled out due to the shattering surface of the specimens. The hope for improving the measurement method can be seen in contactless measurement using DIC.

Acknowledgements

This research has been supported by grant SGS22/089/OHK/2T/11 "Buidings materials and construction under view of experiment".

References

[1] Jaquin P. and Augarde C., *Earth Building: History, science, and conservation.*(IHS BRE Press, Watford, 2012). ISBN 978-1848061927.

[2] Walker P. et al., *Rammed Earth: Design and Construction Guidelines, vol 2010*, (IHS BRE Press, Watford, 2010). ISBN 978-1860817342.

[3] Plaček-Otcovská T., *Composition Analysis of Unfired Earth by a Methylene Blue Test and a Correlation with Final Properties of Rammed Earth*, (Disertation work, 2021).

[4] Otcovská T. and Padevět P., "Dependence of Tensile Bending Strength of Rammed Earth on Used Clay Composition and Amount of Mixture Water," *Advanced Materials Research*, Vol. 1144, pp. 48–53 (2017). https://doi.org/10.4028/www.scientific.net/AMR.1144.48

[5] Otcovská T. and Padevět P., "Dependence of Compressive Strength of Rammed Earth on Used Clay Composition." Západočeská univesita, Faculty of Applied Science, EAN 2016, Srní, Czech Republic.

Peeling Tests of the Human Aortas under the Different Conditions

Zdeněk Petřivý[1], Lukáš Horný[1], Hynek Chlup[1], Jakub Kronek[1], Petr Tichý[1], Tomáš Adámek[2], Alžběta Blanková[3] & Tomáš Suchý[3]

Abstract: Aortic dissection is one of the life-threatening diseases. It is manifested as a delamination process of the arterial layers. The delamination strength is investigated with the help of the peeling experiments, which resemble the mode I crack opening well-known from the fracture mechanics. These experiments were designed to take into account possible effects of material anisotropy and site-specific differences in the mechanical properties that occur along the aortic length. The computational branch of our study is based on the FEM/XFEM model of the peeling experiment built in Abaqus and attempts to find location-dependent parameters of the traction-separation law that would reflect age-related changes in the delamination resistance. The results showed that the delamination strength strongly depends on the age of the donor. It was also found that the delamination properties of the aortic wall differ significantly with respect to the crack tip orientation. It suggests that anisotropy is not only a bulk property, but also a phenomenon, which plays a role in a discontinuity propagation. In contrast to the aortic aneurysm thrombus delamination properties, our results suggest that the delamination properties of the aortic wall itself are not rate-dependent.

Keywords: Aging; Aortic dissection; Delamination; Peeling test; XFEM

1. Introduction

Aortic dissection is a life-threatening disease manifested by a separation of the layers of the aortic wall [1, 2]. During this disease, blood enters the wall and further separation of the layers of the aorta leads to creation of a new false lumen (Fig. 1).

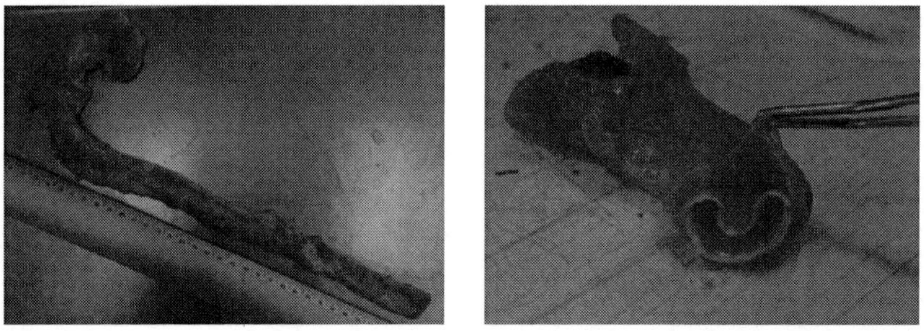

Fig. 1. Aorta and its dissection.

The initiation tear occurs the most frequently in the thoracic part of the aorta, but in the following delamination process it can spread along its entire length. This may cause fatal consequences such as total rupture of the aortic wall.

[1] *Faculty of Mechanical Engineering, Czech Technical University in Prague, Technická 4, 160 00 Prague, Czech Republic, e-mail:* lukas.horny@fs.cvut.cz
[2] *Department of Forensic Medicine and Toxicology, Regional Hospital Liberec, Husova 357/10, 460 63 Liberec, Czech Republic*
[3] *Institute of Rock Structure and Mechanics of the Czech Academy of Sciences, V Holešovičkách 94/41, 182 09 Prague, Czech Republic*

Our study tries to extend the knowledge about the biomechanics of the aortic delamination by means of both experimental and computational approaches. To measure how the delamination strength depends on anatomical site, loading velocity, age, and crack tip direction, we engaged the peeling experiment. The computational branch of our study is based on the FEM/XFEM model of the delamination test built in Abaqus [3].

2. Methods

Segments of the human aortas were obtained in regular autopsies conducted in the Department of Forensic Medicine and Toxicology of the Regional Hospital Liberec. Post-mortem use of human tissue was approved by the Ethics Committee of the Regional Hospital Liberec. Rectangular samples, approximately 8 $\times$ 60 mm, were cut from the thoracic ascending (AAs), the thoracic descending (ADs), the abdominal (AAb) parts of the aorta.

The peeling experiment, which is a method adopted to characterize the delamination properties of the aorta [1,4–7], resembles the mode I crack opening widely used in the fracture mechanics; see Fig. 2. The experiments were carried out with the help of the Zwick/Roell (Messphysik) multipurpose tensile testing machine. The delamination force was measured using HBM U9C $\pm$ 25 N force transducers. The length of the tear was determined from the movement of the clamps. This data was complemented with the recordings carried out by a built-in video-extensometer which measured the distance between marks made on the surface of the samples. To determine whether the delamination strength depends on the loading rate, experiments were carried out with clamps' velocity set to 0.1, 1, 10, 50 mm/s. To study site-specific differences in delamination strength, our study included experiments with strips obtained from AAs, ADs, and AAb. The anisotropy of the delamination strength was studied using samples oriented in the longitudinal and circumferential directions of the aorta.

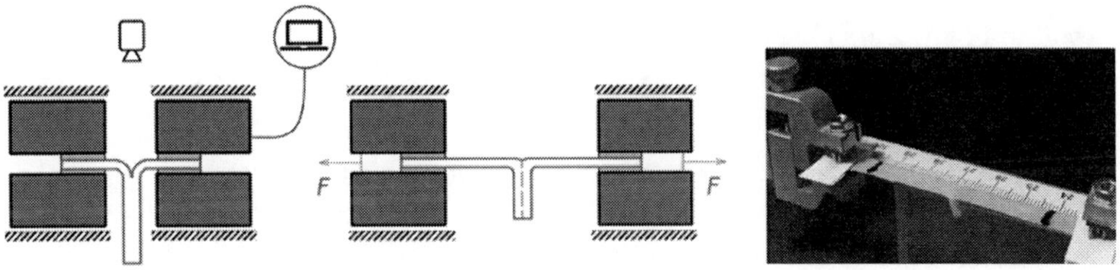

Fig. 2. Schema of the peeling experiment.

In the computational part of the study, a crack propagation is modelled using the traction-separation law, which is implemented in Abaqus [3]. A minimization of the difference between the experimentally measured delamination force and the reaction force calculated in FEM model is used to estimate parameters of the traction-separation law.

3. Results and Conclusions

Our experiments have suggested that there is a strong correlation between age and delamination strength (Fig. 3), which depends on the anatomical location (Fig. 4). It was also found that the delamination properties of the aortic wall differ significantly with respect to the orientation of the tip of the crack (Fig. 5).

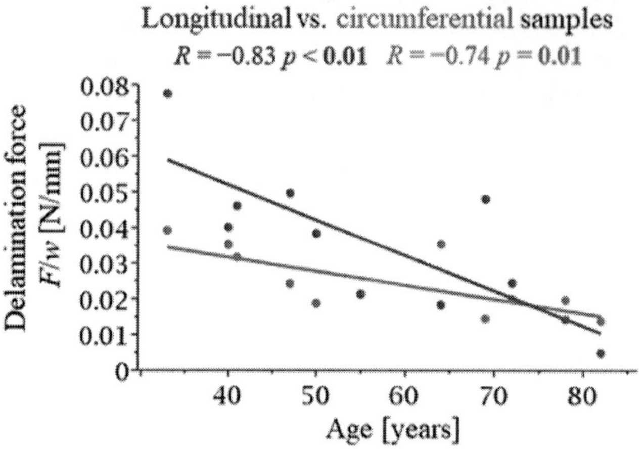

Fig. 3. Results of the age-related peeling test for the abdominial aorta. Points were obtained as medians of recorded forces.

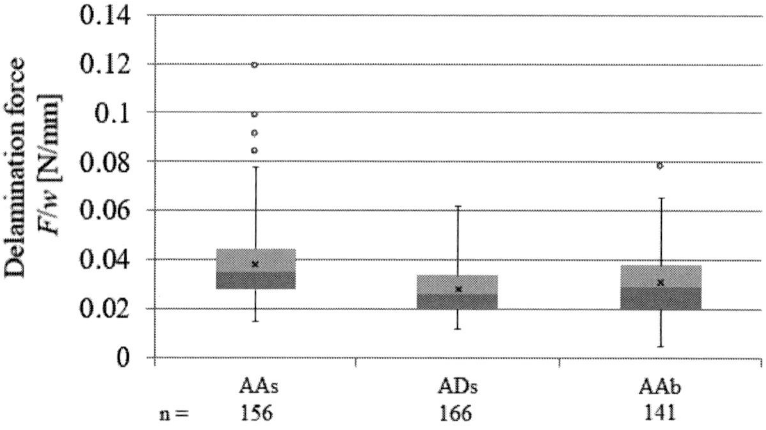

Fig. 4. Effect of location. Kruskal-Wallis test, $p < 0.01$ and also pair Dunn tests $p(As, Ds) < 10^{-6}$, $p(As, Ab) < 10^{-6}$, $p(Ds, Ab) < 0.05$ suggest differences in the delamination strength of each main part of the aorta.

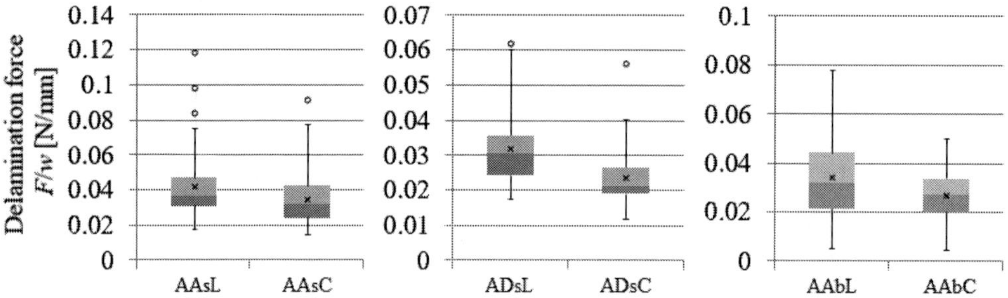

Fig. 5. Delamination force for different locations and both orientations (longitudinal L, circumferential C). Differences in delamination forces medians are examined with the help of the Mann-Whitney U test, $p < 0.01$.

It suggests that anisotropy is not only a bulk property, but also a phenomenon, which plays a role in discontinuity propagation.

On the other hand, data indicate that the delamination strength does not depend statistically significantly on the loading speed.

Regression analysis of the traction-separation law is planned to be carried out with the help of FEM/XFEM model of discretization and traction-separation law for the material description of discontinuity. At this time, it is in the debugging phase.

Acknowledgements

This study has been supported by the Czech Science Foundation in the GA20-11186 project entitled 'Mechanics of Arterial Delamination and Crack Propagation'.

References

[1] Sommer G., Gasser T.C., Regitnig P., Auer M. and Holzapfel G.A., "Dissection properties of the human aortic media: An experimental study," *Journal Biomechanical Engineering*, Vol. 130, Art. no. 021007 (2008). https://doi.org/10.1115/1.2898733.

[2] Yeh T.-Y., Chen C.-Y., Huang J.-W., Chiu C.-C., Lai W.-T. and Huang Y.-B. "Epidemiology and medication utilization pattern of aortic dissection in Taiwan: A population-based study," *Medicine*, Vol. 94, Art. no. e1522 (2015). https://doi.org/10.1097/md.0000000000001522.

[3] Dassault Systemes. *Documentation to Abaqus 2019 version.* (2019), Available online from https://help.3ds.com.

[4] Tong J., Cohnert T., Regitnig P., Kohlbacher J., Birner-Gruenberger R., Schriefl A.J., Sommer G. and Holzapfel G.A., " Variations of dissection properties and mass fractions with thrombus age in human abdominal aortic aneurysms," *J. Biomech.*, Vol. 47, pp. 14–23 (2014). https://doi.org/10.1016/j.jbiomech.2013.10.027.

[5] Wang Y., Johnson J.A., Spinale F.G., Sutton M.A. and Lessner S.M., "Quantitative Measurement of Dissection Resistance in Intimal and Medial Layers of Human Coronary Arteries," *Exp. Mech.* Vol. 54, pp. 677–683 (2014). https://doi.org/10.1007/s11340-013-9836-0.

[6] Kozuń M., Kobielarz M., Chwiłkowska A. and Pezowicz C., "The impact of development of atherosclerosis on delamination resistance of the thoracic aortic wall," *J. Mechan. Behav. Biomed. Mater.*, Vol. 79, pp. 292–300 (2018). https://doi.org/10.1016/j.jmbbm.2018.01.009.

[7] Chung J.C., Wong E., Tang M., Eliathamby D., Forbes T. L., Butany J., Simmons C.A. and Ouzounian M., "Biomechanics of aortic dissection: A comparison of aortas associated with bicuspid and tricuspid aortic valves," *J. Am. Heart Assoc.*, Vol. 9, Art. no. e016715 (2020). https://doi.org/10.1161/jaha.120.016715.

Experimental Dynamic Analysis of the Footbridge over the Jesípek near Hradec Králové Using New Approach

Tomáš Plachý[1], Michal Polák[2], Aleš Mezera[3], Milan Talich[4], Jan Havrlant[5], Lubomír Soukup[6] & Miroslav Čáp[7]

Abstract: The paper presents an experimental dynamic analysis carried out on a footbridge for pedestrians and cyclists, which carries a cycle path over the branch "Jesípek" of the river Elbe near Hradec Králové. Three measurements of its dynamic properties were carried out on this footbridge, using acceleration sensors in 2019, 2020 and 2021, but also using a relatively new method - ground-based radar interferometry - in 2020. In this paper, the dynamic measurements from these three years are compared with each other, as well as the results obtained by radar interferometry with those obtained by classical methods. Finally, the obtained knowledge about the possibilities of using radar interferometry in the measurement of this type of structures are summarized.

Keywords: Footbridge; Radar interferometry; Modal analysis; Dynamic load test; Natural frequency; Acceleration transducers; Mode shape

1. Introduction

Dynamic measurements are a very important part of the evaluation process of building structures, especially those with predominant dynamic loads, such as bridges [1, 2] or footbridges. Therefore, there is an increasing emphasis not only on improving existing methods of measuring dynamic properties, but also on testing new approaches [3].

The paper presents an experimental dynamic analysis carried out using the classical method and ground-based radar interferometry on a footbridge for pedestrians and cyclists, which carries a cycle path over the branch "Jesípek" of the river Elbe near Hradec Králové in the Czech Republic.

2. Description of the footbridge

The load bearing structure of the footbridge is an arch with the lower bridge deck suspended on twenty pairs of stainless-steel rods (see Fig. 1). The length of the footbridge is 50.48 m, the span of the arch is 49.88 m and the rise of the arch is 10.00 m (see Fig. 1). The arch has a rectangular cross-section with dimensions of 412 mm (width) and 420 mm (height) at its apex. Towards

[1] *Ing. Tomáš Plachý, Ph.D.; Department of Mechanics, Faculty of Civil Engineering, Czech Technical University in Prague, Thákurova 7, 166 29 Prague 6, Czech Republic;* plachy@fsv.cvut.cz
[2] *prof. Ing. Michal Polák, CSc.; Department of Mechanics, Faculty of Civil Engineering, Czech Technical University in Prague, Thákurova 7, 166 29 Prague 6, Czech Republic;* polak@fsv.cvut.cz
[3] *Ing. Aleš Mezera; Department of Mechanics, Faculty of Civil Engineering, Czech Technical University in Prague, Thákurova 7, 166 29 Prague 6, Czech Republic;* ales.mezera@cvut.cz
[4] *Ing. Milan Talich, Ph.D.; Institute of Information Theory and Automation, Academy of Sciences of the Czech Republic; Pod Vodárenskou věží 4, 182 00 Prague, Czech Republic;* talich@utia.cas.cz
[5] *Ing. Jan Havrlant, Ph.D.; Institute of Information Theory and Automation, Academy of Sciences of the Czech Republic; Pod Vodárenskou věží 4, 182 00 Prague, Czech Republic;* havrlant@utia.cas.cz
[6] *Ing. Lubomír Soukup, Ph.D.; Institute of Information Theory and Automation, Academy of Sciences of the Czech Republic; Pod Vodárenskou věží 4, 182 00 Prague, Czech Republic;* soukup@utia.cas.cz
[7] *Ing. Miroslav Čáp; Department of Mechanics, Faculty of Civil Engineering, Czech Technical University in Prague, Thákurova 7, 166 29 Prague 6, Czech Republic;* miroslav.cap.1@fsv.cvut.cz

the supports, the arch is divided into two separate parts at about a quarter of the span. The diameter of the stainless-steel ropes is 10 mm. The bridge deck structure (see Fig. 1) consists of a steel main girder, steel cross girders and a reinforced concrete slab. The cross-sectional dimensions of the steel main girder are 400 mm (width) and 400 mm (height). The cross girders are I-shaped with a variable height from 400 mm at the steel main girder to 149 mm at the end of the cross girders where the steinless-steel arch ropes are mounted. The length of the crossbeams is 2.127 m. The bridge deck slab is constructed of C30/37 XF3 reinforced concrete and has a variable cross-sectional thickness that varies from 120 mm at the centre of the bridge deck to 155 mm at the edges of the slab.

Fig. 1. Footbridge over the river branch Jesípek near Hradec Králové – 2019 (on the left) and 2020 (on the right), the radar placed under the footbridge.

3. The 1st measurement – 2019

The 1st measurement in 2019 was realized as the dynamic load test with two goals, measurement of natural frequencies and modes of vibration of the footbridge and measurement of the footbridge response caused by walking pedestrians and vandals.

The eight piezoelectric highly sensitive acceleration transducers Brüel & Kjær Type 8344 were used for measurement (Fig. 3 – on the right). The transducers were connected by cables to the 8-channel data acquisition station SIRIUS Type 6ACC-2ACC+ of the company DEWE-Soft. All channels have their own 2 $\times$ 24-bit A/D converter (so called DualCore) that is able to measure with dynamic range up to 160 dB. All A/D converter are mutually synchronized due to ensuring simultaneous measurement on all active channels.

Fig. 2. The position of the shaker TIRAVIB 5140 on the footbridge.

3.1. *Used measurement system*

The electrodynamic shaker TIRAVIB 5140 was used for vibration excitation of the footbridge. The shaker was situated on the deck eccentrically to the longitudinal axis of the footbridge in the quarter of the span on the steel cross member which is suspended by the sixth pair of stainless-steel rods (Fig. 2). The shaker position was selected based on the theoretical natural modes of vibration as the best position to excite the first seven fundamental natural mode shapes of vibration because of the most of them had some sufficiently high ordinate in this point to excite them. The exciter was set to produce a random excitation force with the frequency range 0.3 to 20 Hz.

Fig. 3. The acceleration transducers attached to the steel weights (on the right) placed in the 12th cross-section (on the left).

A grid of points was placed on the bridge deck and arch of the footbridge to measure its dynamic response. The position of the grid points was determined based on the character of the theoretical mode shapes. The vibration of the bridge deck was measured in 12 cross-sections, at the locations of all even crossbeams. Two points were located in each monitored cross section (see Fig. 3 – left figure). In all of these points, the vibration was measured in three perpendicular directions. Three acceleration transducers were attached by neodymium magnets to each steel weight to enable vibration measurement in three perpendicular directions (Fig. 3 – right figure). The response was measured also on the arch at two points where the arch splits into two parts. On the arch, the vibration was measured in the vertical direction and in the horizontal direction perpendicular to the longitudinal axis of the footbridge.

Two reference acceleration transducers were used in this experiment and were placed in the same cross-section as the exciter but on the opposite side. The first transducer measured the footbridge vibration in the vertical direction and the second one measured the vibration in the horizontal direction perpendicular to the longitudinal axis of the bridge.

During the second part of the dynamic test (measurement of the footbridge response caused by walking pedestrians and vandals) the acceleration was measured in the middle and at the quarters of the footbridge span.

3.2. *Results of the measurement in 2019*

The recorded dynamic responses of the footbridge were normalized to the one measured at the reference point using software ME'scopeVES. Measured time data records were transformed using Fast Fourier Transform (FFT) from time domain to the frequency domain. Frequency response functions for operational modal analysis (OMA FRF) were evaluated from these signals and then the natural frequencies and mode shapes were evaluated (Fig. 4 – Fig. 10).

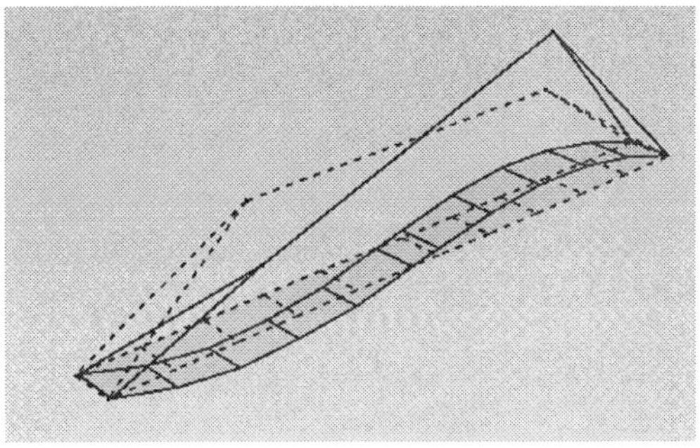

Fig. 4. The 1st natural mode shape of the footbridge, $f_{(1)} = 1.34$ Hz – year 2019.

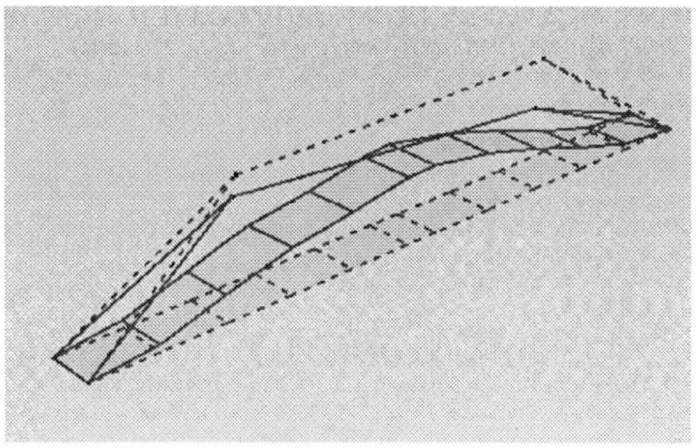

Fig. 5. The 2nd natural mode shape of the footbridge, $f_{(2)} = 1.53$ Hz – year 2019.

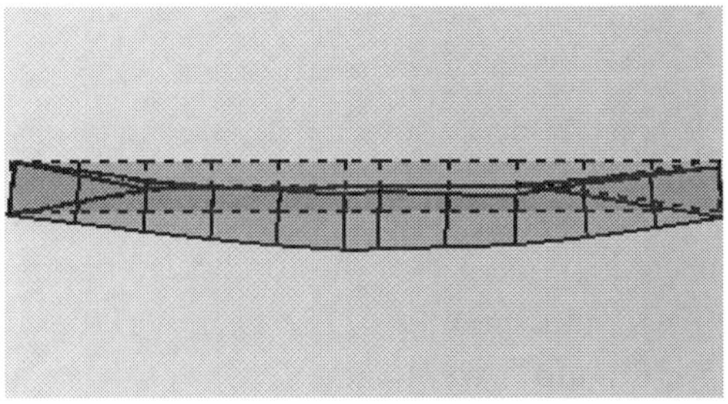

Fig. 6. The 3rd natural mode shape of the footbridge, $f_{(3)} = 1.62$ Hz – year 2019.

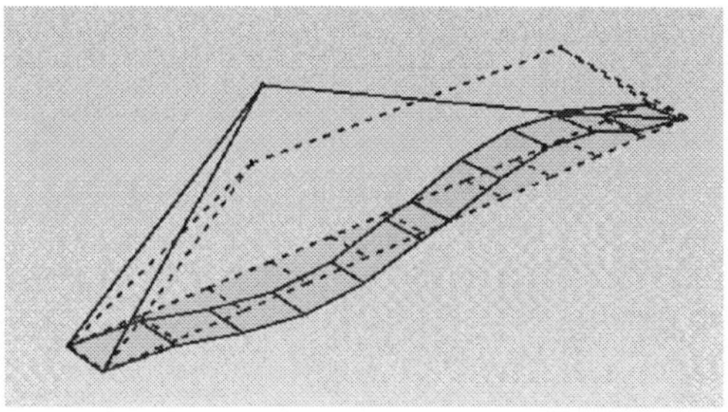

Fig. 7. The 4th natural mode shape of the footbridge, $f_{(4)} = 3.20$ Hz – year 2019.

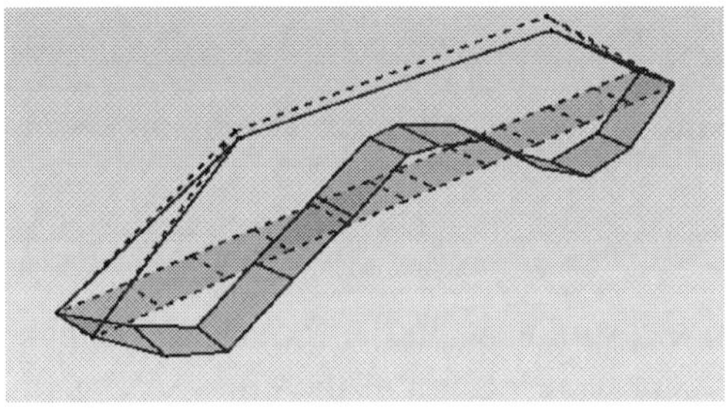

Fig. 8. The 5th natural mode shape of the footbridge, $f_{(5)} = 3.53$ Hz – year 2019.

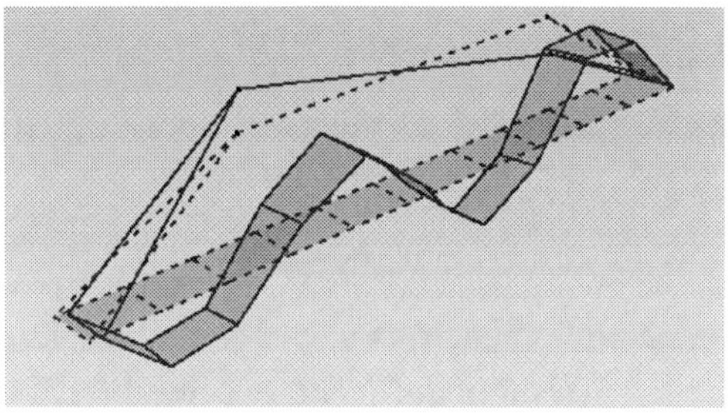

Fig. 9. The 6th natural mode shape of the footbridge, $f_{(6)} = 5.13$ Hz – year 2019.

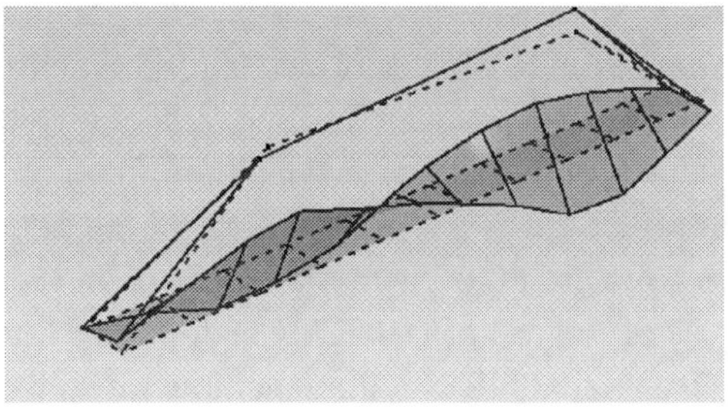

Fig. 10. The 7th natural mode shape of the footbridge, $f_{(7)} = 5.40$ Hz – year 2019.

As an example, the record of the footbridge response caused by 2 pedestrians going synchronously with the step frequency 1.53 Hz measured in the middle of the span is shown in the Fig. 11.

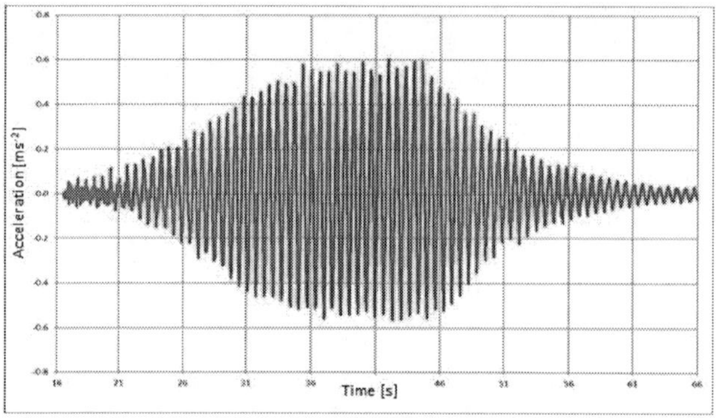

Fig. 11. The record of the footbridge response in the middle of the span caused by 2 pedestrians going synchronously with the step frequency 1.53 Hz – year 2019.

From the suitable records (Fig. 11), the logarithmic decrement δ of the footbridge damping was evaluated. The damping was evaluated using FFT multispectrum. For example, the logarithmic decrement $\delta_{(1)} = 0.034$ was evaluated for the natural frequency $f_{(1)} = 1.34$ Hz, the logarithmic decrement $\delta_{(2)} = 0.033$ for the natural frequency $f_{(2)} = 1,53$ Hz, the logarithmic decrement $\delta_{(4)} = 0.036$ for the natural frequency $f_{(4)} = 3,20$ Hz and the logarithmic decrement $\delta_{(5)} = 0.041$ for the natural frequency $f_{(5)} = 3.53$ Hz.

4. The 2nd measurement – 2020

The main goal of the 2nd measurement in 2020 was to compare data measured by acceleration transducers and by radar interferometry.

4.1. Used measurement system

The same equipment was used for the acceleration measurement on the footbridge Jesípek as for the 1st measurement. The acceleration transducers were placed at the cross-sections No. 3, 4, and 6. They measured vibration in the vertical direction and in the horizontal direction perpendicular to the longitudinal axis of the footbridge. All measurements were realized simultaneously with the second measurement system.

The second measurement system was the ground-based radar interferometry (GBRI) with ground-based real aperture radar (GB-RAR) performed by two synchronized radars (see Fig. 12). The GBRI method has the ability to measure real-time deflections at normal operating conditions with precision better than 0.1mm. Deflections of a monitored object can be simultaneously determined at multiple locations. So it is possible to obtain both general and detailed information on the behaviour of the structure under its dynamic load. In recent literature there are the basic principles and examples of the use of GB-RAR technology for determining deflection of bridges given e.g. in [4–7]. Two Radars IBIS Italian manufacturer Ingegneria Dei Sistemi (IDS) were used for the measurements. Radar R1: IDS Radar IBIS – FS Plus, radar R2: IDS Radar IBIS – RU 172. In both cases, IBIS-ANT3-H17V15 antennas were used. The positions of radars were mapped by detailed terrestrial laser scanning and subsequently a 3D model was created.

Fig. 12. The positions of the Radar R1 (IDS Radar IBIS – FS Plus) and radar R2 (IDS Radar IBIS – RU 172) under the footbridge.

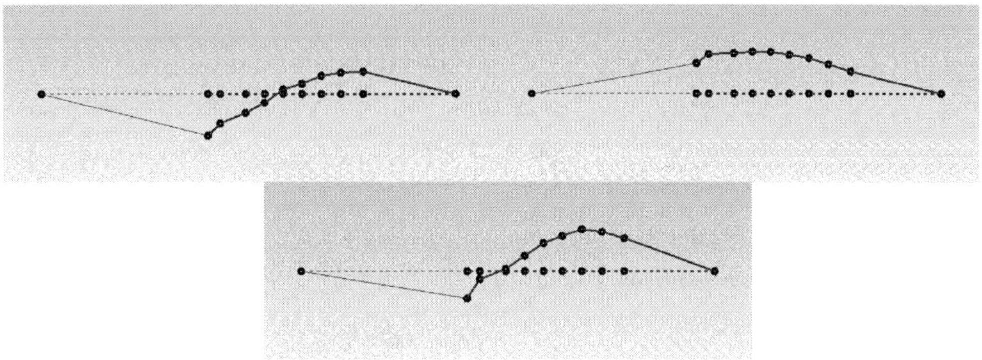

Fig. 13. The first three natural mode shapes of the footbridge vertical vibration evaluated from GBRI, $f_{(1)}$= 1.34 Hz, $f_{(2)}$= 1.49 Hz, $f_{(4)}$= 3.17 Hz – year 2020.

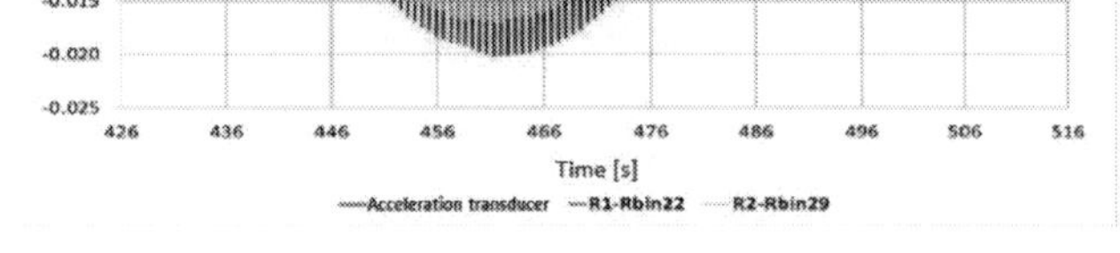

Fig. 14. The comparison of the dynamic deflection of the deck measured by acceleration transducers, radar R1 and radar R2 – year 2020.

4.2. Results of the measurement in 2020

Unfortunately, some bushes grew up under the bridge between the measurements in 2019 and 2020 (see Fig. 1) and thus only the half of the bridge could be measured by radar interferometry.

The first three natural frequencies and mode shapes of the vertical vibration were evaluated from GB-RAR on the half of the footbridge (Fig. 13).

The dynamic deflection recorded by GB-RAR was compared with double integrated acceleration measured by piezoelectric acceleration transducers (Fig. 14).

5. The 3rd measurement – 2021

The main goal of the 3rd measurement in 2021 was to compare new data with data measured in 2019, both measured using acceleration transducers.

5.1. Used measurement system

The same equipment was used for the acceleration measurement of the footbridge Jesípek as for the 1st measurement. The acceleration transducers were placed at the same twelve cross-sections at the same two positions in each cross-section. The only difference in positions of transducers was that the arch vibration was measured not at the points where the arch splits into two parts but at the heights of 2 m from the deck. The difference was also in excitation of the footbridge during experimental modal analysis. The excitation was realized by force impulses induced by jumps of one person in intervals of 5 s at the third quarter of the footbridge span.

5.2. Results of the measurement in 2021

The same modal data evaluation process was realized as in 2019 using the same software ME'scopeVES. Measured time data records were transformed using FFT from time domain to the frequency domain. OMA FRF´s were evaluated from these signals and then the natural frequencies and mode shapes were evaluated (Fig. 15–Fig. 21).

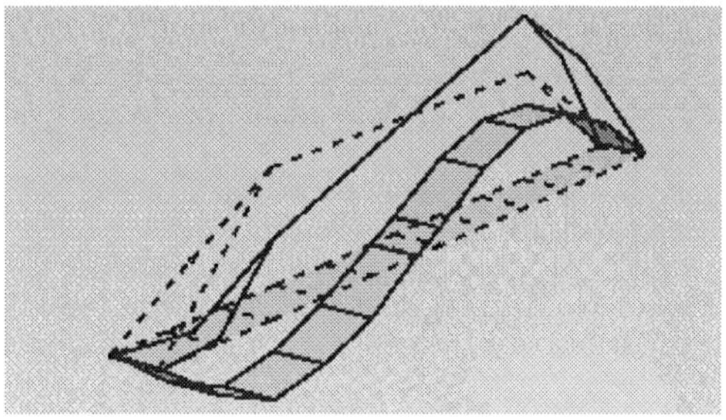

Fig. 15. The 1st natural mode shape of the footbridge, $f_{(1)}$ = 1.36 Hz – year 2021.

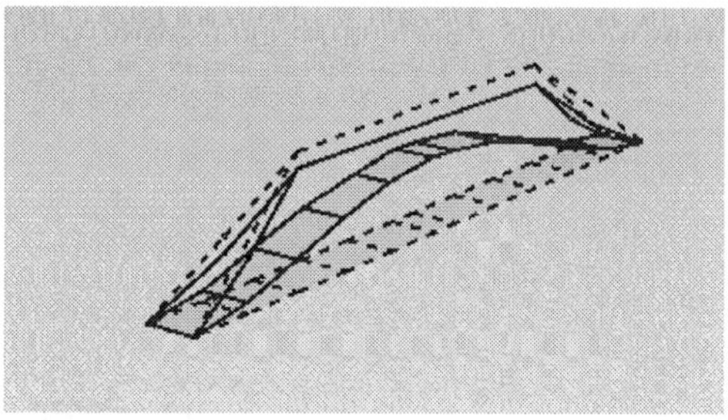

Fig. 16. The 2nd natural mode shape of the footbridge, $f_{(2)}$ = 1.52 Hz – year 2021.

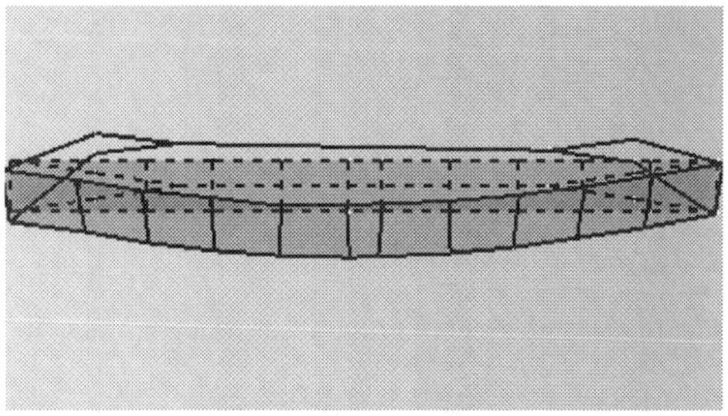

Fig. 17. The 3rd natural mode shape of the footbridge, $f_{(3)}$= 2.42 Hz – year 2021.

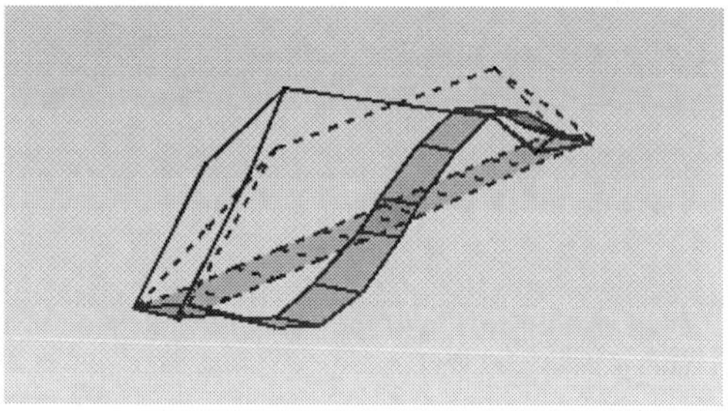

Fig. 18. The 4th natural mode shape of the footbridge, $f_{(4)}$= 3.20 Hz – year 2021.

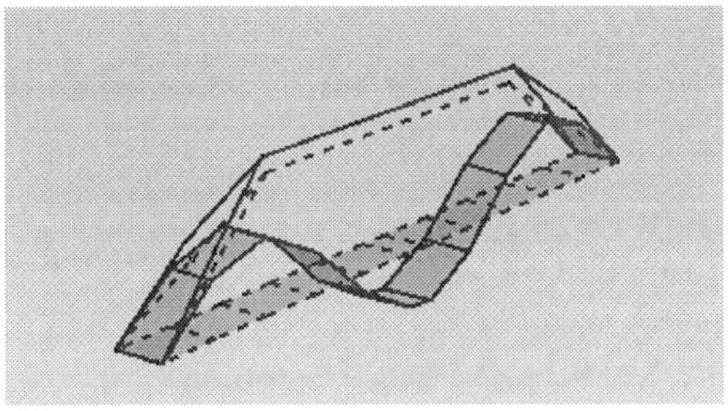

Fig. 19. The 5th natural mode shape of the footbridge, $f_{(5)}$= 3.52 Hz – year 2021.

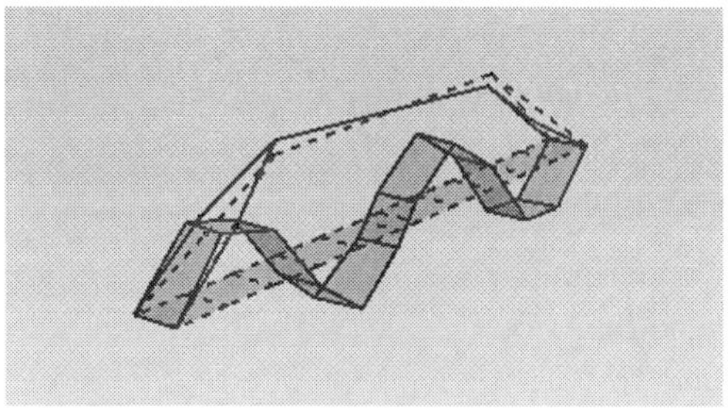

Fig. 20. The 6th natural mode shape of the footbridge, $f_{(6)}$= 5.13 Hz – year 2021.

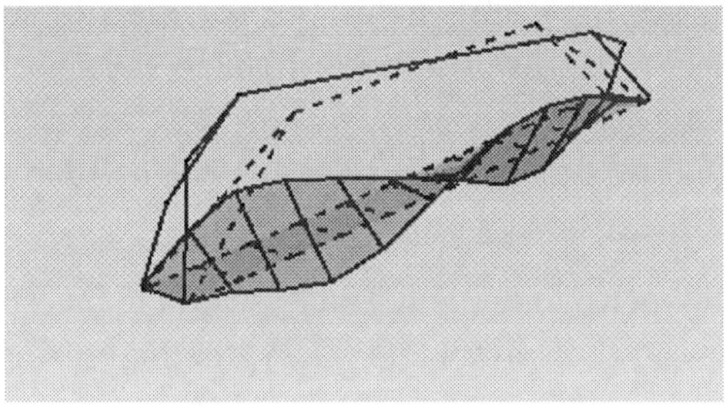

Fig. 21. The 7th natural mode shape of the footbridge, $f_{(7)}$= 5.34 Hz – year 2021.

During the second part of the dynamic test in 2021 (the measurement of the footbridge response caused by walking pedestrians and vandals), the acceleration was measured in the middle and at the quarters of the footbridge span.

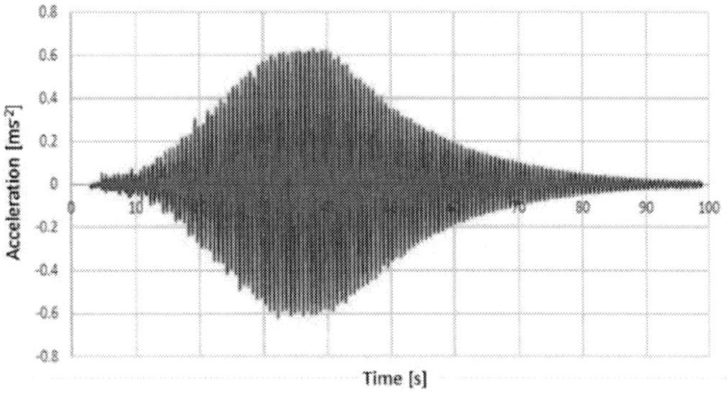

Fig. 22. The record of the footbridge response in the middle of the span caused by 2 pedestrians going with the step frequency 1.52 Hz – year 2021.

6. Conclusions

The paper presents three measurements of dynamic properties on the footbridge carried out using acceleration sensors in 2019, 2020 and 2021, but also using a relatively new method – ground-based radar interferometry – in 2020.

The changes of natural frequencies and modes of vertical vibration evaluated in 2019 and

2020 are very small. For frequencies, they are not bigger than the uncertainty of the measurement (see Fig. 4 – Fig. 10 and Fig. 15 – Fig. 21). Also, the footbridge response caused by pedestrians has not changed practically (Fig. 11 and Fig. 22) between 2019 and 2021. However, the change in the natural frequency of the first horizontal mode of vibration is substantial - $f_{(3)}$ = 1.62 Hz – year 2019, $f_{(3)}$ = 2.42 Hz – year 2021, which indicates the horizontal stiffening of footbridge supports.

Comparing the two different excitation methods applied on this footbridge – the exciter in 2019 and the jumping person in 2021, the second method used was simpler, cheaper and higher resonant frequencies were excited. The substantial disadvantage of this method is the early fatigue of the jumping person.

The dynamic measurement using ground-based real aperture radar (GB-RAR) was also tested on this footbridge in 2020. The advantage of the GB-RAR was the short time to prepare the radar measurements and the simultaneous measurement of deflections in multiple cross sections of the footbridge deck. The disadvantages in the used experimental configuration were that the deflection of each cross section was measured as the average value of the deflections over a whole cross-girder and that there must be no obstacles such as bushes (Fig. 1 – on the right) between the radar and the observed cross-sections.

The use of GB-RAR for experimental modal analysis of the investigated footbridge was also limited compared to the applied accelerometers. Only the three lowest natural frequencies and mode shapes of vertical vibration could be evaluated from measured deflections on this footbridge in a limited number of cross-sections (Fig. 13) and torsional modes cannot be determined at all by this method in the used configuration.

Acknowledgements

The research was supported by the project VI20192022167 “The Advanced Technology of Rapid Determination of Bridges Deformation by Radar Interferometry and its Use in Diagnostics” supported by the Ministry of the Interior of the Czech Republic.

References

[1] Plachý T., Polák M. and Ryjáček P., “Assessment of an Old Steel Railway Bridge Using Dynamic Tests”, *Procedia Engineering*, Vol. 199, pp. 3053–3058 (2017). ISSN 1877-7058.

[2] Plachý T., Polák M., Ryjáček P., “Experimental dynamic analysis of the railway bridge near Žatec,” in *Proceedings of Experimental Stress Analysis - 58th International Scientific Conference, EAN 2020*, Czech Republic, 2020, pp. 417-422. ISBN 978-802484451-0.

[3] Plachý T., Polák M., Ryjáček P., Talich M., Havrlant J., Antoš F., Litoš J., Macho M., Soukup L., “Experimental dynamic analysis of the arch road bridge,” *IOP Conference Series: Earth and Environmental Sc.*, Vol. 906, Art. no. 012061, p. 22 (2021). ISSN 1755-1315. http://dx.doi.org/10.1088/1755-1315/906/1/012061.

[4] Talich M., “The Effect of Temperature Changes on Vertical Deflections of Metal Rail Bridge Constructions Determined by the Ground Based Radar Interferometry Method,” *IOP Conference Series: Earth and Envir. Sc.*, Vol. 221, no. 1, (2019).

[5] Talich M., “Using Ground Radar Interferometry for Precise Determining of Deformation and Vertical Deflection of Structures,” *IOP Conference Series: Earth and Envir. Sc.*, Vol. 95, no.3, (2017).

[6] Lipták I., Erdélyi J., Kyrinovič P., Kopáčik A., "Monitoring of Bridge Dynamics by Radar Interferometry," *INGEO 2014, 6th Int. Conference on Engineering Surveying*, pp. 211–216 (2014).

[7] Liu X., Tong X., Ding K., Zhao X., Zhu L. and Zhang X., "Measurement of Long-Term Periodic and Dynamic Deflection of the Long-Span Railway Bridge Using Microwave Interferometry," *IEEE J. of Selected Topics in Appl. Earth Obser. and Remote Sensing*, Vol. 8, no. 9, pp. 4531–4538 (2015).

Application of Three Different Experimental Methods for Dynamic Analysis Performed on the Railway Steel Arch Bridge

Tomáš Plachý[1], Michal Polák[2], Pavel Ryjáček[3], Milan Talich[4], Jan Havrlant[5], Filip Antoš[6], Jiří Litoš[7], Vojtěch Stančík[8] & Lubomír Soukup[9]

Abstract: In the paper, there are described arrangement and obtained substantial results of an experimental dynamic analysis performed on a new railway bridge using three different methods. The first one was the usual experimental approach realized only by high sensitive piezoelectric accelerometers. The other two methods were new approaches to dynamic measurement – ground-based radar interferometry with ground-based real aperture radar performed by two synchronized radars and digital image correlation, these ones were applied to the measurement of deflection of the horizontal load-bearing structure of the bridge. The experiment was carried out in September 2021 on the new railway steel arch bridge. The main purpose of the realized analysis was to verify applicability of two above mentioned new approaches for measurement of bridge dynamic behavior.

Keywords: Steel bridge; Experimental analysis; Dynamic measurement; Deflection; DIC; Ground-based radar interferometry; Ground-based real aperture radar ; Accelerometer

1. Introduction

In the paper, there are described arrangement and obtained substantial results of an experimental dynamic analysis performed on a new railway steel arch bridge using three different methods. The first one was the usual experimental approach realized only by high sensitive piezoelectric accelerometers because the standard sensors for determination of relative vertical deflection of the bridge deck could not be installed under the bridge due to the deep water. The other two methods were new approaches to dynamic measurement – ground-based radar interferometry (GBRI) with ground-based real aperture radar (GB-RAR) performed by two synchronized radars and digital image correlation (DIC), these ones were applied to the measurement of deflection of the horizontal load-bearing structure of the bridge. The experiment was carried out in September

[1] *Ing. Tomáš Plachý, Ph.D.; Department of Mechanics, Faculty of Civil Engineering, Czech Technical University in Prague; Thákurova 7, 166 29 Prague, Czech Republic;* plachy@fsv.cvut.cz
[2] *prof. Ing. Michal Polák, CSc.; Department of Mechanics, Faculty of Civil Engineering, Czech Technical University in Prague; Thákurova 7, 166 29 Prague, Czech Republic;* polak@fsv.cvut.cz
[3] *doc. Ing. Pavel Ryjáček, Ph.D.; Department of Steel and Timber Structures, Faculty of Civil Engineering, Czech Technical University in Prague; Thákurova 7, 166 29 Prague, Czech Republic;* pavel.ryjacek@fsv.cvut.cz
[4] *Ing. Milan Talich, Ph.D.; Institute of Information Theory and Automation, Academy of Sciences of the Czech Republic; Pod Vodárenskou věží 4, 182 00 Prague, Czech Republic;* talich@utia.cas.cz
[5] *Ing. Jan Havrlant, Ph.D.; Institute of Information Theory and Automation, Academy of Sciences of the Czech Republic; Pod Vodárenskou věží 4, 182 00 Prague, Czech Republic;* havrlant@utia.cas.cz
[6] *Ing. Filip Antoš; Institute of Information Theory and Automation, Academy of Sciences of the Czech Republic; Pod Vodárenskou věží 4, 182 00 Prague, Czech Republic;* antos@utia.cas.cz
[7] *doc. Ing. Jiří Litoš, Ph.D.; Experimental Centre, Faculty of Civil Engineering, Czech Technical University in Prague; Thákurova 7, 166 29 Prague, Czech Republic;* litos@fsv.cvut.cz
[8] *Ing. Vojtěch Stančík, Ph.D.; Department of Steel and Timber Structures, Faculty of Civil Engineering, Czech Technical University in Prague; Thákurova 7, 166 29 Prague, Czech Republic;* vojtech.stancik@fsv.cvut.cz
[9] *Ing. Lubomír Soukup, Ph.D.; Institute of Information Theory and Automation, Academy of Sciences of the Czech Republic; Pod Vodárenskou věží 4, 182 00 Prague, Czech Republic;* soukup@utia.cas.cz

2021 on the bearing structure BS2 which is a part of the new railway bridge across the Váh river at the Púchov town in Slovakia (see Fig. 1 and Fig. 2).

The main purpose of the realized experimental analysis was to verify the dynamic reliability of the bridge structure before its opening and also to verify applicability of two new approaches GB-RAR (see Fig. 6) and DIC (see Fig. 8) for measurement of bridge dynamic behaviour.

The experiment followed on the previous ones performed by the authors. Two experimental analyses carried out on a cable-stayed footbridge and a steel arch road bridge using combination of two methods, which were GB-RAR and the usual experimental approach realized by accelerometers, are described in references [1, 2]. The utilization of various standard experimental methods and the application of the obtained experimental results for advanced analyses of some different bridge structures (of an extremely skew steel arch railway bridge, a steel truss railway bridge, a one hundred years old steel truss railway bridge and a steel-concrete composite road bridge) are published in references [3–7]. The potential of the new experimental approach GB-RAR for its independent application for bridges and other engineering or building structures is described in two next references [8, 9].

There are some other references, for examples [10–13], where their authors described analyses of bridge structures by which experimental data were applied that were obtained using GB-RAR performed by one radar only.

2. Description of the observed bridge

The observed bridge structure is the system of six independent simply supported horizontal load-bearing structures with spans 30.6 m, 30.6 m, 124.8 m, 124.8 m, 30.6 m and 30.6 m long which are separated by dilatation gaps. The experiment was performed only on the third structure named BS2 (see Fig. 1 and Fig. 2) that is practically identical with the fourth one named BS3.

Fig. 1. Side view on the analysed horizontal load-bearing structure BS2. In the background there is the structure BS3.

The steel horizontal load-bearing structure BS2 is a simply supported rigid beam reinforced by an arch with the span length of 124.8 m. This structural system is also called the Langer beam (see Fig. 1). The observed structure BS2 is the double-track railway bridge with the continuous ballast bed (see Fig. 2). The bridge deck is orthotropic with cross beams and stringers. The axial distance of the main girders is 12.60 m and it is constant along the whole length of the structure. The walls of the box rigid beams are 3.0 m height.

Fig. 2. View on the analysed horizontal load-bearing structure BS2 that is oriented in direction of its longitudinal axis. In the background there is the structure BS3.

The both segmental arches are in the shape of a circle. The rise of the arches is 22.0 m. The dimensions 1.15 m and 1.71 m of the welded square box cross section of the both arches are uniform along the whole arch span except the short part at the connection to the rigid beams on the arch ends (see Fig. 2). The rigid beams are stiffened by the vertical steel bar-type arch hangers (see Fig. 1 and Fig. 2).

3. Brief description of used dynamic loading

In the course of the experiment, the observed dynamic response of the bridge bearing structure BS2 was induced by the test train (see Fig. 3) that crossed the bridge at different speeds in both directions. The railway vehicles standardly used in the bridge region were selected for the dynamic load test. The test train was composed of two towing locomotives of type 751 and one freight railway car of type Falns (see Fig. 3). The actual total weight of the individual railway vehicles connected to the test train was verified by weighing and it was 72.0 t (first towing locomotive), 79.4 t (freight railway car) and 72.0 t (second towing locomotive).

The bridge vibrations were measured in the course of twenty-two passages of the test train moving in both directions at its different speeds varying from 5 to 100 km per hour.

4. Used measuring equipment

As was said above, three different experimental techniques were applied during the dynamic experiment. The first one was the standard experimental approach realized only by high sensitive piezoelectric accelerometers. The second one was ground-based radar interferometry (GBRI) with ground-based real aperture radar (GB-RAR) carried out by two synchronized radars and the third one was digital image correlation (DIC).

Fig. 3. View on the test train moving across the bridge.

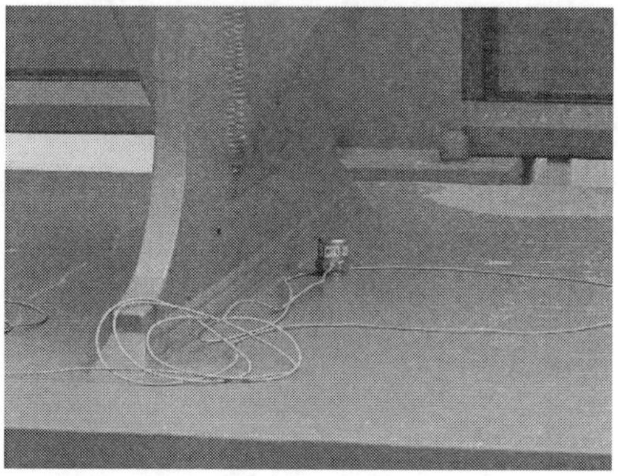

Fig. 4. View on the accelerometer of type 8344 located on the rigid beam.

4.1. Used standard measuring equipment and measurement system

Standard sensors for determination of relative vertical deflection of the bridge deck, that would be most suitable to achieve the goal of the experiment, could not be installed under the bridge due to the deep water in the Váh river (see Fig. 5). Therefore, the vibration of the bridge deck on the rigid beams was measured in the course of the experiment by piezoelectric acceleration transducers Brüel & Kjær Type 8344 (see Fig. 4). The working range of these sensors is from 0.2 Hz to 3 kHz. These transducers are very suitable for the dynamic experiments realized on bridges because the lower frequency limit of their working frequency range is low enough and they have very high sensitivity of approximately 2500 mV/g.

The accelerometers were via cables connected to the 8-channel data acquisition station SIRIUS Type 6ACC-2ACC+ made by DEWESoft. All channels have their own 2×24-bit A/D converter (so called DualCore) that is able to measure with dynamic range up to 160 dB. All A/D converter are mutually synchronized due to ensuring simultaneous measurement on all active channels.

The scope of the realized experimental analysis using standard measuring equipment was greater than is described in this paper. However, the selected set of eight accelerometers of type 8344 was used to fulfilled the goal of the reported experiment. The accelerometers that measured the bridge vibration in the vertical direction were placed on the upper surface of both rigid beams (see Fig. 4) in the first quarter, in the middle and in third quarter of the span of the structure BS2. The last two accelerometers measured vibration on the left rigid beam in the horizontal direction perpendicular to the longitudinal axis of the bridge.

4.2. Used radar interferometry and corresponding measurement system

The new technology for measuring of the dynamic response of engineering structures to dynamic load using ground-based radar interferometry (GBRI) with ground-based real aperture radar (GB-RAR) was verified in the course of the performed experiment. Two synchronized radars were applied for contactless observation of the dynamic deflections on the studied horizontal bearing structure BS2 (see Fig. 5 and Fig. 6). The positions of the both radars are marked in Fig. 7.

During the performed experiment, climatic conditions were recorded using a data logger and video records of the test train passages over the bridge were taken. The local situation and position of both radars were mapped by detailed terrestrial laser scanning and subsequently a 3D model was processed Fig. 7.

Fig. 5. Bottom view on the bridge deck of the structure BS2 with the steel crossbeams from the position of the radar R1.

Fig. 6. View on the radar R2 in its position below the structure BS2.

Two Radars IBIS produced by the Italian manufacturer Ingegneria Dei Sistemi (IDS) were used for the measurements. The radar R1 was of type IDS Radar IBIS – FS Plus with antenna IBIS-ANT3-H17V15 (see Fig. 5) and the radar R2 was of type IDS Radar IBIS – RU 172 with antenna IBIS-ANT5-H12V39 (see Fig. 6).

A bottom view on the bridge deck of the structure BS2 with the steel crossbeams from the position of the radar R2 is shown in Fig. 6. The crossbeams served for both radars as natural signal reflectors.

The sampling frequency was set at 200 Hz on both radars. However, the actual value of the sampling frequency on the radar R2 corrected by the radar control software was 199.2 Hz.

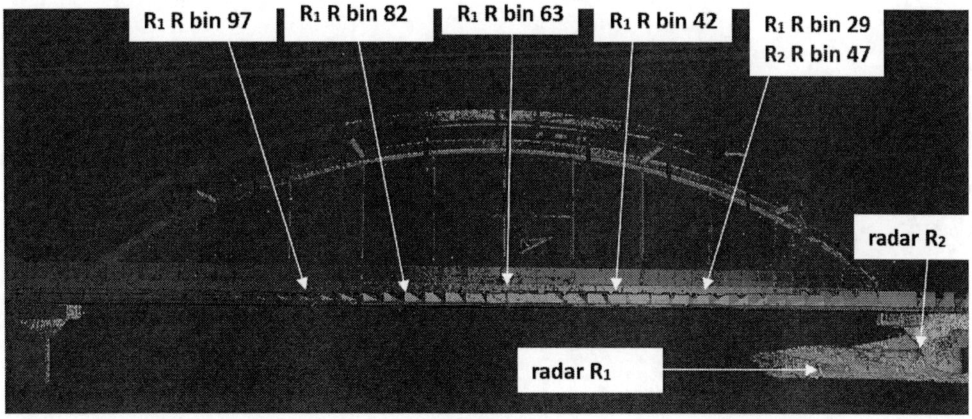

Fig. 7. 3D model of the bridge with colour-highlighted Rbins for radars R1 and R2, which were evaluated, and with marked positions of both radars below the structure BS2.

The basic principles and examples of the application of GB-RAR technology for observation of bridge deflection are described e.g. in recent publications [8] and [12, 13]. More details about application of two synchronized radars for measuring of bridge vibration are described in reference [2].

4.3. Used digital image correlation and corresponding measurement system

The digital image correlation (DIC) is the optical method that was used, as well as GB-RAR, for contactless measurement of the dynamic deflections on the studied horizontal load-bearing structure BS2 (see Fig. 8 and Fig. 9).

Fig. 8. Two active cameras applied during the experiment as part of the 2D-DIC system.

Fig. 9. The steel boards painted by white colour and by a black random pattern fixed at the observed point on the bridge deck.

The DIC system produced by Correlated Solutions, Inc. enables potentially 2D or 3D measurement. In the course of the realized experiment, the 2D-DIC system was applied, two active cameras (see Fig. 8) were used to measure the displacement in vertical plane at three selected points on the bridge deck, which were placed in the first quarter, in the middle and in third quarter of the span of the structure BS2. The steel boards painted by white colour and by a black random pattern (see Fig. 9) were fixed at the selected observed points on the bottom surface of the bridge deck to improve obtained results.

The sampling frequency of DIC was set at 10 Hz. The evaluation of the experimental results was realized in software VIC-2D produce by Correlated Solutions, Inc.

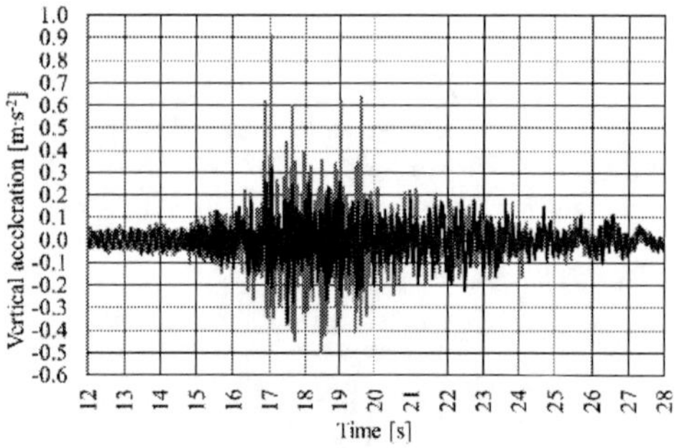

Fig. 10. The vertical acceleration of the rigid beams measured by accelerometers in third quarter of the span at 50 km/h.

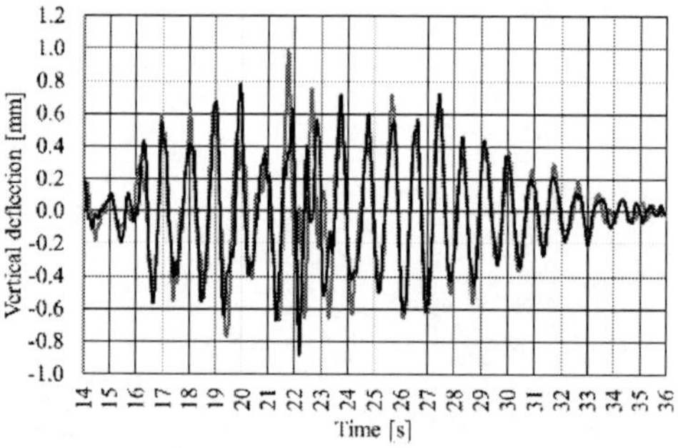

Fig. 11. The dynamic component of the vertical deflection obtained by the second integration of the measured acceleration at 50 km/h that is shown in Fig. 10.

5. Selected results of the dynamic analysis

The vibration of the bridge deck with the maximal peak acceleration of the rigid beams in the vertical direction 0.91 $m \cdot s^{-2}$was recorded during the passage of the test train at 50 km per hour. The corresponding vertical acceleration of the rigid beams measured on the structure BS2 by accelerometers in third quarter of the span is depicted in Fig. 10. The red line is vertical acceleration of the left rigid beam and the black one is vertical acceleration of the right beam.

The dynamic component of the vertical deflection, that is shown in Fig. 11, was obtained by the second integration of the measured acceleration at 50 km/h which is depicted in Fig. 10. The Simpson's rule, which is a numerical method that approximates the value of a definite integral by using quadratic polynomials, was applied for evaluation of resulting functions of the deflection.

The frequency spectrum corresponding to the measured vertical acceleration of the rigid beams in third quarter of the span on the structure BS2, that is depicted in Fig. 10, was evaluated in the frequency range from 0.4 Hz to 100 Hz. The significant part of the frequency spectrum in the frequency range from 0.4 Hz to 50 Hz is shown in Fig. 12.

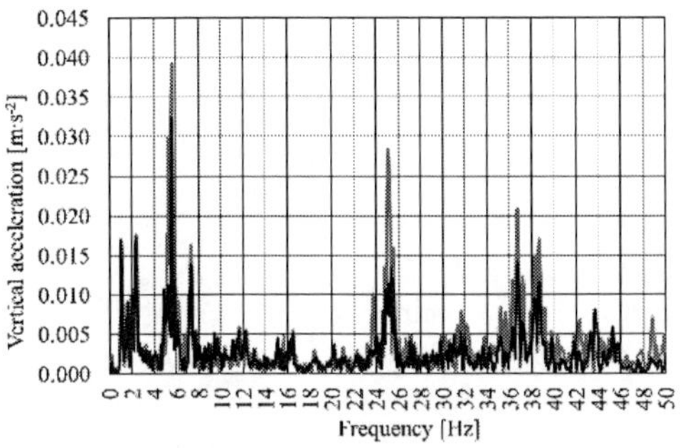

Fig. 12. The frequency spectrum of vertical acceleration of the rigid beams in third quarter of the span at 50 km/h that is shown in Fig. 10.

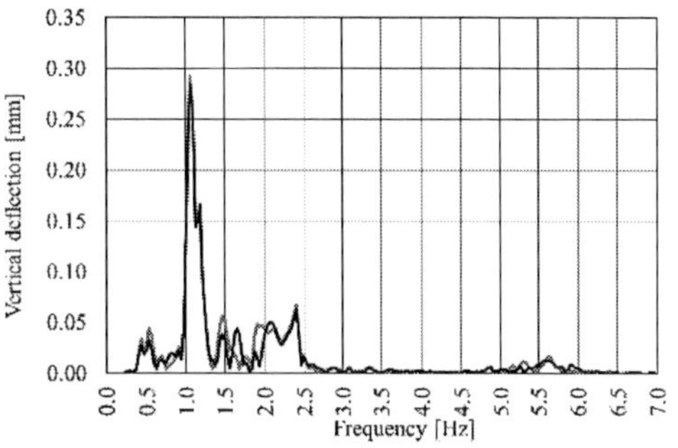

Fig. 13. The frequency spectrum of the dynamic component of the vertical deflection obtained by the second integration of the measured acceleration at 50 km/h that is shown in Fig. 11.

In the Fig. 13, there is depicted the frequency spectrum of the vertical deflection, that is shown in Fig. 11. This spectrum is drawn only for frequency range from 0.4 Hz to 7 Hz because of this interval is the most important one. The effect of the frequencies higher than 7 Hz on the deflection is negligible.

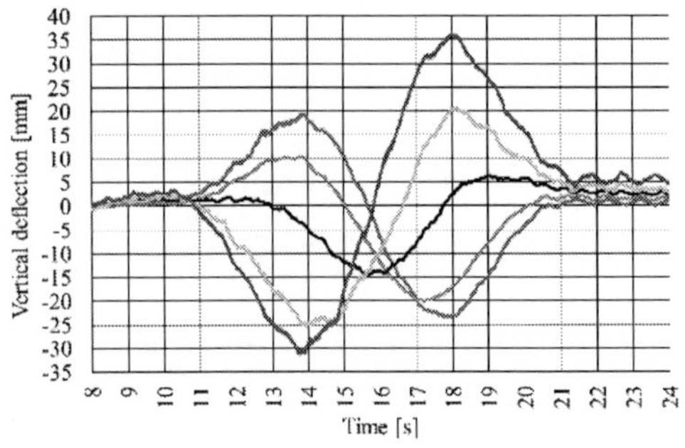

Fig. 14. The vertical deflections of the cross girders measured by GB-RAR in the central part of the bridge deck at 50 km/h.

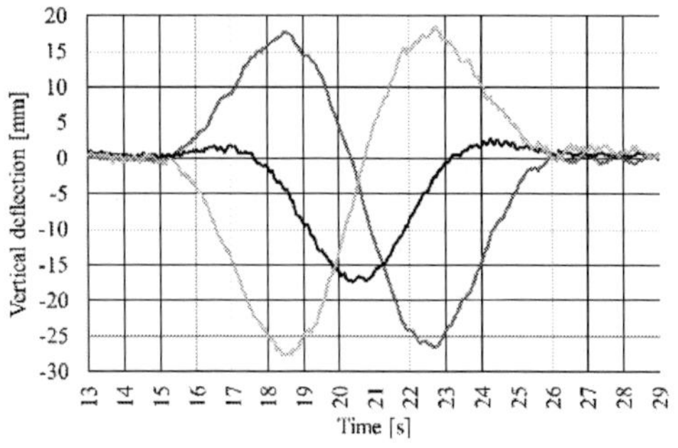

Fig. 15. The vertical deflections of the observed points of the bridge deck measured by DIC at 50 km/h.

The vertical deflections of the cross girders observed by GB-RAR in the central part of the bridge deck that were evaluated during the passage of the test train at 50 km per hour are drawn in Fig. 14 and the vertical deflection measured during the passage at 40 km per hour are shown in Fig. 18.

The comparable vertical deflections at the observed points on the bridge deck measured by DIC that were evaluated in the course of the passage of the test train at 50 km per hour are shown in Fig. 15 and during the passage at 40 km per hour are depicted in Fig. 19. The comparable curves in Fig. 14 and in Fig. 15 are drawn by the same colour. The red curves mark the vertical deflections in first quarter of the span of the structure BS2, the black ones indicate the deflections in the middle and the orange curves designate the vertical deflections in third quarter of the span. The same designation is valid also for the curves in Fig. 18 and Fig. 19.

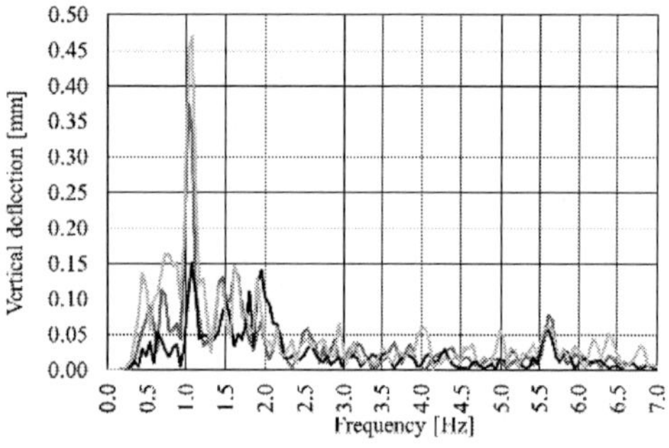

Fig. 16. The frequency spectrum of the dynamic components of the vertical deflections measured by GB-RAR at 50 km/h, which are drawn in Fig. 14.

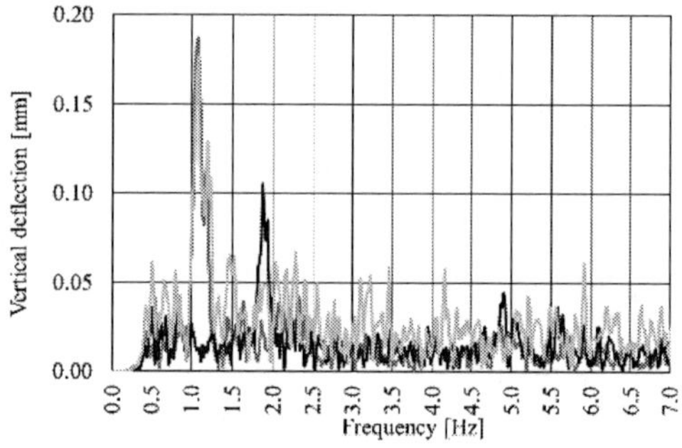

Fig. 17. The frequency spectrum of the dynamic components of the vertical deflections measured by DIC at 50 km/h, which are drawn in Fig. 15.

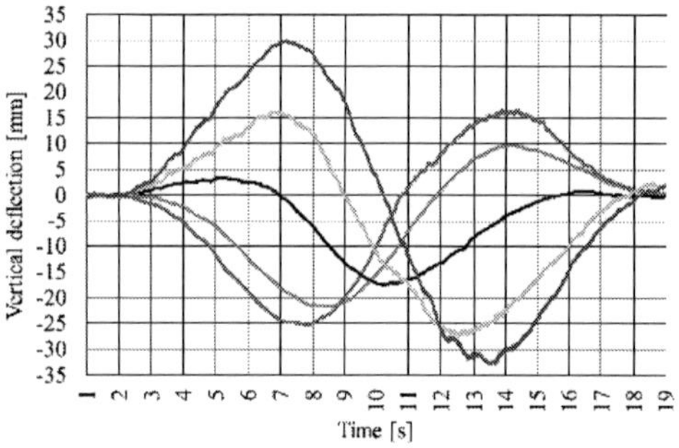

Fig. 18. The vertical deflections of the cross girders measured by GB-RAR in the central part of the bridge deck at 40 km/h.

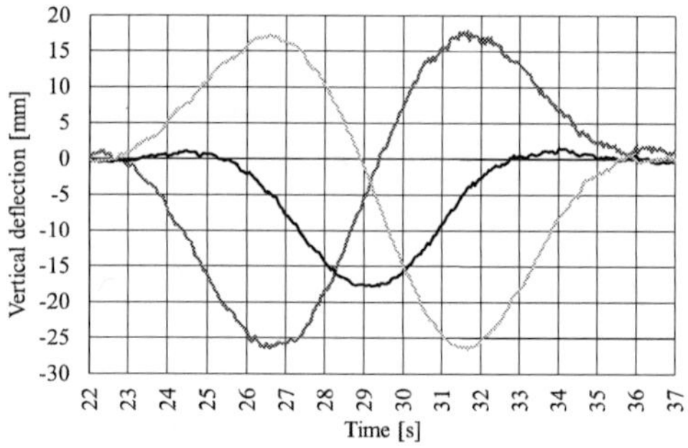

Fig. 19. The vertical deflections of the observed points of the bridge deck measured by DIC at 40 km/h.

As it can be seen from two pairs of the figures (Fig. 14, Fig. 15 and Fig. 18, Fig. 19), the comparable quasi static vertical deflections of the bridge deck measured by GB-RAR and by DIC are very similar.

The form of the frequency spectra shown in Fig. 13, Fig. 16 and Fig. 17, that were evaluated from the dynamic components of the vertical deflections obtained by all three applied approaches during the test train passage at 50 km per hour, is similar for all three spectra.

6. Conclusions

The realized experimental analysis verified the dynamic reliability of the bridge structure before its opening. The dynamic response of the bridge bearing structure BS2, that was induced by the test train that crossed the bridge at different speeds in both directions, was relatively small. The maximum observed peak acceleration in vertical direction for frequency range up to 100 Hz was 0.91 $m \cdot s^{-2}$ and for frequency range up to 30 Hz was 0.63 $m \cdot s^{-2}$. They were evaluated on the left rigid beam in the third quarter of the span and they are both significantly smaller than limit value 3.5 $m \cdot s^{-2}$ that is specified in the standard STN EN 1990 in A2.4.4.2.1 for frequency range up to 30 Hz. The dynamic reliability of the observed structure BS2 was successfully verified.

Any credible experimental information has its meaning in the experimental analysis of bridge structures. However, as expected, each of the compared experimental approaches has its advantages and disadvantages.

In particular, the following benefits were the principal ones of using GB-RAR on the observed bridge. The vibration of the bridge deck was measured simultaneously in multiple cross sections and without installation of additional equipment on the deck bottom surface. It was not necessary to get access to the observed points on the bridge deck. So the time required for installation of the experimental technique was substantially shorter compared to the standard experimental approach. But these two advantages are valid for steel structures only because a special corner radar reflectors made from metal have to be installed at observed points on concrete structures.

If the corner radar reflectors are not used, the measuring signal emitted and measured by the radar is not related to specific point of the investigated bridge deck but to a certain sector of it, that reflects the signal used for evaluation of the results. It means, the specific analysed deflection of the bridge deck should be understood as a certain average deflection in the observed sector. In the case of the observed bridge, the sectors were linked to the cross beams (see Fig. 7). This arrangement may not be used for correct observation of the torsion movement of the bridge deck.

The radars (GB-RAR) measured the component of the displacement on the bridge deck in direction of the transmitted and reflected signal that was oblique to the vertical. The points on the bridge deck usually move especially in vertical direction and so the measured signal has to be converted to the vertical direction. However, if the motion of the observed point is spatial, the conversion error of evaluated vertical deflection is substantial. This problem can be solved by simultaneous application of two or three radars [2]. However, in the case of the studied bridge, the combination of two radars was possible to use only for limited part of the deck (see Fig. 7).

Steel plates with a painted random pattern for measurement by DIC were fastened to three points on the bottom surface of the bridge deck (see Fig. 9). The permanent bridge inspection platform standardly used for the bridge maintenance was used for this activity. In this way, the observed points of the bridge deck were located more precisely compared with GB-RAR.

The instability of the tripod positions on the ground (see Fig. 5 and Fig. 8) and weather effects (e.g. changes of air temperature, potential changes in insolation of the tripods or the used devices, wind gusts) may influence the quality of the obtained experimental results of both new approaches especially in the case of a long term monitoring of the observed structure. However, similar instability and weather effects may affect negatively also the results obtained via the standard sensors for determination of relative vertical deflection of the bridge deck.

Acknowledgements

The results presented in this paper are outputs of the research project VI20192022167 "The Advanced Technology of Rapid Determination of Bridges Deformation by Radar Interferometry and its Use in Diagnostics" supported by Ministry of the Interior of the Czech Republic.

References

[1] Polák M., Plachý T., Čítek A., Berková K., Hájková P. and Čáp M., "Experimental dynamic analysis of the existing footbridge in Dobřichovice town," *MATEC Web of Conferences*, Vol. 313, Art. no. 0002, (2020). ISSN 2261-236X. http://dx.doi.org/10.1051/matecconf/202031300002.

[2] Plachý T., Polák M., Ryjáček P., Talich M., Havrlant J., Antoš F., Litoš J., Macho M. and Soukup L., "Experimental dynamic analysis of the arch road bridge," *IOP Conference Series: Earth and Environmental Science*, Vol. 906, no. 012061, p. 22 (2021). ISSN 1755-1315. http://dx.doi.org/10.1088/1755-1315/906/1/012061.

[3] Kašpárek P., Ryjáček T., Rotter T., Polák M. and Calcada R., "Long-term monitoring of the track-bridge interaction on an extremely skew steel arch bridge," *J. Civil. Struct. Health. Monit.*, Vol. 10, pp. 377387 (2020). ISSN 2190-5452. https://doi.org/10.1007/s13349-020-00389-1.

[4] Ryjáček P., Polák M., Plachý T. and Kašpárek J., "The dynamic behavior of the extremely skew railway bridge "Oskar"," *Procedia Structural Integrity*, Vol. 5, pp. 10511056 (2017). ISSN 2452-3216. https://doi.org/10.1016/j.prostr.2017.07.069.

[5] Plachý T., Polák M. and Ryjáček P., "Experimental dynamic analysis of the railway bridge near Žatec," in *Proceedings of the 58th International Scientific Conference on Experimental Stress Analysis EAN 2020*, Sobotín, 2020 (Czech Society for Mechanics, Brno, 2020), pp. 417422. ISBN 978-80-248-4451-0.

[6] Plachý T., Polák M. and Ryjáček P., "Assessment of an Old Steel Railway Bridge Using Dynamic Tests," *Procedia Engineer.*, Vol. 199, pp. 30533058 (2017). ISSN 1877-7058. https://doi.org/10.1016/j.proeng.2017.09.555.

[7] Polák M., Rotter T. and Plachý T., "Long-time Monitoring of the Bridge Response Caused by Heavy Traffic and Temperature Changes," in *Proceedings of the 49th International Scientific Conference on Experimental Stress Analysis EAN 2011*, Znojmo, 2011 (Brno University of Technology Faculty of Mechanical Engineering, Brno, 2011), pp. 341348. ISBN 978-80-214-4275-7.

[8] Talich M., "The Effect of Temperature Changes on Vertical Deflections of Metal Rail Bridge Constructions Determined by the Ground Based Radar Interferometry Method," *IOP Conference Series: Earth and Environmental Science*, Vol. 221, no. 1, Art. no. 012076 (2019). ISSN 1755-1307. http://dx.doi.org/10.1088/1755-1315/221/1/012076.

[9] Talich M., "Using Ground Radar Interferometry for Precise Determining of Deformation and Vertical Deflection of Structures," *IOP Conference Series: Earth and Environmental Science*, Vol. 95, no. 3, Art. no.032021 (2017). ISSN 1755-1307. http://dx.doi.org/10.1088/1755-1315/95/3/032021.

[10] Lamperová K., Sokol M. and Timková B., "Identification of Bearings State on the Bridge Checked by Dynamic Tests," *Strojnicky Casopis*, Vol. 70, no. 2, pp. 6776 (2020). ISSN 0039-2472. http://dx.doi.org/10.2478/scjme-2020-0021

[11] Sokol M. and Lamperová K., "Dynamic response of bridges tested by radar interferometry," *Vibroengineering Procedia*, Vol. 23, pp. 138143 (2019). ISSN 2345-0533. http://dx.doi.org/10.21595/vp.2019.20674

[12] Gentile C., "Dynamic investigation of a suspension footbridge using accelerometers and microwave interferometer," *MATEC Web of Conferences*, Vol. 24, Art. no. 07004 (2015). ISSN 2261-236X. http://dx.doi.org/10.1051/matecconf/20152407004.

[13] Liu X., Tong X., Ding K., Zhao X., Zhu L. and Zhang X., "Measurement of Long-Term Periodic and Dynamic Deflection of the Long-Span Railway Bridge Using Microwave Interferometry," *IEEE Journal of Selected Topics in Applied Earth Observations and Remote Sensing*, Vol. 8, no. 9, pp. 45314538 (2015). ISSN 1939-1404. http://dx.doi.org/10.1109/JSTARS.2015.2464240

Measured and Calculated Stiffness and Damping Coefficients of Water-Lubricated Tilting-Pad Journal Bearing

Jan Pokorný[1], Tomáš Návrat[2] & Michal Vajdák[3]

Abstract: This paper deals with the measurement and computational modelling of the dynamic performance of a hydrodynamic water-lubricated tilting-pad journal bearing. The measurement is performed on a developed test rig, which contains a special bearing housing with integrated piezo-actuators enabling dynamic excitation of the bearing. A reported computational model is based on the generalized dynamic Reynolds equation coupled with the 2D energy equation in the bearing midplane. This model involves the lubricant mixing, deformation and tilting of the pads, and establishing the static equilibrium position of the bearing. Results of the dynamic performance dependent on rotational speed are given for a particular water-lubricated journal bearing. These results are compared with the measured data.

Keywords: Tilting pad journal bearing; Measurement; Numerical analysis; Stiffness; Damping

1. Introduction

The dynamic performance of hydrodynamic bearings is represented by their stiffness and damping coefficients, written in the stiffness and damping matrices [1]. These matrices usually have a shape of 2×2 and contain the direct terms on the diagonal and cross-coupled terms off the diagonal [2]. In our case, the dynamic performance is determined experimentally and by computational modelling.

The dynamic performance is determined by very small displacements of the journal from the equilibrium position and is therefore dependent not only on the bearing geometry but also on speed and load [3]. In computational modelling, the equilibrium position of the bearing must first be determined [4]. This is a position at which the hydrodynamic forces resultant is in equilibrium with the external load on the bearing and all tilting pads are in equilibrium. This position is established iteratively by solving the Reynolds equation together with the energy equation numerically [5]. Furthermore, the mixing of hot and cold lubricant in the groove between the pads must also be solved [6]. The cold lubricant is supplied from the reservoir, while the hot lubricant exits from the lubricant film between the pads and the rotor. Other effects considered in the establishment of equilibrium are the deformation of the pads due to hydrodynamic forces and also the very important tilting of the pads [7]. Once the above-mentioned equilibrium conditions are satisfied, the equilibrium position is determined for a given geometry, load and speed [8].

To identify the stiffness and damping coefficients, it is necessary to apply small dynamic motions to the bearing journal from its equilibrium position and determine the reaction forces to these motions [9]. The applied motions must be sufficiently small because the hydrodynamic bearing behaves non-linearly [10]. By using very small-amplitude motions we try to linearize its properties around the equilibrium position.

[1] *Ing. Jan Pokorný; Institute of Solid Mechanics, Mechatronics and Biomechanics, Faculty of Mechanical Engineering, Brno University of Technology; Technická 2, 616 69 Brno, Czech Republic;* jan.pokorny4@vut.cz
[2] *doc. Ing. Tomáš Návrat, Ph.D.;* navrat@vut.cz
[3] *Ing. Michal Vajdák;* michal.vajdak@vut.cz

2. Methods

2.1. Test Rig and Instrumentation

Fig. 1 shows a device for measuring the dynamic performance of water-lubricated hydrodynamic tilting-pad journal bearings. This device consists of three bearing housings that are mounted on a base frame made of precision aluminium profiles. The outermost bearing housings are used to support the rotor to be measured, while the middle housing allows dynamic excitation of the bearing by very small forces in two mutually perpendicular directions independently. For this purpose, the housing contains two piezo-actuators in the horizontal and vertical directions. To determine the dynamic excitation, the device is equipped with two force transducers. The response to the excitation is measured by four eddy current proximity probes, two of which are in the horizontal direction and two in the vertical direction. The drive is provided by a high-speed electric motor up to 15,000 rpm, which is connected to the measured rotor by a flexible coupling. This coupling compensates for any misalignment between the rotating parts and allows radial movement of the rotor in the bearings due to hydrodynamic forces. The speed is measured by an optical tachometer and the safety of the device is ensured by a protective cover.

Fig. 1. Test rig for measuring the dynamic performance of water-lubricated tilting-pad journal bearings.

2.2. Measurement procedure

The device was started up to the required rotational speed during which the rotor and tilting pads establish the equilibrium position. Subsequently, the test bearing was excited with a very small harmonic force at a selected frequency of 40 Hz in the vertical direction. The waveform of the excitation force was recorded by a force transducer, while the waveform of the relative displacements of the bearing with respect to the rotor was measured by proximity probes. Due to the non-linear behaviour of the bearing, the excitation force in one axial direction caused displacement in the general direction. Furthermore, excitation was performed in the horizontal direction with force and displacement recording. The evaluation of the measurements takes advantage of the fact that the excitation and response have a harmonic waveform [11]. The measured signals were processed in Matlab where bandpass filtering was performed to remove unwanted frequencies. The stiffness coefficients were determined based on the ratio of the excitation and response amplitudes, while the damping coefficients were identified based on the phase shift of the response relative to the excitation [1]. To determine the dependence of the stiffness and damping coefficients on speed, this procedure was repeated for different speeds.

2.3. Computational Model

The hydrodynamic pressure in the journal bearing under dynamic operating conditions is described by the generalized dynamic Reynolds equation of the following form [12, 13]:

$$\frac{\partial}{\partial\theta}\left(\frac{H^3}{K_x}\bar{\boldsymbol{G}}\frac{\partial\bar{p}}{\partial\theta}\right)+\frac{\partial}{\partial\bar{y}}\left(\frac{H^3}{K_y}\bar{\boldsymbol{G}}\frac{\partial\bar{p}}{\partial\bar{y}}\right)=\frac{\partial}{\partial\theta}\left(\bar{\boldsymbol{G}}_{\boldsymbol{R}}H\right)+\frac{2}{\omega}\frac{\partial}{\partial t}\left(\bar{\boldsymbol{G}}_{\boldsymbol{R}}H\right) \tag{1}$$

where $H = h/c_b$ is the dimensionless lubricant film thickness, $c_b = R - R_J$ is the radial assembly clearance, R is the bearing radius, R_j is the journal radius, $\bar{p} = \frac{pc_b^2}{U\eta_0 R}$ is the dimensionless pressure, U is the sliding velocity, η_0 is the initial dynamic viscosity, $\theta = x/R$ is the angular coordinate, $\bar{y} = y/R$ is the dimensionless axial coordinate, ω is the rotational angular speed, t is the time, $\bar{\boldsymbol{G}}$ and $\bar{\boldsymbol{G}}_{\boldsymbol{R}}$ characterize the variable lubricant viscosity and density, and K_x and K_y models effect of local turbulent flow and Taylor vortices. (More details are provided in [8].)

The temperature distribution in the bearing midplane in the circumferential and radial directions is given by the energy equation:

$$\frac{\partial}{\partial\theta}\left(\bar{U}\bar{T}\right)+\frac{\partial}{\partial\bar{z}}\left(\bar{W}\bar{T}\right)=\frac{\partial}{\partial\theta}\left(\Lambda_\theta\frac{\partial\bar{T}}{\partial\theta}\right)+\frac{\partial}{\partial\bar{z}}\left(\Lambda_z\frac{\partial\bar{T}}{\partial\bar{z}}\right)+K_T\phi \tag{2}$$

where $\bar{T} = T/T_0$ is the dimensionless temperature, T_0 is the water supply temperature, $\bar{U}$ and $\bar{W}$ are the modified dimensionless circumferential and radial velocities, Λ_θ and Λ_z are the modified dimensionless thermal diffusivities, K_T is the dissipation number, ϕ is the dissipation function, and $\bar{z} = z/h$ is the dimensionless radial coordinate. Detailed information is presented by the author in [8].

In the case of a water-lubricated bearing, the temperature change of the lubricant film is not very significant, a greater effect can be observed in bearings lubricated with higher viscosity lubricants such as oil. The change in temperature affects mainly the viscosity and density of the lubricating medium. The computational model used also takes into account the mixing of the lubricant in the grooves between the pads, the deformation of the pads due to pressure and temperature loads, and, in particular, the tilting of the pads [8]. The pad tilting has a significant effect on the bearing properties, especially in terms of stability during operation and resistance to vibration [14]. Furthermore, the static equilibrium position of the bearing is established using the bounded Anderson method [8].

The static equilibrium position of the bearing is used to determine the stiffness and damping coefficients Fig. 2. These coefficients are obtained by a small-amplitude dynamic excitation of the rotor about its equilibrium position. The mathematical notation of the relationship between the reaction forces d$\boldsymbol{F}$ and the motions of the journal d$\boldsymbol{U}$ is given below using the stiffness and damping matrix [1]:

$$\begin{bmatrix}\mathrm{d}F_x\\ \mathrm{d}F_y\end{bmatrix}=-\begin{bmatrix}K_{xx} & K_{xy}\\ K_{yx} & K_{yy}\end{bmatrix}\begin{bmatrix}\mathrm{d}x\\ \mathrm{d}y\end{bmatrix}-\begin{bmatrix}B_{xx} & B_{xy}\\ B_{yx} & B_{yy}\end{bmatrix}\begin{bmatrix}\mathrm{d}\dot{x}\\ \mathrm{d}\dot{y}\end{bmatrix} \tag{3}$$

where $\boldsymbol{K}$ is the stiffness matrix and $\boldsymbol{B}$ is the damping matrix of the bearing.

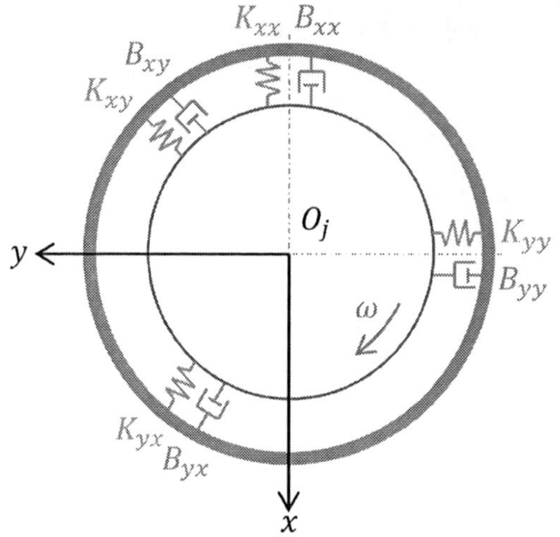

Fig. 2. Scheme of the stiffness and damping coefficients represented by a mechanical spring and a viscous dashpot [15].

To determine the stiffness and damping coefficients, it is advantageous to use isolated excitation in only one of the axial directions, while the excitation in the other direction is zero. If we excite the journal with a small displacement in one direction so that the displacement in the other direction is zero, then we can express two coefficients of the stiffness matrix from the reaction forces obtained [1]. This can be written as follows:

$$K_{xx} = -\frac{\mathrm{d}F_x}{\mathrm{d}x}; K_{yx} = -\frac{\mathrm{d}F_y}{\mathrm{d}x} \tag{4}$$

where $\mathrm{d}x$ is the displacement in the x direction and the corresponding reaction forces are.

Similarly, if the displacement $\mathrm{d}x$ in the vertical direction is zero, then:

$$K_{xy} = -\frac{\mathrm{d}F_x}{\mathrm{d}y}; K_{yy} = -\frac{\mathrm{d}F_y}{\mathrm{d}y} \tag{5}$$

where $\mathrm{d}y$ is the displacement in the y direction and are the reaction forces on this displacement.

The damping coefficients can be determined analogously. The difference is that the infinitesimal displacement is not used for excitation, but its time derivative.

3. Results

Table 1 shows the parameters of the investigated water-lubricated bearing.

Table 1. Characteristic data of the test bearing.

Number of tilting pads	3, load between pad
Bearing radius [mm]	9
Bearing length [mm]	11
Radial assembly clearance [µm]	37.5
Radial machined clearance [µm]	50
Pivot offset [-]	0.5
Rotational speed [rpm]	0–30 000
Bearing load [N]	2.7
Water supply temperature [°C]	25

Fig. 3 shows the distribution of the hydrodynamic pressure field in the tilting-pad journal bearing under different operating conditions. These are speeds of 5,000 rpm (Fig. 3a) and 30,000 rpm (Fig. 3b). The magnitude of hydrodynamic pressure increases with speed, which can be seen especially in the upper pad, which is not loaded at 5000 rpm. The maximum pressure is 45.7 kPa and 58 kPa, respectively. At higher speeds, the lubricant film thickness is also greater, with minimum thicknesses of 15.3 μm (Fig. 3a) and 28.1 μm (Fig. 3b). For the sake of clarity, the film thickness is enlarged 50 times in Fig. 3. A more precise view of the spatial pressure distribution in the bearing is given in Fig. 4. The pressure field is shown here for 30,000 rpm.

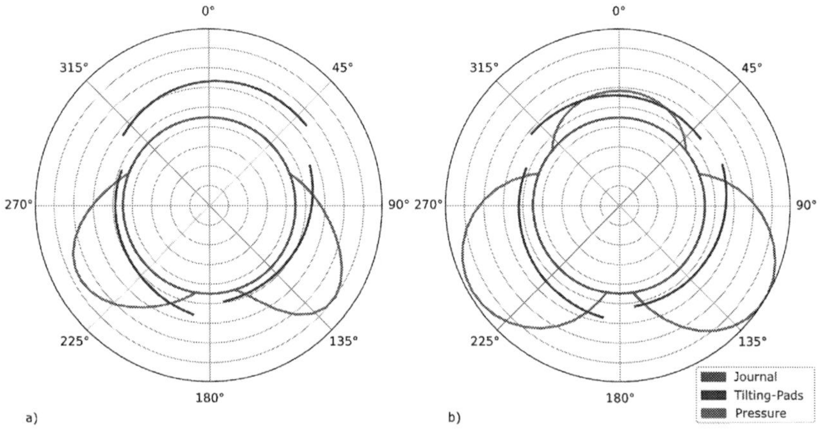

Fig. 3. Hydrodynamic pressure field in the bearing midplane and lubricant film thickness (50 times magnified), rotational speed 5,000 rpm a), 30,000 rpm b).

The increase in lubricant film thickness can also be seen in Fig. 5, which shows the dimensionless eccentricity as a function of speed. The eccentricity decreases with increasing speed, showing a shift of the journal towards the centre of the bearing. This results in all pads being loaded more similarly at higher speeds.

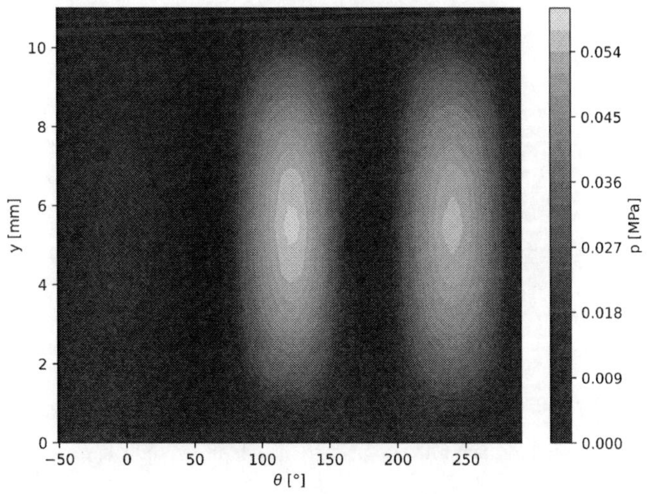

Fig. 4. Distribution of the hydrodynamic pressure, rotational speed 30,000 rpm.

Fig. 6 shows the temperature change in the bearing midplane at 30,000 rpm, showing the temperature rise of the shaft, lubricant film and pads. Due to the very small size of the lubricant film, its thickness is increased to the same size as the thickness of the pads. The greatest temperature increase occurs near the rotating surface of the shaft. The temperature at the leading edge of the pads is lower than at the trailing edge due to the cold lubricant supply. Because of the relatively low viscosity of the water, the heat-up is not very high.

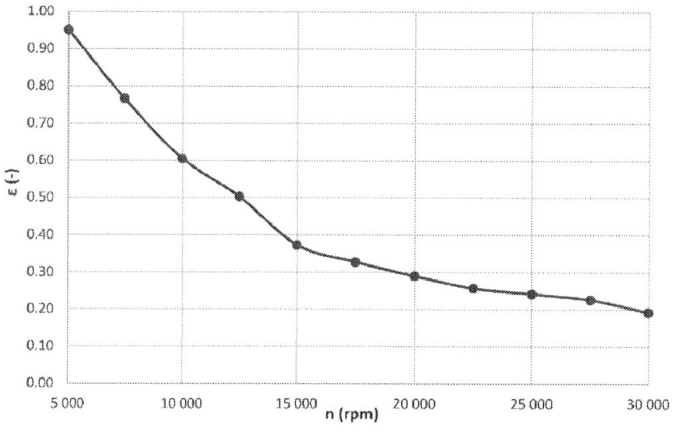

Fig. 5. Bearing dimensionless eccentricity in dependence on rotational speed, radial assembly clearance 37.5 μm.

The dependence of the stiffness coefficients on the speed is shown in Fig. 7. These coefficients obtained by computational modelling are shown in comparison with the measurement. The direct coefficients reach the largest values, the cross-coupled ones are much smaller. This property is generally true for hydrodynamic bearings with tilting pads, where the cross-coupled stiffness and damping terms are minimal due to the tilting of the pads [15]. In this case, the direct stiffness coefficient is larger in the y direction than in the x direction. This means that the largest direct term is in the direction perpendicular to the load, which is generally not the case.

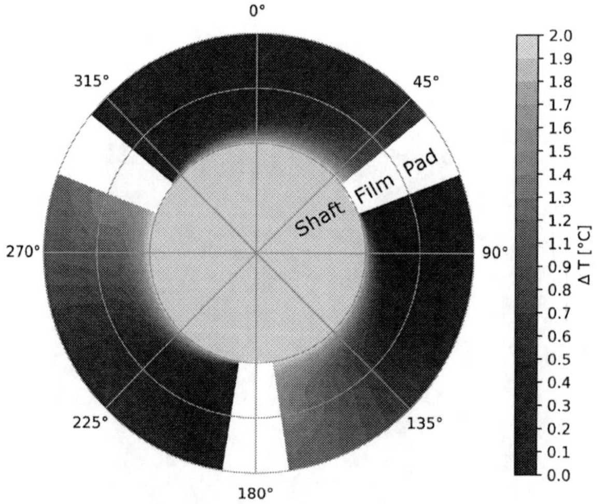

Fig. 6. Temperature change in the lubricant film and pads in the bearing midplane, rotational speed 30,000 rpm.

Measurements were limited to 15,000 rpm due to available equipment. Considering the nature of the measured quantities, a fairly good agreement between measurement and calculation can be reported, especially for speeds higher than 10,000 rpm. The response of the bearing to dynamic excitation was in the order of micrometers. Only the values of the direct stiffness coefficients are compared since the values of the cross-coupled coefficients are already below the resolution level of the measurement.

The values of the damping coefficients decrease globally with speed (Fig. 8). As with the stiffness coefficients, the largest direct term is in the y direction, and similarly, the cross-coupled terms of the damping coefficients are significantly smaller than the direct terms.

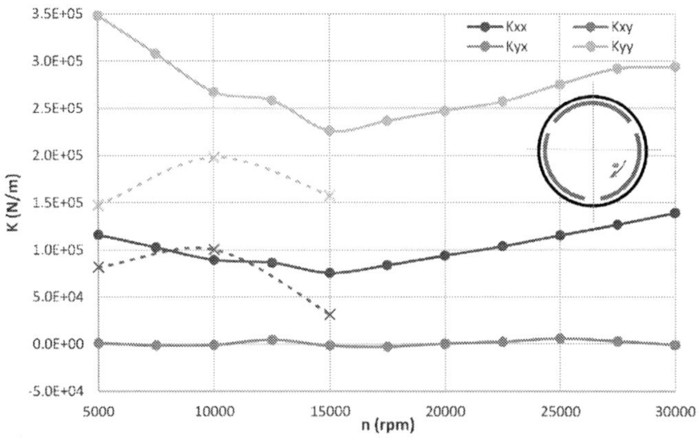

Fig. 7. Stiffness coefficients in dependence on speed, $K_{xy} = K_{yx}$, computation (the solid curve) and measurement (the dashed curve).

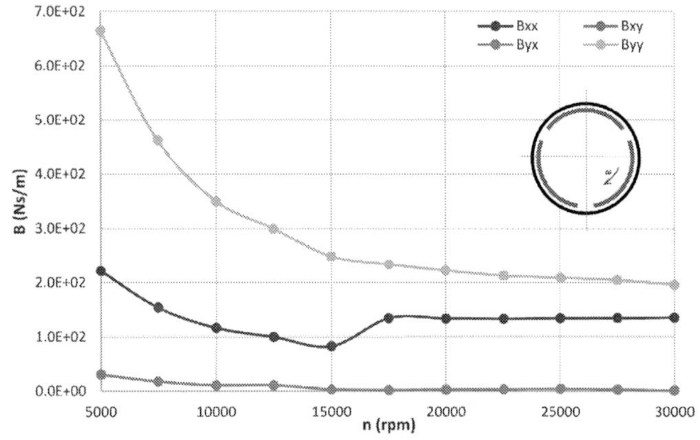

Fig. 8. Damping coefficients in dependence on speed, $B_{xy} = B_{yx}$, computation.

The measured results of the damping coefficients are not reported here because they do not fall within the expected range of values. The experimentally obtained damping values are generally much larger than expected. This may be due to the additional damping that may be caused by the bearing being supported in the housing by two rubber O-rings. They are used here to prevent the leakage of the lubricant. These O-rings can influence the measured force during dynamic excitation. However, measurement deviations can also be caused by other influences.

4. Conclusion

The dynamic performance of the water-lubricated tilting-pad journal bearing was measured, calculated, and then compared. The reported measurement required dynamic excitation of the bearing about the static equilibrium position by a small-amplitude force. The displacement response to this excitation was measured by proximity probes. The calculation performed was based on the generalized dynamic Reynolds equation coupled with the energy equation. Lubricant mixing in the bearing, deformation and tilting of the pads were also solved. After establishing the static equilibrium position, the bearing stiffness and damping matrices were calculated.

To conclude, the values of the direct coefficients of the stiffness matrix were compared with the calculation up to 15,000 rpm. The values of the cross-coupled coefficients were below the resolution level of the measurement. For speeds higher than 10,000 rpm, a fairly good agreement between measurement and calculation can be reported. The damping coefficients obtained by measurement were not presented because they were probably negatively influenced by the mounting of the bearing to the housing using rubber O-rings. The experimental determination of bearing stiffness and damping places high demands on the excitation of the bearing with small forces, on the measurement of very small displacements and on the subsequent processing of the measured data. Further research might be focused on eliminating the mentioned parasitic effects in the measurement of the dynamic performance of journal bearings.

Acknowledgements

This work is an output of the project „Research of an electrically powered fuel pump for the next generation rocket engines“ (FV40386) created with financial support from the Ministry of Industry and Trade of the Czech Republic under the TRIO program.

References

[1] Tiwari R., *Rotor Systems: Analysis and Identification*, (CRC Press, Boca Raton, 2017). ISBN 978-113-8036-284.

[2] Shapiro W. and Colsher R., "Dynamic Characteristics Of Fluid-Film Bearings," in Proceedings of the Sixth Turbomachinery Symposium, A&M University, Texas, 1977. https://doi.org/10.21423/R1RH4S.

[3] Zhang X., Yin Z., Gao G. and Li Z., "Determination of stiffness coefficients of hydrodynamic water-lubricated plain journal bearings," *Tribology International*, Vol. 85, pp. 37–47 (2015). https://doi.org/10.1016/J.TRIBOINT.2014.12.019.

[4] Pokorný J. and Návrat T., Measured and Calculated Static Equilibrium Position of Water-Lubricated Tilting-Pad Journal Bearing, in: Experimental Stress Analysis, Praha, 2021: pp. 186–192. ISBN 978-80-01-06885-4.

[5] Dowson D., "A generalized Reynolds equation for fluid-film lubrication," *International Journal of Mechanical Sciences*, Vol. 4, pp. 159–170 (1962). https://doi.org/10.1016/S0020-7403(62)80038-1.

[6] Heshmat H. and Pinkus O., "Mixing Inlet Temperatures in Hydrodynamic Bearings," *Journal of Tribology*, Vol. 108, pp. 231–244 (1986). https://doi.org/10.1115/1.3261168.

[7] Ettles C.M.M., "The analysis and performance of pivoted pad journal bearings considering thermal and elastic effects," *Journal of Tribology*, Vol. 102, pp. 182–191 (1980). https://doi.org/10.1115/1.3251465.

[8] Pokorný J., "An efficient method for establishing the static equilibrium position of the hydrodynamic tilting-pad journal bearings," *Tribology International*, Vol. 153, pp. 106641 (2021). https://doi.org/10.1016/j.triboint.2020.106641.

[9] Flack R.D., Kostrzewsky G.J. and Taylor D.V., "A Hydrodynamic Journal Bearing Test Rig with Dynamic Measurement Capabilities," *Tribology Transactions*, Vol. 36, pp. 497–509 (1993). https://doi.org/10.1080/10402009308983190.

[10] Szeri A.Z., *Fluid film lubrication*, second edition, (Cambridge University Press, 2010). https://doi.org/10.1017/CBO9780511782022.

[11] Wygant K., Flack R. and Barrett L., "Measured Performance of Tilting-Pad Journal Bearings over a Range of Preloads—Part II: Dynamic Operating Conditions," *Tribology Transactions*, Vol. 47, pp. 585–593 (2004). https://doi.org/10.1080/05698190490504208.

[12] Mittwollen N. and Glienicke J., "Operating conditions of multi-lobe journal bearings under high thermal loads," *Journal of Tribology*, Vol. 112, pp. 330–340 (1990). https://doi.org/10.1115/1.2920261.

[13] Huang P., *Numerical Calculation of Lubrication: Methods and Programs*, (John Wiley & Sons, Singapore, 2013). ISBN 978-1-118-45119-9.

[14] Boyce M.P., *Gas Turbine Engineering Handbook, Butterworth-Heinemann*, (Boston, 2012). https://doi.org/10.1016/C2009-0-64242-2.

[15] San Andrés L., "Modern Lubrication Theory: Gas Film Lubrication", *Notes*, Vol. 15, Texas, (2010). http://repository.tamu.edu/handle/1969.1/93197 (accessed September 3, 2020).

[16] Synnegård E., Gustavsson R. and Aidanpää J.O., "Influence of cross-coupling stiffness in tilting pad journal bearings for vertical machines," International Journal of Mechanical Sciences, Vol. 111–112, pp. 43–54 (2016). https://doi.org/10.1016/J.IJMECSCI.2016.03.017.

Lightweight Gypsum-based Masonry Block: Mechanical Properties

Zdeněk Prošek[1], Jan Trejbal[2] & Pavel Tesárek[3]

Abstract: The paper deals with the use of recycled gypsum for the production of lightweight masonry block for cladding. The effects of different percentages of foaming additives on the mechanical properties are discussed. In addition, the paper discusses the use of recycled fibres which positively affect the foam structure by stabilizing it. The results were obtained on 100×100×100 mm specimens and always compared with a reference mixture. The resulting gypsum-based composite can be used in the form of insulating material.

Keywords: Compressive strength; Experimental; Gypsum; Lightweight masonry block

1. Introduction

Gypsum is one of the most commonly used materials in civil engineering. It is a natural material whose resources are limited. For this reason, efforts are being made to replace it with other sources. One of the possibilities is energy-gypsum. It is a chemically pure material, produced by the desulphurization of flue gases from thermal power plants and heating plants where the main fuel is brown coal containing sulphur [1, 2]. Energy-gypsum is a suitable option for the protection of natural resources, but due to the decreasing share of coal in energy production, it is necessary to find a replacement for this source as well. One option is recycling, as gypsum-based products contain large amounts of pure gypsum with minimal admixture (up to 1%) [3].

Within the framework of the project FV30359 of the Ministry of Industry and Trade, experiments were carried out at the FCE, CTU in Prague, where the result was plasterboard recyclate with a high purity of up to 95% [4]. In the follow-up project TA ČR Trend FW03010054, the aim is to calcinate this recyclate and create a new product, unhydrated gypsum, which can form a binder in new building materials. Such a material can be used for the production of lightweight blocks, and this issue is the focus of this paper. New building products are nowadays subject to higher demands for a wider range of performance properties such as compressive strength, bulk density, thermal conductivity, etc. Basically, an optimum compromise between the parameters under consideration is sought.

Lightweight building elements for envelopes are conceived/designed in a similar way, where a compromise is sought between compressive strength and thermal conductivity to keep the thickness of the envelope as small as possible relative to the built-up area. In most cases, the envelope is then externally insulated with thermal insulation.

One option to create a lightweight block is to use foam to lighten the structure in a controlled manner, thereby improving some of the performance characteristics of the resulting material. The important thing is to use the right foam and other chemical additives so that the foam does

[1] *Ing. Zdeněk Prošek, Ph.D.; Department of Mechanics, Czech Technical University in Prague, Faculty of Civil Engineering, Thákurova 7, 166 29 Prague 6, Czech Republic;* zdenek.prosek@fsv.cvut.cz

[2] *Ing. Jan Trejbal, Ph.D.; Department of Physics s, Czech Technical University in Prague, Faculty of Civil Engineering, Thákurova 7, 166 29 Prague 6, Czech Republic;* jan.trejbal@fsv.cvut.cz

[3] *doc. Ing. Pavel Tesárek, Ph.D.; Department of Mechanic, Czech Technical University in Prague, Faculty of Civil Engineering, Thákurova 7, 166 29 Prague 6, Czech Republic;* pavel.tesarek@fsv.cvut.cz

not “collapse” and, ideally, a dense foam made up of small regular bubbles is created. The fresh mixture created must be adjusted at the beginning of production so that it holds its shape, does not collapse, and does not form inclusions, i.e. areas of different bulk density. For this purpose, plasticizing or accelerating additives are added to the mixture, which can positively influence the behavior of the mixture with respect to the materials used and the resulting performance properties [5]. Another option to maintain the stability of the foam and light weight structure of material in the initial stages of setting and hardening is the addition of microfibers.

2. Materials and Specimens

Specimens of lightweight gypsum-based composites are made using a foaming additive and the application of recycled fibers. Unhydrated waste gypsum produced by calcination of secondary waste material was used as the binder. The gypsum used is a very pure chemical material with a purity of 97 to 98%. By comparison, mined gypsum (natural quarries) has a purity of at most 70 to 75%. An amide- and sulfonic acid-based foaming agent was used to foam the structure of the lightweight gypsum-based composite. The foam itself was produced by an "ex-situ" foam generator and then mixed into the fresh mix. The foaming agent used in the foam generator was at a concentration of 50% solution with a foaming capacity of 35 ml/g and a foam stability of 465 minutes. The selection of foaming agent type and its concentration can be found in previous research [6]. To stiffen the foamed structure, Trevos, Ltd. microfibres were used, which were made from 100% recycled polypropylene. The microfibres used were 32 microns in diameter and 4 mm long. Specimens of 100 100100 mm were made from these materials. The different mixtures were varied to determine the optimum lightening with respect to the resulting performance properties, in this case mechanical and thermal properties, and to determine the effect of the use of microfibres on the "stiffening" of the structure. All the mixtures produced had the same water coefficient (coefficient water to gypsum 0.7). The composition of each mixture can be seen in Table 1.

Table 1. Composition of tested mixtures.

Set	Gypsum [kg]	Water [kg]	Fibers [g]	Foaming agent [g]
A	3.0	2.1	0	0
B	3.0	2.1	15	0
C	3.0	2.1	0	30
D	3.0	2.1	15	30
E	3.0	2.1	15	15

3. Experimenthal methods

The thermal conductivity (λ) coefficient was tested for all specimens with an age of 7 days. For these purposes, heat transfer analyzer ISOMET 2104 (Applied Precision) was employed, equipped with two surfaces probes API210411 and API210412 with measuring range from 0.04–0.3 W/mK and 0.3–2.0 W/mK, respectively. Such a device applies a dynamic measurement method which is based on response monitoring of examined material on heat flow impulses. Temperature of specimens was equal to constant 27 °C. Each specimen was measured three-times in a different position in order to obtain statistically relevant data. Specimens marked as A1 and B1 were analyzed with API210412 probe as their thermal conductivity coefficient exceeded 0.3 W/mK. The other specimens (C, D, E) were measured with API210411 probe.

After testing the thermal properties, the specimens were destructively tested to determine the compressive strength on a FHF Strassentest hydraulic press with a load area of 100×100 mm.

The test was controlled by a constant speed displacement of 1 mm/min. The compressive strength was determined from 5 specimens.

4. Results

The results of each experiment follow the same trend and are directly dependent on the value of the bulk density of the material. Specimens of sets A and B had a comparable bulk density, namely 1300±65 kg/m^3. Set C had a bulk density 325±2 kg/m^3, set D 335±1 kg/m^3 and set E 372±1 kg/m^3.

The results of compressive strength (Fig. 1) correspond to the values of the bulk weights; i.e., the specimens with lower bulk weights have higher porosity and also lower compressive strengths. In the case of the use of fibers, a noticeable effect can be seen, reaching a higher compressive strength of approximately 15% in the case of the mixture containing fibers. Specimen C had the lowest compressive strength, nominally 250 kPa. This value is very low and such material is not suitable for structural purposes and can only form a fill insulation material.

The results of the thermal moisture parameters (Fig. 2) are also consistent with the bulk density. With a lower bulk density, the specimens contain more pores with air, which is an excellent insulator. For this reason, specimen C has the best thermal-moisture properties. Specimen C had a thermal conductivity coefficient value of 0.077 W/mK, which is comparable to conventional thermal insulation products.

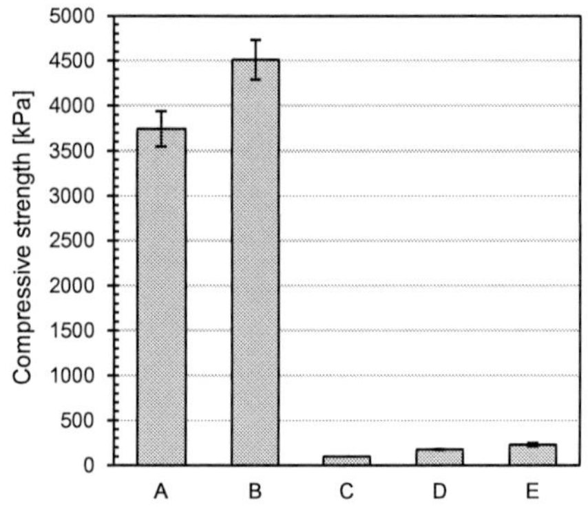

Fig. 1. Compressive strength of test mixtures with standard deviation.

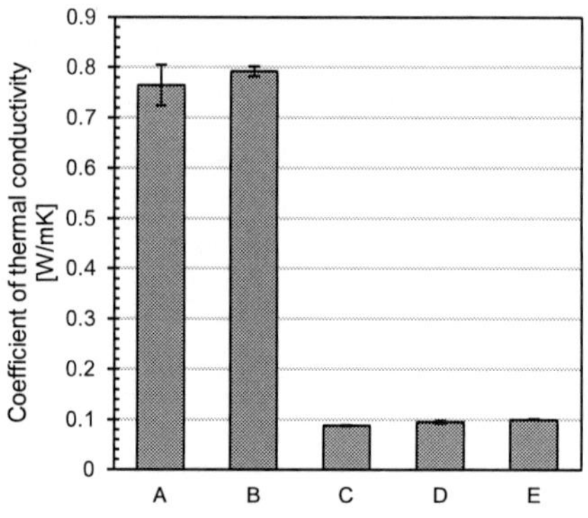

Fig. 2. Coefficient of thermal conductivity of tested mixtures with standard deviation.

5. Conclusion

This work focuses on the design of a lightweight masonry block mix based on waste gypsum. The different mixes were composed of waste gypsum fibers and different amounts of foaming agent. Based on the results, it can be concluded that:

- Creation of foam using a foam generator within the application to the gypsum composite is possible.
- The foaming in this way is too high and the specimens lose their mechanical properties.
- The use of fibres has a positive effect on the stability of the foam, resulting in an increase in compressive strength.

Future research will be based on the results presented in this paper. Future work will test gypsum mixtures lightened with foam created in-situ during mixing of the mixture. This should result in a reduction in foaminess.

Acknowledgements

This paper was financially supported by by Technology Agency of the Czech Republic (TA ČR) research project FW03010054 and Czech Technical University in Prague under No. SGS project SGS22/089/OHK1/2T/11. The authors also thank KNAUF Prague Ltd. for the supplied materials.

References

[1] Černý R., Rovnaníková P. and Keršner Z., *Vlastnosti modifikované sádry* (VUT, Brno, 2009). ISBN 978-80-214-3988-7.

[2] Tesárek et P. al., "Flue Gas Desulfurization Gypsum: Study of Basic Mechanical, Hydric and Thermal Properties," *Construction and Building Materials* Vol. 21, pp. 1500–1509 (2007). ISSN 0950-0618. http://dx.doi.org/10.1016/j.conbuildmat.2006.05.009.

[3] Singh M. and Garg M., "Making of anhydrite cement from waste gypsum," *Cement and Concrete Research*, Vol. 30, pp. 571–577 (2000). ISSN 0008-8846.

[4] Trejbal J. et al., "Recyklace sádrových výrobků pomocí vysokorychlostního mikromletí," in Proceeding of RECYCLING, 2019 (VUT Brno, Brno, 2019), pp. 113–116.

[5] Tesárek P. and Prošek Z., "Možnosti využití jemné frakce odpadního betonu v lehčených cementových kompozitech," in Proceeding of 18. konference Speciální betony, 2021 (SEKURKON, Ostrava, 2021), pp. 45-48.

[6] Ďureje J. and Prošek Z., "Foamed cement paste with micronized recycled concrete-measuring of compressive strength and thermal conductivity," in Proceeding of Experimental Stress Analysis - 57th International Scientific Conference, 2019 (Czech Society for Mechanics, Luhačovice, 2019), pp. 66–69.

EAN 2022 60th annual conference on experimental stress analysis

Using the Electrical Tomography Method to Determine Damage of the Surface Layer

Paweł Pyrzanowski[1] & Marek Stepnowski[2]

Abstract: The paper describes the method, which allows to find and measure parameters of damage that appears on the surface of the body. Changes of the resistivity of very thin conductive layer applied into the surface allow to determine localization, shape and dimensions of the damage. In measurement only 8 electrodes connected to the conductive layer were used. Thanks to supplementing the measurements with preliminary FEM simulations, it was enough to obtain satisfactory measurement accuracy.

Keywords: Electrical tomography; Conductive layer; Damage; Hybrid method

1. Introduction

A lot of damage to elements begins on their surface, or after some time is noticeable there. It can be seen as cracks, scratches or material losses. For this reason, monitoring only the surface of components is often a sufficient method of alerting of possible catastrophic damage. Proposed method can be used in Structural Health Monitoring (SHM) system, because information about possible damage can be acquired and analysed in real time.

Electrical Tomography Method (ETM), also referred to as Electrical Impedance Tomography (EIT) is useful in structural health and condition monitoring [1–3]. Majority of the methods are based on experimental measurements only and require the connection of a large number of electrodes. It allows to create maps of the analysed surface with the indication of resistance changes presence, which is interpreted as the occurrence of damage. The method is used to monitoring carbon composites [4, 5], but due to their low resistance, the accuracy of the method is low. For this reason, most research today focuses on using the method to test thin conductive layers [6, 7].

2. Material and methodology

2.1. Basics of the electrical tomography method

ETM measures the difference between two states - before and after the event changing the resistance. In this case a thin layer applied to the surface works as the conductor through which the current flows. The change in resistance resulting from the decrease in the cross-section of the layer as a consequence of its breaking by mechanical damage is directly measured (Fig. 1).

A simple measurement of the resistance between two contacts gives information that the state has changed - e.g. damage to a layer has caused a change in its resistance. In order to determine the location and magnitude of this damage, multiple measurements between numerous contacts surrounding the damage site need to be made. This method is called electrical tomography.

[1] *Prof. Paweł Pyrzanowski, Ph.D., D.Sc.; Warsaw University of Technology, Institute of Aeronautics and Applied Mechanics, Nowowiejska 24, 00-665 Warsaw, Poland;* pawel.pyrzanowski@pw.edu.pl

[2] *Marek Stepnowski, M.Sc.; Warsaw University of Technology, Institute of Aeronautics and Applied Mechanics, Nowowiejska 24, 00-665 Warsaw, Poland;* marek.stepnowski.dokt@pw.edu.pl

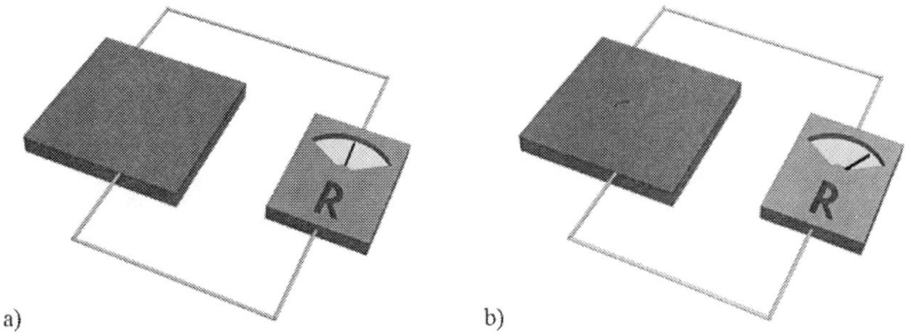

Fig. 1. Basics of the changing electrical resistance method: a) surface layer without the damage – small resistance; b) surface layer with the damage – high resistance.

An example of the use of this method is presented in [8]. It was employed to locate a 5 mm diameter semi-circular layer damage using a sample with 72 electrodes – 18 on each side of a square specimen (Fig. 2).

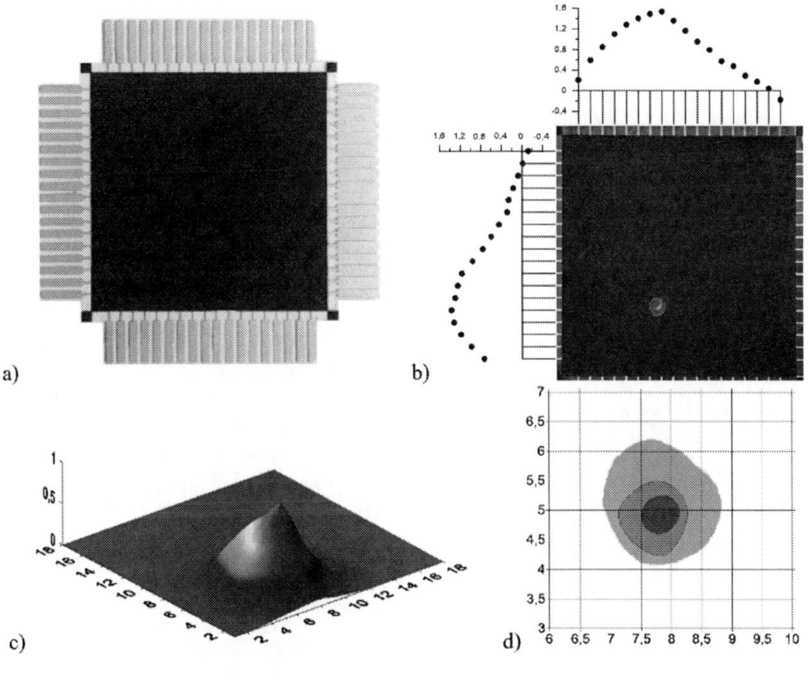

Fig. 2. An example of the use of the changing electrical resistance method: a) specimen without the damage; b) resistance changes in vertical and horizontal direction; c) 2-D resistance change map; d) localisation of the damage: grey area – damage, green ang red area – 2-D resistance change map.

2.2. Sensitivity maps obtained from the Finite Element Method

Finite Element Method can be used to obtain sensitivity maps, that help to increase the accuracy of the method, especially in the case of small damage and a small number of electrodes [9].

These maps are created taking into account the change in the resistance calculated between a pair of electrodes as a consequence of the appearance of a small damage (in this case with a diameter of 1 mm) in various places within the simulated conductive layer. Fig. 3 presents an example of one of 28 maps.

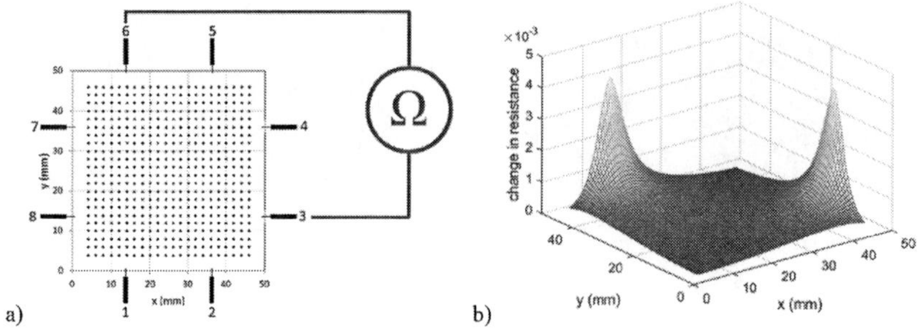

Fig. 3. FEM simulation: a) scheme of the simulation of the measurement between electrodes 3 and 6, dots on the sample show the locations of the simulated damage locations; b) smoothed sensitive map.

2.3. *Hybrid method*

The hybrid method proposed in this paper consists of two main stages: numerical simulations of sensitivity maps using real sample geometry and variable location of slight layer damage and performing experimental measurements and calculation of damage parameters with the use of sensitivity maps.

Initially, for each pair of contacts, the relative change in resistance is measured:

$$\Delta R_n = \frac{R_{f,n} - R_{0,n}}{R_{0,n}} \tag{1}$$

where: n – pair of the measurement.

$R_{0,n}$ – resistance between n-pair before damage.

$R_{\mathrm{f,n}}$ – resistance between n-pair after damage.

Then the $Z_{\mathrm{n}}(x, y)$ function is calculated, comparing the measurement result with the value from the sensitivity map:

$$\begin{aligned} Z_{\mathrm{n}}(x,y) &= \left[\frac{\Delta R_{\mathrm{n}}}{\Delta R_{\mathrm{sn}}(x,y)}\right]^k \\ & k = 1 \text{ for } \Delta R_{\mathrm{sn}}(x,y) \geq \Delta R_{\mathrm{n}} \\ & k = -1 \text{ for } \Delta R_{\mathrm{sn}}(x,y) < \Delta R_{\mathrm{n}} \end{aligned} \tag{2}$$

where $\Delta R_{\mathrm{sn}}(x, y)$ is the change of the resistance between n-pair of electrode from the sensitive map.

Finally, the sum of the function Z is calculated:

$$Z_{\mathrm{sum}}(x,y) = \sum_{n=1}^{m} [Z_n(x,y)]^p \tag{3}$$

where: m – number of pair of electrodes.

p – power exponent, usually taken as 10.

Fig. 4 shows an example of the use of the proposed method for the damage of diameter 1 mm, located at x = 27 mm, y = 20 mm.

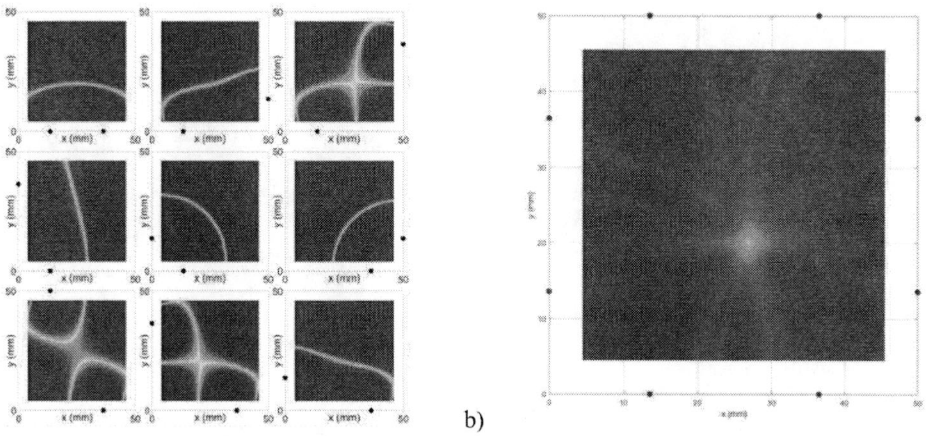

a) b)

Fig. 4. Use of the hybrid method: a) example of the Z_n maps; b) diagram of the Z_{sum} function.

3. Monitoring of the real surface

Presented method was used for damage monitoring on real specimen with the same shape as in the simulations shown earlier.

The specimen was made using carbon-based Loctite EDAG PF 407A E&C paste (for conductive layer) and silver-based Loctite EDAG PM 460A E&C paste (for electrodes) polymer composite using AUREL Mod. C902 printer. The layers were applied to a polycarbonate foil Autostat CT5 made by MacDermid. The measured value of resistivity of the carbon coating after drying was $\rho = 70.23\ \Omega\text{m}$.

In the sample made in this way, 3 holes in a shape similar to circular were made one after the other. Before making holes, and after making each one, resistance measurements were made between all pairs of electrodes (28 measurements each). Unfortunately, the method is not able to correctly detect a change as a result of more than one damage at the same time, therefore the distribution of measurements after the occurrence of the previous damage is always taken as the zero state.

The use of an additional iterative algorithm makes it possible to calculate the most probable size of a damage.

Obtained results are presented in Fig. 5.

Conclusions

The presented example shows that the ETM with the use of a small number of electrodes can be used for the correct location of even very small damage of the surface layer. This was achieved thanks to the use of a hybrid method combining experimental measurements with numerical simulations. The coordinates of the centre of damage were found with an error not exceeding 3% of the specimen dimensions. The use of special iterative algorithms allows to find the size of the damage and the angle of its major axis.

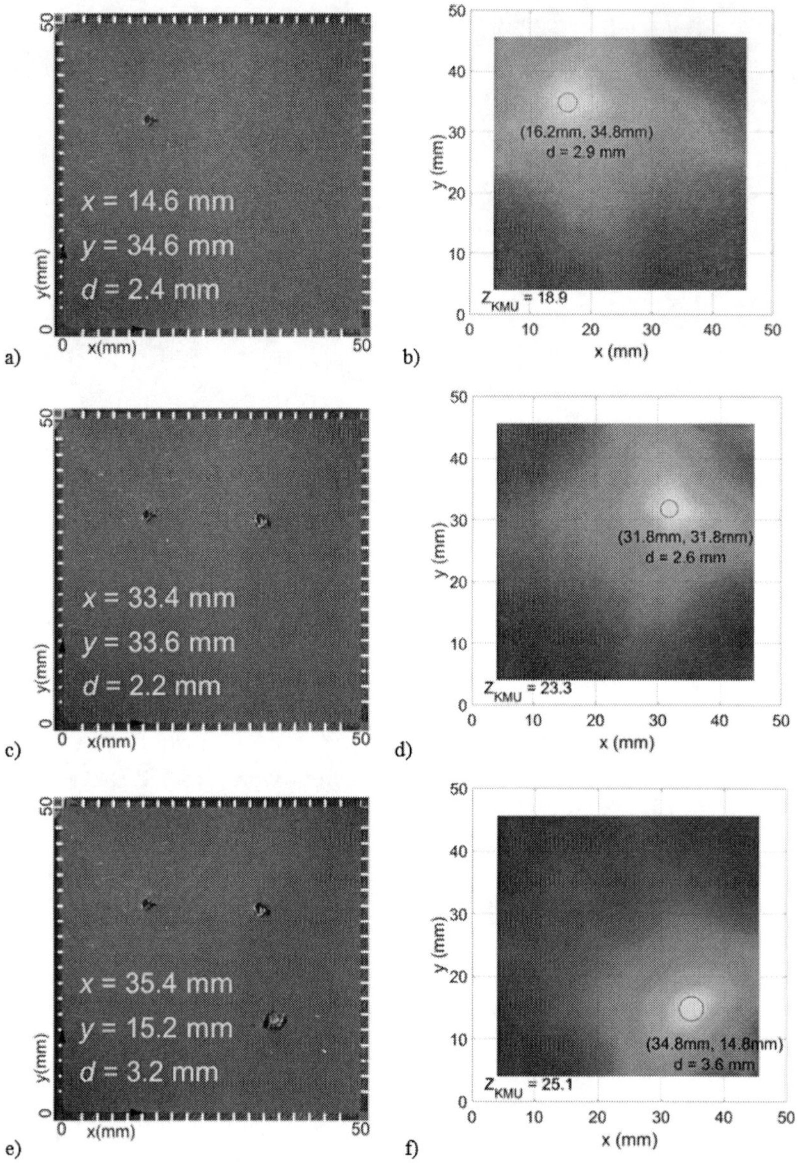

Fig. 5. Experimental results: a) 1st damage – photograph; b) 1st damage – detected location and size; c) 2nd damage – photograph; d) 2nd damage – detected location and size; e) 3rd damage – photograph; f) 3rd damage – detected location and size.

References

[1] Todoroki A., Ueda M. and Hirano Y., "Strain and Damage Monitoring of CFRP Laminates by Means of Electrical Resistance Measurement," *Journal of Solid Mechanics and Materials Engineering*, Vol. 1, no. 8, pp. 947–974 (2007). DOI: 10.1299/jmmp.1.947.

[2] Nonn S. et al. "Application of Electrical Impedance Tomography to an Anisotropic Carbon Fiber-Reinforced Polymer Composite Laminate for Damage Localization," *Composites Science and Technology*, Vol. 160, pp. 231–236 (2018). DOI: 10.1016/j.compscitech.2018.03.031.

[3] Gao, X et al., "Damage Detection in 2.5D C/SiC Composites Using Electrical Resistance Tomography," *Journal of the European Ceramic Society*, Vol. 39, no. 13, pp. 3583–3593 (2019). DOI: 10.1016/j.jeurceramsoc.2019.04.046.

[4] Thomas A.J. et al., "Damage Detection in Self-Sensing Composite Tubes via Electrical Impedance Tomography," *Composites. Part B, Engineering*, Vol. 177, 2019, pp. Composites. Part B, Engineer-

ing, 15 November 2019, Vol.177. DOI: 10.1016/j.compositesb.2019.107276.

[5] Gadomski J. and Pyrzanowski P., "Experimental Investigation of Fatigue Destruction of CFRP Using the Electrical Resistance Change Method," Measurement : Journal of the International Measurement Confederation, Vol. 87, 2016, pp. 236–245 (2016). DOI: 10.1016/j.measurement.2016.03.036.

[6] Hallaji M. et al., "Electrical Impedance Tomography-Based Sensing Skin for Quantitative Imaging of Damage in Concrete," *Smart Materials and Structures*, Vol. 23, no. 8, pp. 1–13 (2014). DOI: 10.1088/0964-1726/23/8/085001.

[7] Dai H. et al., "A Novel Methodology for Spatial Damage Detection and Imaging Using a Distributed Carbon Nanotube-Based Composite Sensor Combined with Electrical Impedance Tomography." *Journal of Nondestructive Evaluation*, Vol. 35, no. 2, pp. 1–15 (2016). DOI: 10.1007/s10921-016-0341-0.

[8] Janczak D. et al., "Screen-Printed Resistive Pressure Sensors Containing Graphene Nanoplatelets and Carbon Nanotubes," *Sensors* (Basel, Switzerland), Vol. 14, no. 9, pp. 17304–17312 (2014). DOI: 10.3390/s140917304

[9] Stepnowski M. et al., "Surface Damage Assessment By Analysis Of Electrical Resistance Changes In Graphite-Based Sensing Skin", *Metrol. Meas. Syst.*, Vol. 29, no. 2, (2022). DOI: 10.24425/mms.2022.140037.

Fatigue Life Assessment of the Agitator Mixer Shaft

Milan Růžička[1], Karel Doubrava[1], Martin Nesládek[1] & Jiří Kuželka[1]

Abstract: A crack was detected on the drive shaft of the horizontal chemical agitator mixer and based on experimental and numerical analysis, the crack propagation mechanism was determined and the possible operation time of the critical equipment was predicted. The actual operating loading of the shaft was determined using strain gauges, the fracture mechanical properties were determined based on tests of the material obtained from the original failed shaft and numerical analysis using fracture surfaces was used to determine the stress intensity factor.

Keywords: Agitator mixer shaft; Fatigue crack; Crack growth models; Stress intensity factor calculation; Strain gauge measurement

1. Introduction

Equipment failures in the chemical industry caused by fatigue cracks can result in large economic losses. Their early detection and determination of the propagation mechanism allows the use of the equipment's service life and replacement during technological shutdowns.

The subject of the assessment was a drive shaft of the horizontal chemical agitator mixer, which is used to produce polypropylene powder. The scheme of the equipment as well as the beam model of the assembly is depicted on the Fig. 1 (left). The Zapex gear coupling is mounted to the shaft using keyed connection on the agitator drive shaft. This joint during operation proved to be a critical point of the structure and showed the occurrence of fatigue cracks and insufficient service life. The detail of the shaft with the keyway and the area of an initiated crack is shown in Fig. 1 (right). The operation of the previous shaft was terminated due to an accident caused by the propagation of a fatigue crack. The question of how long it will be possible to operate the current shaft was answered based on the analysis performed.

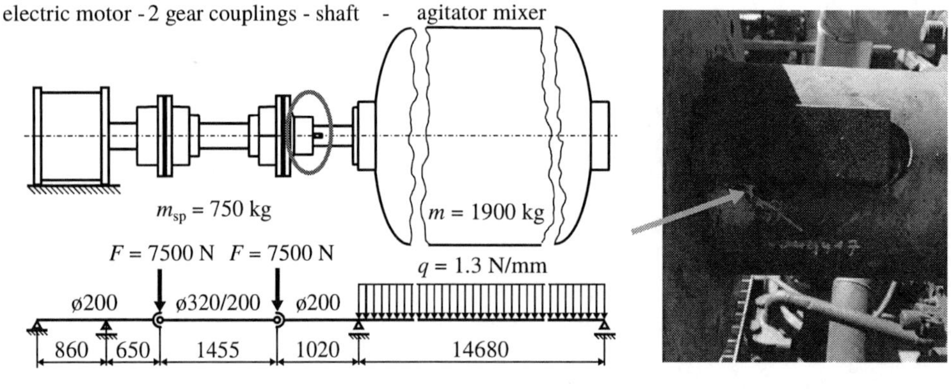

Fig. 1. The drive assembly of the mixer as well as the beam model of the assembly (left) and detail of the shaft with the key and the keyway and the area of the initiated crack (right).

[1] *Prof. Ing. Milan Růžička, CSc.; Ing. Karel Doubrava, Ph.D., Ing. Martin Nesládek, Ph.D., Ing. Jiří Kuželka, Ph.D.: Department of Mechanics, Biomechanics and Mechatronics, Fac. of Mech. Engineering, Czech Technical University in Prague; Technická 4, 166 07 Prague, Czech Republic;* milan.ruzicka@fs.cvut.cz, karel.doubrava@fs.cvut.cz, martin.nesladek@fs.cvut.cz, jiri.kuzelka@fs.cvut.cz

2. Experimental analysis

The strain-gauge analysis on the shaft was performed with the aim to obtain a correlation relationship between the long-time monitored electrical power input of the electromotor against the nominal stress on the agitator shaft, or also against the local stress in an area of occurrence of the fatigue crack. The strain gauge placement is shown in Fig. 2. Strain gauges installed in the $\pm 45°$ direction were used to determine the torque (CH1), and strain gauges installed in two sections in the direction of the shaft axis were used to determine the bending moment components (CH2, CH3). A strain gauge rosette installed near the crack tip was used to map the stresses around the tip of the crack (CH4–CH6). A GSV6BT data acquisition unit was used to record the data during device operation.

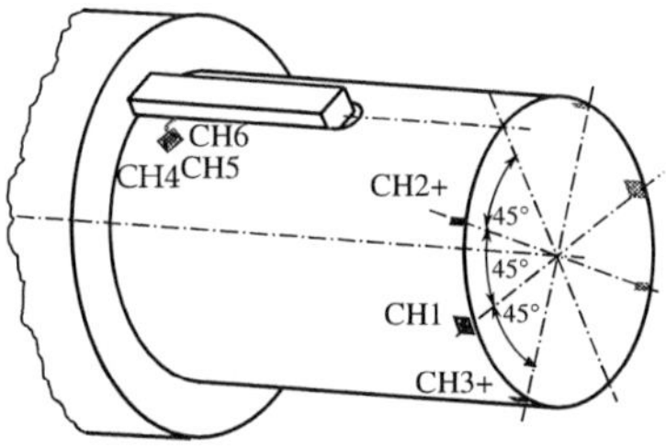

Fig. 2. Schematic of strain gauge placement on the shaft.

Fig. 3. Strain gauges and measuring unit installed on the shaft during operational measurement.

The load monitoring was carried out in technological modes of gradual filling of the mixer drum with raw materials without the start of the catalytic reaction. The mixer drive thus worked at different power levels, which caused different measured strains. Since real operation then also takes place at different powers levels, this data was used to compile calibration dependence, which relates the measured signal to the drive power. The mixer filling took two days, during which data was recorded throughout. The situation during the measurement is shown in Fig. 3.

The recording of the torque and bending moment components evaluated from channels CH1–CH3 at an electric motor power level of 270 kW is shown in Fig. 4. The superimposed oscillations in the torque recording correspond to the engagement of the eight rows of mixer blades. The range of bending stress for a given level of electric motor power was determined based on the equation (1):

$$\Delta\sigma_b = \frac{32\Delta M_b}{\pi \cdot D^3} = E \cdot \Delta\varepsilon_b, \tag{1}$$

where D is diameter of shaft, E is Young's modulus of elasticity and $\Delta\varepsilon_b$ is measured strain range.

Similarly the range of shear stress was determined using the equation (2):

$$\Delta\tau = \frac{16\Delta T}{\pi \cdot D^3} = G \cdot \Delta\gamma, \tag{2}$$

where G is shear modulus and $\Delta\gamma$ is measured shear strain range.

The strain-gauge measurement on four different power levels was realised to evaluate the calibration lines, which were linearized as is shown in Fig. 5.

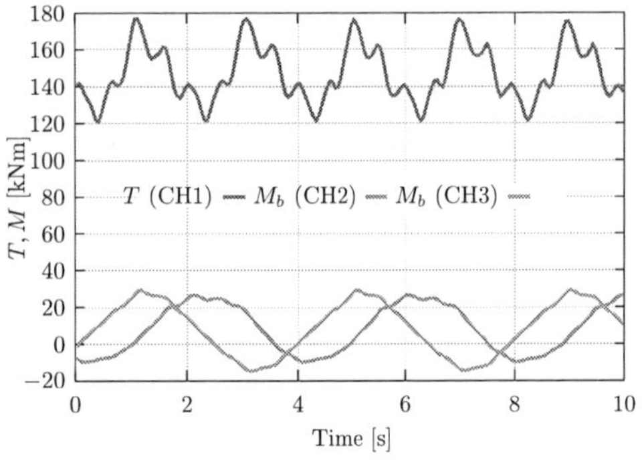

Fig. 4. Recording of the torque and bending moment components at an electric motor power level of 270 kW.

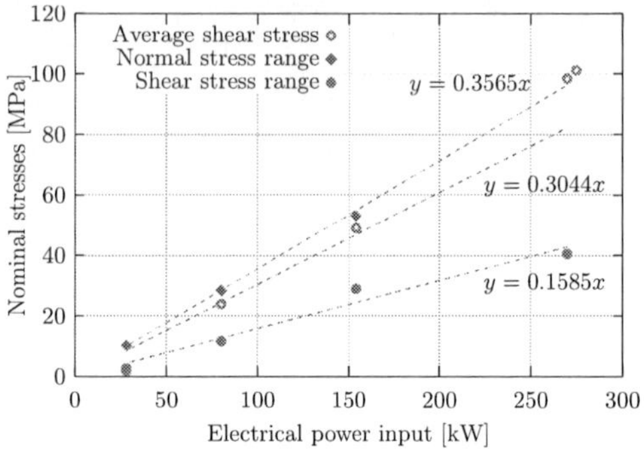

Fig. 5. Calibration of the power input and nominal stresses on the shaft.

3. Numerical analysis

Knowledge of the shape of the fracture surface and crack front is crucial for the correct determination of the stress intensity factor, or the geometric shape factor in given load conditions using a numerical (FEM) approach. The geometric shape factor of a crack include both the influence of the crack shape and the influence of the loading regime. Their knowledge is necessary for the prediction of the crack propagation rate, or the residual service life of the structure.

Numerical keyed joint model was created based on documentation provided by the mixer operator. The model consists of linear tetrahedral and hexahedral elements with full integration. The boundary conditions are shown in Fig. 6. The magnitudes of the torsional and bending moments were chosen based on operational measurements described in section 2. A total of eight calculations of the given model were performed, differing only in the considered crack length. The analysed cracks are shown in the Fig. 7. The shapes of the considered crack faces and the fracture surfaces were determined based on the 3D reconstruction and photographs of the fracture surface of the already fractured shaft.

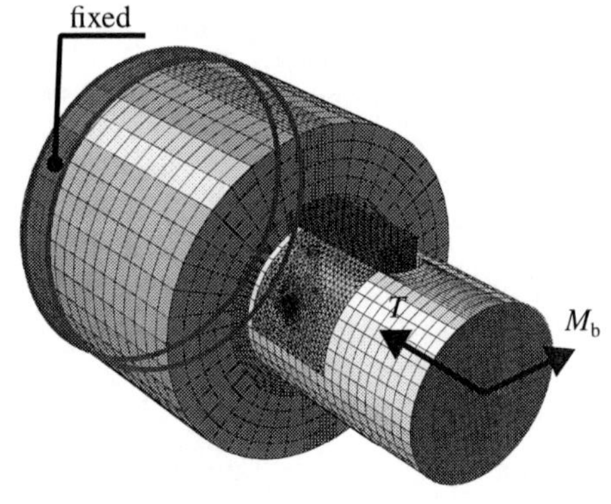

Fig. 6. Numerical model of keyed joint.

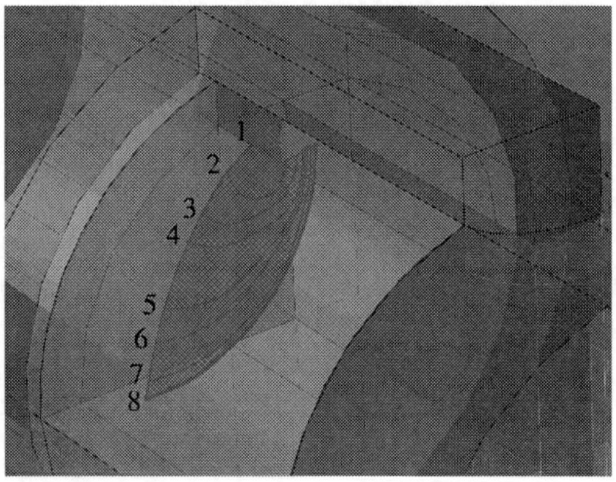

Fig. 7. Analysed crack faces.

Along the crack face, the stress intensity factors were numerically calculated in 240 positions, both for torsional and for bending loading. The position on the crack face is considered from the shaft surface towards the surface of the keyway. For bending load, the envelope of maximum values from all calculated positions of the bending moment vector was considered for evaluation.

Since the numerically determined value of the stress intensity factor directly on the surface is inaccurate, extrapolation from values near it was used to determine K_{II} and K_{III}. The K_I values were determined as maximum near the surface.

4. Results

The residual fatigue life analysis was done according the Paris-Erdogan, Klesnil-Lukáš and NASGRO crack growth models. Material parameters of these models were investigated for the used material (steel 36NiCrMo16) experimentally on the small specimens [1]. Fig. 8 (top) shows the time history of the electrical power input of the machine for which a fatigue crack growth of the initiated crack was simulated. Predictions of further crack growth were subsequently calculated for repeated typical three-months working loading until the limited crack length before formation of a quasi-brittle fracture in the whole shaft cross-section. The crack growth curve in Fig. 8 shows the predicted propagation of the detected crack (12 mm) and a limited service life for further operation of the agitator mixer.

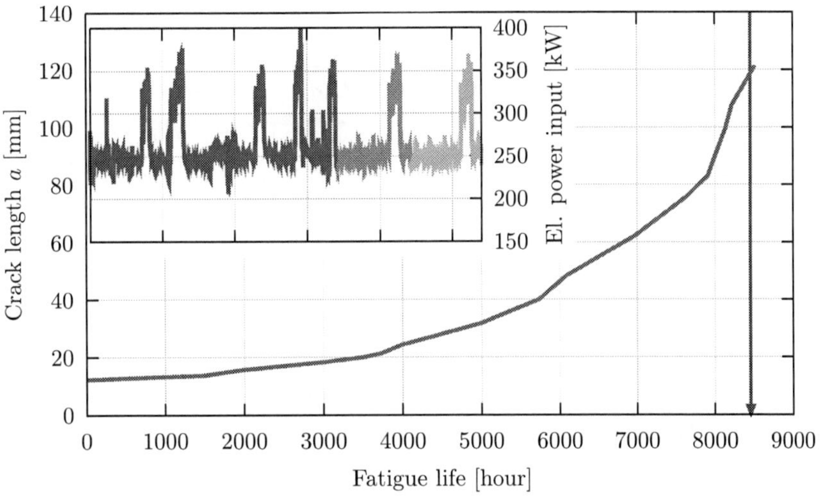

Fig. 8. The time history of the electrical power input of the machine (top) and predicted crack length growth curve according the NASGRO model (bottom).

5. Conclusion

The crack was discovered on a shaft replaced after the failure of the previous shaft. The operating load of the shaft was determined using strain gauges. The fracture surface on the original shaft was scanned and used for creating a numerical model for calculating the stress intensity factor. A part of the original shaft was used to experimentally determine the fracture mechanical properties of the material used. Based on the obtained data, the residual fatigue life analysis was made. The accuracy of the results was demonstrated during the planned shaft replacement, when the achieved operating time and crack growth corresponded well to the analysis.

References

[1] Kovářík O. and Kunz J., *Study of the crack grow rates in steel 1.6773* (CTU in Prague, Faculty of Nuclear Sciences and Physical Engineering), internal report, 2020.

Modelling and Testing of 3D Printed Flexible Structure With Spatial Pattern

David Rybanský[1], Martin Šotola[1], Pavel Maršálek[1] ,
Lukáš Drahorád[1], Juraj Hronček[1] & Ondřej Kušnír[1]

Abstract: The objective of this work is to evaluate stiffness of 3D spatial structures using experimental data and numerical modelling. In this paper, we present possible design of such 3D spatial structures which stiffness can be set by changing parameters of pattern. For evaluation of the stiffness, we used previous methodology from previous research. These structures were 3D printed using selective laser sintering (SLS) method and made of PA12 material. Measured data were used to set computational model such as its resulting stiffness would be comparable to the experimental ones. With prepared computational model, we performed parametric study to determine effect of the thickness of spatial structures to its stiffness.

Keywords: Flexible Structures; Experimental; Stiffness; Spatial structure

1. Introduction

Flexible structures (FS) are patterns of repeating units called cells, which have specific properties such as stiffness-mass ratio or greater stiffness in one direction. Main idea of such structures is to control its mechanical properties by changing geometry of such pattern (including cells) rather than changing material to achieve same properties. These structures can be used in many field such as biomechanics, prosthetic, protective equipment and medical gadgets, etc [1–3]. For example it can help with manufacturing proper cranial orthoses with predefined properties. Or it can be used as infilling for helmets which will be light weighted but strong enough to protect rider.

Spatial structures offer large number of options compare to planar pattern structures. Also spacial structures usually comes with larger numbers of parameters which can be optimized for specific problems. The objective of this work is to evaluate stiffness of 3D spatial structures using experimental data and numerical modelling. In this paper, the possible design of such 3D spatial structures which stiffness can be set by changing parameters of pattern is presented.

This work is based on previous research of flexible structures, with planar patterns. That pattern were created by repeating identical two dimensional cells. This pattern was projected onto curved surface and then thickness was added creating 3D model which can be printed using additive manufacturing. These structures were 3D printed using selective laser sintering method and made of PA12 material.

2. Specimens and measurement

With certain simplifications, large amount of application is made of curved surfaces which are nearly hemisphere (for example already mentioned helmets). Therefor, measured specimen has hemisphere shape with defined inner radius. The rest of dimension depends on chosen pattern.

[1] *Ing. David Rybanský, Ing. Martin Šotola, Ing. Pavel Maršálek, Ph.D., Ing. Lukáš Drahorád, Ing. Juraj Hronček, Bc. Ondřej Kušnír; Deparment of Applied Mechanics, Faculty of Mechanical Engineering, VSB - Technical University of Ostrava; 17. listopadu 2172/15, 708 00 Ostrava, Czech Republic; david.rybansky@vsb.cz, martin.sotola@vsb.cz, pavel.marsalek@vsb.cz, lukas.drahorad@vsb.cz, juraj.hroncek@vsb.cz, ondrej.kusnir.st@vsb.cz*

Measuring was carried with same equipment as in articles [4] and [5] to be able to compare result of planar and spatial pattern structures.

For experiment following equipment was used: testing device (1), specimen (2) and plunger (3), scheme of testing assembly is shown on Fig. 1. All parts were 3D printed. Plunger has shape of ball cap with radius R_2 and height h. The cell of the pattern is represented by its middle surface, its parameters are also shown on Fig. 1. These parameters are a = 7.4 mm, b = 1.8 mm, c = 2.5 mm, r_0 = 0.9 mm and r_1 = 1.9 mm. By repeating of the cells, the planar geometry was obtain, which was projected onto hemisphere with radius R_1, using advanced modeling tools. Whole assembly was put in universal testing device TESTOMETRIC 500-50CT, load was displacement u driven and response force F was measured.

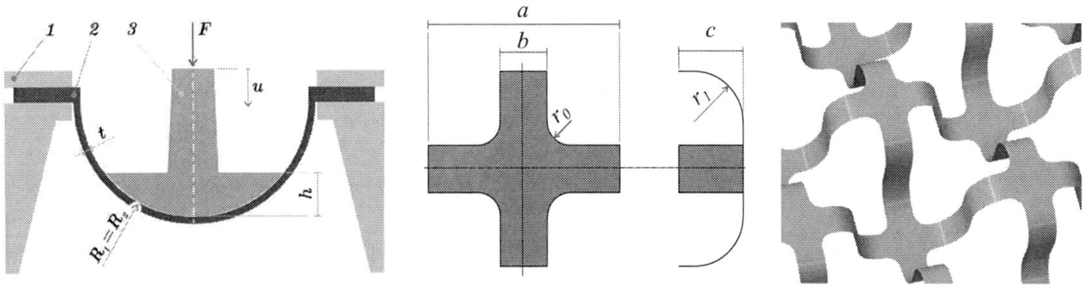

Fig. 1. Designed pattern (right), cell of pattern (middle) and scheme of the testing device (left): 1 – testing device, 2 – specimen, 3 – plunger [5].

For experimental measurement two specimens were used. These specimen were 3D printed using SLS method with designed thicknesses 1.2 mm and 1.6 mm. Velocity of plunger displacement was set to 3 mm per minute and the specimen was loaded until first cracks occurred. The measured relationships between displacement u and response force F of both specimens is depicted on Fig. 2. For determining of the stiffness, the linear approximation of measured data is used, nonlinear parts in the beginning of loading were excluded.

Measured data were used for validation of nonlinear computational model such as its resulting stiffness would be comparable to the experimental ones. The computational model was created in Ansys Workbench 2021 R2 and it consists of the parametric geometrical model which represent specimen and model of plunger. The computational model took into account large displacements and deformations. Model also included frictional contact between the structure and the plunger. To capture specimen behaviour during load, bilinear elastoplastic material model was used with Young's modulus E = 1 224 MPa, Poisson ratio μ = 0.39, Yield stress σ_Y = 21 MPa and Tangent modulus E_T = 334 MPa [6,7]. A material model describing an absolutely rigid body was assigned to the plunger because its rigidity was significantly higher than in the case of the specimen. Frictional contact with a coefficient of friction of f_s = 0.3 [5] was considered between the specimen and the plunger. The geometric model was discretized by shell finite elements (SHELL181), with an element size of 0.4 mm. As a result of SLS procces, real thicknesses t_r are decreased by 0.20 mm [5]. The boundary conditions were set to correspond most closely to the experimental ones. The edges of the FS were fixed and the load was applied using a plunger displacement function.

The main result is the relationship between the plunger displacement and the resistance force generated by FS. Comparsion of the force-displacement relationships between experimental and simulation is shown on Fig. 2. In the direction of force, stiffness is determined by approximating linear part of curve and the slope of approximated line is the seeked stiffness. High quality calibration between experiment and simulation is achieved, see Table 1.

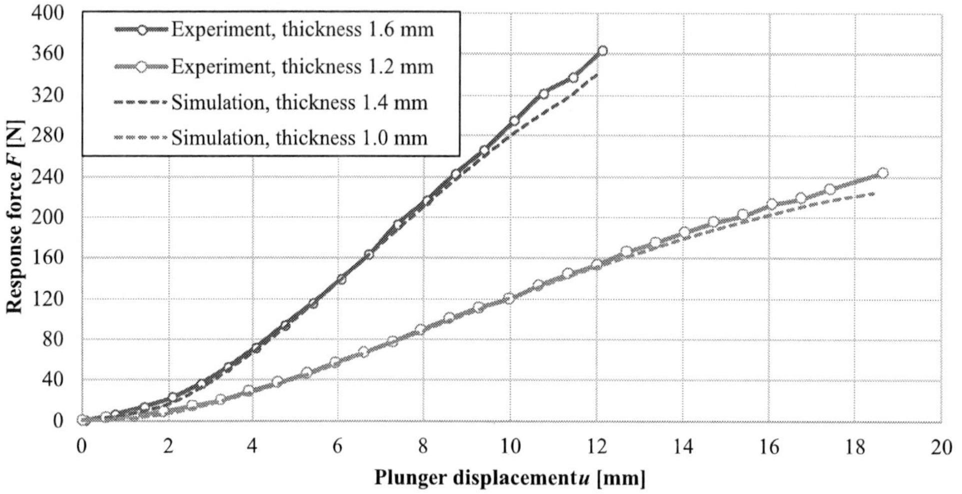

Fig. 2. Comparsion of the force-displacement relationships between experiment and simulation.

Table 1. Comparsion of the stiffness k between experiment and simulation.

Designed thickness t [mm]	Real thickness t_r [mm]	Experiment k [N/mm]	Simulation k [N/mm]	Difference [%]
1.6	1.4	36.82	35.89	2.53
1.2	1.0	15.06	14.68	2.52

With prepared computational model, parametric study was performed to determine effect of the thickness of spatial structure to its stiffness. The results are shown on Fig. 3.

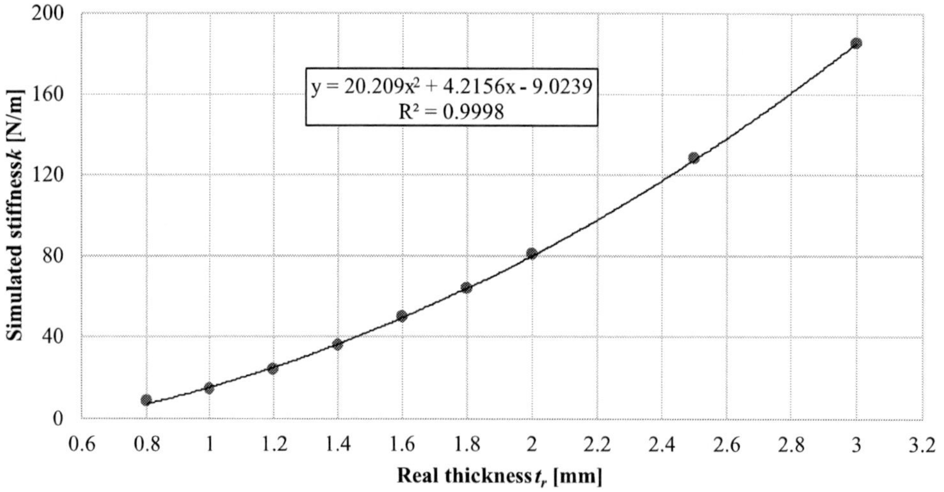

Fig. 3. Parametric study of the stiffness of FS.

3. Conclusions

The objective of this work was to evaluate stiffness of 3D spatial structures using experimental data and numerical modeling. Even though all parts of assembly were 3D printed, they were more stiffer than tested specimen. Therefor influence of these parts was neglected. The testing device was also analyzed using finite element method which proved how much was deformation of specimen influence by compliance of these parts. Deformation of testing device was smaller compared to deformation of spatial structure.

Stiffness overall cannot be described as linear, however after some initial deformation force-displacement diagram became linear. Initial deformation was different for each pattern and was excluded for evaluation of stiffness. Value of stiffness was slope of linear regression of non excluded data.

Nonlinear computational model was tuned by measured data. Material model was bilinear elastoplastic material. Also computational model took into account another nonlinearities such as large displacements and contacts. It should be noted that mathematical model had slightly decreased thickness to corresponde to experimental specimens. This modification resulted into quality calibration witf difference 2.53% at most. Therefor, computational model was used for parametric study which was perfomed to determine effect of the thickness to specimens stiffness. In this case, data were not acting linear like in case of planar patterns instead best fit curve was second order polynominal curve.

Acknowledgements

This work was supported by The Ministry of Education, Youth and Sports from the Specific Research Project (SP2022/26).

References

[1] Panetta J., Zhou Q., Malomo L., Pietroni N., Cignoni, P. and Zorin, D., "Elastic textures for additive fabrication," *Materials, ACM Transactions on Graphics*, Vol. 34 no. 4, pp. 1–12 (2015). https://doi.org/10.1145/2766937.

[2] Schumacher C., Thomaszewski B. and Gross M.,"Stenciling: Designing Structurally-Sound Surfaces with Decorative Patterns," *Computer Graphics Forum*, Vol. 35 no. 5, pp. 101–110 (2016). https://doi.org/10.1111/cgf.12967.

[3] Schumacher C., Marschner S., Gross M. and Thomaszewski B., "Mechanical characterization of structured sheet materials, " *ACM Transactions on Graphics*, Vol. 37 no. 4, pp. 1–15 (2018). https://doi.org/10.1145/3197517.3201278.

[4] Řepa V., Maršálek P., Prokop J., Šotola M., Stareczek D., Rybanský D. and Halama R., "Modelling and testing of 3D printed flexible structures with three-pointed star pattern used in biomedical applications," in Proceedings of Experimental Stress Analysis - 58th International Scientific Conference, EAN 2020, Ostrava, 2020, pp. 451-455.

[5] Maršálek P., Šotola M., Rybanský D., Řepa V., Halama R., Fusek M. and Prokop J., "Modeling and testing of flexible structures with selected planar patterns used in biomedical applications," *Materials*, pp. 1–16 (2021). ISSN. 19961944, DOI: 10.3390/ma14010140.

[6] Lammens N., De Baere I. and Van Paepegem W., "On the orthotropic elasto-plastic material response of additively manufactured polyamide 12," 7th bi-annual international conference of Polymers & moulds innovations, (Ghent, Belgium) (2016).

[7] Stoia D.I., Linul E. and Marsavina L., "Influence of Manufacturing Parameters on Mechanical Properties of Porous Materials by Selective Laser Sintering," *Materials*, Vol. 12, Art. no. 871 (2019). https://doi.org/10.3390/ma12060871.

Mechanical Behavior of a Geopolymer Composite System with 3D Fibre Reinforcement

Martina Ryvolová[1] & Totka Bakalova[1]

Abstract: The third dimension of the spacer fabrics is a benefit that significantly improves their resistance to compression. During reinforcement loading, good cohesion is assumed between the top and the reinforcement component of the fabric. Such a replacement brings both material and production cost savings. The composite system with 3D textile reinforcement has the potential for civil engineering in producing sandwich panels, facade elements, or cladding. The research aims to determine the mechanical properties of a geopolymer composite reinforced with 2D (mineral fiber mesh) and 3D textile (glass fiber spacer fabric). The premise is that the reinforcing fabric will significantly improve the mechanical properties of the geopolymer composite. Evaluating and comparing the flexural load tests of a geopolymer system reinforced with several layers of geogrid and spacer fabric will contribute significantly to the knowledge of the behavior of composite systems with 3D reinforcement.

Keywords: Mineral fibres; Spacer fabric; Geopolymer matrix; Flexural strength

1. Introduction

Spacer fabrics are the summary name for fabrics that consist of two layers connected by vertical threads that are part of both cover layers. The function of vertical threads is to maintain a constant distance between the two layers. Textiles can be produced using warp and weft knitting technology and weaving technology. Combining weft-knit spacer fabric with epoxy resin [1] and woven spacer fabric and epoxy resin [2] brings several advantages in various applications. The material is strong, stiff, and lightweight. The use of three-dimensional reinforcement eliminates the layering of flat reinforcements, resulting in financial savings. The authors investigated the effect of the structure on the tensile and compressive strength and impact resistance [1, 2]. Warp knitted spacer fabrics (WKSF) are successfully introduced in building structures as thin reinforcement for wall panels, exterior cladding, roof tiles, floor tiles, pressure pipes, and other products. Their design advantages are a highly oriented reinforcement fabric system and easy bonding with a cementitious matrix in manufacturing the composite. The compression resistance of the spacer fabric is a significant advantage; It provides performance and simplifies the composite manufacturing process [3]. The authors [4] investigated the effect of the geometry of the WSF bond on the reinforced strength behavior of the WSF concrete and compared the lattice model of the behavior of the cement matrix with actual measured values. They developed a model based on the fundamental WKSF structure based on the theoretical findings. Study investigates the effect of WKSF bond geometry on the tensile behavior of concrete canvas (CC). They described the crack propagation patterns of CC in the 0° and 90° directions due to tensile force. The experimental results showed that the samples with one solid outer fabric substrate showed an increase in tensile strength.

The study compares the flexural properties of the composite material with the woven spacer reinforcement and the composite material with the basalt geogrid. The aim is to obtain a de-

[1] *Ing. Martina Ryvolová, Technical University of Liberec, Faculty of Mechanical engineering, Department of Materials, Studentská 1402/2, 461 17 Liberec 1, Czech Republic,* martina.ryvolova@tul.cz

scription of the mechanical behavior of the panels with 2D and 3D reinforcement, compare the results, and specify the partial differences in behavior.

2. Materials and Methods

The composite samples were made of a geopolymer matrix (Baucis LK cement – ČLUZ, a.s., CZ) activated with potassium water glass, with silica sand as filler and reinforcement with mineral fibers. Three layers of 2D reinforcement (basalt fiber geogrid) is commonly used in concrete mixes. The interlacing points are formed by rovings crossed in the 0 ° and 90 ° directions. Mesh size 10×14 mm, weight 375 g/m^2. 3D reinforcement - woven spacer fabric (WoSF) from glass fibers type E, distance 3 mm, weight 780 g/m^2 and distance 5 mm, weight 840 g/m^3. The flow of the geopolymer matrix during spacer fabric was ensured by means of a vibrating device. Samples was tested after standard maturation period – 30 days. The four-point bending test for composite samples was performed on a tensile tester. The support span was 300 mm, the loading span was 100 mm.

The four-point bending test for composite samples was conducted on a tensile tester. The support span was 300 mm, the loading span was 100 mm and testing speed was 6 mm/min.

3. Results

A box plot was created from the force output values to exclude outlying measurement values, Fig. 1. The sample parameters and statistically processed data present Tab. 1.

Table 1. Results of four-point bending test.

Specimen	Specimen size [mm]	Maximal force [N]	Flexural stress [MPa]	Deflection [mm]
Basalt geogrid	400×100×10	898.3±141.1	26.95±4.23	18.2±2.1
3D glass 3 mm	400×100×15	–	–	–
3D glass 5 mm	400×100×15	338.2±26.3	4.51±0.35	7.0*

The table contains only measured data for samples with basalt reinforcement (the test record you can see in Fig. 2). Data for samples with 3 mm 3D glass reinforcement could not be measured. The joining between the matrix and the reinforcement was insufficient, the reinforcement separated from the matrix and the geopolymer matrix did not flow inside the spacer fabric. It was stated that the density of the geopolymer material and the thickness of the 3 mm material are not suitable material combination to make the geopolymer composit. The geopolymer material, even after using the vibrating device, did not get inside the fabric and thus did not ensure good adhesion between the materials.

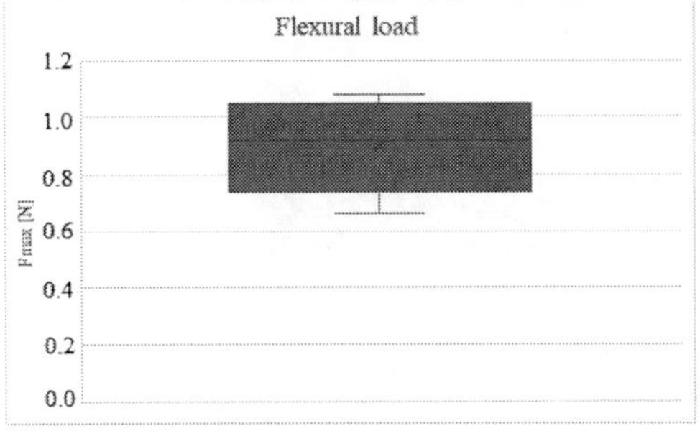

Fig. 1. Box plot – samples with basalt grid.

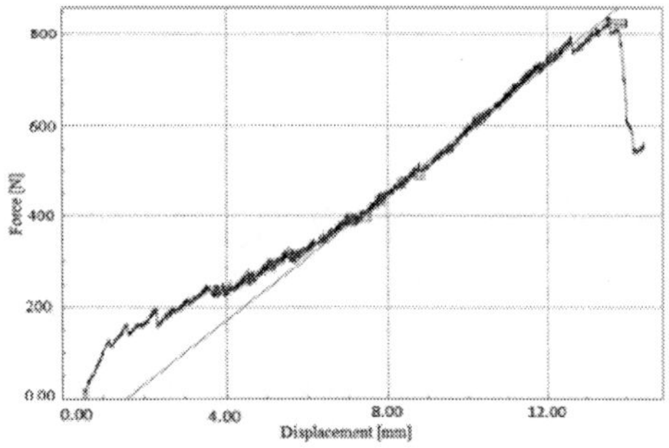

Fig. 2. Test record – samples with basalt grid.

Only very low values of the maximum bending force could be measured for the new series of specimens. A comparison of the two test runs is shown in Fig. 3 and Fig. 4 ; the basalt geogrid reinforced specimens show very high flexural and deflection strength values. The failure of the specimens was due to bending in the center region of the specimen between the loading elements (Fig. 3). For the specimens with spacer reinforcement, the flexural strength was very low, and the deflection could not be measured (*the value in the table corresponds to the path of movement of the jaw of the device). The failure of the samples occurred at the point of contact with the loading elements, Fig. 4.

Fig. 3. Deflection of samples with 2D grid

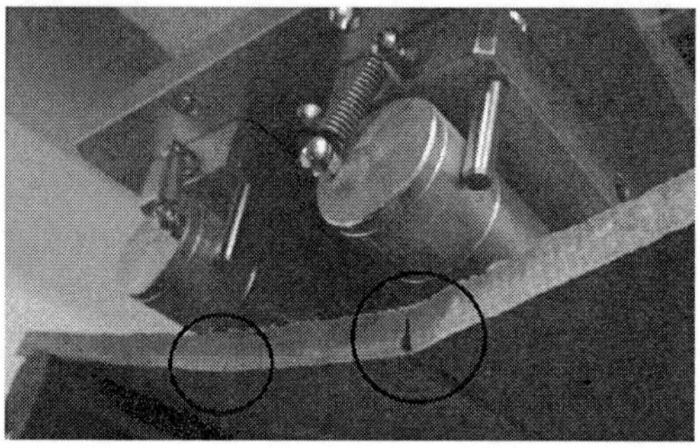

Fig. 4. Failure of samples with WoSF.

The cause of the low bending strength and delamination of composite system is insufficient flow of the matrix through the reinforcing fabric (Fig. 5). Subsequently, it was found that even 3D textiles with a thickness (5 mm) did not meet expectations. This type of textile cannot be used for this purpose because the fiber density does not allow the geopolymer material to flow into the inner region of the textile material. The woven spacer fabric consists of two boundary layers with a plain weave made of warp and weft yarns. The plain weave is thickened by spacer threads (pilots) which are part of the weave in both layers. The upper layer of the spacer woven fabric is subjected to compressive loading (application of a pair of forces) during bending, while the lower layer is subjected to tensile loading. The neutral axis lies in the middle in the area that is filled with pilots. The geopolymer matrix flowed only through the reinforcement borders. Without the presence of a matrix in the central part of the reinforcement between the pilots, there is no force transfer and the specimen fractures under the influence of the external load, followed by delamination of the layers, Fig. 6.

As a result of the very low strength values, the glass spacer fabric reinforced geopolymer matrix was evaluated as unsuitable for applications where a certain degree of flexural strength is required. However, it is suitable for those applications where the material is not subjected to bending forces.

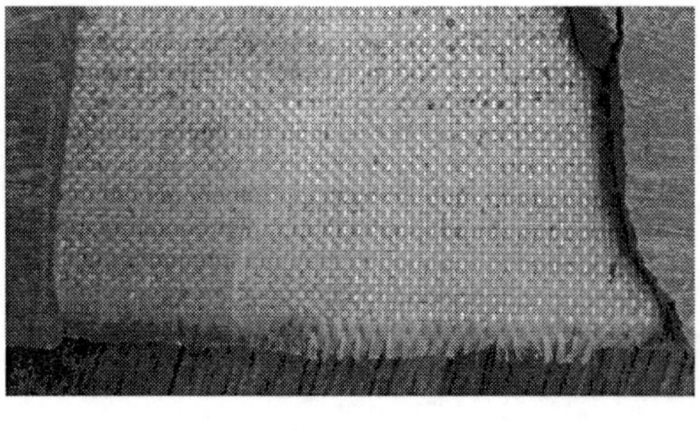

Fig. 5. WoSF reinforcement without matrix.

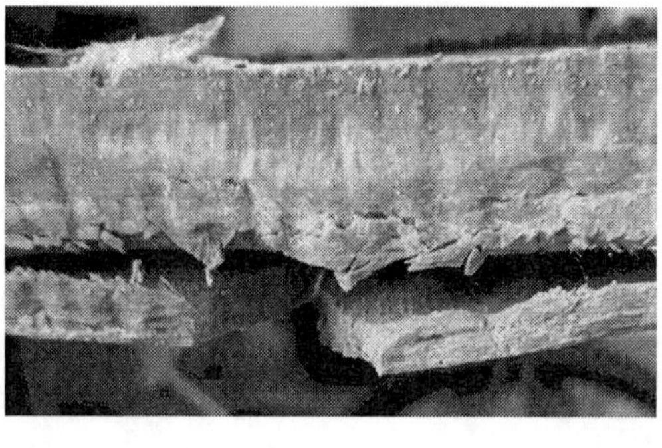

Fig. 6. Composite delamination.

4. Conclusions

The experiment has not yet produced the expected result. The combination of geopolymer matrix material with WoSF of 3 mm thickness proved to be inappropriate. For this study, a set of new samples with WoSF with a spacing of 5 mm was prepared. In this case, the experiment also did not produce satisfactory results. The flexural strength of the WoSF-reinforced geopolymer matrix is minimal. A building slab with such a composition would only withstand very small load values.

The importance of composite materials prepared in this way is primarily to reduce heat dissipation, ie as panels for thermal insulation. The creation of an air bubble in the 3D fabric in the middle of the composite material will significantly reduce the thermal conductivity of, for example: the walls and ceilings of residential buildings.

In the following period, the experiment will be directed towards the use of warp knitted spacer fabrics (WKSF) and the investigation of the properties of geopolymer composite reinforced with WKSF.

Acknowledgements

The results of the project "Heat nanoisolation for automotive, aviation and aerospace" reg. number TH04020189, were obtained through the financial support of the TACR in the program "Epsilon" 4th call.

References

[1] Pamuk, G. et al., "Mechanical properties of spacer fabric-reinforced composites knitted from glass yarn," *Journal of Composite Materials*, 45 (17), (2011). DOI: 10.1177/0021998310386262

[2] Li, M. et al., "Effect of Structure on the Mechanical Behaviors of Three-Dimensional Spacer Fabric Composites," *Applied Composite Materials*, 16 (1), (2009). DOI: 10.1007/s10443-008-9072-4

[3] Li, H. et al., "Lattice modeling for the influence of geometrical patterns of 3D spacer fabric on tensile behavior of concrete canvas," *Journal of Sandwich Structures & Materials*, (2021). DOI: 10.1177/10996362211020430

[4] Han, F. et al.,"Influences of geometric patterns of 3D spacer fabric on tensile behavior of concrete canvas," Construction and Building Materials, 65, (2014) DOI: 10.1016/j.conbuildmat.2014.05.041

The Composite Component with Self-detection of Damage

Martina Ryvolová[1] & Martin Mazač[2]

Abstract: The aim of the paper is to describe development, design and testing of a carbon-glass-epoxy composite for automobile part. This modified composite material can provide a feedback about possible arising damages.

Keywords: Composite push-rod; Self-diagnostic composite; NiTi wire

1. Introduction

This paper describes self-damage indication part application and testing. The aim of the paper is to describe development, design and testing of a carbon-glass-epoxy composite for automobile part. This modified composite material can provide a feedback about possible arising damages. Testing of the final designed automobile component by the universal loading machine equipped with special fixing parts were realised. The goal of the paper is recognition of possible application of self-damage indication material. Design, mechanical and indication material properties are described.

2. Self-diagnostic composite part

2.1. *Chosen part for modification*

Part of the automobile axle was chosen as a sample for our research. This part called push rod is typical easy machine component. For its shape is it a component with wide range of other possible application. Presently used pushrod is common steel tube with two ball pins mounted on the ends. Our modification used carbon and glass composite tube with two special ends for mounting.

2.2. *Used composite material*

The load-bearing part of the push-rod consists of a composite tube made from carbon fabric with plain weave and epoxy matrix. This part is the most stressed during mechanical loading. The tube is wrapped with a braided glass roving sleeve, with NiTi (nickel, titanium) wires added as a mechanically sensitive element. The wire formed a network on the surface of the tube to capture the mechanical response in the form of electrical resistance as sensitively as possible to measure. Another layer of braided fabric finished the entire composite part and the final lamination was done with epoxy resin.

Fig. 1. Push-rod composition: NiTi wire net on glass fibres layer.

[1] *Ing. Martina Ryvolová; Technical University of Liberec; Studentská 1402/2, 46001 Liberec 1, Czech Republic;* martina.ryvolova@tul.cz

[2] *Ing. Martin Mazač, Ph.D.; Technical University of Liberec; Studentská 1402/2, 46001 Liberec 1, Czech Republic;* martin.mazac@tul.cz

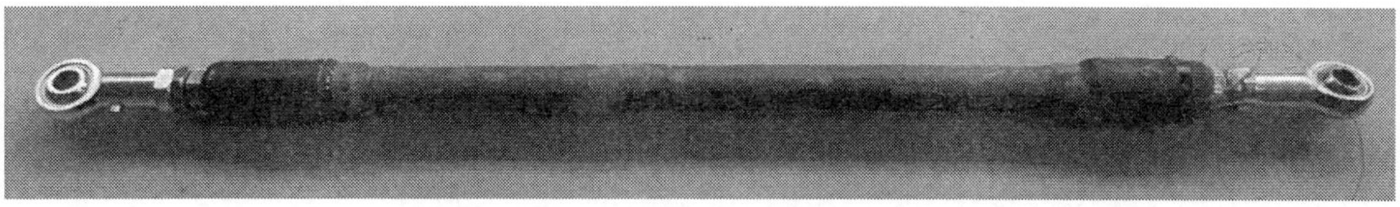

Fig. 2. Final composite push-rod with added NiTi wire sensor.

2.3. Self-diagnostic composite part testing

The composite push-rod was fixed to the static testing machine and loaded after. Loading was measured by integrated sensor of a laboratory loading system. Resistivity of the sensitive wire inside composite material of push-rod by the accurate digital multimeter (Rigol DM3068) was measured. Four-wire resistivity measuring method was applied. Measured data (load, resistivity) by the computer were recorded with 1 Hz sampling frequency. Loading cycle is very slow and sampling frequency is sufficient.

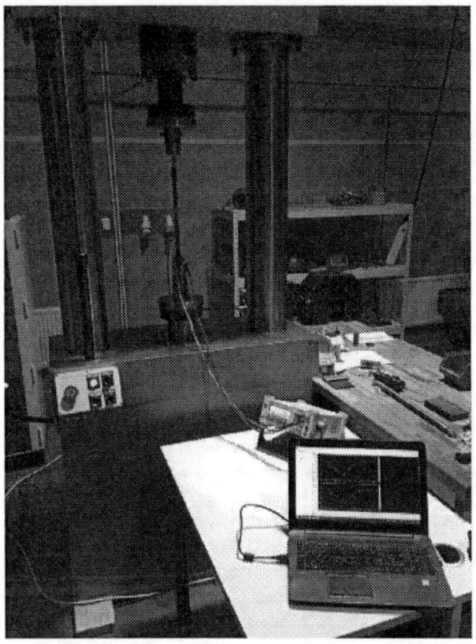

Fig. 3. Loading of composite part – loading and measurement equipment.

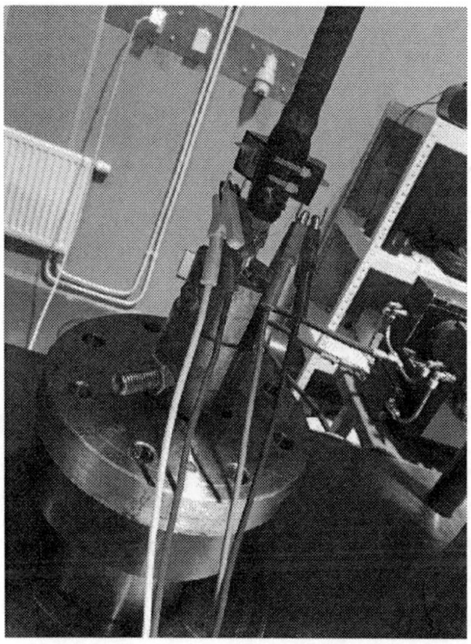

Fig. 4. Parts for composite push rod fixing and wiring for resistivity measurement by the four-wire method.

The testing equipment for loading composite part and measurement resistivity of sensitive wire inside is on Fig. 3 depict. Resistivity data loading by the PC with UltraSensor for DM3000 series multimeter software was realized. PC with software widow on the screen is possible to see on the figure too.

The tested composite element with special ends is equipped. The special end as a part for mounting and fixing of the main tube is necessary. Special ends are working by the principle of press joint. Ball joints in the ends is applied for final composite part fixing. The composite part by the ball joints to the special part is fixed during loading. Fixing of the composite push rod end on the Fig.4 is depict. Tested composite part have already absolved mounting and riding during real operation in experimental car. Real operation condition operation of the tested composite push rod was verified.

Tested part was pulled up to 1 kN force. First section of loading cycle is tested part preloading by the 0.1 kN force. During this preloading suppression of all mechanical clearances occurs. Second section is slow loading (0.05 kN/s) up to 1 kN. Third section is load decreasing to zero.

The real loading on the Fig. 5 and measured resistivity on Fig. 6 is depict. The measured resistivity of sensitive wire inside of push-rod material is depend on the external load. The external load from the measured resistivity to determine is possible.

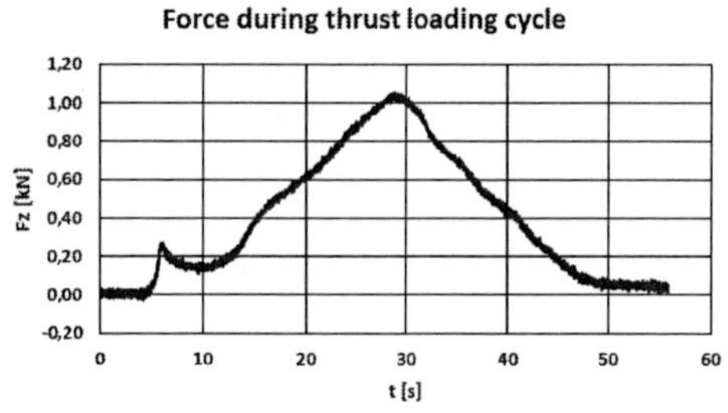

Fig. 5. Loading force measured by integrated sensor of equipment.

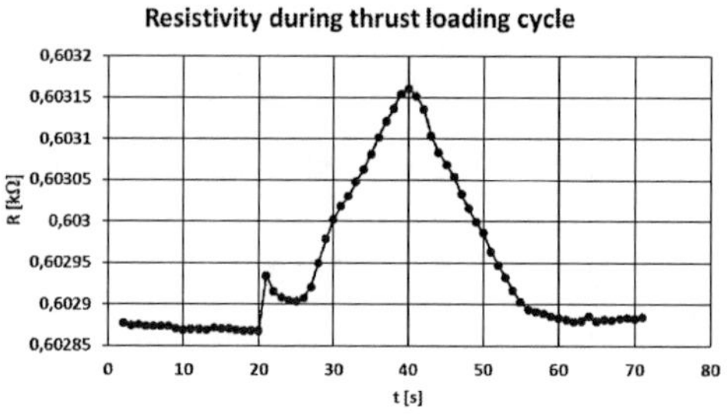

Fig. 6. Resistivity of sensitive wire applied in composite structure of developed part; resistivity measured by multimeter (device) with high accuracy.

The loading direction is possible to specify from measured resistivity. Similar loading cycle for press loading were used for push rod testing. The results of this mirrored test cycle on next figures are depict. The loading force measured by integrated sensor during loading cycle on the Fig. 7 is depict. Measured resistivity of sensitive wire implemented to the composite part is on Fig. 8.

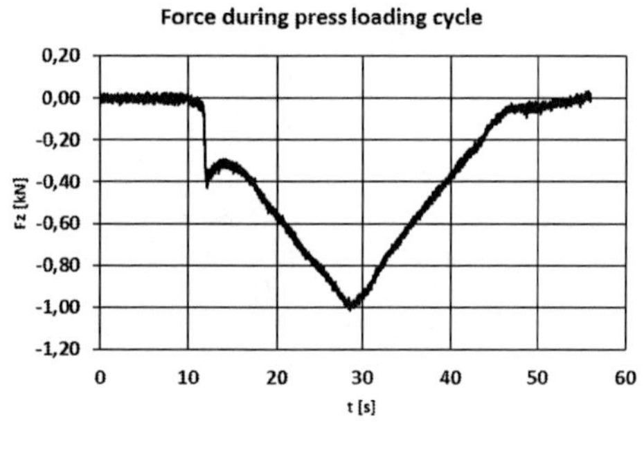

Fig. 7. Loading force (press) measured by integrated sensor of equipment.

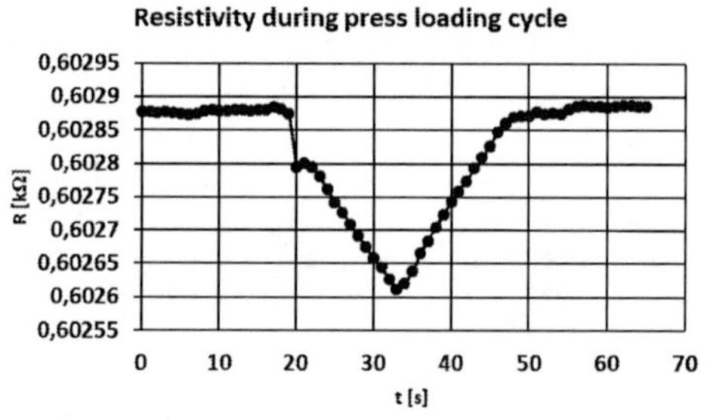

Fig. 8. Resistivity of sensitive wire applied in composite structure of developed part; resistivity measured by multimeter (device) with high accuracy.

Resistivity of sensitive wire applied in composite material of unloaded modified push rod is R_o = 602.87 Ω. Resistivity is changed, changes are dependent on loading of all composite part what is very good appreciable from the stated graphs. The resistivity depends on loading direction- tension means higher resistance, stress means lover resistivity against zero load resistance. When the sensitive wire will be broken the resistance is not possible to measure and that is a symptom of mechanical composite part damage. Similar system is possible to use in many composite parts for broke diagnostics- the broken can be diagnostics before fatale accidents of observed parts. Resistance of sensitive wire depends on load of the composite part and that is why the real time load measurement during common operation is possible.

3. Conclusions

This short paper describes manufacturing of self-diagnostic composite automobile component. The sequence of composite layers is described. Prototype of the composite part was manufactured for next testing. Testing of prototype functionality is described. It has been shown that the composite part can be applied to real car- part was aplied during real operation condition. Static loading cycles were realized. Self-diagnostics feature of composite part during static loading shown that the damage is possible to detect. Next significant output is proved possibility to real time measure load of composite part- resistivity of the sensitive wire depends on load.

This work was created in cooperation with Entry Engineering s.r.o. company. Used push rod is a part of experimental car developed by mentioned company.

Acknowledgements

Publishing of the results was financially supported by the Ministry of Education, Youth and Sports of the Czech Republic and the European Union (European Structural and Investment Funds – Operational Programme Research, Development and Education) in the frame of the project "International Research Laboratories", Reg. No. CZ.02.2.69/0.0/0.0/18_054/0014685.

References

[1] Gatto E. et al., "Load–deflection characteristics of superelastic and thermal nickel–titanium wires," *The European Journal of Orthodontics*, Vol. 35(1) pp. 115–123 (2013). https://doi.org/10.1093/ejo/cjr103.

[2] Gravina M. A. et al., "Mechanical properties of NiTi and CuNiTi shape-memory wires used in orthodontic treatment. Part 1: stress-strain tests," *Dental Press Journal of Orthodontics*, Vol. 18 (4), pp. 35–42 (2013). https://doi.org/10.1590/s2176-94512013000400007.

[3] Amin S. A., and Hasan A. Y., "Numerical and Experimental Study of Shape Effect Behaviour of Nitinol Wire," *Journal of Engineering and Sustainable Development*, Vol. 23.02, pp. 1–11, (2019).

[4] Kapoor D. "Nitinol for medical applications: a brief introduction to the properties and processing of nickel titanium shape memory alloys and their use in stents," *Johnson Matthey Technology Review*, Vol. 61, pp. 66–76 (2017). http://dx.doi.org/10.1595/205651317X694524.

Measurement of Selected Types of Oils Properties for Conical Friction Variator

Vít Řípa[1], Jakub Štětina[2], Karel Petr[3], Marek Štádler[4]

Abstract: To increase the service life of the friction variators, traction oil is fed to the contact point of the friction gear. The right choice of oil will not only increase the service life, but also affect the transmitted power and the overall efficiency of the friction gear. The article deals with several common types of oils that are tested on a special stand. It consists of a simple gear with a smooth surface, where one wheel is conical for the possibility of changing the gear ratio. Several types of oils were selected, according to their use, i.e., transmission, hydraulic, bearing and also oils intended directly for variators (CVT). Based on the measured results, the most suitable type of oil will be selected and will be used for the prepared design of a continuously variable planetary gear. Decisive factors in selecting a suitable oil include wear on the friction wheel surface and transferable torque at the same contact stress.

Keywords: CVT; Friction gears; Traction oils; Wear

1. Introduction

Friction members are used in friction gears. The Fig. 1 shows a simple friction gear with a constant gear ratio. Today, friction gears are used mainly in continuously variable transmissions (CVT). The gear ratio can be changed continuously. This is usually achieved by rolling the wheels on different diameter (one of the wheels is conical).

The power transmission of the friction wheels is realized due to the action of frictional forces, which arise as result of the development of a sufficient pressing force of the wheels and the existence of a slip in their contact surface [1] . However, this results in a partial wheel slip. The wheel surfaces are smooth and can have a cylindrical or conical shape (or the front surface of the plate). Friction gears transmit low to medium powers for parallel or multi-parallel shafts with small axial distances. The need to develop the pressure force of the friction wheels and the resulting friction for power transmission brings several specific pros and cons [2].

Positive advantages include quiet running, smooth and flexible starting and the ability to work at high speeds. Friction gears can absorb shocks well and reduce small variations in variable load differences (shocks are compensated by slippage). Cylindrical and conical friction gears are less demanding on production and assembly accuracy. The service life of the friction surfaces can be extended by using a replaceable friction ring, but this design increases the production complexity. By using wheels with different diameters, it is possible to achieve a different gear ratio, which can be continuously variable (CVT). With some special friction gear designs, continuous speed reversal is possible, it is called infinite variable transmissions (IVT).

[1] *Ing. Vít Řípa; Department of Designing and Machine Components, Czech Technical University in Prague; Technická 4, 166 07 Prague, Czech Republic; vit.ripa@fs.cvut.cz*

[2] *Ing. Jakub Štětina; Department of Designing and Machine Components, Czech Technical University in Prague; Technická 4, 166 07 Prague, Czech Republic; Jakub.Stetina@fs.cvut.cz*

[3] *Ing. Karel Petr, Ph.D.; Department of Designing and Machine Components, Czech Technical University in Prague; Technická 4, 166 07 Prague, Czech Republic; karel.petr@fs.cvut.cz*

[4] *Ing. Marek Štádler; Department of Designing and Machine Components, Czech Technical University in Prague; Technická 4, 166 07 Prague, Czech Republic; marek.stadler@fs.cvut.cz*

The need for a relatively large normal pressing force of the wheels (proportional to the transmitted friction force) is one of the main disadvantages. This leads to high stresses on the shafts and bearings. Another disadvantage is the high contact stress in contact area between two friction wheels. It limits the amount of transferable power and has a major impact on the service life and durability of the wheel surfaces. A smaller contact area leads to less spin losses (relative slip at the edges of contact area [3]), but it increases the contact stress. Excessive load will cause completely wheel slip and leading to permanent destruction of the friction wheel surfaces.

Several parameters affect power transmission of the friction gears. There is always an effort to increase the friction coefficient. Increasing the pressing force results in an increase in the transferable torque, but increases the contact stress and leads to surface degradation. Thus, there is an effort to increase the coefficient of friction through various surface roughness [4], materials, surface treatments and traction oils used [5]. The following article will be devoted to the selection of traction oils. Traction oils help to prolong the life of the contact surfaces, improve heat dissipation and increase the friction coefficient. The resulting selected oil will be used in upcoming planetary CVT [6] and it will increase the power torque transmission.

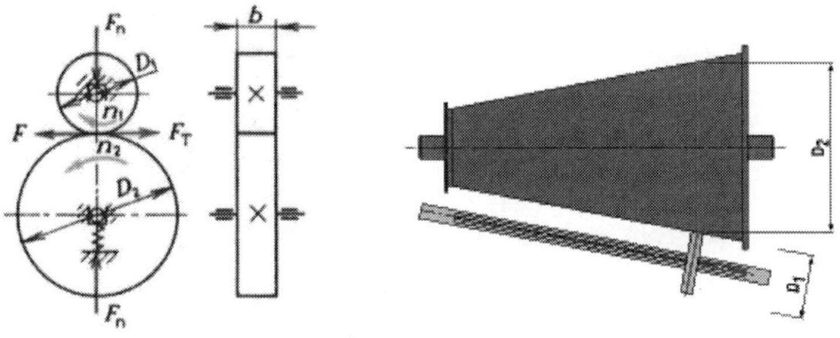

Fig. 1. Friction gears with constant gear ratio (left) and continuously variable transmission (right) [2, 7].

The geometric gear ratio is given by the diameters of the engaging wheels. For friction gears, a kinematic gear ratio is considered. This differs from the geometric due to the partial slip of the wheels. This is due to the fact that the friction surfaces are elastically deformed.

$$i_K = \frac{i_G}{1-\xi} \tag{1}$$

The value of the slip factor ξ depends mainly on the wheel material and increases with the load [2].

2. Test rig description

A test rig stand shown in Fig. 2 was designed as a simple friction gear to test transferable performance. It consists of an input drive electric motor (Siemens, 4 kW, 1450 rpm) and a load braking electric motor (Siemens, 7.5 kW, 975 rpm). Between them is the tested variator (friction gearbox). The variator is designed so that measurements can be made for different gear ratios. It consists of a pair of smooth wheels. The first wheel has the shape of a cylindrical disc, and the second wheel is conical. The gear ratio can be changed when the electric motor is stopped. It is realised by changing axial position of the output disc and it also necessary to change radial position of input bearing housing (change the axial distance).

The amount of contact wheel pressure is exerted by pair adjusting screws (Fig. 2 position 5). The normal force from screws passes through axial bearing and presses the disk to the conical wheel. A strain gauge is placed between the screw and the housing to indicate the magnitude of the pressing force.

A computer with LabVIEW is connected to the test stand. A controlling VI program has been developed for measuring and testing various operating conditions. Frequency converters connected to electric motors can be used to simulate operation at different loads and speeds. The evaluation is performed using speed and torque sensors on the input and output shafts.

For example, the maximum transferable torque at constant normal force (contact pressure) and speed can be determined. This is realized by gradually increasing the load on the braking electric motor. When the wheel slips, i.e., when the output speed differs from the input speed, considering calculated gear ratio, the maximum torque reached is automatically recorded. When the maximum torque is reached, the braking torque is stopped, the wheels stop slipping and the speed is equalized for the given gear ratio. This prevents the wheels from being subjected to excessively slippage and does not damage the surface of the friction wheels.

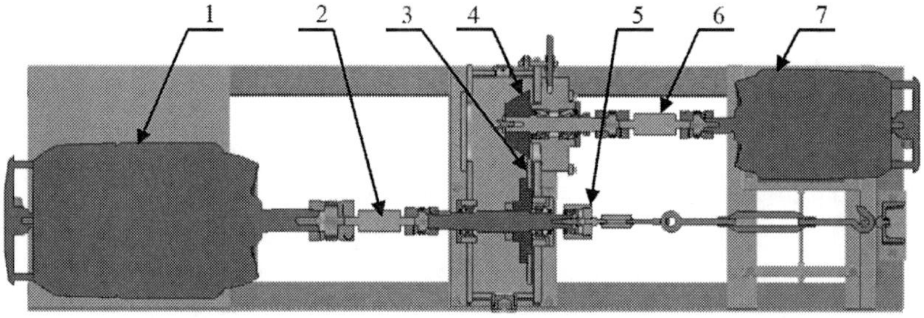

Fig. 2. Description of CVT test stand: 1 – Output breaking electromotor (load), 2 – Output torque and rpm sensor, 3 – Disc, 4 – Cylindrical wheel, 5 – Axial bearing housing with tension adjusting, 6 – Input torque and rpm sensor, 7 – Input drive electric motor.

Another automated program is surface testing of the engaging wheels for constant operating conditions. The wheels are pressed together with a constant axial force, the highest possible braking torque is set before the wheels slip (determined from previous measurements) and the input speed is set at 1500 rpm. The test is allowed to run for a longer period of time, on the order of several days, and then the subjective damage to the wheel surface is evaluated. If the oil temperature in the gear-case (upper and lower level), in the bearing housing or outside is exceeded from the set limits, the program will stop automatically.

The test stand at Fig. 3 is used to verify the analytically calculated moments that the gear unit is able to transmit. Already during the calculations, it was found that the accuracy of the calculation is affected by several important parameters. In addition to gear geometry (wheel diameters, edge fillet, angle of conical wheel), it is the effect of surface roughness (micro and macro structure), wheel material, surface layer and type of oil used. It was the selection of a suitable type of traction oil that was the subject of the first measurement.

3. Experimental measurement of oils

The use of lubrication oils in the friction gears helps to increase the life of the friction wheels. With the right choice of lubricant oil used, the variator can transmit high power and at the same time we can ensure that the friction wheels do not touch directly (contact metal-to-metal is supressed), but the frictional force in the gear will be transmitted through a layer of oil. This will prevent wheel wear due to gear slippage.

Fig. 3. Test stand of cone friction variator (CVT) at Czech technical university in Prague.

3.1. Selection of suitable oils for experimental measurements

For the first series of measurements, we chose several types of oils (Table 1), from highly viscous transmission oils, which expected good power transmission, low wear but high wading losses, through hydraulic oils, low viscosity bearing oils to oils developed specifically for variator gearboxes (CVT traction oils).

Table 1. Characteristic parameters of oils.

Oil typ	Kinematic viscosity at 100°C [mm^2/s]	Viscous index [–]	Point of ignition [°C]	Pour point [°C]
Gear oil Mogul Trans 85W-140H	24.3	95	200	-18
Hydraulic oil Paramo HM 10	10.06	–	150	-42
Variator traction oil Ravenol CVT Fluid	–	–	–	-45
Bearing oil Paramo OL-J3	4.33	–	116	-24
Bearing oil Paramo OL-J22	20.23	–	200	-27
Haldex gearbox oil Ravenol AWD-H	5.4	166	1952	-63
According to the standard	*ČSN EN ISO 3104*	*ČSN 65 6218*	*ČSN EN ISO 2592*	*ČSN ISO 3016*

3.2. Measurement of the coefficient of friction and losses in the variator

To determine the maximum coefficient of friction, several measurements were made for each oil at different input rpm. All measurements were performed at a constant axial force, ie. at constant contact pressure. For an axial force of 1000 N applied, the Hertzian contact pressure is 1202 MPa.

The measurement of the coefficient of friction was such that the axial force was set to 1000 N and the input shaft speed 300 rpm, then the load torque was increased until the wheels slipped completely. The maximum slip moment and other parameters were recorded in the program. The same procedure was repeated for 600, 900, 1200 and 1500 rpm and for all oils.

The resulting coefficient of friction was then determined from the magnitude of the output torque. However, during the measurement, the torque at the output sensor is still affected by the wading wheel losses and the losses in the bearings and seals. Input and output shaft losses were also measured to correctly determine the magnitude of the coefficient of friction.

To be able to separate the input and output losses during the measurement, the losses when the variator in disconnected state were measured. This is a condition where the input friction cone is not in contact with the friction output disk. When measuring losses, the speed on the input shaft was gradually increased and the torque was recorded. We repeated the same for the output shaft. In this measurement, the bearings are not loaded by radial or axial forces that arise from wheel contact and thus are not considered in the losses. The result of the loss measurement is plotted on graphs Fig. 4 and Fig. 5.

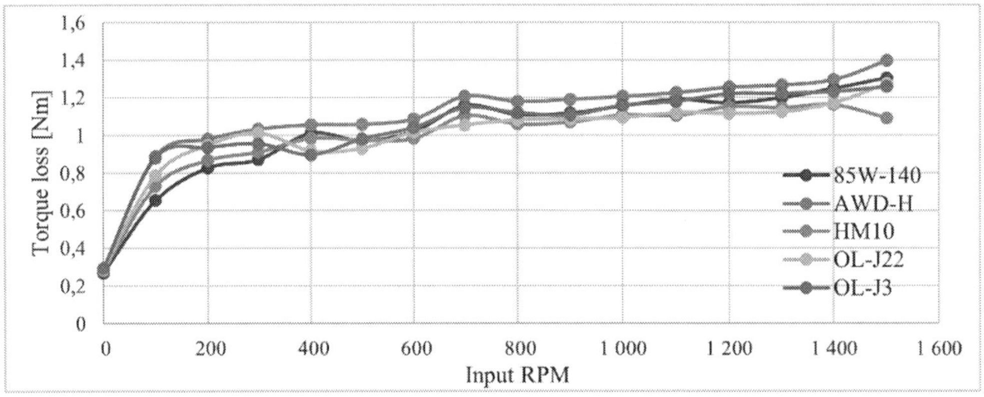

Fig. 4. Torque loss on output shaft.

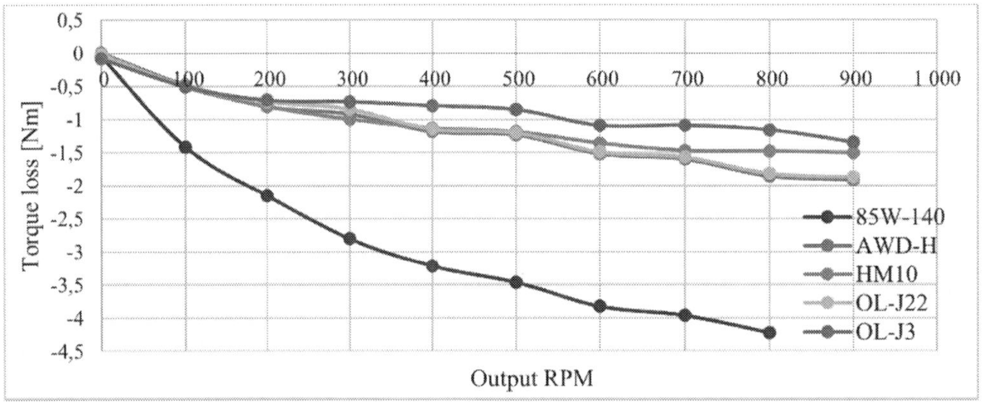

Fig. 5. Torque loss on output shaft.

When input and output losses are known, it is possible to calculate the coefficient of friction in the friction gear. This can be calculated from the measured nominal torque on the output shaft. This torque is reduced by the losses on the output shaft from the transmitted torque to the friction disc wheel, so that the output torque must be added to the measured torque. The resulting at graph Fig. 6 shows the maximum transferable torque by the test stand (average of three consecutive measurements) and the graph Fig. 7 shows the maximum coefficient of friction. The maximum transferable torque is not added to the torque, so it is the actual amount of torque read from the torque sensor.

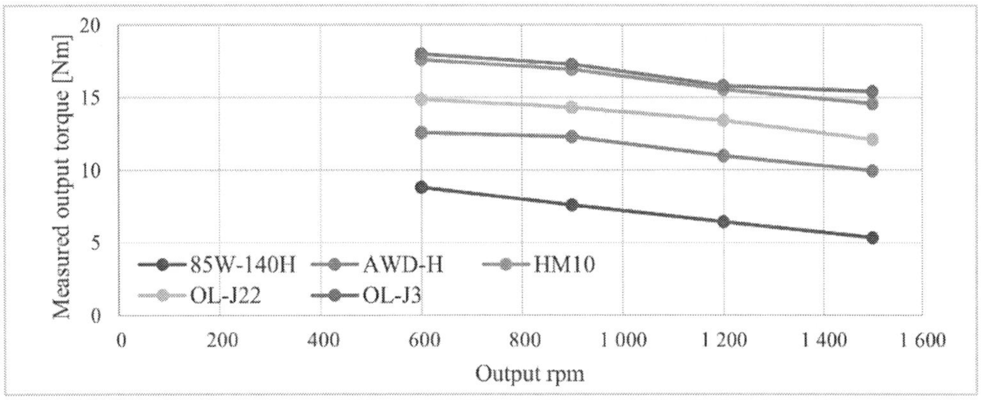

Fig. 6. Maximum measured output torque of test stand.

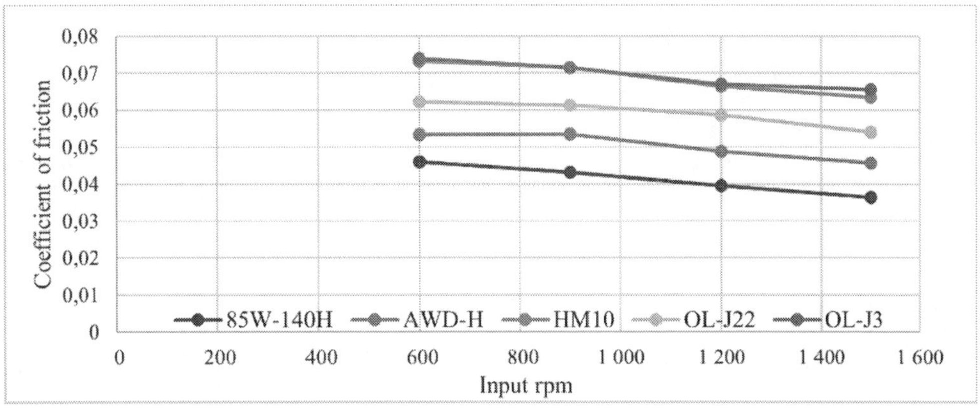

Fig. 7. Maximum coefficient of friction.

Ravenol CVT Fluid was not yet available at the time of measuring coefficient of friction (and torque losses), but during the experimental measurement of the coefficient of friction we could not cause total slippage during the measurement even at the maximum load of the measuring sensor on the output shaft, which is 20 Nm.

3.3. Measurement of contact surface wear

Contact surface wear measurement is used as one of the standard parameters when selecting a suitable oil type for friction gears. This will be further used in the upcoming planetary friction variator (planetary CVT). Testing was performed under operating conditions corresponding to the maximum achievable load torque. Unlike previous shorter measurements, the test stand was allowed to run for several days.

To ensure the same conditions for different oils, a constant contact pressure value of the engaging wheels has been set to 1400 MPa. For each measurement, a new worn track was created on the input cone friction wheel. Since the input wheel is conical and each new measurement runs on a different diameter on the cone (different gear ratio), it was also necessary to ensure the value of the contact pressure for each track (gear ratios). The geometry of both wheels, i.e., their diameters, plays a big role in the relationship for the calculation of the contact pressure. It was necessary to change the value of the axial force for each track, which was indicated by the strain gauge. Thus, the same contact pressure was achieved for each track.

After each measurement, the stand was rebuilt to prepare new measurement at different gear ratio and different oil specimen. This consisted of moving the output disk along the shaft (in axial direction) and changing the axial distance by moving the bearing housing of the input shaft and the entire drive electric motor. For each measurement, we chose a new unworn track on the input cone wheel. Then the original oil was drained, and more tested oil was filled into gear-case. We tested all oil specimens at the same volume in the gear-case.

There was an effort for approximately the same measurement time for all types of oil specimens. Initially, the program shut down automatically after approximately 30 hours. This was caused by filling the memory of one of the rpm counters, which was corrected by an additional modification of the program by frequent resetting of counters.

Fig. 8. Wear of the input cone friction wheel.

After performing long-term tests for all selected oils, an image with worn surfaces was taken (Fig. 8). The individual paths are numbered, and the individual oil types are assigned in Table 2. For each track, the measurement times, the set axial force, the corresponding contact pressure and the gear ratio are listed here.

Table 2. Evaluation of testing of different types of oils.

Trac Nr.	Oil type	Gear ratio	Measurement time [s]	[h]	Initial axial force [N]	Contact pressure [MPa]
1	Variator traction oil Ravenol CVT Fluid	2.55	701 921	195	1500	1400
2	Bearing oil Paramo OL-J3	2.36	381 000	106	1560	1400
3	Previous slip measurements	2.28	–	–	–	–
4	Gear oil Mogul Trans 85W-140H	2.18	458 290	127	1630	1400
5	Hydraulic oil Paramo HM 10	2.11	430 380	119.5	1650	1400
6	Haldex gearbox oil Ravenol AWD-H	2.04	599 532	166.5	1690	1400

The final choice of a suitable oil, including a brief comment on surface wear, will be described in the following conclusion part of this paper.

4. Conclusions

All measurements on the test stand were used to predict the behaviour of the prepared planetary CVT. The main goal was to find out which oil would be most suitable for future use. The most important criteria for selecting the oil were the achieved coefficient of friction and the assessment of the degree of wear of the contact surface (worn track) on the friction wheels.

The wear of the contact surface was evaluated by comparing the individual worn tracks for individual oils (Fig. 8). The most common cause of wear was pitting, which can also be observed on toothed gear surfaces. This involves pulling particles out of the wheel surface under high contact pressure through the lubricating oil layer. The result is the formation of small dimples or craters on the surface of the friction wheel. Furthermore, it can be seen that the individual tracks are not abrasively worn ("abraded"), so it can be stated that there was no complete wheel slip, which would cause the measurement to stop due to programmed slip protection.

As described, the contact pressure was the same for all measurements, but the degree of wear is different for different oil types. From the measurements we can state that the more viscous the oil, the smaller the wear of the surfaces, but there is a significantly lower coefficient of friction. Which confirms the results according to [8, 9]. This leads to a reduction in the transferable torque of the friction gears.

In the end, the Ravenol CVT Fluid was chosen as the most suitable oil. In the picture of worn tracks (Fig. 8), this track appears to be worn, but it must be considered that the measurement took place for the longest time. Furthermore, when looking at the actual worn surface in more detail, the wear is not so significant. However, this type of oil has the greatest benefit in the achieved coefficient of friction. This was significantly higher than the other samples. The coefficient of friction of this oil could not be precisely determined, because the total wheel slip did not occur even when the maximum torque of the measuring sensor was reached. It can therefore be considered that we can set a lower contact pressure for the selected Ravenol CVT Fluid oil, without such wear of the wheel surface at the same transmitted power. An alternative to this oil (the second option) is Paramo J3 bearing oil, which has a significantly lower degree of wear, and the coefficient of friction is the highest of the measured oils.

The planned planetary CVT will have hardened surface 62 to 64 HRC. This is much higher than in the tested stand, where a hardness of 45 HRC was prescribed. Therefore, the wear of the wheel surfaces should not be so significant.

Acknowledgements

This work was supported by the grant TH03010053 – Planetary continuously variable transmission (https://starfos.tacr.cz/en/project/TH03010053).

References

[1] Švec, V., *Části a mechanismy strojů: Mechanické převody*. Praha: Vydavatelství ČVUT, 2003.

[2] Kříž R., *Stavba a provoz strojů II: Převody.* Praha: Státní nakladatelství technické literatury (1978).

[3] Loewenthal S. H. and Rohn D.A., *Regression analysis of traction characteristics of traction fluids*, NASA Lewis Research Center Cleveland, OH, United States: NASA Technical Paper, (May 1 1983).

[4] Nanbu T., Yasuda Y., Ushijima K. et al., "Increase of Traction Coefficient due to Surface Microtexture," *Tribol Lett*, Vol. 29, pp. 105–118 (2008), http://dx.doi.org/10.1007/s11249-007-9287-9.

[5] Vengudusamy B., Enekes C. and Spallek, R. "EHD friction properties of ISO VG 320 gear oils with smooth and rough surfacesL," *Friction*, Vol. 8, pp. 164–181 (2020), https://doi.org/10.1007/s40544-019-0267-5.

[6] Hampl J., Vítek J., "A New Concept of Infinitely Variable Transmission for Mechanically Powered Gyronuses," *MECCA 1 + 2 2006*, (ČVUT Praha,2006).

[7] Transmission Technic, 2014 [online], *Automatické převodovky. Available from:* http://www.transmission-technic.cz/variator.html *Accessed:* 2021-09-20.

[8] Guegan J., Kadiric A., Gabelli A. et al., "The Relationship Between Friction and Film Thickness in EHD Point Contacts in the Presence of Longitudinal Roughness," *Tribol Lett*, Vol. 64, Art. no. 33 (2016), https://doi.org/10.1007/s11249-016-0768-6.

[9] Nanbu T., Yasuda Y., Ushijima K. et al. "Increase of Traction Coefficient due to Surface Microtexture," *Tribol Lett* Vol. 29, pp. 105–118 (2008) http://dx.doi.org/10.1007/s11249-007-9287-9.

On Uniaxial Ratcheting of AlSi10Mg Prepared by SLM Technology

Fatih Sari[1], Radim Halama[2] & Jiří Hajnyš[3]

Abstract: This study examines benchmark experiments for a cyclic plasticity model. The aim was to investigate the influence of mean stress and stress amplitude on uniaxial ratcheting behavior. Data were collected from low cycle fatigue tests that were applied to the AlSi10Mg aluminum alloy manufactured by selective laser melting technology. Low cycle fatigue tests were performed in stress-controlled mode as a block loading in three different conditions. These conditions are in order, first, to keep the loading coefficient the same and change the stress amplitude and mean stress, second, keep the maximal stress the same and change the stress amplitude and mean stress, and third, keep the stress amplitude the same and change the mean stress. All fatigue tests were simulated using the Chaboche model to verify its ability to predict uniaxial ratcheting of the material. All simulations were done using Python code. It is difficult to describe the effect of the mean stress and the stress amplitude by the Chaboche model in the whole range of stress amplitude / mean stresses correctly.

Keywords: Uniaxial loading; Ratcheting; Cyclic plasticity; Chaboche model

1. Introduction

Selective laser melting (SLM) is an additive manufacturing technique (AM) developed to melt and solidate metallic powders using a high-energy laser beam [1]. Recently, Selective laser melting (SLM) is being used for producing the geometrically complex or customized structures as a promising method in the various industries that require manufacturing of light-weighting parts. In addition to its contribution to the lightweight production of parts with unique shapes, it adds functionality to parts and has advantages such as cost-effective production and high precision of parts compared to traditional production techniques [2, 3]. Additive manufacturing used produced parts are already widely used in various fields, such as the automotive [4], medical, aviation [5] and military industries [6]. Manufacturing load bearing components via SLM, such as turbine parts [7] or aerospace parts [8], is becoming more common and attractive; therefore, research of their mechanical performance is increasingly focused on coping with the growing popularity of the process and confirming its reliability [9]. Compared to the static tensile properties of the parts produced, fatigue performance is mainly ignored in recent investigations [10]. The fatigue behavior of material is a significant property that should be taken into account when evaluating materials for industrial use. Phenomena such as pores, cracks, weak surface roughness, and high residual stresses [11] caused by the SLM process can be observed in 3D printed products, and these can cause an accelerated damage of SLM parts under cyclic loading [12, 13]. SLM products also have a distinctive microstructure [14] in comparison to conventional processing products. All heterogeneities affect the mechanical properties of the SLM components and distinguish them from conventional processing [15]. The AlSi10Mg alloy is one of the most

[1] *Ing. Fatih Sari; Kyndryl Client Center, s.r.o., Technická 2995/21, 616 00 Brno, Czech Republic; Fatih.Sari@kyndryl.com*

[2] *prof. Ing. Ph.D. Radim Halama; Dep. of Applied Mechanics, Fac. of Mech. Engineering, VSB-Technical University of Ostrava; 17.listopadu 2172/15, 70800 Ostrava, Czech Republic; radim.halama@vsb.cz*

[3] *Ing. Ph.D. Jiří Hajnyš; Dep. of Machining, Assembly and Eng. Metrology, Fac. of Mech. Engineering, VSB-Technical University of Ostrava; 17.listopadu 2172/15, 70800 Ostrava, Czech Republic; jiri.hajnys@vsb.cz*

popular materials used in light structures produced via SLM and commonly used in the aviation [16] and automotive industries due to its high resistance and fatigue properties [17].

In this paper, the influence of mean stress and stress amplitude on the uniaxial ratcheting behavior of the AlSi10Mg aluminum alloy manufactured by SLM technology is studied. For this purpose, low cycle fatigue (LCF) incremental tests were performed in a stress-controlled mode.

2. Experiment description

For fabricating specimens, 3D printer from company Renishaw namely type AM500E was used. Specimens were printed with the power of laser 350 W, scanning speed of 1150 mm·s^{-1} and the layer thickness of 30 μm. Scanning strategy "Meander" (see Fig. 1) was applied, which is the most used strategy for specimens with small cross-sections. During fabrication, the chamber was filled with argon with purity 5.0 to prevent oxidation.

Fig. 1. Used scanning strategy "Meander"; Reproduced with permission from [18].

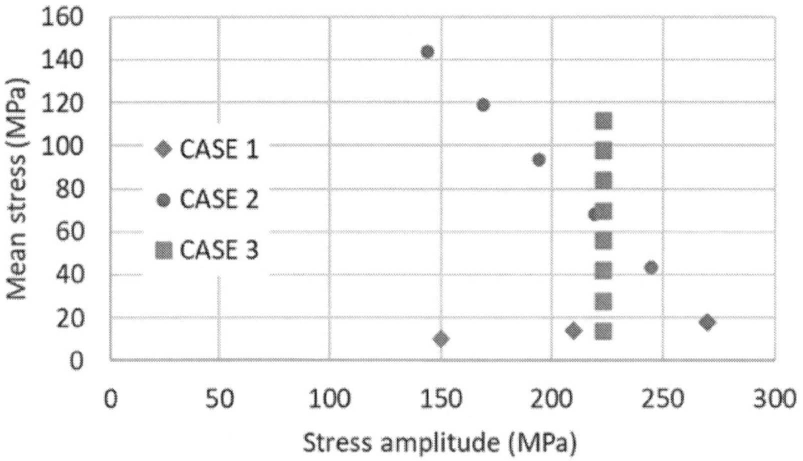

Fig. 2. Definition of stress amplitude and mean stress for particular loading blocks in the three low cycle fatigue tests.

All LCF tests were performed on vertically printed tubular specimens. The LabControl 100kN/1000Nm hydraulic fatigue testing machine was used in connection with the Model 3550 extensometer manufactured by Epsilon Technology Corp with a gauge length of 25 mm. Uniaxial ratcheting was investigated for a loading frequency of 0.25 Hz. The three ratcheting tests consisted of different number of loading blocks. There are three loading blocks for the first test, six loading blocks for the second test, and eight loading blocks for the third test, see Fig. 2.

The loading conditions can be seen in Fig. 2, and are in order, first, keeping the loading coefficient (R) the same and changing the stress amplitude and mean stress, as shown in Fig. 3, second, keeping the maximal stress the same and changing the stress amplitude and mean stress, as shown in Fig. 4 and third, keeping the stress amplitude the same and changing the mean stress, as shown in Fig. 5. For the first and second ratcheting tests, the stress amplitude is increased abruptly from one load block to another. In the third ratcheting test, the step change is performed only for the mean stress value.

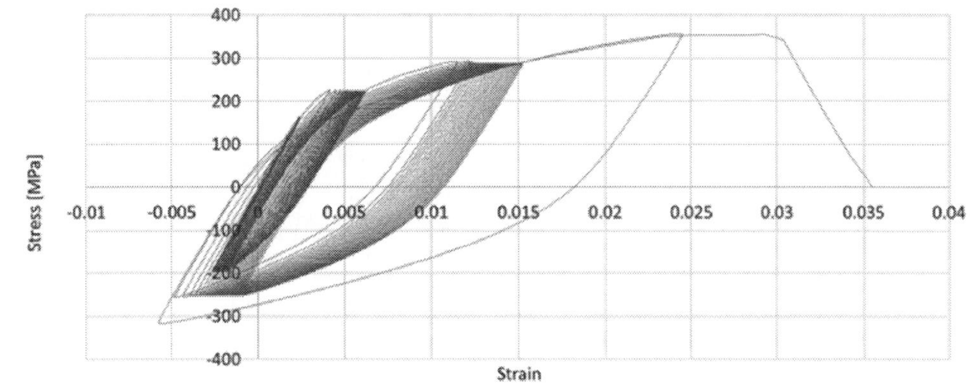

Fig. 3. Loading case 1; keeping the loading coefficient(R) the same and changing the stress amplitude and mean stress.

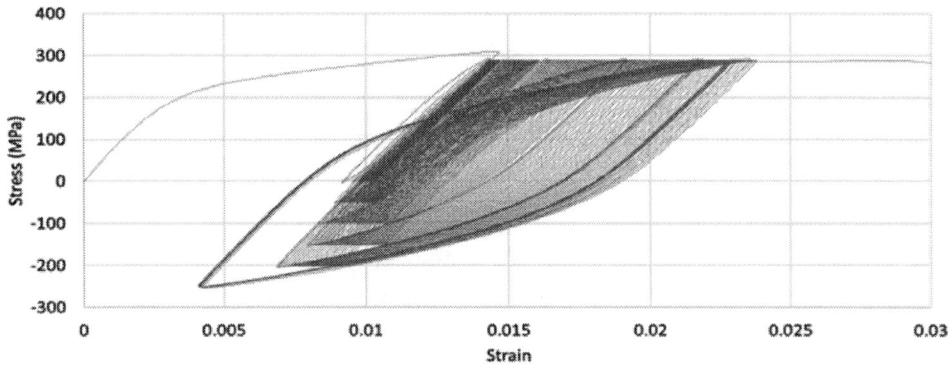

Fig. 4. Loading case 2; keeping the maximal stress the same and changing the stress amplitude and mean stress.

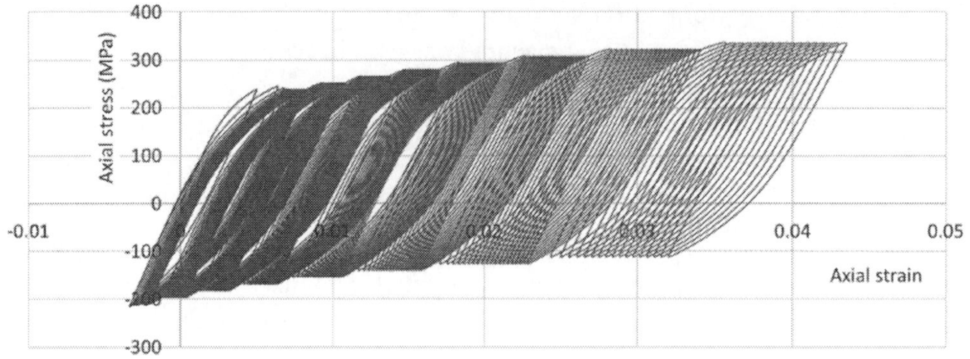

Fig. 5. Loading case 3; keeping the stress amplitude the same and changing the mean stress.

The number of cycles in the last blocks of loading is shown in Table 1. Additionally, all the previous loading blocks contain 25 cycles for case one, and 20 cycles for the case two and three.

Table 1. Number of cycles in the last block for each ratcheting test.

Test number	Number of cycles
1	1
2	11
3	16

3. Numerical prediction

In the description of constitutive equations, all equations will be written for uniaxial loading only. Axial stress $\sigma$is calculated from the back stress $\alpha$and the yield stress σ_Y, thus

$$\sigma = \sigma_Y + \alpha. \tag{1}$$

The Chaboche model is the cyclic plasticity model with pure kinematic hardening that establishes a linear superposition of several back-stress parts keeping the Armstrong-Frederick evolution equation as follows

$$\alpha = \sum_{i=1}^{M} \alpha^{(i)}, \tag{2}$$

whereas the evolution of the i-th back-stress part is given by Armstrong-Frederick kinematic hardening rule

$$\mathrm{d}\alpha^{(i)} = C_i \mathrm{d}\epsilon_p - \gamma_i \alpha^{(i)} \mathrm{d}p, \tag{3}$$

where C_i, $\gamma_i$are material parameters of Chaboche model and $\mathrm{d}p$ is the increment of accumulated plastic strain. In this study, two parts of the back-stress are considered ($M = 2$). The second part of the back-stress is considered without the memory term $(\gamma_2 = 0)$ in the initial phase of calibration procedure. Then, Young modulus E was obtained from a large stable hysteresis loop [19]. The identification of other material parameters was done using the cyclic stress strain curve of the AlSi10Mg aluminum alloy considering the following equation

$$\sigma_a = \sigma_Y + \sum_{i=1}^{M-1} \frac{C_i}{\gamma_i} \tanh(\gamma_i \cdot \epsilon_{ap}) + C_M \epsilon_{ap}, \tag{4}$$

where σ_a is the stress amplitude and ϵ_{ap} is the amplitude of plastic strain.

The optimization of material parameters was performed using a non-linear least-squares method. For the prediction of ratcheting, the calibration of the cyclic plasticity model was done using the data of the first test, thus the parameter γ_2 was identified. The same value of γ_2 was used for both other tests as well. The resulting material parameters of the optimization task are presented in Table 2. All simulations were performed using a Python code described elsewhere [19, 20].

Table 2. Optimized parameters of the Chaboche model with two back-stress parts.

Parameter	Value
σ_Y [MPa]	143
C_1 [MPa]	105532
γ_1 [-]	1293
C_2 [MPa]	9206
γ_2 [-]	8

4. Comparison of numerical and experimental results

A comparison of numerical predictions using the Chaboche model with experimental results of case one is shown in the Fig. 6. The prediction fits the experiment well for two loading blocks. For the last loading block, there is an overprediction of ratcheting.

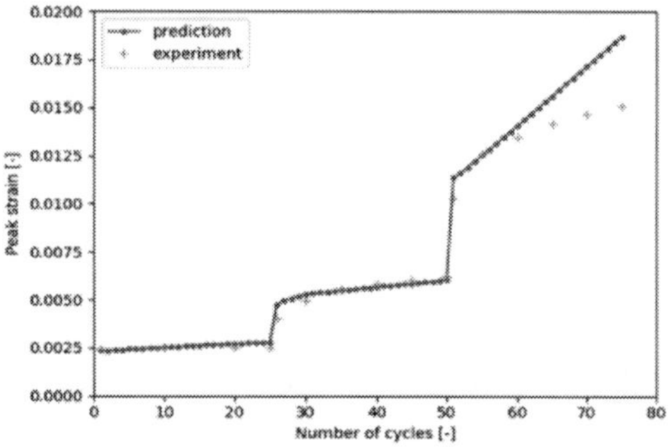

Fig. 6. Results of uniaxial ratcheting prediction on AlSi10Mg by Chaboche model, case 1.

The second test prediction results are presented in Fig. 7. An underprediction was observed for the first loading block, but an overprediction was obtained in the simulation for all other loading blocks.

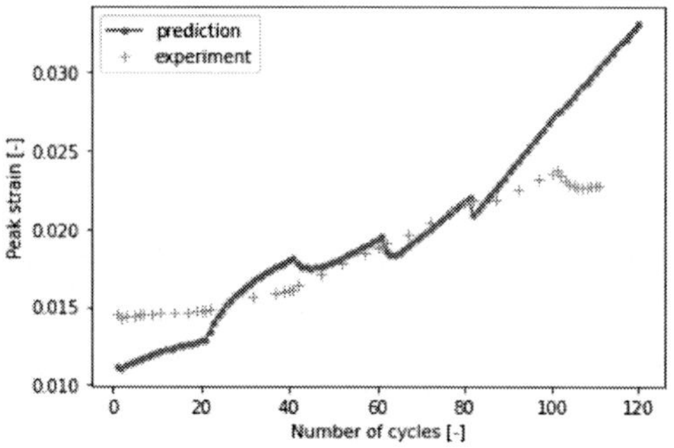

Fig. 7. Results of uniaxial ratcheting prediction on AlSi10Mg by Chaboche model, case 2.

For the third case, the prediction of the Chaboche model is excellent for loading blocks number four to seven, which can be seen in Fig. 8. There is an over-prediction of the ratcheting rate in the first three blocks of loading and an under-prediction for the last one. This means that the influence of mean stress is captured very well in the 50MPa to 100MPa interval.

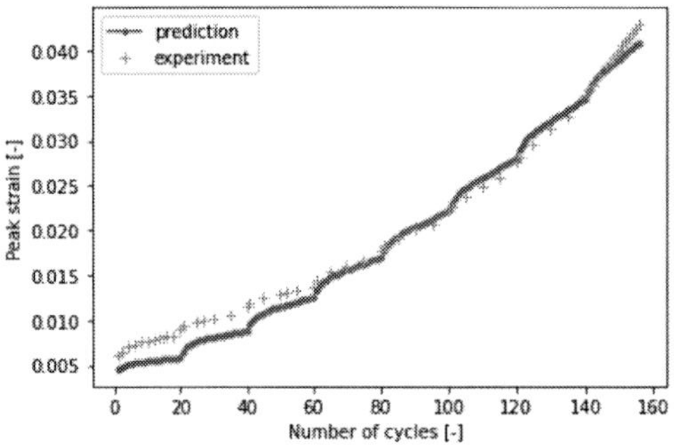

Fig. 8. Results of uniaxial ratcheting prediction on AlSi10Mg by Chaboche model, case 3.

5. Conclusion

The experimental study focused on the investigation of mean stress and stress amplitude on the ratcheting behavior of the AlSi10Mg aluminum alloy manufactured by SLM technology. The material shows steady ratcheting after few cycles, which is in good correlation with almost cyclically stable behavior observed under the strain-controlled loading.

All low cycle fatigue tests were simulated using the Chaboche model to verify the prediction of uniaxial ratcheting. The conclusion of the comparative study is that the Chaboche model is not capable of describing the mean stress and stress amplitude effects in the whole range of loading conditions. Therefore, a possible solution is to apply an advanced cyclic plasticity model, e.g. the AbdelKarim-Ohno model with memory surface [21], which will be done in future work.

Acknowledgements

This work was supported by the Ministry of Education, Youth and Sports from the Specific Research Project (SP2022/26) and has been done in connection with the DMS project reg. no. CZ.02.1.01/0.0/17_049/0008407 is financed by the Structural Funds of the European Union.

References

[1] Song X., Zhai W., Huang R., Fu J., Fu M.W. and Li F., "Metal-Based 3D-Printed Micro Parts & Structures," *Encyclopedia of Materials: Metals and Alloys,* Vol. 4, pp. 448–461, (2022).

[2] Yadroitsev I. and Smurov I., "Selective laser melting technology: From the single laser melted track stability to 3D parts of complex shape," *Physics Procedia*, Vol. 5, pp. 551–560, (2010). http://dx.doi.org/10.1016/j.phpro.2010.08.083.

[3] Yadroitsev I., Bertrand Ph. and Smurov I., "Parametric analysis of the selective laser melting process," *Applied Surface Science*, Vol. 253, no. 19, pp. 8064–8069, (2007). https://doi.org/10.1016/j.apsusc.2007.02.088-

[4] Matilainen V., Piili H., Salminen A., Syvänen T. and Nyrhilä O., "Characterization of Process Efficiency Improvement in Laser Additive Manufacturing," *Physics Procedia*, Vol. 56, pp. 317–326, (2014). https://doi.org/10.1016/j.phpro.2014.08.177.

[5] Wycisk E., Solbach A., Siddique S., Herzog D., Walther F. and Emmelmann C., "Effects of Defects in Laser Additive Manufactured Ti-6Al-4V on Fatigue Properties," *Physics Procedia,* Vol. 56, pp. 371–378, (2014). https://doi.org/10.1016/j.phpro.2014.08.120.

[6] Tian Q., "The Development Status of Selective Laser Melting Technology (SLM)," *Journal of Physics: Conference Series,* Vol. 1798, p. 012045, (2021). http://dx.doi.org/10.1088/1742-6596/1798/1/012045.

[7] Buchbinder D., Schleifenbaum H., Heidrich S., Meiners W. and Bültmann J., "High Power Selective Laser Melting (HP SLM) of Aluminum Parts," *Physics Procedia,* Vol. 12, pp. 271-278, (2011). https://doi.org/10.1016/j.phpro.2011.03.035.

[8] Seabra M., Azevedo J., Araújo A., Reis L., Pinto E., Alves N., Santos R. and Mortágua J.P., "Selective laser melting (SLM) and topology optimization for lighter aerospace componentes," *Procedia Structural Integrity,* Vol. 1, pp. 289-296, (2016). https://doi.org/10.1016/j.prostr.2016.02.039.

[9] Dai D., Gu D., "Thermal behavior and densification mechanism during selective laser melting of copper matrix composites: Simulation and experiments," *Materials & Design,* Vol. 55, pp. 482–491, (2014). https://doi.org/10.1016/j.matdes.2013.10.006.

[10] Brandl E., Heckenberger U., Holzinger V. and Buchbinder D., "Additive manufactured AlSi10Mg samples using Selective Laser Melting (SLM): Microstructure, high cycle fatigue, and fracture behavior," *Materials & Design,* Vol. 34, pp. 159–169, (2012). https://doi.org/10.1016/j.matdes.2011.07.067.

[11] Shiomi M., Osakada K., Nakamura K., Yamashita T. and Abe F., "Residual Stress within Metallic Model Made by Selective Laser Melting Process," *CIRP Annals,* Vol. 53, pp. 195–198, (2004). https://doi.org/10.1016/S0007-8506(07)60677-5.

[12] Divya V.D., Muñoz-Moreno R., Messé O.M.D.M., Barnard J.S., Baker S., Illston T. and Stone H.J., "Microstructure of selective laser melted CM247LC nickel-based superalloy and its evolution through heat treatment," *Materials Characterization,* Vol. 114, pp. 62–74, (2016). https://doi.org/10.1016/j.matchar.2016.02.004.

[13] Aboulkhair N.T., Everitt N.M., Ashcroft I. and Tuck C., "Reducing porosity in AlSi10Mg parts processed by selective laser melting," *Additive Manufacturing,* Vols. 1-4, pp. 77–86, (2014). https://doi.org/10.1016/j.addma.2014.08.001.

[14] Chakrabarty A., Sahu V.K, Das A., Mukherjee S., Gurao N.P., Chakraborty P., Bar H.N., Khutia N., "Study of the effect of two separate tilt angles of laser scanning lines on the microstructure and mechanical properties in direct metal laser sintered AlSi10Mg alloy," *Metals and Materials International*, Vol. 28, pp. 250–268, (2022). https://doi.org/10.1007/s12540-021-01080-w.

[15] Read N., Wang W., Essa K. and Attallah M.M., "Selective laser melting of AlSi10Mg alloy: Process optimisation and mechanical properties development," *Materials & Design (1980-2015),* Vol. 65, pp. 417–424, (2015). https://doi.org/10.1016/j.matdes.2014.09.044.

[16] Raja A., Cheethirala S.R., Gupta P., Vasa N.J. and Jayaganthan R., "A review on the fatigue behaviour of AlSi10Mg alloy fabricated using laser powder bed fusion technique," *Journal of Materials Research and Technology,* Vol. 17, pp. 1013–1029, (2022). https://doi.org/10.1016/j.jmrt.2022.01.028.

[17] Aboulkhair N.T., Maskery I., Tuck C., Ashcroft I. and Everitt N.M., "Improving the fatigue behaviour of a selectively laser melted aluminium alloy: Influence of heat treatment and surface quality," *Materials & Design,* Vol. 104, pp. 174–182, (2016). https://doi.org/10.1016/j.matdes.2016.05.041.

[18] Mesicek J., Ma Q.-P., Hajnys J., Zelinka J., Pagac M., Petru J. and Mizera O., "Abrasive Surface Finishing on SLM 316L Parts Fabricated with Recycled Powder," *Appl. Sci.*, Vol. 11, p. 2869 (2021). https://doi.org/10.3390/app11062869.

[19] Halama R, Šmach J, Govindaraj B. and Čep R., "Implementation of the Chaboche model in numerical analysis of fatigue damage," in *Fatigue Damage in Metals - Numerical Based Approaches and Applications*, ed. (Elsevier, in press).

[20] Halama R, Šmach J, Govindaraj B. and Čep, R. , "Chaboche model with memory implemented in Python code," in *Fatigue Damage in Metals - Numerical Based Approaches and Applications*, ed. (Elsevier, in press).

[21] Halama R, Markopoulos A, Šmach J, and Govindaraj B., "Theory, Application and Implementation of Modified Abdel-Karim-Ohno Model for Uniaxial and Multiaxial Fatigue Loading," in *Fatigue Damage in Metals - Numerical Based Approaches and Applications*, ed. (Elsevier, in press).

Monitoring of Integrated Loop Technology Joints under Flexural Loading Using CFS Attached to the Surface

Nikola Schmidová[1], Tomáš Pońížil[2], Bohumil Kropík[1], Karel Doubrava[1], Milan Dvořák[1] & Milan Růžička[1]

Abstract: New technology of joining wounded tubes (carbon fiber and polymer-matrix composite) has been developed. During the development of Integrated Loop Technology (ILT) joints, possible ways to monitor structural health of these components were also investigated. This paper presents results of damage detection of ILT joints obtained by the Carbon Fibre Sensors (CFS) attached to the surface.

Keywords: Damage detection; Polymer-matrix composites; Carbon fiber sensor

1. Introduction

The ILT joints have been developed to overcome joint design issues of composite tubes such as hole machining and hand lamination. Production of composite frames from subassembly parts with fully integrated joints is rapid and cost effective.

Composite frames are used in a construction of production machines, where collisions and thus overloads can occur while operating. Accidental collision can cause damage to the construction, which can be overlooked, and subsequent loading of the construction can have fatal consequences. In the framework of our project, we have investigated several possibilities to monitor damage of profiles with ILT joints. Damage and deformation detection using CFS attached to the surfaces, CFS integrated in the structure, conductive epoxy applied to the surface, electrical resistance change measurement of the structure itself, optical fibres with Bragg gratings integrated in the structure and attached to the surface. Detailed information on measurement using CFS attached to the surface is presented in this paper.

2. Experimental procedure

2.1. Specimen preparation

The specimens were manufactured using an automated filament winding procedure of carbon fibre tows T700 (Toray Composite Materials America, Inc.) with epoxy resin. The length of the specimen was 175 mm, Diameter of clamping mandrel 45 mm. The configuration of the composite layers corresponds to specimen type SN-09 described in [1].

[1] *Ing. Nikola Schmidová; Department of Mechanics, Biomechanics and Mechatronics, Fac. of Mech. Engineering, Czech Technical University in Prague; Technická 4, 160 00 Prague, Czech Republic;* nikola.schmidova@fs.cvut.cz, bohumil.koropik@fs.cvut.cz, karel.doubrava@fs.cvut.cz, milan.dvorak@fs.cvut.cz, milan.ruzicka@fs.cvut.cz

[2] *Ing. Tomáš Pońížil; CompoTech PLUS, spol. s.r.o.,Nová 1316, 34201, Sušice, Czech Republic;* tomas.ponizil@compotech.com

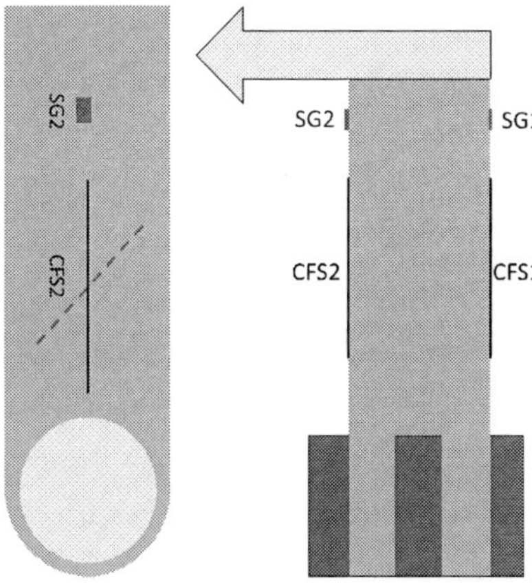

Fig. 1. Sensor configuration on the specimen.

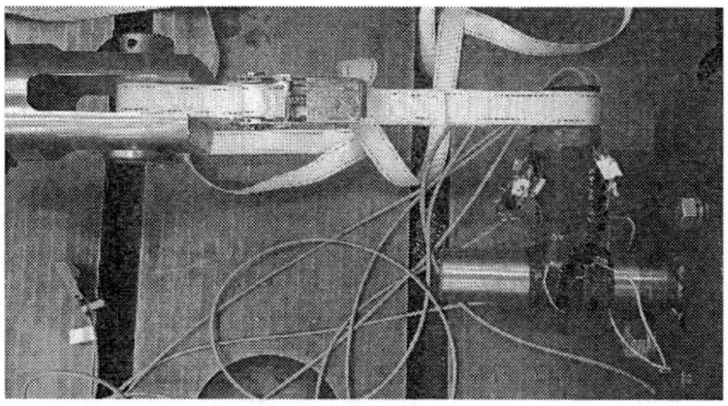

Fig. 2. Configuration of the flexural test.

2.2. *Carbon Fiber Sensors (CFSs)*

The CFSs were manufactured using carbon fiber roving ex-PAN T300 ((Toray Composite Materials America, Inc.) according to procedure published in [2] in a length of 80 mm. The distance between the eyelet and the sensor was 10 mm. The CFSs were attached to the surface of 7 specimens along the longitudinal axes of the specimen according to Fig. 1 using epoxy adhesive CC-33a (Kyowa Electronic Instruments Co., Ltd.). The CFSs on the specimen's number 8 and 9 are positioned so that the angle between the longitudinal axes and the sensor is 45°, see Table 1. The epoxy adhesive was used also as the electrical insulation between the sensor and the specimen.

Table 1. Specimen overview, configuration of the CFSs and position of the crack.

Specimen number	CFS1 tension side	CFS2 compression side	Crack on the compression side
8	45°	45°	through the sensor
9	45°	45°	through the sensor
13	0°	0°	next to the sensor
14	0°	not installed	–
15	0°	0°	next to the sensor
34	0°	0°	through the sensor
35	0°	0°	next to the sensor

3. Experimental Setup

The flexural test of the composite profile with the ILT joint was performed in the configuration depicted in Fig. 2. All specimens were tested using an IST Instron PL63N servo hydraulic cylinder, with a loading speed 8mm/min. The change in electrical resistance of the CFSs was measured using a strain gauge transducer (QuantumX, HBM).

The changes in electrical resistance were calculated using the following equations, where U_O is output voltage and U_I is input voltage.

$$\frac{U_\mathrm{O}}{U_\mathrm{I}} = \frac{1}{4} \cdot \frac{\Delta R}{R} \left[\frac{\mathrm{mV}}{\mathrm{V}}\right] \rightarrow \left[\frac{\mathrm{m}\Omega}{\Omega}\right] \tag{1}$$

$$\Delta R = 4 \cdot \frac{U_\mathrm{O}}{U_\mathrm{I}} \cdot R_0 \; [\mathrm{m}\Omega] \tag{2}$$

$$R_0 = R_{\text{CFS}} + R_{\text{ad}} \tag{3}$$

$$\text{Relative change in electrical resistance} = \frac{\Delta R}{R} = \frac{\Delta R}{R_{\text{CFS}}} \tag{4}$$

$$\text{Relative change in electrical resistance in \%} = \frac{\Delta R}{R} \cdot 100 \tag{5}$$

4. Results and discussion

The specimen fracture occurred on the compression side of the specimens. We observed two types of fracture. First one was observed in distance approx. 30 mm from the eyelet (specimens 8, 9, 34). The crack was in the area of installed CFS and was caused probably by the buckling of the specimen, see Fig. 3. The second type of the fracture was observed closer to the eyelet in the distance approx. 10 mm (specimens 13, 14, 15, 35), see Fig. 4. The second type of damage occurred next to the installed CFS.

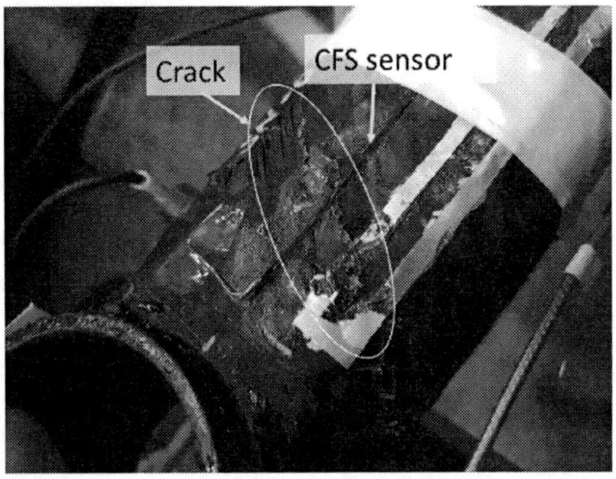

Fig. 3. Spec. 34 with the crack in the area of CFS.

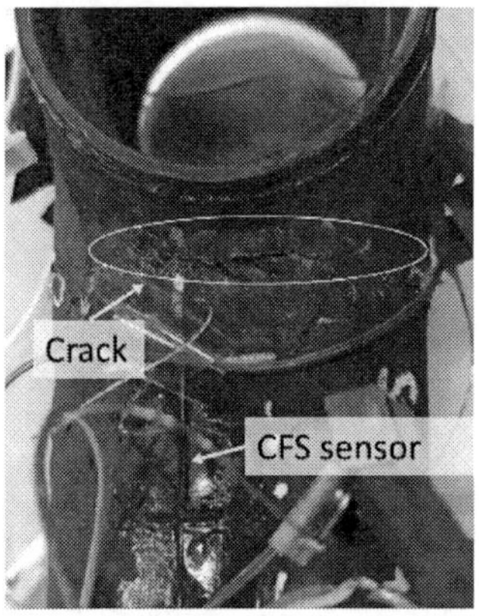

Fig. 4. Spec. 13 with the crack next to the CFS.

Table 2. Measured step change in electrical resistance for CFS installed on the compression side.

Specimen number	ΔR [%] during loading	ΔR [%] after final fracture	Position of the fracture
8	11	overload	In the area of the sensor
9	0.05	overload	
34	1.7	20	
13	0.12	1	Next to the sensor
15	0.1	overload	
35	0.05	0.3	

The electrical resistance of installed CFS were measured during the loading. There are two possible ways damage detection scenarios, which are damage detection during continual measurement and damage detection during production line shutdowns. In the Table 2 you can see value of measured step change in measured electrical resistance of the CFS installed on the compression side during loading and after final fracture. The measured step change of about 0.05% could be possibly caused by a temperature change of 1.5 °C, see [3], therefore it is necessary to take into account the change of the operational temperature for the practical application.

5. Conclusion

Possible damage modes of the composite profiles with the ILT joints were described. Profiles with ILT joints underwent flexural test during which possibility of damage detection using CFS attached to the surface was experimentally investigated. The CFSs attached to the surface were sensitive to damage in the form of cracks.

According to conducted experiments – it seems possible to detect damage of profiles with ILT joints using CFS attached to the surface during continual measurement and also during production line shutdowns

In case, when the fracture occurred close to the eyelet – the step change in measure el. res occurred earlier than for the fracture farer to the eyelet. The above conclusions should be verified on a larger number of experiments.

References

[1] Kropík B. et al., "Tension and Compression Optimization of Integrated Loop Technology Joint" in EAN 2021 59th conference on experimental stress analysis, Litomyšl, 2022 (ČVUT v Praze, Praha, 2022), pp. 99 – 106. ISBN 978-80-01-06885-4.

[2] Horoschenkoff A. and Christner C., "Carbon Fibre Sensor: Theory and Application," in *Composites and Their Applications*, Hu N., ed. (IntechOpen, 2012), online. DOI: 10.5772/50504.

[3] Schmidová N., Macken J., Horoschenkoff A., Sedláček R., Kostroun T., Šimota J. and Růžička M., "Impact damage detection of a glass fabric composite using carbon fiber sensors with regard to mechanical loading," *Applied Sciences*, Vol. 12, no. 3, Art. no. 1112 (2022). https://doi.org/10.3390/app12031112.

Measurement and FEM Simulation of Process 3D Printing by Direct Metal Laser Melting

Radek Skácelík[1] & Miroslav Španiel[2]

Abstract: Printing is accompanied by problems usually caused by large temperature gradients, such as temperature expansion, residual stress, porosity, delamination, local overheating and recoater crash or recoater impact. These defects need to be eliminated by choosing the appropriate technological parameters, layer thickness, laser speed, laser power, scanning strategy. These parameters are currently tuned experimentally, which is time and money consuming.

Keywords: Laser powder bed fusion; FEM; 3D prinitng simulation

1. Introduction

Laser powder bed fusion (LPBF) is one of the additive technologies based on melting metal powder. The principle consists in applying thin layers of metal powder that melts at the sites of the printed part. This creates the entire geometry of the printed part (Fig. 2). The printed part must normally be complemented by supporting structures that anchor the part to the platform and improve heat dissipation from the printed part.

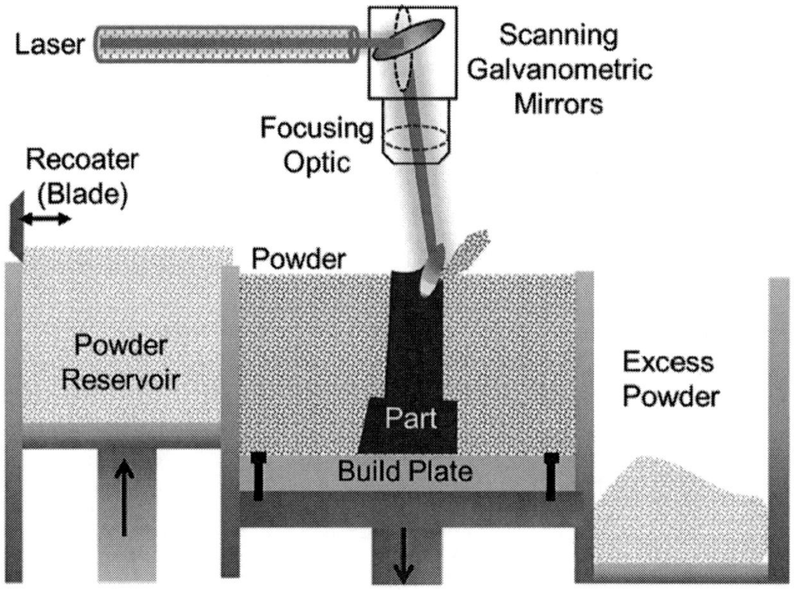

Fig. 1. A schematic of the laser powder bed fusion (LPBF) process [1].

Printing is accompanied by problems usually caused by large temperature gradients, such as temperature expansion, residual stress, porosity, delamination, local overheating and recoater

[1] *Ing.Radek Skácelík; Department of Mechanics, Biomechanics and Mechatronics, Faculty of Mechanical Engineering, Czech Technical University in Prague. Technická 4, 166 07 Praha 6, Czech Republic* radek.skacelik@fs.cvut.cz

[2] *doc. Ing. Miroslav Španiel,CSc.; Department of Mechanics, Biomechanics and Mechatronics, Faculty of Mechanical Engineering, Czech Technical University in Prague. Technická 4, 166 07 Praha 6, Czech Republic* miroslav.spaniel@fs.cvut.cz

crash or recoater impact [1]. These defects need to be eliminated by choosing the appropriate technological parameters, layer thickness, laser speed, laser power, scanning strategy. These parameters are currently tuned experimentally, which is time and money consuming.

Numerical simulation of the printing process can reduce the risk of some defects and thus increase the efficiency of printing. Therefore, it is important to incorporate finite element method (FEM) simulation into the printing design itself. Currently, there are several simulation approaches that usually take into account only one type of printing problem. Moreover, the simulation itself is complicated by the difficult validation caused by the technological process itself.

2. Simulations method and validation

The printing simulation consists of a sequential calculation of temperature fields and subsequent mechanical response [2]. The aim of the simulation is to predict the behavior of the part and thereby predict the problems arising from the thermal expansion and deformation of the part during and after the printing itself. One of the very complex tasks is to predict the collision with the coating device. In such a simulation, it is also necessary to verify sufficient anchoring of the part by calculating the stiffness in individual cuts, depending on the rotation of the part in relation to the coating device.

2.1. Thermo-mechanical model

The print simulation is based on a calculation by the final element method. The calculation is designed as a sequential calculation, so that the temperature field is first calculated, which is then used to calculate the mechanical response. The model heater is caused by laser action. The variation in model size and laser zone of action complicates the whole simulation process. Material undergoes complex processes, such as thermal conduction, heat loss due to convection and radiation, phase transformation, and melting and cooling solidification. Laser heat input is represented by Gaussian distribution of heat source (1) for simulate the interaction between the laser and material [3].

$$q\left(x,y,z\right)=\frac{3c_sQA}{\pi H(1-\frac{1}{e^3})}\exp\left[\frac{-3c_s}{\log\left(\frac{H}{z}\right)}(x^2+y^2)\right] \tag{1}$$

Where A is the laser absorptivity of materials affected by the wavelength, Q is the laser power, H is the height of heat source and c_S is the concentric coefficient of heat-flux distribution of the cross section. The enthalpy (H_e) is defined as a function of temperature (2). Where ρ is density and c is thermal capacity of based material

$$H_e=\int \rho c \mathrm{d}T. \tag{2}$$

Heat energy losses from part through surface and conduction into the support structure and substrate. Boundary conditions for heat transfer should be considered, described as

$$q_{cond}=q\left(x,y.z\right)-q_c-q_r, \tag{3}$$

where the heat losses through conduction q_{cond}, convection q_c and radiation q_r.

$$q_{cond}=\ -k\Delta T \tag{4}$$
$$q_c=h_c(T-T_o) \tag{5}$$
$$q_r=\sigma\in(T^4-T_o^4) \tag{6}$$

A variable k is for thermal conductivity, h_c is the heat convection coefficient, σ is the Stefan-Boltzman constant, $\in$ is the emissivity of the powder bed, T is the temperature of the molten pool, and T_0 is the ambient temperature.

The structural analysis assumes a small strain and small deformation formulation using an elasto-plastic constitutive model with the standard relation. The residual stress is formed after the cooling solidification stage. If the residual stress is greater than the yield strength of the material, it will produce local deformation. The material strain rate $\dot{\varepsilon}$ is influenced by external force and temperature including the elastic strain rate $\dot{\varepsilon_e}$, plastic strain rate $\dot{\varepsilon_p}$, creep strain rate $\dot{\varepsilon_c}$, and the strain rate caused by temperature change $\dot{\varepsilon_T}$.

$$\dot{\varepsilon} = \dot{\varepsilon_e} + \dot{\varepsilon_p} + \dot{\varepsilon_c} + \dot{\varepsilon_T} \tag{7}$$

2.2. *Experimental validation*

The aim of the experiment was to discover the cause of the collision of the coater behind the device, so the experimental printing was designed to affect as many possible influences as possible. The test sample was chosen according to [4], because in the absence of the predicted collision, it would be possible to evaluate the degree of displacement of individual layers due to temperature expansion. The layout of the sample was chosen over the entire width and length of the printing area, so that the influence of the loading blade deflection and the influence of the flowing gas can be excluded (Fig. 2). The rotation of the piece was chosen at an angle of 0, 45, 90, 135, 180 degrees around the vertical axis. Different angles of rotation change the size of the contact area and the way the friction stresses the part.

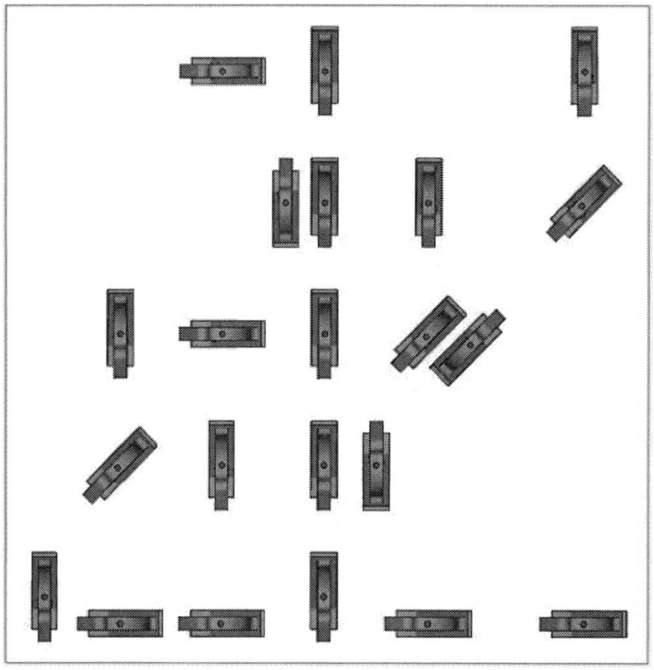

Fig. 2. Distribution of samples on the platform.

During the printing process, parts oriented with a curved side against the deposition device were destroyed (Fig. 3). The sample rotated 180 degrees did not show a collision, this is most likely due to a deformation that increases the gap between the piece and the deposition device. The other sample did not close this gap due to deformation. The experiment thus revealed as a key factor in the collision of the application device the rigidity of the component itself.

3. Conclusions

The measurement can only be performed indirectly due to the technology. Therefore, the measurement was carried out on the already printed part to detect residual stresses. The temperature of the part was monitored using a thermal camera. The verification of the calculation was carried out using a test printing, an experimental body distributed over the area of the printing area so that the influence of the part position on the printing result could be excluded. The experiment demonstrated the influence of the rotation of the part and the independence of the placement on the platform (Fig. 3).

Fig. 3. The result of the experiment after collision with the coater.

References

[1] Yavari R., Smoqi Z., Riensche A., Bevans B., Kobir H., Mendoza H., Song H., Cole K. and Rao P., "Part-scale thermal simulation of laser powder bed fusion using graph theory: Effect of thermal history on porosity, microstructure evolution, and recoater crash, " *Materials & Design*, Vol. 204, p. 109685 (2021). ISSN 0264-1275. https://doi.org/10.1016/j.matdes.2021.109685.

[2] Bayat M., Dong W., Thorborg J., To A.C. and H. Hattel J.H., "A review of multi-scale and multi-physics simulations of metal additive manufacturing processes with focus on modeling strategies," *Additive Manufacturing*, Vol. 47, p. 102278 (2021). ISSN 2214-8604. https://doi.org/10.1016/j.addma.2021.102278.

[3] Jamshidinia M., Kong F. and Kovacevic R., "Numerical modeling of heat, distribution in the electron beam melting of Ti-6Al-4V," *J. Manuf. Sci. Eng.*, Vol. 135, no. 6, p. 061010 (2013). http://dx.doi.org/10.1115/1.4025746.

[4] Santos L.S., Gupta S.K. and Bruck H.A., "Simulation of buckling of internal features during selective laser sintering of metals," *Additive Manufacturing*, Vol. 23, pp. 235–245 (2018). ISSN 2214-8604. https://doi.org/10.1016/j.addma.2018.08.002.

Identification of Constitutive Parameters of the Human Vena Saphena Magna in an Inflation-extension Experiment

Zbyněk Sobotka[1], Lukáš Horný[1], Hynek Chlup[1], Miroslav Špaček[2] & Pavel Měřička[3]

Abstract: In the present biomechanics, computational simulations are used to model the adaptation of the venous wall to arterial conditions, which is a consequence of the bypass surgery carried out with venous grafts. However, these simulations can only be as accurate as the constitutive models used within the calculations. In this study, the results of constitutive modelling of the human great saphenous vein are presented. A hyperelastic material model was fitted to data from an inflation-extension experiment carried out with a sample of the vessel. The model was able to describe the behaviour of the vein subjected to interior pressure and axial forces.

Keywords: Great saphenous vein; Hyperelastic material model; Inflation-extension experiment

1. Introduction

Vascular grafts are used in both coronary and peripheral arterial bypass surgery. However, the structural and mechanical properties of veins differ from those of arteries, as each type of blood vessel is optimised to work under different loads. As a result of this mismatch, remodelling of the vein is triggered immediately after implantation in order for the graft to adapt to its new environment. Computational simulations can be used to model the adaptation of the venous wall to arterial conditions. A constitutive model of the wall is necessary for simulations to be carried out.

It is known that arteries are prestretched and subjected to multiaxial stress states induced by internal pressure and axial forces in vivo. To ensure that a model can describe the mechanical behaviour of a vein in the arterial environment, it is beneficial to fit the model to data from experiments that accurately simulate these in vivo conditions. A method of mechanical testing that allows both internal pressure and axial forces to be applied is the inflation-extension experiment.

In this study, the results of the constitutive modelling of the human great saphenous vein are presented. A hyperelastic material model was fitted to data from an inflation-extension experiment carried out with a sample of the vessel. The model was able to describe the behaviour of the vein when subjected to interior pressure and axial forces.

2. Methods

An inflation-extension experiment was carried out with a sample of the human vena saphena magna. The experiment consisted of three series of cyclic pressurizations, each with a different axial force applied. The sample was positioned in a vertical configuration during the experiment and weights were suspended from its lower end to induce axial forces. The pressure-strain relationships of the sample with weights of 0, 10 and 50 g were obtained in the experiment. Then, a constitutive model was fitted to the experimental data.

[1] *Faculty of Mechanical Engineering, Czech Technical University in Prague, Technická 4, 160 00 Prague, Czech Republic*
[2] *2nd Department of Surgery – Department of Cardiovascular Surgery, 1st Faculty of Medicine, Charles University in Prague and General University Hospital in Prague, U nemocnice 2, 120 00 Prague, Czech Republic*
[3] *Tissue Bank, University Hospital Hradec Králové, Sokolská 581, 500 05 Hradec Králové, Czech Republic*

2.1. Mathematical model

In the constitutive modelling, the sample is considered to be a hyperelastic, incompressible, thin-walled cylindrical tube. By accepting these assumptions, it is possible to express mathematical relations between the applied loads (pressure and axial force) and the deformations observed in the experiment.

If shear deformations are neglected, as is the case in this study, the deformation of the sample can be described by the deformation gradient $\mathbf{F}$ as

$$\mathbf{F} = \begin{bmatrix} \lambda_r & 0 & 0 \\ 0 & \lambda_\theta & 0 \\ 0 & 0 & \lambda_z \end{bmatrix} \tag{1}$$

where λ_r, λ_θ and λ_z are radial, circumferential and longitudinal stretches of the sample, respectively. λ_θ and λ_z are known from the experimental data. λ_r can be expressed from the assumption of incompressibility as $\lambda_r = 1/(\lambda_\theta \cdot \lambda_z)$.

The stress-strain relationship in an incompressible hyperelastic material model is defined as

$$\sigma = \frac{\partial W}{\partial \mathbf{F}} \mathbf{F}^T - p\mathbf{I} \tag{2}$$

where W is the elastic potential characterizing the specific material model, $\mathbf{I}$ denotes second-order identity tensor, and p is an indetermined multiplier, which has to be determined from the boundary conditions. On the basis of the equilibrium equations in the radial and axial directions, one can express the mathematical relationships between applied loads and deformations of the sample as read in Eq. (3). For a detailed description of the mathematical operations leading to the resulting equations, see, for example, [1].

$$P = \frac{H}{R_i} \frac{1}{\lambda\theta\lambda_z} \frac{\partial W}{\partial \lambda_\theta}, \qquad F_{RED} = \pi H R_i \left(2\frac{\partial W}{\partial \lambda_z} - \frac{\lambda_\theta}{\lambda_z} \frac{\partial W}{\partial \lambda_\theta} \right) \tag{3}$$

Here, P is the internal pressure, F_{RED} denotes the axial force, H and R_i are the reference thickness and inner radius of the sample. $\hat{W}$ denotes the elastic potential formulated as a function of circumferential and longitudinal stretches $\hat{W} = W(\lambda_\theta, \lambda_z)$, which is allowed by adopting the condition of incompressibility.

2.2. Material Model

The hyperelastic model used in this study is based on the HGO model [2] and combines two of its modified forms [3,4]. The specific form of the elastic potential is adopted as follows.

$$W = \frac{c_0}{2}(I_1 - 3) + \sum_{i=1}^{4} \frac{c_1^i}{4c_2^i} \left(e^{c_2^i (\kappa^j I_1 + (1-3\kappa^i) I_4 - 1)^2} - 1 \right) \tag{4}$$

The first term is a classical neo-Hookean model. c_0 is a stress-like material parameter and I_1 is the first principal invariant of the right Cauchy-Green strain tensor. The second term assumes the existence of four preferred directions within the venous wall, which cause the material to be anisotropic. c_1^i is a stress-like parameter, c_2^i and κ^i are dimensionless parameters of the i-th direction. I_4^i is an invariant defined as $I_4^i = \lambda_\theta^2 \cos^2(\beta^i) + \lambda_z^2 \sin^2(\beta^i)$. β^i is a parameter that represents the angle between the circumferential direction of the sample and the i-th preferred direction.

The β and κ parameters are based on the general structural description of blood vessel walls [3]. In the arterial and venous walls, collagen fibre bundles are responsible for the vessels' anisotropy. Therefore, β may be interpreted as the angle between the bundles and the circumferential direction of the vessel. The κ parameter was proposed in order to incorporate the fact that the collagen bundles are not oriented uniformly but rather are dispersed along the direction specified by β. Note that if $\kappa^i = 0$, the isotropic part of the exponent becomes equal to 0 and only the anisotropic part linked to I_4^i remains. This would mean that all collagen fibres are oriented in the preferred direction. On the other hand, if $\kappa^i = 1/3$ then the anisotropic invariant I_4^i no longer contributes to the strain energy density function. This would mean that the collagen fibre bundles are oriented evenly in all directions and the material becomes isotropic. However, although the concept of β and κ originates from the general architecture of blood vessel walls, the parameter values identified are not necessarily consistent with the histology of the specific vessel. Rather, β and κ should be regarded as mere constitutive parameters used to incorporate anisotropy into the model.

In this study, several restrictions are applied to the values of the constitutive parameters. In accordance with [4], it is assumed that the first two preferred directions are oriented circumferentially and axially, respectively, e.g.: $\beta^1 = 0$, $\beta^2 = 90°$. The third and fourth directions are of unknown orientation, but they are assumed to be symmetric with respect to the vein's axis, e.g.: $\beta^3 = -\beta^4 = \beta \in \langle 0; 90° \rangle$. Also, the remaining parameters of the third and fourth direction are constrained to be of the same value, e.g.:$c_1^4 = c_1^3$, $c_2^4 = c_2^3$, $\kappa^4 = \kappa^3$. Furthermore, it is assumed that $\kappa^1 = \kappa^2 = 0$, which leaves only a single κ parameter: $\kappa^3 = \kappa^4 = \kappa \in \langle 0; 1/3 \rangle$. Upon applying all the restrictions, there are a total of 9 constitutive model parameters: c_0, $c_1^{1,2,3}$, $c_2^{1,2,3}$, β, κ.

2.3. *Parameter Identification*

An optimization algorithm formulated as a minimization of a loss function was used to identify the values of the constitutive parameters. Eq. (3) allows one to express model pressure P and axial force F_{RED} corresponding to a given deformation state. Therefore, it is possible to define the loss function as a weighted square error function of the loads as measured in the experiment and predicted by the model:

$$\sum_{i=1}^{m} \left(\frac{1}{\overline{P}_{\text{exp}}^2} \cdot \left(P_{\text{exp}}^i - P_{\text{mod}}^i\right)^2 + w_F \frac{1}{\overline{F}_{RED\text{exp}}^2} \cdot \left(F_{RED\text{exp}}^i - F_{RED\text{mod}}^i\right)^2 \right) \tag{5}$$

Here, exp and mod subscripts denote experimental and model-predicted loads, respectively, overbar denotes mean value. w_F is a weight parameter chosen for the force contribution to the overall error function. The summation is done over m deformation states measured in the experiment. The loss function was minimized using genetic algorithms available in the Maple 2021 computing environment to optimize the 9 constitutive parameters of the model.

3. Results

The optimal constitutive parameter values for the sample were identified as shown in Table 1. The corresponding pressure-stretch relations of the model can be seen in Fig. 1.

Table 1. Material parameters of the human great saphenous vein.

c_0 [kPa]	c_1^1 [kPa]	c_2^1	c_1^2 [kPa]	c_2^2	c_1^3 [kPa]	c_2^3	β [°]	κ [-]
0	38.87	100.09	469.37	13.08	3000	259.54	28.2	0.27

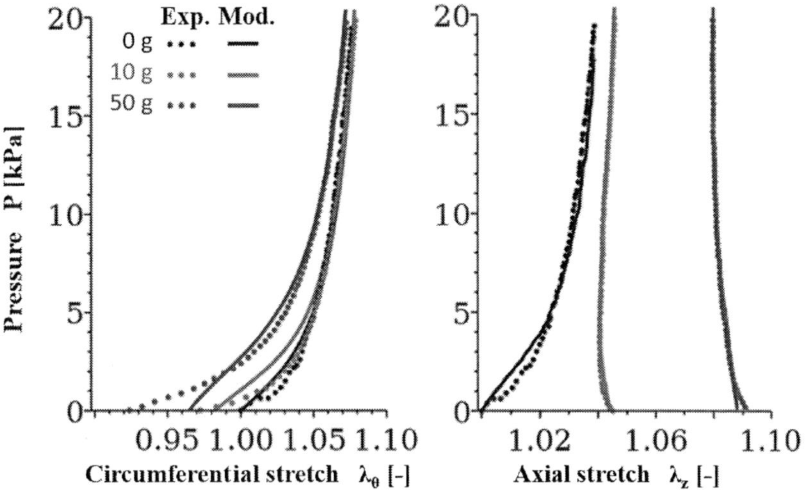

Fig. 1. Experimental data (dots) and model predictions (lines) of the pressure-stretch behaviour of the vena saphena magna with three different suspended weights (0, 10, 50 g).

To measure the goodness of fit of the model, the coefficients of determination R^2 and standardized root-mean-square errors were used. Root-mean-square errors were defined per degree of freedom of the model as follows:

$$RMSE(P) = \frac{\sqrt{\frac{\sum_{i=1}^{m}(P^i_{\text{exp}} - P^i_{\text{mod}})}{m-9}}}{\overline{P}_{\text{exp}}}, \quad RMSE(F_{RED}) = \frac{\sqrt{\frac{\sum_{i=1}^{m}(F^i_{RED\text{exp}} - F^i_{RED\text{mod}})}{m-9}}}{\overline{F}_{\text{REDexp}}} \tag{6}$$

The resulting error measures are shown in Table 2:

Table 2. Error measures of the model.

$R^2(P)$	$R^2(F_{RED})$	$RMSE(P)$	$RMSE(F_{RED})$
0.97	0.93	0.1	0.28

4. Conclusions

The constitutive parameter values of the human great saphenous vein were determined by fitting a hyperelastic model to data from an inflation-extension experiment. The model was a combination of the GOH and Baek constitutive models adopted from [3, 4]. Surprisingly, neither of the original models were able to describe the behaviour of the sample satisfactorily. There were 9 constitutive parameters defined in the strain energy density function of the model (c_0, $c_1^{1,2,3}$, $c_2^{1,2,3}$, β, κ). The parameter values were determined by adopting an optimization method based on the minimization of an error function.

The model was able to describe the mechanical response of the saphenous vein when subjected to internal pressure and axial forces. The experimental loading conditions were similar to those expected in the arterial part of the human circulatory system. Therefore, the model should be suitable for simulations concerning the initial mechano-biology of a vascular graft exposed to an arterial environment.

Acknowledgements

The Czech Science Foundation has supported this study via project GA18-26041S entitled "Effect of axial prestretch on mechanical response of nonlinearly elastic and viscoelastic tubes".

References

[1] Horný L. and Petřivý Z., "Inversion point and internal volume of pressurized nonlinearly elastic tube," *International Journal of Non-Linear Mechanics*, Vol. 125, p. 103530 (2020). https://doi.org/10.1016/j.ijnonlinmec.2020.103530.

[2] Holzapfel G.A., Gasser T.C. and Ogden R.W., "A new constitutive framework for arterial wall mechanics and a comparative study of material models," *Journal of elasticity and the physical science of solids*, Vol. 61, pp. 1–48 (2000). https://doi.org/10.1023/A:1010835316564

[3] Gasser T.C., Ogden R.W. and Holzapfel G.A., !Hyperelastic modelling of arterial layers with distributed collagen fibre orientations," *Journal of the royal society interface*, Vol. 3.6, pp. 15–35 (2006). https://doi.org/10.1098/rsif.2005.0073.

[4] Baek S. et al., "Theory of small on large: potential utility in computations of fluid–solid interactions in arteries," *Computer methods in applied mechanics and engineering*, Vol. 196.31-32, pp. 3070–3078 (2007). https://doi.org/10.1016/j.cma.2006.06.018.

Artificial Intelligence in Material Testing with Digital Image Correlation

Bořek Ščerba[1], Tomáš Návrat[2], Michal Vajdák[3] & Jan Čepela[4]

Abstract: Artificial intelligence improves user experience and work efficiency in digital image correlation (DIC) systems by automating certain tasks, thus improving the competitive ability of local DIC system developers in the global market. The main objective is to detect the type of sample used for material testing, choose the proper measurement tool in the DIC system, and find the correct placement of the tool in relation to the sample. This leads to solving problems of computer vision and image processing, such as object localization and classification, using convolutional neural networks. The open-source PyTorch framework is used for machine learning activities. Tasks like data selection and labeling, model selection, training and validation, or hyperparameter optimization are dealt with. As a result, about 95% of the images were detected successfully.

Keywords: Artificial intelligence; Digital image correlation; YOLOv5; DCNN; PyTorch; Machine learning; Deep convolutional neural network

1. Introduction

Digital image correlation (DIC) "is an optically-based technique used to measure the evolving full-field 2D or 3D coordinates on the surface of a test piece throughout a mechanical test" [1]. Since the early 1980s, when the method was invented, DIC algorithms have made incredible progress and have become one of the most powerful non-destructive surface deformation measurement techniques in wide variety of experimental mechanics applications [2]. Compared to conventional contact displacement or strain gauges, which are suitable for small and rather uniform deformations, digital image correlation offers a full-field approach better suited to describe more common heterogonous deformations. Confronted with other full-field optical methods for deformation measurements, for example, holographic interference, electric speckle pattern or speckle photography, DIC offers advantages such as relatively simple preparation of the measuring system, lower requirements to the experimental environment enabling easy outside-of-the-laboratory testing, it does not require a laser light source, it offers a high degree of automation in measurement and data processing, and it can be coupled with various magnifying instruments or other devices [2]. The technique is also independent of the tested material and the scale of testing [1].

Recently, the equipment needed to acquire quality images for DIC has become widely available in terms of pricing. Also, with DIC algorithms presenting increased accuracy by subpixel interpolation [3], relatively low resolution (e.g. 5 megapixels (MP)) cameras are sufficient for precise measurements offering a cost-effective solution. Together with many open-source codes

[1] *Bořek Ščerba; Brno University of Technology, Faculty of Mechanical Engineering, Institute of Solid Mechanics, Mechatronics and Biomechanics; Technická 2896/2, Královo Pole, 616 69 Brno, Czech Republic; E-mail:* Borek.Scerba@vutbr.cz

[2] *Tomáš Návrat; Brno University of Technology, Faculty of Mechanical Engineering, Institute of Solid Mechanics, Mechatronics and Biomechanics; Technická 2896/2, Královo Pole, 616 69 Brno, Czech Republic; E-mail:* navrat@fme.vutbr.cz

[3] *Michal Vajdák; X-Sight s.r.o., Staňkova 557/18a, 602 00 Brno, Czech Republic; E-mail:* michal.vajdak@xsight.eu

[4] *Jan Čepela; X-Sight s.r.o., Staňkova 557/18a, 602 00 Brno, Czech Republic; E-mail:* jan.cepela@xsight.eu

(AL-DIC, Ncorr, PReDIC) or commercial software (VIC from Correlated solutions, GOM from Zeiss, and Alpha from Czech developer X-Sight), it enables rapid spreading in the industry and scientific community.

As DIC measuring systems become more popular, users are seeking intuitive and easy-to-use environments. The main objective of this article is consistent with the current industry 4.0 and aims to increase the competitive ability of local DIC system developers X-Sight in the global market by automating part of standard laboratory workflows. Such automation includes detection of the sample type used for material testing, suggesting proper software tool for 2D measurement in the DIC system Alpha and placing it into proper position in the image.

2. Methods

Considering tasks on hand that include the recognition of a sample class and the assessment of its position, in machine learning called classification and localization, it leads to their combination called object detection, which falls under computer vision problems [4].

The success of traditional methods for object detection (e.g. deformable part-based machines – DPMs) depends on precise definition of features, which describe the object to be recognized. This manual approach could not capture complex structures embedded within the image. Amongst the most notable limitations of these detectors were: Many redundant false positive detections, detection windows did not match objects well, the detection pipeline could not be optimized as one system [5].

Great accomplishments in object detection were achieved by applying deep convolutional neural networks (DCNN), which are nature-inspired structures and processes that resemble neurons connected by synapses in the brain. Among other attributes, DCNN includes a technique called feature learning, where feature maps are extracted by convolutional filters (or kernels), which are optimized during network training to properly describe the object in question [6].

2.1. Selecting deep convolutional neural network for object detection

Generally, they can be divided into two groups: a) two-stage detectors, for instance, Region-based CNN (R-CNN) and its modifications, and b) one-stage detectors, for example, YOLO (You Only Look Once) and its variations [5]. "Two-stage detectors first use a proposal generator to generate a sparse set of proposals and extract features from each proposal, followed by region classifiers which predict the category of the proposed region. One-stage detectors directly make categorical prediction of objects on each location of the feature maps without the cascaded region classification step. Two-stage detectors commonly achieve better detection performance and report state-of-the-art results on public benchmarks, while one-stage detectors are significantly more time-efficient and have greater applicability to real-time object detection" [5].

In our case, real-time detection is demanded. Considered objects are not so complicated and in such a complex environment as in public object detection benchmarks, thus the performance of one-stage detector will be sufficient, and YOLO is chosen to be used for this purpose. Furthermore, detectors for a small number of classes can be highly optimized [7].

Since 2015, when Redmon et al. presented the first version of YOLO [7], four versions of YOLO have been presented: YOLOv2, YOLOv3, YOLOv4 and YOLOv5. These versions progressively improve YOLO weak areas compared to two-stage object detectors: Accuracy and detection of small objects; with further improvement in speed [8]. From these available versions, YOLOv5 was chosen to use here, as it contains all improvements from previous versions with further enhanced performance, it offers more model sizes, and it implements the PyTorch framework, which is more user-friendly and easier to put into production than previous versions working with Darknet framework [8].

2.2. *On the YOLOv5*

YOLOv5 is composed from Backbone, Neck, and Head, similarly as other modern object detection models. It uses the CSPDarknet53 backbone, which is fed by images of specific size (usually 640×640 pixels) and which is responsible for creating feature maps. The PANet (path aggregation network) neck consisting of several bottom-up and top-down layers improves information flow through the model and helps feature propagation, thus improving localization accuracy. Finally, the anchor-based head that generates predictions [9]. Fig. 1 shows the architecture of the model.

The CSP (Cross Stage Partial) backbone consists of Convolutional (Conv), Spatial Pyramid Pooling (SPP – later SPPF as a faster version [10]) and C3 blocks. Convolutional blocks apart from convolution also perform batch normalization and includes activation function (SiLU – Sigmoid weighted Linear Unit). SPP(F) processes features using pooling layers of different sizes in parallel and then concatenates (Concat) them. C3 blocks are combination of three convolutional operations [11] combined with several bottleneck layers. The number of bottleneck layers together with the number of convolutional filters per layer affect the model size resulting in different variations of the model: YOLOv5n, YOLOv5s, YOLOv5m, etc. The YOLOv5s was chosen as the final model file has a reasonable size, but it still has more than 10% of the Alfa installation file size.

The Focus structure of the backbone was later replaced by convolutional layer, which turned out to be more efficient [10]. Upsampling blocks in the architecture ensures correct data dimensions when moving bottom-up in the model.

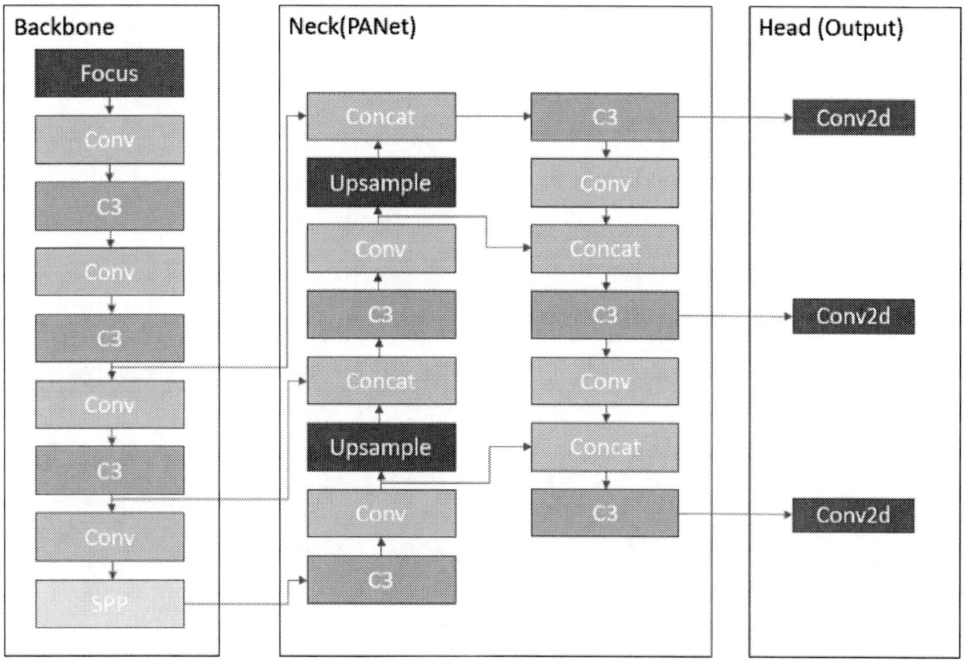

Fig. 1. YOLOv5 architecture [9].

The Head consists of 3 convolutional layers (Conv2d), each one predicting object of different size (scale). The top layer uses the grid 80×80, having the highest resolution and thus being suitable for detecting small objects. The middle layer with a 40×40 grid is suitable for inference of medium objects, and bottom layer with 20×20 grid detects large objects. The output from the model is vector containing information about predefined number of bounding boxes, for each according to the equation (4).

YOLOv5 uses for training the loss function which consists of three parts: a) classification loss using Binary Cross Entropy (BCE) [12]; b) objectness loss which characterizes the confidence of object presence also uses BCE; c) bounding box regression loss using Complete Intersection over Union loss (CIoU) [13], which evaluates overlap, central points distance and aspect ratio between training bounding box (anchor) and ground truth bounding box (defined by label – see the equation 4 in subsection 2.4 named Data selection and labeling). ADAM method for stochastic optimization [14] is used to minimize the loss function.

Hyperparameters in machine learning control various aspects of training. Optimal values can significantly enhance the model performance, but optimization is generally time-consuming, and their finding is not guaranteed. YOLOv5 has about 30 hyperparameters for various training settings and implements genetic algorithm as the method of their retrieval [15]. These hyperparameters modify ADAM optimizer (e.g. learning rate), they scale and balance loss functions or control aspects of data augmentation (described later).

2.3. Performance metrics

To better understand results of the model, let us summarize YOLOv5 metrics. There are several basic types of detections [16]:

- True positive (TP) – correct detection of ground truth (label) bounding box
- False positive (FP) – incorrect detection, either of nonexistent object or misplace detection of existing object
- False negative (FN) – undetected ground truth bounding box

Correct and incorrect detections are evaluated using IOU (Intersection Over Union) of the detected bounding box and ground truth box. The name already describes the calculation formula:

$$IOU = \frac{area\ of\ overlap}{area\ of\ union} \tag{1}$$

When the value of IOU exceeds defined threshold value t, the detection is considered correct, otherwise it is considered incorrect [16]. Then, basic performance metrics Precision (P) and Recall (R) can be defined by the following statements [16]:

$$P = \frac{TP}{TP + FP} = \frac{correct\ detections}{all\ detections}, \tag{2}$$

$$R = \frac{TP}{TP + FN} = \frac{correct\ detections}{all\ ground\ truths}. \tag{3}$$

This is often used to form the PR curve stated as $f(R) = P$ for certain threshold value of IOU in the following manner: All the predictions are sorted by confidence in descending order. Precision and Recall values are evaluated progressively from top to bottom of the list. Recall values are increasing monotonically, and precision values are overall decreasing. After interpolation to remove possible zigzag behaviour, the area under the curve (AUC) is calculated, which is called average precision (AP) [16]. It is usually reported together with the IOU threshold value. One of the most common metrics is the mean average precision (mAP) calculated for all classes using the 0.5 IOU threshold value. It is denoted as $mAP@.5$. Average value from evaluation using different IOU threshold values ranging from 0.5 to 0.95 with the 0.05 step is usually denoted as $mAP@.5 : .95$ [16].

2.4. Data selection and labeling

The image database, which was available for the task, contained a variety of different standardized and other tests. For example, tensile, compression, three- or four-point bending, shear, torsion, fracture, or crack growth tests, where materials such as steel, copper, and aluminum alloys, concrete, carbon fiber composites, polymers, textiles, or rubbers were used. Standard tensile specimens with flat or round cross sections, smooth bars, reinforcing bars (rebar), or bolts were present among other nonstandard samples.

For good results in machine learning, there must be a sufficiently large dataset to reliably learn typical features of classes in question and to make accurate predictions. For YOLOv5, it is recommended to have at least 1500 images per class with adequate variety, which should represent the environment and different conditions of the scene [17] (in this case different lighting conditions, sample positions, surface textures, or image qualities).

Software tools to be automatically matched with proper sample are: (i) Extreme or Trans Line used for tensile or compression tests, (ii) ROD Line used for rebar testing, (iii) Bend Line used for three- or four-point bending tests, (iv) Torsion Line used for torsion tests, and (v) Crack Probe for fracture or crack propagation tests.

Above mentioned enabled creation of following classes suitable for machine learning: a) standard tensile samples regardless of the cross section (named for the purpose of labeling SAMPLE) to be matched with Extreme or Trans Line and b) reinforced bars (named REBAR) to be matched with ROD Line. Other classes that would match the rest of tools were not possible to use mainly due to insufficient size of the dataset.

Each one of selected classes contains significant features. For the SAMPLE class, it is a typical radius between its shoulders and gauge, causing its distinct shape, often called "dog bone". For the REBAR class, these are ribs on its surface.

In total, 75 900 images from 33 different measurements were available for the REBAR class and 85 000 images from 54 different measurements were accessible for SAMPLE class. Final dataset for machine learning was uniformly chosen from it in the count of 2850 images per class. These were divided into training, validation and testing set in the 70/20/10 ratio. It helps prevent overfitting, meaning the situation in which the machine learning model 'memorizes' training data and loses the ability to generalize on previously unseen data. The validation dataset gives some sense on how the model is performing on the unseen data and indicates when the model hits the best performance. The test dataset is used to evaluate the model after training to obtain an unbiased view of model performance [18].

YOLOv5 also offers a novel tool to fight overfitting called data augmentation, where the input for machine learning is formed into so-called mosaics, created randomly from training images using operations like rotation, flipping, scaling, shearing, translating, and color space adjustments by altering the hue, saturation, or intensity [11] and thus increasing the variability of the dataset.

Another necessary step in data preparation for machine learning is labeling. The label is an output from machine learning model, which must be identified beforehand, usually by the user. In the task at hand, the label is bounding box (*bb*), which can be defined as the following vector

$$bb = (x, y, w, h, c), \tag{4}$$

where (x, y) are coordinates of the bounding box center, (w, h) are its width and height, and c is the class of the object within.

Machine learning labels must be created consistently and accurately [19] to achieve good results. Consistency means that obejcts of all classes in all images must be considered. Accuracy means that the object must be enclosed by the bounding box as precisely as possible. Microsoft VoTT and Roboflow tools available online were used for labels definition.

For the REBAR class, the bounding box was selected so that it defines part of the sample between grips (Fig. 2 top). For the SAMPLE class, the bounding box is defined so that it covers only the gage length of the sample (Fig. 2 bottom).

Fig. 2. REBAR class member with defined bounding box (top); SAMPLE class member with defined bounding box (bottom).

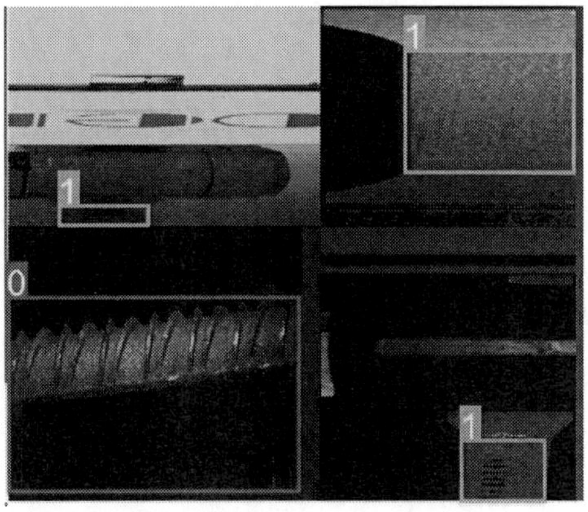

Fig. 3. Mosaic – training image after data augmentation.

2.5. Model training

Before training, all labeled images were resized to the 640×640 pixels format. Then, data augmentation was performed to prevent overfitting of the model. Fig. 3 depicts resulting mosaic, which is fed in the batches of 5 into the training engine.

The training of the YOLOv5s [17] model consisted of two runs. Each run consisted of hyperparameter optimization and training in this order (Fig. 4).

For the first run, pretrained model parameters and hyperparameters obtained from COCO dataset were used as initial guess for optimization. Parameters in first ten layers of the model (backbone), containing basic features, were frozen for the first run, and only the rest of the model was optimized. For the second run, all layers were unfrozen and the whole model was trained. Final set of parameters, in the figure 4 called PAR_2, was obtained.

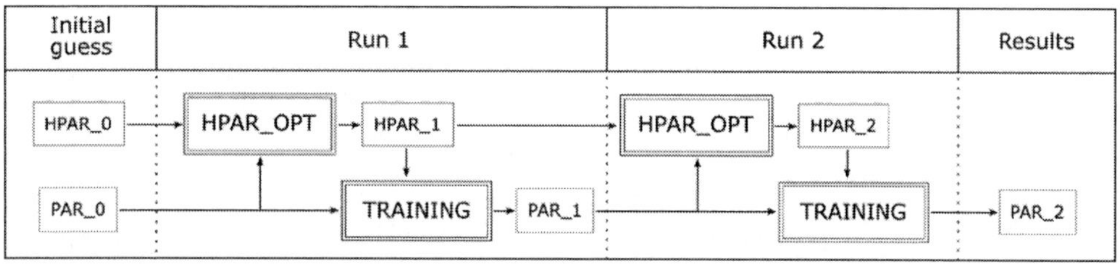

Fig. 4. Scheme of the training process; HPAR stands for hyperparameters, PAR stands for parameters, HPAR_OPT stands for hyperparameters optimization; Single frame marks parameters set, double frame marks process.

Hyperparameters were optimized using the genetic algorithm for 20 generations, each generation having 50 epochs of training. The fitness function to maximize by the optimization was a linear combination of $mAP@.5$ with weight 0.1 and $mAP@.5 : .95$ with weight 0.9.

The training was performed for 500 epochs with the patience option, which is meant to stop the training after 50 epochs without improvements in fitness function value. Fitness function was the same as for hyperparameters optimization. Its development during both training runs is shown in the Fig. 5.

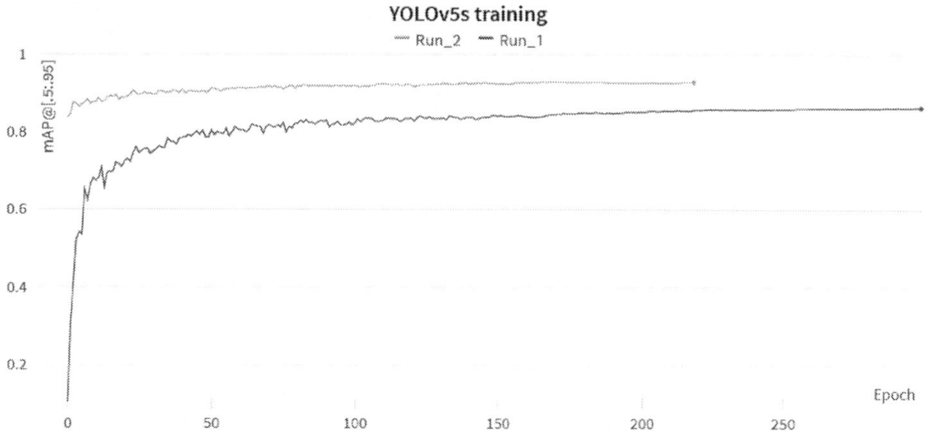

Fig. 5. Fitness function development during training runs.

3. Results and Discussion

The performance of the trained model was evaluated using the test dataset. The inference took 15.6 ms in average per image on the NVIDIA GeForce GTX 1650, 4096 MiB graphics processing unit. Table 1 contains a summary of the model inference on the testing dataset.

Table 1. Summary of results.

Class	Precision [–]	Recall [–]	$mAP@.5$ [-]	$mAP@.5:.95$
All	0.989	0.953	0.973	0.935
REBAR	0.989	0.989	0.994	0.959
SAMPLE	0.989	0.918	0.952	0.912

In the Fig. 6, there are few examples of good inference results from testing dataset (high confidence, correct class, and location).

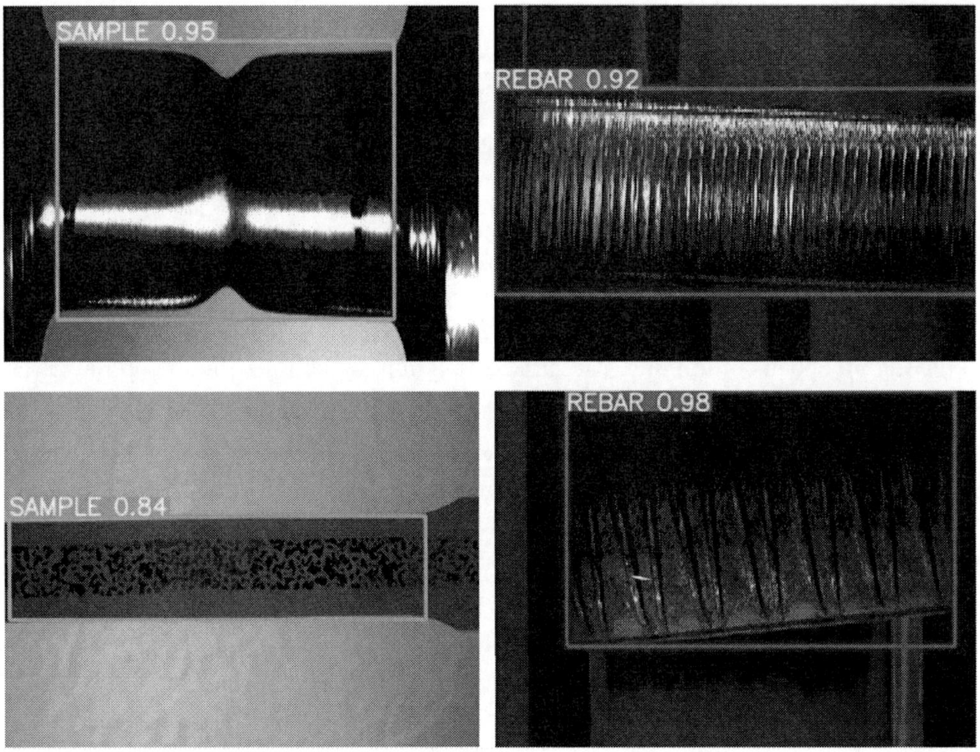

Fig. 6. Good inference results – examples.

It is obvious from results, that the smaller object in the image the worse inference results. It leads to lower confidence of the inference (Fig. 7 top), undetected object (Fig. 7 middle) or wrong detection (Fig. 7 bottom).

Although it is preferable to cover as many pixels as possible for the DIC measurement to achieve the best correlation results and detected objects tend to appear large in the image, there are cases when the object in the image seems rather small, as shown in Fig. 7. It happens most often when measuring samples with high strain values (e.g. from hyperelastic material). Such cases are present in the SAMPLE class causing its lower Recall value.

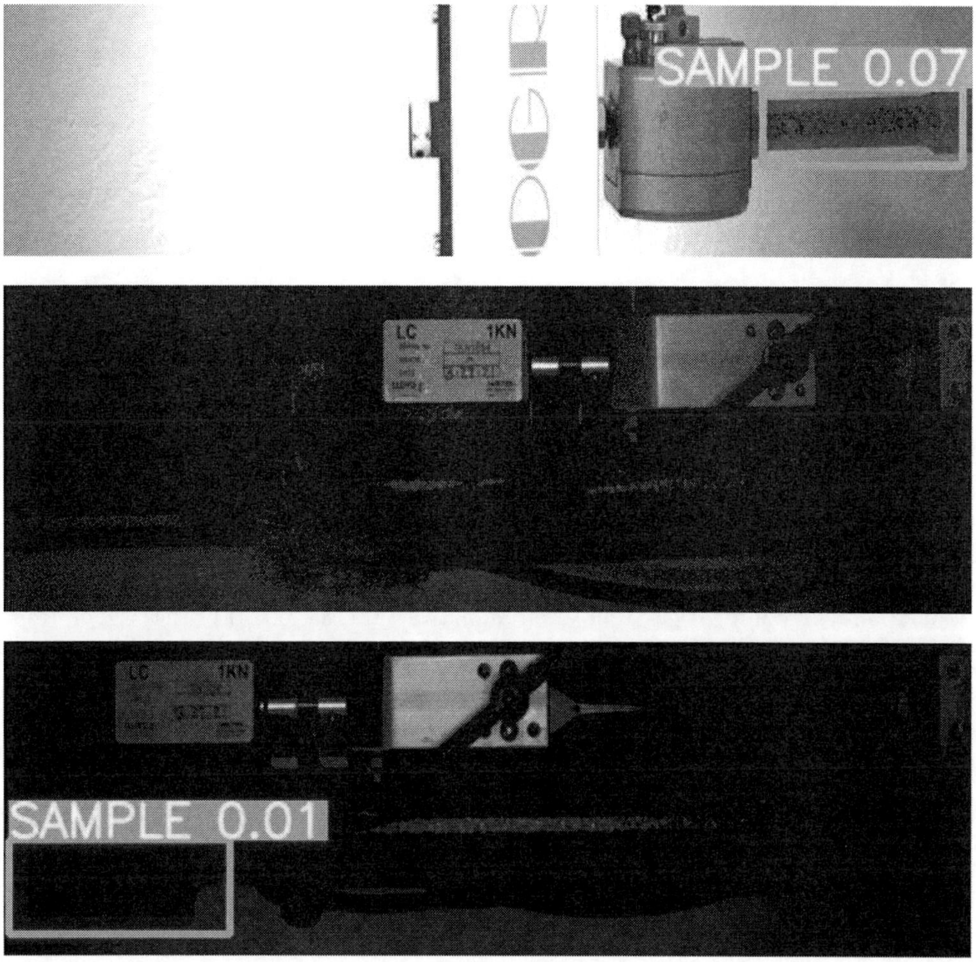

Fig. 7. Examples of small object inference; from top to bottom: small inference confidence, wrong detection, undetected object.

In the REBAR class, drop in the Precision and Recall values was produced by class misinterpretation due to the excessive input image dimension change resulting in hardly distinguishable ribs (the main feature of this class). It is demonstrated in the Fig. 8, where in the left part, inference is not correct, but in the right part, after some deformation occurred in the test, the distance between ribs increased enabling correct detection.

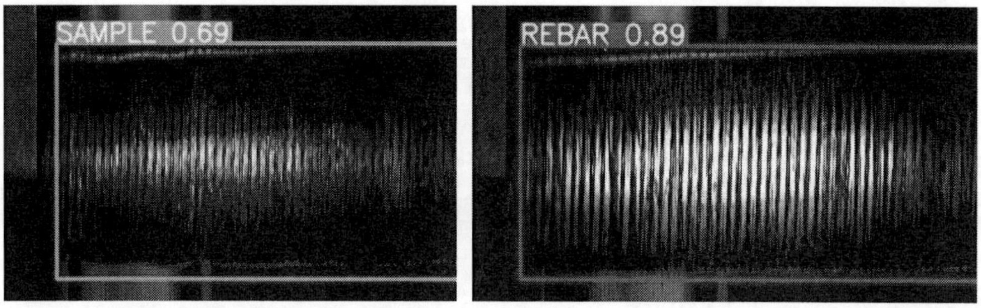

Fig. 8. Left – incorrect inference of the REBAR class object; right – correct inference after some deformation of the bar.

The detection of small objects could be improved by adding a detection layer to the head of the model with the 160×160 inference grid to the current layers with 20, 40 and 80 grid sizes. Together with increased input image dimensions to 1280×1280 pixels, it could also improve the REBAR class problem shown in the Fig. 8. On the other hand, it would be at the cost of enlarging the model, which is not desired, leading to the need of making compromises.

4. Conclusions

Selected model shows excellent results for this application, although it offers room for improvement in the future research, especially in the area of small object detection that occurs e.g. in experiments with samples from hyperelastic material.

Also, in case of dataset enlargement and the need/possibility of adding more classes to be recognized by the model, it would not be complicated to implement these classes and retrain the model.

Acknowledgements

The cooperation with company X-Sight s.r.o., Staňkova 557/18a, 602 00 Brno, Czech Republic; namely Jan Čepela, Petr Stolař and Petr Gajdoš, who provided consultations and expertise in the field of digital image correlation, is gratefully acknowledged. They also provided a database full of various measurements to be used for machine learning. This work was also supported by Technology Agency of the Czech Republic through the project Optical analysis of deformations using artificial intelligence (FW03010653).

References

[1] Jones E.M.C. and Iadicola M.A., "A good practices guide for digital image correlation," *Available from* https://doi.org/10.32720/idics/gpg.ed1 *Accessed:* 2022-04-13.

[2] Zhao J., Sang Y., and Duan F., "The state of the art of two-dimensional digital image correlation computational method," *Available from* https://doi.org/10.1002/eng2.12038 *Accessed:* 2022-04-13.

[3] Zhou J., Liu D., Shao L., and Wang Z., "Application of Digital Image Correlation to Measurement of Packaging Material Mechanical Properties," *Available from* https://doi.org/10.1155/2013/204875 *Accessed:* 2022-05-21.

[4] Wu Y., et al., "Rethinking classification and localization for object detection," *Available from* https://doi.org/10.48550/arXiv.1904.06493 *Accessed:* 2022-05-21.

[5] Wu X., Sahoo D. and Hoi S.C.H., "Recent advances in deep learning for object detection," *Available from* https://doi.org/10.1016/j.neucom.2020.01.085 *Accessed:* 2022-04-15.

[6] Chen M., et al., "Deep Feature Learning for Medical Image Analysis with Convolutional Autoencoder Neural Network," *Available from* http://doi.org/10.1109/TBDATA.2017.2717439 *Accessed:* 2022-05-21.

[7] Redmon J., et al., "You only look once: Unified, real-time object detection," *Available from* https://doi.org/10.48550/arXiv.1506.02640 *Accessed:* 2022-05-21.

[8] Jiang P., Ergu D., Liu F., Cai Y., and Ma B., "A Review of Yolo Algorithm Developments," *Available from* https://doi.org/10.1016/j.procs.2022.01.135 *Accessed:* 2022-04-15.

[9] Nepal U., and Eslamiat H., "Comparing YOLOv3, YOLOv4 and YOLOv5 for Autonomous Landing Spot Detection in Faulty UAVs," *Available from* https://doi.org/10.3390/s22020464 *Accessed:* 2022-05-23.

[10] Zhe W., "YOLOv5 (6.0/6.1) brief summary," *Available from* https://github.com/ultralytics/yolov5/issues/6998 *Accessed:* 2022-05-25.

[11] Dadboud F., et al., "Single-Stage UAV Detection and Classification with YOLOV5: Mosaic Data Augmentation and PANet," *Available from* http://doi.org/10.1109/AVSS52988.2021.9663841 *Accessed:* 2022-05-25.

[12] Torch Contributors, "BCELOSS," *Available from* https://pytorch.org/docs/stable/generated/torch.nn.BCELoss.html *Accessed:* 2022-05-25.

[13] Arie L., "The practical guide for Object Detection with YOLOv5 algorithm," *Available from* https://towardsdatascience.com/the-practical-guide-for-object-detection-with-yolov5-algorithm-74c04aac4843 *Accessed:* 2022-05-25.

[14] Kingma D., and Ba J., "Adam: A Method for Stochastic Optimization," *Available from* https://doi.org/10.48550/arXiv.1412.6980 *Accessed:* 2022-05-25.

[15] Jocher G., "Hyperparameter Evolution," *Available from* https://github.com/ultralytics/yolov5/issues/607 *Accessed:* 2022-05-25.

[16] Padilla R., Netto S., and Silva E., "A Survey on Performance Metrics for Object-Detection Algorithms," *Available from* http://doi.org/10.1109/IWSSIP48289.2020.9145130 *Accessed:* 2022-05-26.

[17] Jocher G. et al., "ultralytics/yolov5: v6.1 - TensorRT, TensorFlow Edge TPU and OpenVINO Export and Inference," *Available from* https://doi.org/10.5281/zenodo.6222936 *Accessed:* 2022-04-15.

[18] Solawetz J., "Train, Validation, Test Split for Machine Learning," *Available from* https://blog.roboflow.com/train-test-split/ *Accessed:* 2022-05-27.

[19] Jocher G.., "Tips for Best Training Results," *Available from* https://github.com/ultralytics/yolov5/wiki/Tips-for-Best-Training-Results *Accessed:* 2022-04-17.

Ply-by-Ply model Application in Analyses of Composite Structures by Finite Elements

Jakub Šedek[1], Jarmil Vlach[2] & Jan Horňas[3]

Abstract: The stress analysis of a composite structure is mostly carried out by finite element (FE) models utilizing shell elements. All composite layers are usually integrally included in element properties via homogenization and only the reference face is discretized in the model. Such process is computationally efficient, but if results at meso-scale are required, the FE model representing individual plies is necessary. The PbPModGen plug-in intended for ABAQUS/CAE created by the authors facilitates the preparation of multi-layered (ply-by-ply) FE models and enables more detailed analyses of impact damage, delamination or separation of individual layers. In this paper the PbPModGen plug-in is introduced and examples of multi-layered models application is shown.

Keywords: Composite; Finite elements; Ply-by-Ply

1. Introduction

Composite materials are used in primary structure of transport vehicles increasingly. Several component tests with damage or defect included are generally conducted to investigate the impact of an imperfection or a damage on the structural integrity. The tests shall be supported by computational analyses. A stress-strain response of a composite structure is mostly analysed by FE model utilizing the theory of shells. Nevertheless, if some local imperfections should be included into, or between layers, the FE model representing individual plies is necessary. The requirement to include the detail on meso-scale leads to building of multi-layered models [1–3].

The pre-processing of multi-layered FE model is time consuming task and precise systematic work with elements categorization into groups, sections and their interactions is fundamental. To simplify, speed up and improve the pre-processing of multi-layered model, the algorithms for model build are being applied. The ABAQUS is advanced product for implicit and explicit FE solutions and through CAE module offers the preprocessing of the model using graphical user interface (GUI) [4]. CAE is also suitable for the extension of its functionality and was therefore utilized for the semi-automatic generation of multi-layered or alternatively ply-by-ply FE models at the Aviation division of the Czech Aerospace Research Centre. The solution was released as the PbPModGen plug-in (***P**ly-**b**y-**P**ly **Mod**el **Gen**erator*) and is introduced in this paper together with its applications.

2. PbPModGen plug-in description

The PbPModGen plug-in is intended for ABAQUS/CAE. Plug-in facilitates the preparation of multi-layered FE models and enables more detailed analyses of impact damage, delamination or separation of individual layers. The basic function of the plug-in is to generate several layers

[1] *Ing. Jakub Šedek, Ph.D.; Czech Aerospace Research Centre; Beranových 130, 19905 Prague-Letňany, Czech Republic;* sedek@vzlu.cz
[2] *Ing. Jarmil Vlach; Czech Aerospace Research Centre; Beranových 130, 19905 Prague-Letňany, Czech Republic;* vlach@vzlu.cz
[3] *Ing. Jan Horňas; Czech Aerospace Research Centre; Beranových 130, 19905 Prague-Letňany, Czech Republic;* hornas@vzlu.cz

of continuum shell elements representing plies of a composite. A composite detail in the form of a sub-model can also be generated and connected to the simple model utilizing the sub-modelling capability of ABAQUS.

The GUI of the plug-in is divided into several thematic sections concerning modelling, material, mesh, interactions and job definition (see Fig. 1). Faces for modelling should be the first user's input to the plug-in. Faces are defined in imported model, or they can be constructed in CAE. They can be arbitrary in a shape and can also contain information about number of plies, its orientation or material in sections or composite layup property. If no associated information is available, then plies parameters can be specified in the plug-in. In the material section, only stacking sequence and elastic material property are supported, but it does not obstruct to edit advanced material definition in the CAE database after detailed model generation.

The Standard and Explicit solvers require different definition of the job and therefore the appropriate solver choice is required from the user. Elements representing plies can be connected by cohesive elements or cohesive contact interaction. The initial stiffness of the connection and damage initiation and evolution can be specified.

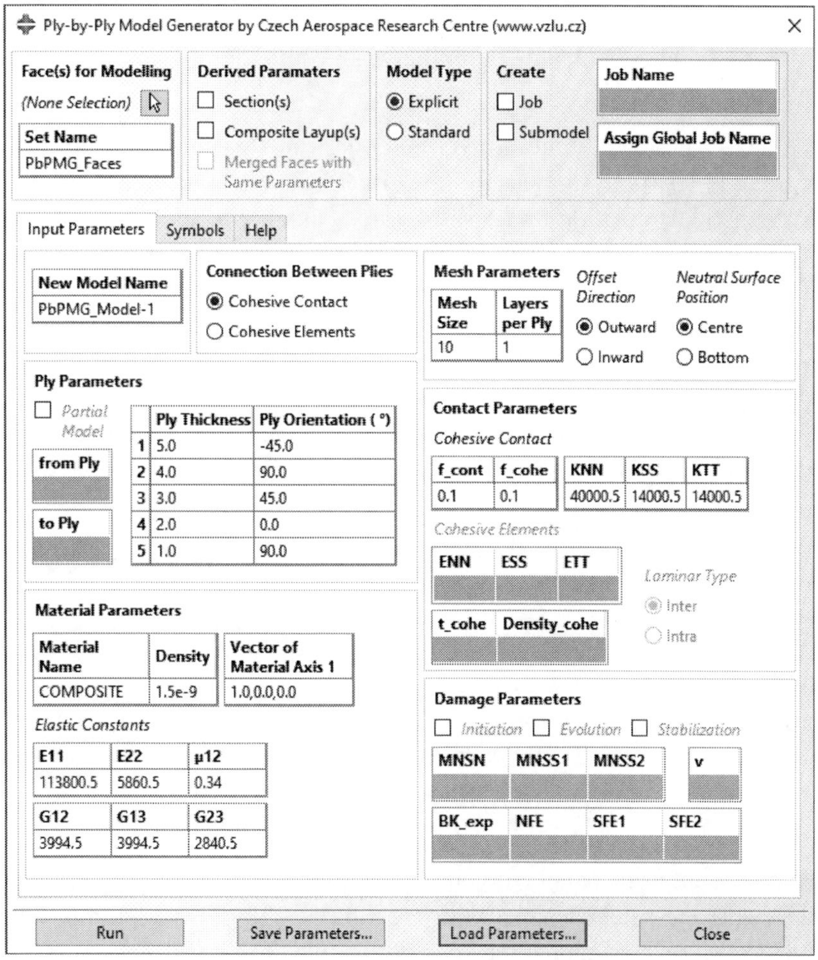

Fig. 1. PbPModGen plug-in GUI.

The great advantage of the plug-in is the capability to generate coincident mesh of several layers, although each layer is represented as independent part in assembly. It allows correct plies connection via interaction or inserting cohesive elements. The mesh is controlled by mesh size parameter and the number of elements through the ply thickness can be specified. The mesh pattern is based on the underlaying shell mesh of the base face. Finally, after setting of all input parameters, layers of elements are generated together with their properties and interactions.

3. PbPModGen plug-in applications

The plug-in simplifies the pre-processing task when multi-layered models are utilized. It covers the problems from adhesive connection of two isotropic plates up to detailed ply-by-ply analysis of the composite part. Some applications of the plug-in are presented below.

The bending of a composite causes high interlaminar stresses. The interlaminar strength can be reached and plies can delaminate while the collapse of the structure does not need to occur. The delamination of internal plies due to bending of a L-shaped composite specimen is shown in Fig. 2. The model represents 50 mm long, 10 mm wide and the 3 mm thick part consisting of 15 unidirectional carbon fibre reinforced plies. Only the internal layers are represented ply-by-ply, the rest is homogenized into side layers of elements from simplicity reason. The cohesive zone modelling is used between neighbouring layers so that delamination initiation and extension is allowed. Prior to ply failure, the delamination between centre plies occurs and the crack can be clearly visible as shown in Fig. 3.

Similarly, to above problem, a multi-layered model can also be utilized to model the interlaminar shear strength test used to measure plies interface resistance [5]. During the bending, the capability against delamination under shear forces parallel to the layers of the laminate is tested, but the delamination is not always clearly visible since the shear is dominant [6].

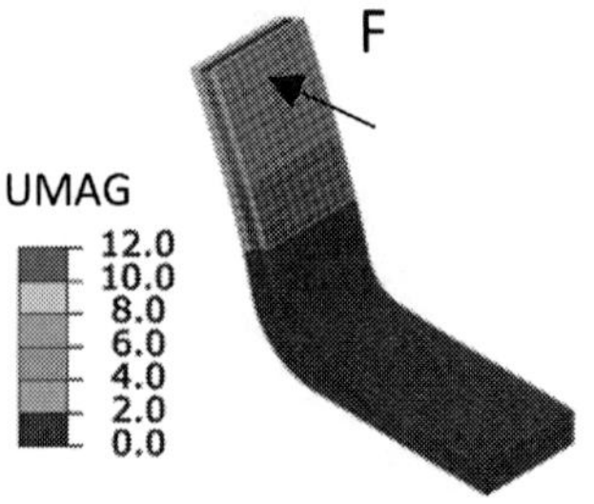

Fig. 2. Displacement of L-shaped specimen; Isometric view.

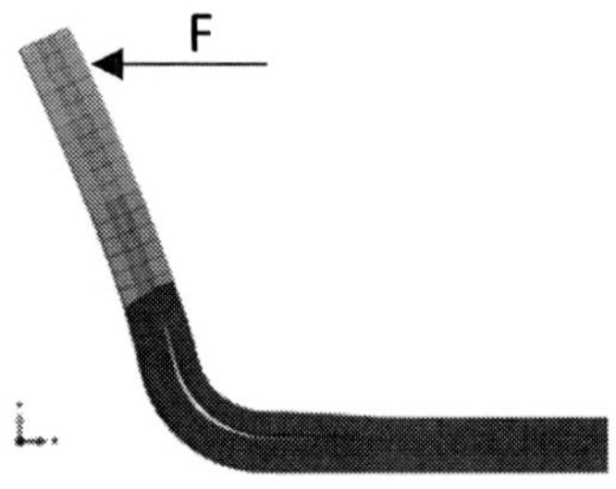

Fig. 3. Displacement and delamination of L-shaped specimen; Side view.

The analyses of damage caused by impact often require meso-scale modelling. Especially barely visible impact damages (BVID) cannot be simulated without detailed model of a composite, since only matrix damage and delamination often appear. The dependence of the damage severity on impactor energy can be observed [7].

The analysis of a composite panel with large hole impacted by half-sphere is shown in Fig. 4. The model represents the squared panel with the side of 230 mm and centre hole of 40 mm in diameter. Impacts were located near the edge of the hole. The resistance against the impact in relation to its energy was studied. The impact with energy of 15 J and 20 J resulted in structural damage, which can be detected only using special non-destructive testing (NDT) equipment, since the signs of this damage are invisible or are barely visible. On the contrary, the 50 J impact penetrated the composite and created visible impact damage, which is well detectable by observing a damaged region with a naked eye from both sides. Delamination induced by impact also occurred.

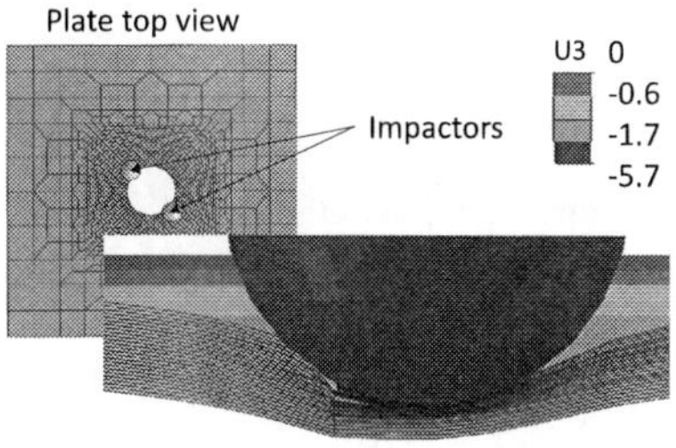

Fig. 4. Displacement during impact of the panel with large hole at reverse velocity point; Cross-section view.

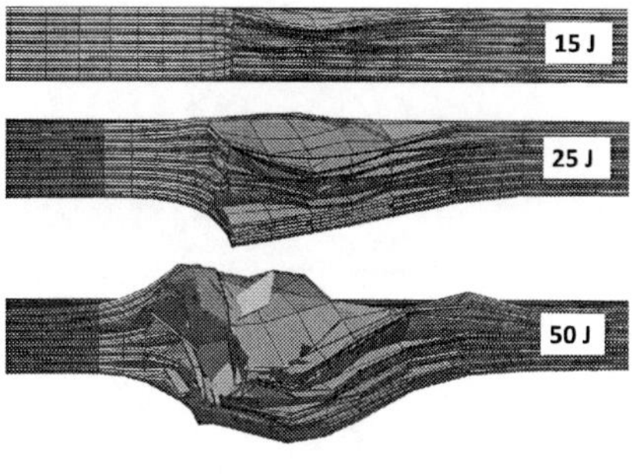

Fig. 5. Damage of the panel after impact at energies of 15 J, 25 J and 50 J; Cross-section view.

4. Conclusions

Multi-layered models are often used in meso-scale FE analyses of composites. Their pre-processing can be efficient using PbPModGen plug-in for ABAQUS/CAE. Such analysis brings more detailed view into stress-strain fields or failure processes resulting in plies damage and delamination.

Acknowledgements

These works were carried out from funds provided for the "Long-term development of the research organization" (DKRVO) in the form of institutional support of the Ministry of Industry and Trade of the Czech Republic.

References

[1] Signetti S., Bosia F., Ryu S., Pugno N. M., "A combined experimental/numerical study on the scaling of impact strength and toughness in composite laminates for ballistic applications", *Composites Part B: Engineering*, Vol.195 (2020) 108090, ISSN 1359-8368, DOI: 10.1016/j.compositesb.2020.108090.

[2] Xu Z.W., Chen Y.H., Cantwell W.J., Guan Z.W., "Multiscale modelling of scaling effects in the impact response of plain woven composites," *Composites Part B: Engineering*, Vol.188 (2020) 107885, ISSN 1359-8368. DOI: 10.1016/j.compositesb.2020.107885.

[3] Teimouri F., Heidari-Rarani M., Aboutalebi F. H.,"Finite element modeling of mode I fatigue delamination growth in composites under large-scale fiber bridging," *Composite Structures*, Vol. 263 (2021), 113716, ISSN 0263-8223, DOI:10.1016/j.compstruct.2021.113716.

[4] Dassault Systemes Simulia Corp., "SIMULIA User Assistance 2017," *Available from* https://abaqus-docs.mit.edu/2017/English/DSSIMULIA_Established.htm *Accessed:* 2022-04-15.

[5] ASTM D2344/D2344M-16, "Standard Test Method for Short-Beam Strength of Polymer Matrix Composite Materials and Their Laminates, " ASTM International, PA, USA, (2016). DOI:10.1520/D2344_D2344M-16.

[6] Lopes C. S., et al. "Influence of Porosity on the Interlaminar Shear Strength of Fibre-Metal Laminates," *Key Engineering Materials*, vol. 383, Trans Tech Publications, Ltd., pp. 35–52 (2008). DOI: 10.4028/www.scientific.net/kem.383.35.

[7] Šedek, J. "Modelling the failure of near-edge impacted carbon fibre-reinforced composite subjected to shear loading," *IOP Conference Series: Materials Science and Engineering* (2018), pp. 369, 012035. ISSN 1757-8981. DOI: 10.1088/1757-899X/369/1/012035.

Analysis of Clinched Joints Manufacturing Residual Stress

Marek Šťádler[1] & František Lopot[2]

Abstract: Specific designs of clinched joint groups are mostly engineered on the basis of empirical experience of designers and the accuracy of the design is often verified in the testing laboratories of manufacturers of clinching tools. Therefore, the problem of clinching is often mastered for a one specific type of joint. The designs of joint groups are based on the properties of an individual joint in two basic directions [1] without considering the possibility of their mutual interaction. This contribution is based on the measurement of possible parasitic stress arising in the groups of clinched joints during their production. Within the work, experiments are designed and performed in order to determine the presence and possible size of the above-mentioned interaction.

Keywords: Clinched joint; Residual stress; Array of joints; DX51D

1. Introduction

Clinching is a high-performance and rapidly developing method for reliable mechanical joining of sheet metal parts using local plastic deformation of the material. It is a modern technology that speeds up the production of sheet metal assemblies and in some areas replaces traditional bolted or welded joints. The clinching mechanism has been known for many years, but it is only in the last 30 years that clinched joints have become more widely used. These joints are successfully used in areas of industry where is a strong pressure to make lightweight construction and hence to reduce primary energy or fuel consumption [3].

The principle of clinching is to join two or more layers of material together, which is performed by a special tool (punch and die). Very often, clinching is used for heterogeneous materials, where the use of traditional joining methods brings increased costs.

This is especially in the case of the automotive, aerospace and electronics industries, where clinching allowed a more advantageous matching of the material to the part, with less regard to the joining technology.

In industrial research at the authors' workplace, specimens of joints of sheet metal parts were investigated in terms of their load capacity.

The large disproportion between the theoretical and measured load distribution between the individual connections, together with the characteristic deformation of the connection, led the authors to believe that there was an internal prestressing field in the groups of clinched joints. This prestress is theoretically generated by the differently sized plastic deformation region of the connected parts.

The possible pre-stressing of clinch joints is marginally addressed by the authors [1], investigating the properties of single joints or two joints located close to each other. Only a partial correlation of theoretical values of the load capacity of the connections with experiments is explained by the authors [2] by the combined stresses of the connection.

[1] *Ing. Marek Šťádler; Department of Designing and Machine Components, Faculty of Mechanical Engineering, Czech Technical University in Prague; Technická 4, 166 07 Prague, Czech Republic;* marek.stadler@fs.cvut.cz

[2] *Ing. František Lopot, Ph.D.; Department of Designing and Machine Components, Faculty of Mechanical Engineering, Czech Technical University in Prague; Technická 4, 166 07 Prague, Czech Republic;* frantisek.lopot@fs.cvut.cz

To confirm the existence and magnitude of the mentioned prestressing, the following experiments were designed and performed.

2. Methods

The examined specimens with joints in a row in the longitudinal axis of symmetry of the specimen were formed by joining galvanised strips of 3 mm thick from one sheet of sheet metal. The material of the specimens is low carbon steel DX51D (1.0226) suitable for cold forming. The material properties are summarised in Table 1.

Table 1. Samples material properties.

Material	Tensile strength [MPa/mm^2]	Ductility [%]	Chemical composition [%]
			C 0.18
			Si 0.5
DX51D (1.0226)	270-500	22	Mn 1.2
			S 0.045
			Ti 0.3

The joints were made with one set of ToxxRound 10 clinching tools. The dimensions of the samples and the size of the formed joints can be seen in Table 2.

2.1. Deformations

The existence of the aforementioned prestress can be inferred indirectly from Fig. 1. The figure shows a sample (03-40-40) from the research set. In all samples a significant deformation is evident, which can be considered as a key manifestation of strain in the joint array.

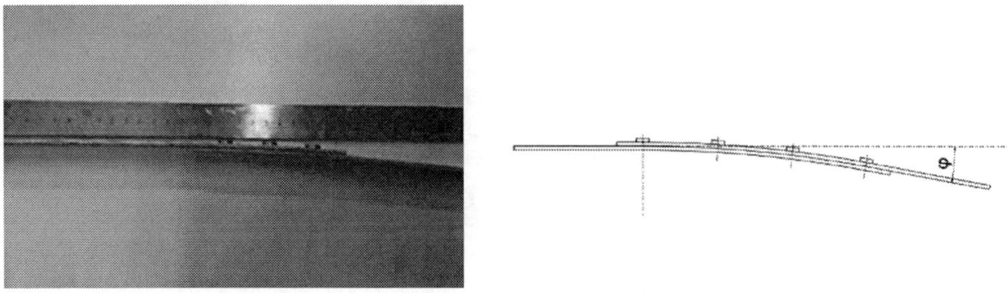

Fig. 1. Sample angular deformations.

The deformation can be best described by the angular deviation from the original axis of the connected parts. The measured angle is shown in Fig. 1.

In the following table (Table 2) the measured deviation of the individual samples is given.

Table 2. Dimensions of tested samples.

Sample	Cross-section [mm]	Number of joints [1]	Spacing [mm]	Measured deformation [°]	Deformation per joint [°]
02-40-20		2	20	2.4	1.20
02-40-40		2	40	1.2	0.6
03-40-20	40×3	3	20	3.1	1.03
03-40-40		3	40	2.1	0.7
04-40-20		4	20	4.5	1.13
04-40-40		4	40	3.2	0.8

2.2. *Measurement of residual stress*

The arrays of clinch joints were further subjected to residual stress testing using a destructive method, drilling out the joints. Sample 03-40-40 (Table 2) was chosen for illustration. The area around each joint was fitted with uni-axial strain gauges (3/120LY11) HBM [4] oriented in the axis of the joint so that the strain gauges were not damaged during the measurement. Fig. 2 shows the location of the individual strain gauges (labelled T1–T6).

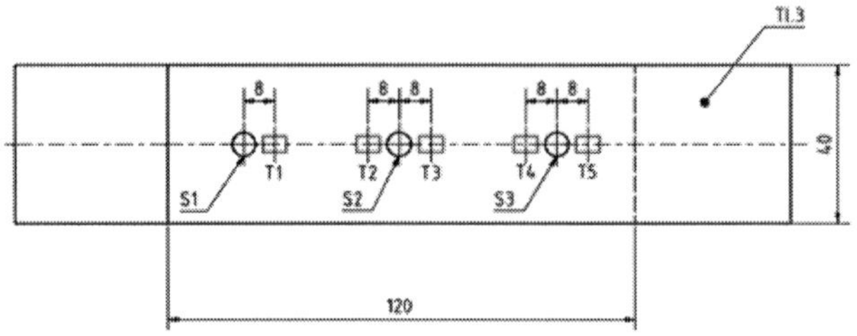

Fig. 2. Location of strain gauges.

During the measurements, all the clinched joints of the samples were drilled out one by one, see Fig. 3 and the response of the strain gauge signals was monitored, and converted to normal stress values see Fig. 4 for illustration.

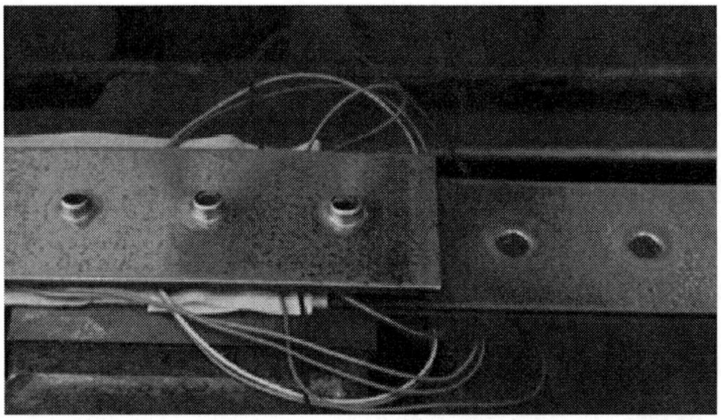

Fig. 3. Drilled out joints.

As can be seen (Fig. 3) drilling out of the joint did not result in an immediate and perfect break of the neck of the joint due to manufacturing inaccuracies of the sample.

This phenomenon was reflected in the signal changes of the strain gauges. For example, during the drilling of joint S2 see Fig. 2, there is still a change of stress on strain gauges T1 and T2 near joint S1, even though joint should not carry load any longer. An explanation can be found in the following consideration. When the neck of the joint is imperfectly broken, the two connected parts are still held together. This, together with the specific shape of the surrounding area of the neck itself, results in a residual shape interaction of the parts.

The measured strain gauge stresses during the joint testing are summarised in Table 3.

Table 3. Measured strain gauge stresses.

Joint	T1-S_{11} [N/mm^2]	T2-S_{11} [N/mm^2]	T3-S_{11} [N/mm^2]	T4-S_{11} [N/mm^2]	T5-S_{11} [N/mm^2]
S1	-93.28	-22.75	-5.5	-1.83	-3.81
S2	-8.65	-96.86	-139.47	7.00	-5.08
S3	-6.60	-16.44	-27.27	-111.36	-46.09
Resulting stress change	-108.53	-136.05	-172.24	-106.19	-54.98

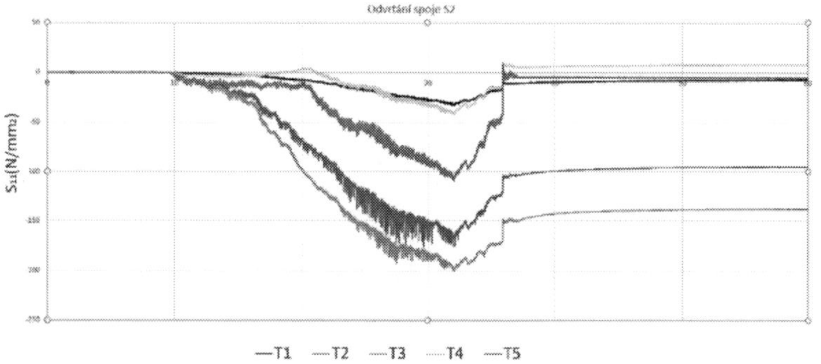

Fig. 4. Change of mechanical stress of the individual strain gauges during joint S2 was drilled out.

All the resulting stress changes are apparently compressive, however, due to the nature of the test where the internal stresses of the joint are released, the real stresses are of an inverse nature, i.e. tensile.

To determine the influence of the stress change caused by drilling of the neck of the joint itself, control samples with just one clinched joint were designed and tested.

These samples were made of the same material and by the same set of clinching tools as samples in Table 2.

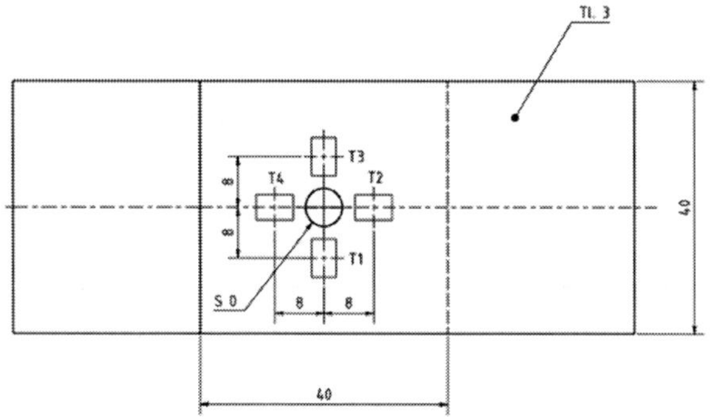

Fig. 5. Location of strain gauges- control samples S0.

The area around joint was fitted with uni-axial strain gauges (3/120LY11) HBM [4] oriented in the two perpendicular axes of the joint. Fig. 5 shows the location of the individual strain gauges (labelled T1–T4).

During the measurements, clinched joints of the control samples were drilled out and the response of the strain gauge signals was monitored, and converted to normal stress values see Fig. 6 for illustration.

The obtained stress values were used for superposition of the data measured at samples with the linear array of joints for elimination of the effect of joint neck breakage and subsequent calibration of the FEA model.

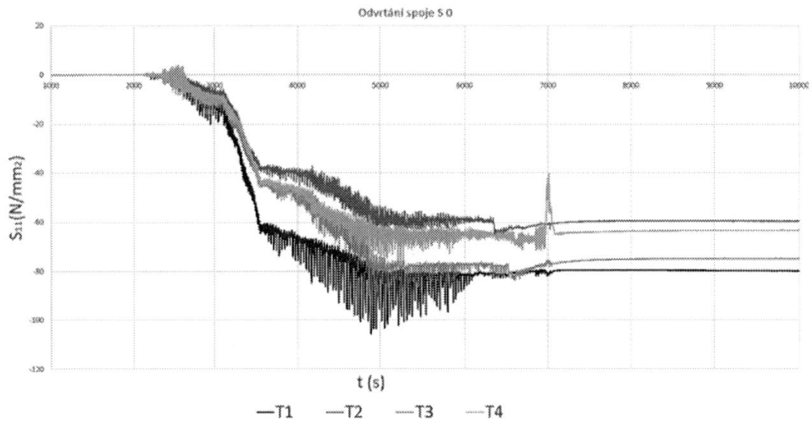

Fig. 6. Change of mechanical stress of the individual strain gauges during joint S0 was drilled out.

2.3. *Finite element models*

A finite element model of the clinched joint was created and calibrated to produce the same strain and stress response as strain gauges measurements according to subsection 2.2. All drilled out joints were simulated.

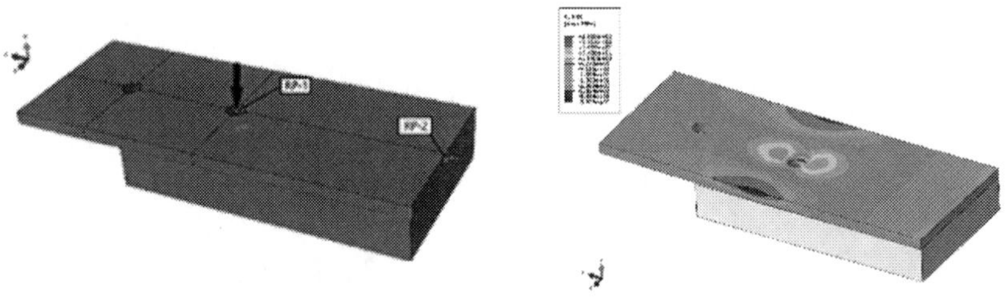

Fig. 7. Clinched joint FEM modelling.

In the simulations, the reaction force was monitored, at point RP-2, in the Z-axis direction, i.e. in the longitudinal axis of the sample, see Fig. 7. The given value was then compared with the design load capacity of the individual clinched joint.

3. Conclusions

From the measurements of samples deformation (subsection 2.1) can be seen, that the value of the angular deformation of the specimen normalised to one clinched joint, see Table 2 is inversely proportional to the spacing of each joint.

When producing an array of joints in line, where the parts to be joined are already bonded to each other by the first joint of the array, the angular deformation of each subsequent joint is greater the greater the stiffness of the intermediate section of the parts. If the deformation of the specimens is dependent on the stiffness of the sections of the connected parts between the joints, the existence of axial prestressing between the parts can be assumed.

The results of simulations with 3D solid models shows that the selected simulation model is suitable for further work and evaluation of strain gauge measurements in destructive testing of clinched joints arrays. The model is easy to calibrate and sufficiently captures the behavior of the connection in the assumed stress distributions.

Initial simulations show, (see Table 4) that in the extreme case up to 24% of the load carrying capacity of the connections is used up by the pre-stressing of the connection array due to manufacturing.

Table 4. Load capacity reduction.

Joint	Design load capacity [N]	FEM reaction [N]	Load capacity reduction [%]
S1	3000	596	19.87
S2	3000	738	24.60

Acknowledgements

This work was supported by SGS CTU grant no.: SGS21/109/OHK2/2T/12, „Metodika návrhu polí clinchových spojů pro použití v konstrukční praxi“.

References

[1] Breda A. et al., “Mechanical Behaviour of Clinched Joints in Configurations,” in Proceedings of ICEM 2018: vol. 2 (8), Paper No. 414, (Brussels, 2018).

[2] Breda A. et al., “Experimental validation of an equivalent modelling strategy for clinch configurations,“ *Materials & Design*, Vol. 157, (2018). doi: 10.1016/j.matdes.2018.07.035

[3] Peng H. et al.,“Recent development of improved clinching process,” *The International Journal of Advanced Manufacturing Technology*, Vol. 110, pp. 1–31 (2020). https://doi.org/10.1007/s00170-020-05978-4.

[4] Linear straingauges with 1 measurement grid [online], *Available from*: https://www.hbm.com/en/4561/ly-linear-straingauges-with-1-measurement-grid/ *Accessed:* 2020-12-01.

Numerical Analysis of the Subgrade while Strengthening with the Micropiles

Oleksii Tiutkin[1], Larysa Neduzha[2] & Jan Kalivoda[3]

Abstract: The paper analyzes the advantages and disadvantages of strengthening a railway track subgrade with micropiles. To optimize the parameters of subgrade strengthening technologies a finite element model of a subgrade, which is reinforced by vertical micropiles, has been created. The results of the numerical analysis without reinforcement and two alternatives of subgrade strengthened by micropiles in various placements are presented in the paper. In order to find the most effective option for track subgrade reinforcement, the plots in a deformed state have been created and vertical and horizontal displacements have been analyzed at four characteristic points of the ballast prism.

Keywords: Numerical analysis; Subgrade; State of Deformation; Finite-element Model; Micropile; Strengthening; Jet-grouting

1. Introduction

Recently, chemical soil consolidation, in particular cementation, has been widely used to strengthen the subgrade. The essence of this method is to inject the cement mortar into the pores of the soil, the hardening of which significantly increases the strength of the base. Jet Grouting technology and Boring and Mixing Technology are the most common types of cementation [1–3].

The fundamental difference between Boring and Mixing Technology and Jet Grouting could be observed in the pressure of the cement mortar, which is ten times less for the Jet Grouting. The destruction of the soil is occurred not by jet, but by a special chisel or impeller. The cement-sand mortar acts as a binder, not a crushing force. Accordingly, the strength and deformation characteristics of micropiles created by Boring and Mixing Technology are lower compared to the Jet Grouting. Therefore, in order to obtain the desired reduction in deformation a larger number of micropiles should be created by Boring and Mixing Technology.

Another advantage of Jet Grouting technology is also that in the soil massif it is possible to form vertical elements with a diameter from 0.2 up to 1.5 m. In the case of micropiles the minimum diameter is 0.2–0.3 m with a length up to 6.0 m [4–6], which will give the desired effect [1, 3, 8] at the lowest cost of material and funds [7–10].

2. Numerical Analysis of the Subgrade while Strengthening with the Micropiles

Three finite-element models were developed for the numerical analysis [11–15] of the subgrade: Option 0 – without strengthening; Option 1 – strengthening with jet-grouting micropiles located near the ends of sleepers (Fig. 1, top); Option 2 – strengthening with jet-grouting micropiles located inside the rail-and-sleeper grid near the rail (Fig. 1, bottom). In both options of as strengthening, the jet-grouting micropile with a diameter of 0.3 m and a length of 4 m is adopted.

[1] *prof. Oleksii Tiutkin, D.Sc.Tech; Faculty of Bridges and Tunnels, Ukrainian State University of Science and Technologies; Lazaryan Str. 2, 49010 Dnipro, Ukraine;* alexeytutkin@gmail.com

[2] *prof. (Associate) Larysa Neduzha, Ph.D.; Faculty of Transport Engineering, Ukrainian State University of Science and Technologies; Lazaryan Str. 2, 49010 Dnipro, Ukraine;* nlorhen@i.ua

[3] *Ing. Jan Kalivoda, Ph.D.; Faculty of Mechanical Engineering, Czech Technical University in Prague; Technická 4, 166 07 Prague 6, Czech Republic;* jan.kalivoda@fs.cvut.cz

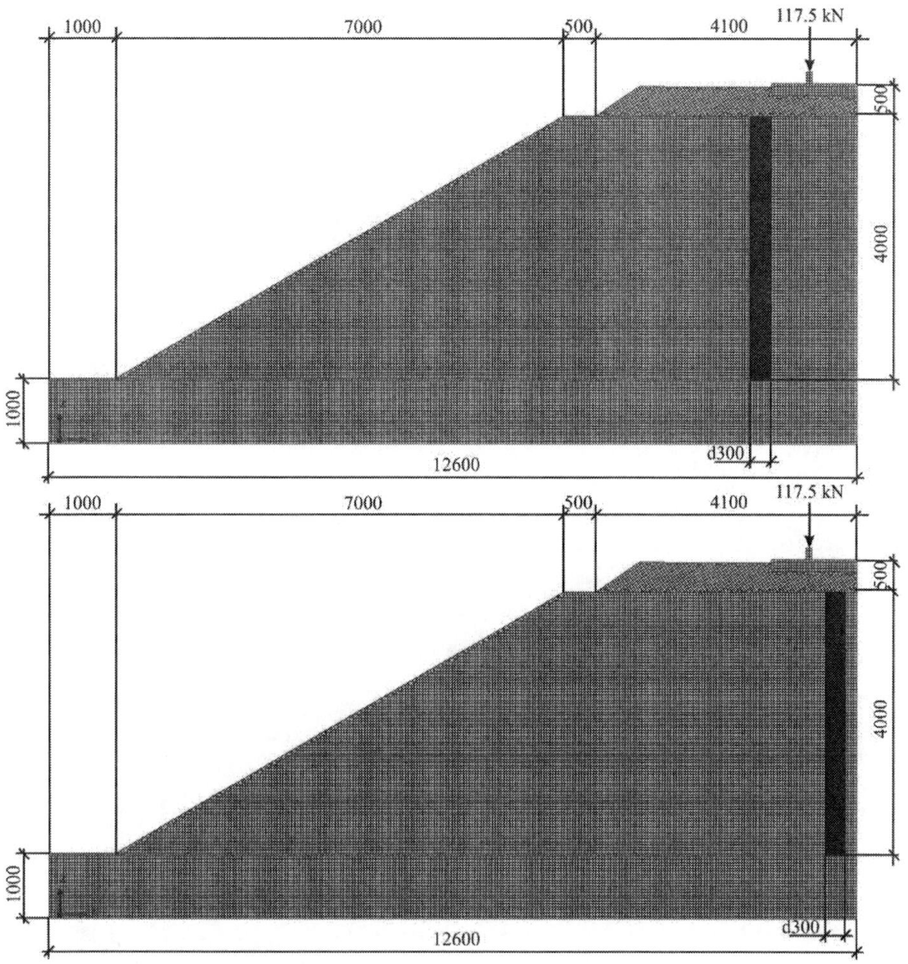

Fig. 1. The finite element model of the subgrade with jet-grouting micropile strengthening: top – Option 0, bottom – Option 1.

A static train weight was adopted as the model's load. The vertical axle load is assumed to be equal to the limit value of standard axle load $P = 235$ kN per axle, that is, 117.5 kN per rail.

Fig. 2 shows the results of numerical analysis of the subgrade loaded by a train (isolines of vertical displacements). The illustrations of the deformed state present the values of the characteristic isolines for comparing the level of vertical displacements.

The results on Fig. 2 shows that the lower deformation limit (isoline with a value of -4.0 mm) is almost unchanged for all three options. However, the next characteristic isoline (with a value of -16.0 mm) in Option 1 moves up, and in Option 2 it is located at the interface between the subgrade and ballast. The last characteristic isoline (with a value of -32.0 mm) is absent in both options of strengthening, which indicates a significant reduction in deformation by micropiles.

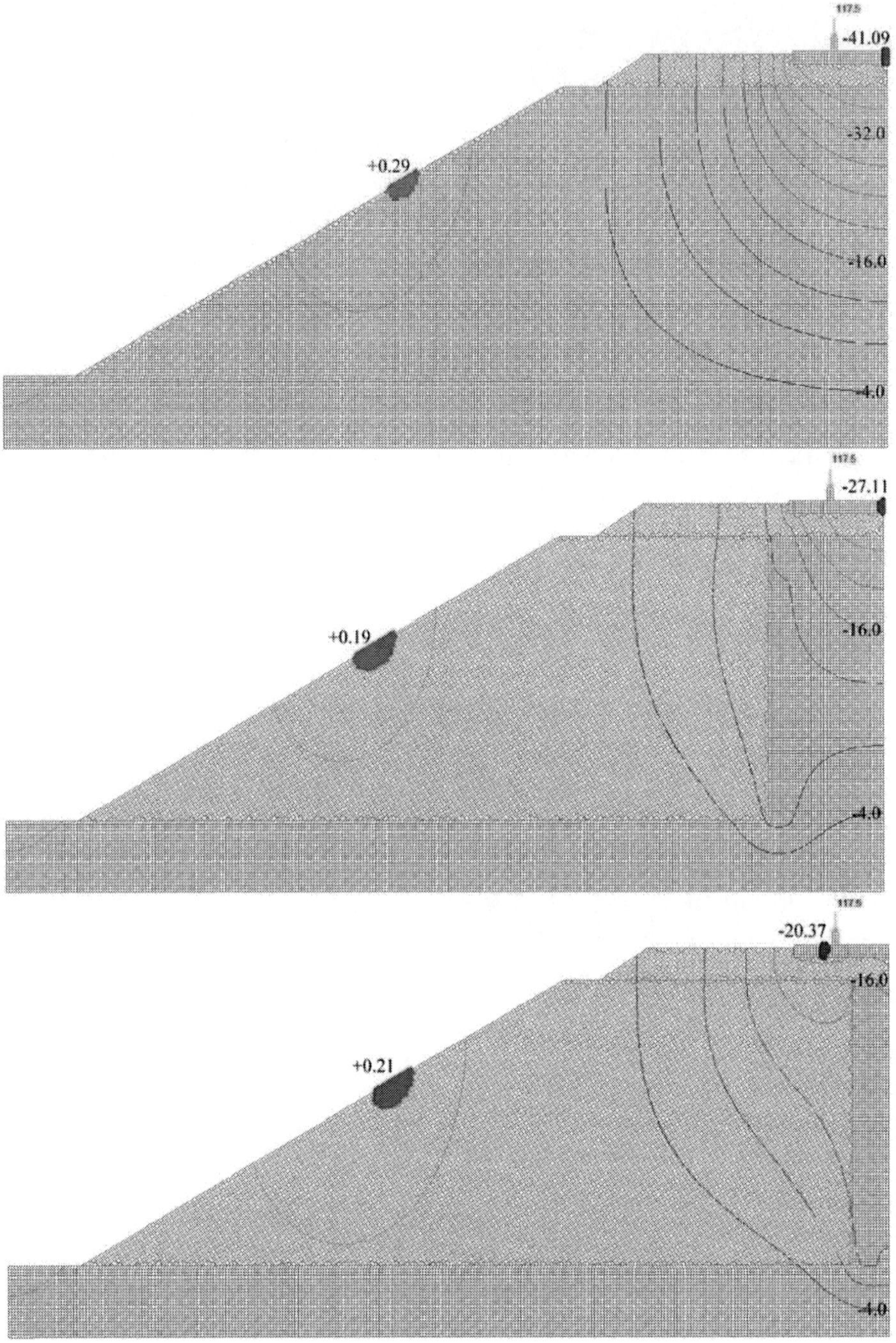

Fig. 2. Isolines of vertical displacements with a constant step (4.0 mm) between isolines: top – Option 0, middle – Option 1, bottom – Option 2.

For the quantitative analysis of the parameters of obtained deformed state, the scheme of characteristic points was applied (Fig. 3). In points 1–4, indicated on the scheme, the components of displacements were analyzed.

Fig. 4 shows a diagram of vertical displacements, in four characteristics points.

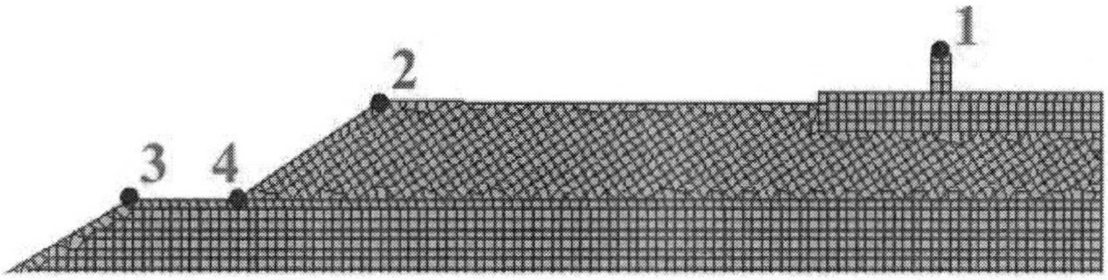

Fig. 3. Points at which the values of the components of the deformed state are measured.

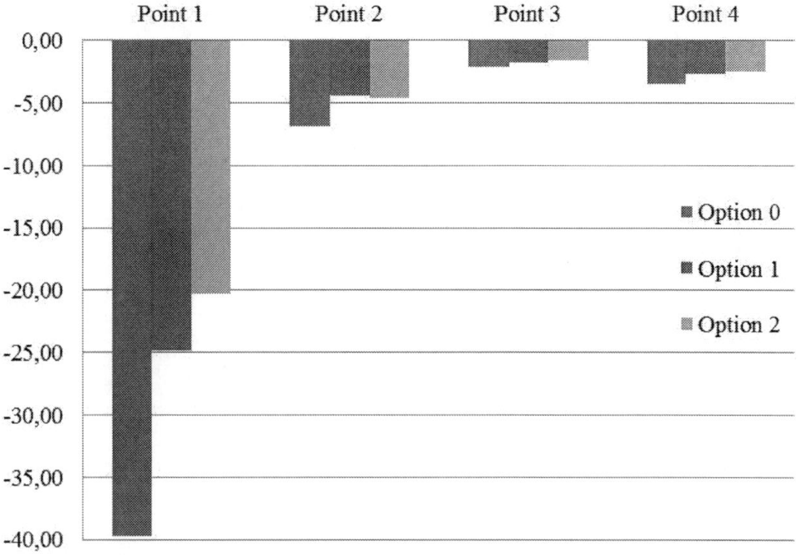

Fig. 4. The diagram of distribution of the vertical displacements (mm) at characteristic points of the model.

Analysis of the diagram on Fig. 4 allows concluding that both strengthening options significantly reduce the level of displacements. If we compare the vertical displacements of Option 0 (not strengthened subgrade) with options of strengthening, it could be noted that at point 1 (rail head) they are reduced by 1.60 times for Option 1 and 1.96 times for Option 2.

In the points 2 – 4 the level of vertical displacements is reduced from 1.20 to 1.58 times for Option 1 and from 1.34 to 1.51 for Option 2. This is due to the fact that immersion of jet-grouting micropile, which leads the deformed state to inhomogeneous, most affects in the zone of maximum development of deformations – the "core", that is being formed in the area of the rail-and-sleeper grid.

Therefore, Option 1 (jet-grouting micropiles located near the end of a sleeper) is less effective in reducing vertical displacement compared to the Option 2 where the stiff jet-grouting micropile is outside the zone of active deformation.

3. Conclusions

Analysis of subgrade strengthening technologies demonstrates that the most rational option is jet-grouting micropile. Micropiles as strengthening elements, in contrast to others, are less demanding both in material sources and in money funds. This is justified by the length (4 ... 6 m) and small diameter (200 ... 300 mm) of the jet-grouting micropile.

Option 2 (micropiles located inside the rail-and-sleeper grid) affect the deformed state of the track subgrade more efficiently compared to the Option 1 (micropiles located near the ends of sleepers). The Option 2 reduce vertical displacements by 1.96 to 1.34 times in selected points, whereas the reduction of vertical displacements for Option 1 is by 1.60 to 1.20 times.

The maximal vertical displacement at the rail head is 20 mm for the Option 2 and 25 mm for the Option 1, whereas without strengthening it reaches 40 mm. Values of vertical displacements up to 20 mm allow testifying to the normal operation of the subgrade.

References

[1] Tiutkin O., Neduzha L. and Kalivoda J., "Changing the Stress State of the Track Superstructure while Strengthening the Subgrade," in *Proceedings of the 58th International Scientific Conference on Experimental Stress Analysis*, 2020, pp. 533–539.

[2] Tiutkin O., Neduzha L. and Kalivoda J., "Finite-Element Analysis of Strengthening the Subgrade on the Basis of Boring and Mixing Technology," *Transport Problems*, 2021, Vol. 16, iss. 2, pp. 189–197. doi: 10.21307/tp-2021-034.

[3] Severino A., Wahrhaftig A.M., Tiutkin O., Gubashova V. and Neduzha L., "Effective Jet-Grouting Application for Improving the State of Deformation of Landmarks," *Buildings* Vol. 12, no. 3, Art. no. 368 (2022). https://doi.org/10.3390/buildings12030368.

[4] Petrenko V., Bannikov D., Kharchenko V. and Tkach T., "Regularities of the deformed state of the geotechnical system soil base – micropile", in *Proceedings of III International Conference Essays of Mining Science and Practice, IOP Conference Series: Earth and Environmental Science*, Vol. 970, Art. no. 012028 (2022). https://doi.org/10.1088/1755-1315/970/1/012028.

[5] Sun S., Wang J. and Bian X., "Design of micropiles to increase Earth slopes stability," *J. Cent. S. Univ*, Vol. 20, pp. 1361–1367 (2013) https://doi.org/10.1007/s11771-013-1623-7.

[6] Mollaali M., Alitalesh M. and Yazdani M., "Soil Improvement Using Micropiles," Conference Paper, 2014. doi: 10.1201/b17017-96.

[7] Alnuaim A.M., El Naggar M.H. and El Naggar H., "Performance of micropiled rafts in clay: Numerical investigation," *Comput Geotech*, Vol 99, pp. 42–54 (2018). https://doi.org/10.1016/j.compgeo.2018.02.020.

[8] Qian Z.Z. and Lu X.L., "Behavior of micropiles in soft soil under vertical loading," *Advanced Materials Research*, Vol. 243, pp. 2143–2150 (2011). http://dx.doi.org/10.4028/www.scientific.net/AMR.243-249.2143.

[9] Hwang T.-H., Kim K.-H. and Shin J.-H., "Effective installation of micropiles to enhance bearing capacity of micropiled raft," *Soils and Foundations*, Vol. 57, pp. 36-49 (2017). http://dx.doi.org/10.1016/j.sandf.2017.01.003.

[10] Reddy K.M. and Ayothiraman R., "Experimental Studies on Behavior of Single Pile Under Combined Uplift and Lateral Loading," *Geotechnical and Geoenvironmental Engineering*, Vol. 1, pp. 1–10 (2015). http://dx.doi.org/10.1061/(ASCE)GT.1943-5606.0001314.

[11] Zolfegharifar S.Y., Daram H. and Rahimi M., "Static Analysis, Using Finite-Element Method (FEM) for Micropiles Application Evaluation: Empirical and Numerical Perspective," http://dergi.cumhuriyet.edu.tr/ojs/index.php/fenbilimleri.

[12] Abdelaziz A., Hafez D. and Hussein A., "The effect of pile parameters on the factor of safety of piled-slopes using 3D numerical analysis," *HBRC Journal*, Vol. 13, pp. 277–285, (2017). http://dx.doi.org/10.1016/j.hbrcj.2015.06.002.

[13] Xu K.J. and Poulos H.G., "3D elastic analysis of vertical piles subjected to "passive" loadings," *Computers and Geotechnics*, Vol. 28, no. 5, pp. 349–375 (2001). https://doi.org/10.1016/S0266-352X(00)00024-0.

[14] Dubinchyk O., Petrenko V., Ihnatenko D. and Kildieiev V., "Comprehensive analysis of the retaining pile structure with the determining the stability factor by numerical methods," in International Conference Essays Of Mining Science And Practice, E3S Web of Conferences, Vol. 109, Art. no. 00020 (2019). https://doi.org/10.1051/e3sconf/201910900020.

[15] Yang Z. and Jeremic B., "Numerical analysis of pile behaviour under lateral loads in layered elastic-plastic soils," *Int. J. Numer. Anal. Methods Geomech.* Vol.26, no. 14, pp. 1385–1406 (2002). https://doi.org/10.1002/nag.250.

Investigation of Residual Stresses of Thick Laser Welded Steel Plates Using X-Ray Diffraction

Karel Trojan[1], Jiří Čapek[2], Jan Kec[3], Nikolaj Ganev[4], Ivo Černý[5] & Stanislav Němeček[6]

Abstract: X-ray diffraction, due to its small penetration depth into metallic materials, can describe the state of residual stress only in a thin surface layer. However, not only for thick welds, it is necessary to describe the course of residual stresses within the whole thickness of the material. As the presence of tensile residual stresses is detrimental in physical processes such as, for instance, fatigue or, in combination with defects, promotes brittle fractures. This research describes the possibility of X-ray diffraction to characterize the state of residual stress beneath the surface of thick laser welded steel plates. High tensile stresses along the weld have been characterized in the entire thickness of the laser-welded plate.

Keywords: Laser welding; X-ray diffraction; Residual stresses

1. Introduction

Welding is a widely used method in mechanical engineering, including ships, trains, steel bridges, and pressure vessels. Therefore, high demands are laid on the mechanical properties and durability of the welds used to connect two or more components [1]. However, not only for thick welds, it is necessary to describe the course of residual stresses (RS) within the entire thickness of the material. Because of the heterogeneous application of energy and localized fusion, which occur during the welding process, high RS can be present in the region near the weld and in the weld itself. These RS are a superposition of thermal and transformation stresses and could reach high values [2]. The presence of tensile residual stresses is detrimental in physical processes such as, for example, fatigue or, in combination with defects, promotes brittle fractures [3].

X-ray diffraction, due to its small penetration depth into metallic materials, can describe the state of residual stress only in a thin surface layer. Neutron diffraction, which provides diffraction from a greater penetration depth, can be used to determine residual stresses within thick samples, but its use is more demanding and not commonly available [4]. Therefore, this research described the possibility of X-ray diffraction to characterize the state of residual stress beneath the surface of thick laser-welded steel plates. The only way to obtain sub-surface residual stress values using X-ray diffraction is to remove some of the material. Two methods were compared, where, in order to describe the residual stresses within the welded plate, a part of the upper side was milled. The second option was to cut the plate perpendicular to the weld and

[1] *Ing. Karel Trojan; Department of Solid State Engineering, Faculty of Nuclear Sciences and Physical Engineering, Czech Technical University in Prague; Trojanova 13, 120 00 Prague, Czech Republic;* karel.trojan@fjfi.cvut.cz
[2] *Ing. Jiří Čapek, Ph.D.; Department of Solid State Engineering, Faculty of Nuclear Sciences and Physical Engineering, Czech Technical University in Prague; Trojanova 13, 120 00 Prague, Czech Republic;* jiri.capek@fjfi.cvut.cz
[3] *Ing. Jan Kec; SVÚM a.s.; Tovární 2053, 250 88 Čelákovice, Czech Republic;* kec@svum.cz
[4] *prof. Ing. Nikolaj Ganev, CSc.; Department of Solid State Engineering, Faculty of Nuclear Sciences and Physical Engineering, Czech Technical University in Prague; Trojanova 13, 120 00 Prague, Czech Republic;* nikolaj.ganev@fjfi.cvut.cz
[5] *Ing. Ivo Černý, Ph.D.; SVÚM a.s.; Tovární 2053, 250 88 Čelákovice, Czech Republic;* ivo.cerny@seznam.cz
[6] *doc. Ing. Stanislav Němeček, Ph.D.; Department of Material and Engineering Metallurgy, Faculty of Mechanical Engineering, University of West Bohemia; Univerzitní 22, 301 00 Plzeň, Czech Republic;* snemecek@kmm.zcu.cz

to analyse the cross-section, which assumes that there is no significant redistribution of residual stresses in the plane of the cut. This assumption also appears in articles in highly impacted journals, e.g. [5], and may not always be true.

2. Experiment

The residual stress state across the butt laser weld made of P355 steel 10 mm thick plates for high-pressure vessels was examined. This fine-grained steel is widely used for the fabrication of high-pressure vessels, steam boiler parts, pressure piping, compressors, etc. Plates were firstly annealed for stress relieving in an inert gas atmosphere. After this treatment, the double-sided square butt weld (the upper side was welded first) was performed using a 3 kW fibre-coupled diode Laserline LDM laser with the wavelength 980 nm equipped with a Laserline's welding optics with a beam diameter of 0.6 mm.

RS were described in two mutually perpendicular directions (the L direction is longitudinal with the weld and the T direction is transversal with the weld) up to a distance of 30 mm from the weld axis. In the first phase, the residual stresses on the upper side (marked as 10 mm) and the lower side (marked as 0 mm) of the plate were analysed. In the next step, 1.3 mm of material was milled from the upper side of the analysed area and subsequently 200 μm was electropolished. The same analyses were performed on the new surface and the results were compared. For comparison, the plate was cut perpendicular to the weld and residual stress values were obtained after electropolishing the affected layer on the resulting cross-section.

3. Results

From the obtained profiles, it could be said that the RS have mainly a tensile character in the longitudinal direction, see Fig. 1. This is in agreement with the RS forming mechanisms during the welding process [2]. On the other hand, the character of RS is compressive in the transversal direction (which is more important from a technological point of view when welding pressure vessels) only on the lower side, see Fig. 2. As the presence of tensile RS is detrimental in physical processes such as, for instance, fatigue, it is unfavourable that tensile stresses occur not only on the upper surface but also below it. From the results, it could be further stated that the longitudinal RS obtained are, with respect to errors, sufficient to describe the longitudinal residual stress state over the entire thickness of the weld.

Nevertheless, the measurements show that the transversal RS change significantly with depth. On the upper side (10 mm), there are high tensile RS; on the contrary, on the lower side (0 mm), there are compressive ones. This course can be explained by the shrinkage of the weld during cooling and subsequent warping of the plate, this effect is most pronounced on the surface. At a depth of 1.5 mm below the upper surface, a slight stress drop can be seen at a distance of 4 mm from the weld axis. In this area, the effect of the volume-restricted phase transformation of austenite to ferrite, which results in compressive RS, is noticeable. The need to properly describe the state of residual stress is complemented by the fact that the fatigue cracks of these welds were always initialized on the side, where the tensile RS were analysed [3].

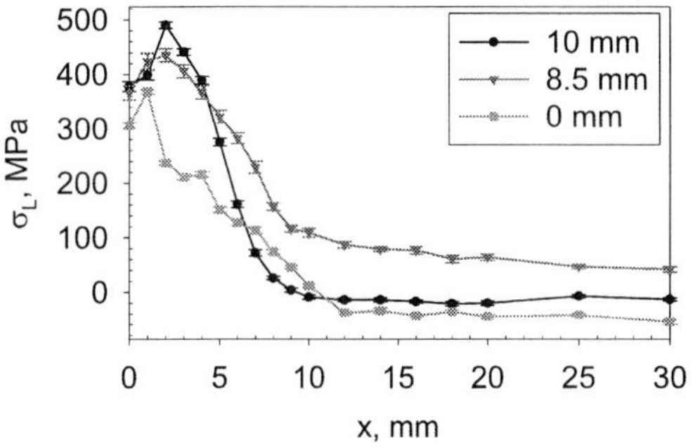

Fig. 1. Longitudinal (parallel to the weld) residual stresses with respect to distance perpendicular to the weld.

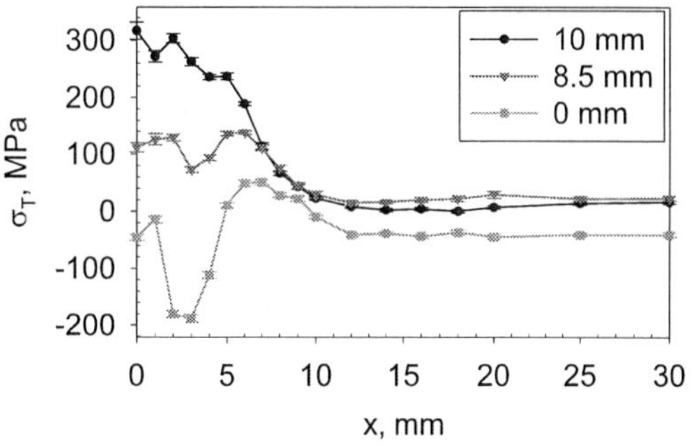

Fig. 2. Transversal (perpendicular to the weld) residual stresses with respect to distance perpendicular to the weld.

Fig. 3 compares RS obtained from the top (using milling) and on the cross-section (using cutting) at a depth of 1.5 mm below the upper surface, when a different course is clearly visible. After cutting the plate, the weld slightly dropped and thus caused compressive RS in the plane perpendicular to the weld.

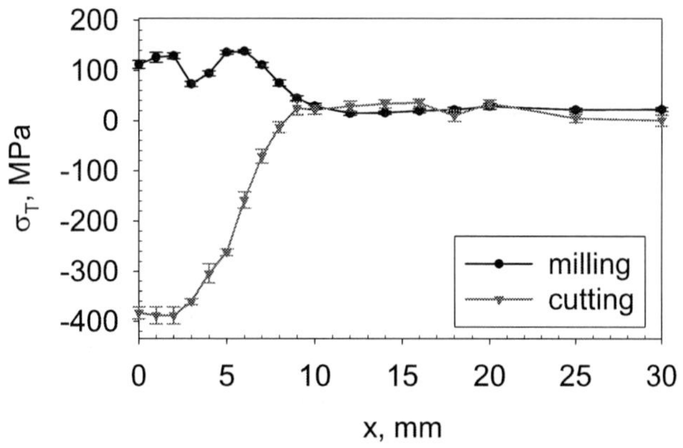

Fig. 3. Comparison of RS obtained from the top (milling) and on the cross-section (cutting) at a depth of 1.5 mm below the upper surface.

The FEM model was used to describe the differences in the comparison of milling and cutting. The model should verify the hypothesis that redistribution after cutting the sample probably produced compressive RS on the cross-section. ANSYS Mechanical APDL 19.0 finite element software was used, where the interpolated RS state obtained from the previous results was used as the initial state for the modelled volume of $300 \times 150 \times 10$ mm^3. The linear static approach was used in the model with SOLID185 elements, when the elastic modulus E (220 GPa) and the Poisson constant (0.3) were used. Constraints were applied on the plate edge y = 150 mm and x = 0 mm in all directions. Fig. 4 shows the results of the model showing a deformed plate with a scale of approx. 200 to highlight the deformations after applying the initial stress state and redistribution with RS in T direction. It is clear that RS have been redistributed and the weld and the surroundings on the cross-section show high compressive RS.

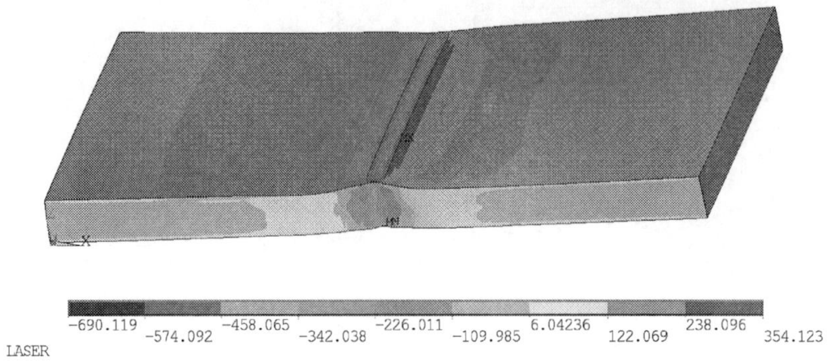

Fig. 4. Residual stresses in MPa in T direction for sample after applying initial stress state and redistribution.

4. Conclusion

High tensile stresses along the weld have been described in the entire thickness of the laser welded plate and also on the upper side in the transversal direction. In [3], it was shown that the fatigue cracks of these welds were always initialized on the side, where the tensile RS were analysed. Furthermore, it was shown that even with X-ray diffraction it is possible to reliably determine not only the surface values of residual stresses of thick laser welded steel plates, but it is necessary to pay attention to the preparation of the sample. Significant redistribution of residual stresses in the cross-section plane occurred due to high tensile residual stresses perpendicular to this plane (parallel to the weld).

Acknowledgements

Measurements were supported by the project TH02010664 of the Technology Agency of the Czech Republic. Research was funded by the Center for Advanced Applied Science, grant number CZ.02.1.01/0.0/0.0/16_019/0000778. "Center for Advanced Applied Science" within the Operation-al Program Research, Development and Education supervised by the Ministry of Education, Youth and Sports of the Czech Republic. K.T.'s work was supported by the Grant Agency of the Czech Technical University in Prague, grant No. SGS22/183/OHK4/3T/14.

References

[1] Oyyaravelu R., Kuppan P. and Arivazhagan N., "Metallurgical and mechanical properties of laser welded high strength low alloy steel,“ *Journal of advanced research*, Vol. 7 (3), pp. 463–472 (2016). https://doi.org/10.1016/j.jare.2016.03.005.

[2] Nitschke-Pagel T. and Dilger K., "Sources and consequences of residual stresses due to welding,“ in *Materials Science Forum Vols. 783-786*, Mishra B., Ionescu M. and Chandravol T. eds. (Trans Tech Publications, Switzerland, 2014), pp. 27772785. http://dx.doi.org/10.4028/www.scientific.net/MSF.783-786.2777.

[3] Čapek J., Trojan K., Kec J., Černý I., Ganev N. and Něměček S., "On the Weldability of Thick P355NL1 Pressure Vessel Steel Plates Using Laser Welding,“ *Materials*, Vol. 14, no. 1, Art. no. 131 (2021). https://doi.org/10.3390/ma14010131.

[4] Holden T.M., "Neutron diffraction,“ in *Practical Residual Stress Measurement Methods*, Schajer G.S., ed. (John Wiley & Sons, Chichester, UK, 2013). pp. 195223. ISBN 9781118342374.

[5] McDonald E. J., Hallam K. R., Bell, W. and Flewitt, P. E. J., "Residual stresses in a multi-pass CrMoV low alloy ferritic steel repair weld," *Materials Science and Engineering: A*, 325(1-2), pp. 454-464 (2002). https://doi.org/10.1016/S0921-5093(01)01510-6.

On-line Measurement and Analysis of the Fan Blade Stress

Jaroslav Václavík[1], Radka Musilová[2], Hynek Purš[3] & Vojtěch Rulc[4]

Abstract: The object of the article is the parametric study of an operational stress on the radial 1000 mm industrial fan. The measurement system as well as the FEM model system validation was based on on-line monitoring with virtual models, called OLIN-4. The measurement system and obtained results are described in more detailed way.

Keywords: Fan; Blade; Stress; Telemetry system

1. Introduction

Within the solved project, advanced systems for on-line measurement and analysis of the operational load of structures are being developed, for which virtual numerical models are created in parallel. The OLIN-4 system is adapted for monitoring operating loads on rotating parts. To verify the function of the system, a fan test was performed on a test stand.

This system was developed within the project of the Ministry of Industry and Trade FV40260, aimed at supporting the development of transport means and other equipment which are loaded in operation by a variable load and must therefore be properly dimensioned for such loads.

2. Measuring system

OLIN-4 is the four-channel Bluetooth telemetry system for strain measurement on rotational parts, consisting of transmitter, receiver and software written in NI LabVIEW. The sampling frequency of the system is 40 ks/s. The supply of the system is made with the battery. Only full bridges can be connected to the system; for quarter bridge measurement the other gauges have to be substitute with precise resistors with low coefficient of thermal expansions. The receiver is equipped with two antennas which enables the signal passage through several mm thickness tested machine wall. The transmitted data are processed in stationary unit and send to the server. For this purpose the system has a web connection through LE VPN access. Here the data are used for on-line validation of the virtual model. The scheme of the system and laboratory realisation is given in Fig. 1 and Fig. 2.

After starting the module, the program saves the measured data to short files on the computer disk. At the same time, it calculates statistics from the measured files: maximum and minimum value, average value and standard deviation. It sends statistics to the cloud, alternatively to the VZÚ Plzeň server. It leaves the measured data on the disk. At the same time, it calculates the spectral power densities from the measured sections, which it leaves on the measuring computer and at the same time sends them to the cloud.

[1] *Ing. Jaroslav Václavík; Dynamic Testing Laboratory, Research and Testing Institute Pilsen; Tylova 1581/46, 301 00 Pilsen, Czech Republic;* vaclavik@vzuplzen.cz

[2] *Ing. Radka Musilová, Ph.D.; Dynamic Testing Laboratory, Research and Testing Institute Pilsen; Tylova 1581/46, 301 00 Pilsen, Czech Republic;* musilova@vzuplzen.cz

[3] *Ing. Hynek Purš, Ph.D.; Advanced Engineering s.r.o.; Na Ostrohu 2405/16, 160 00 Prague, Czech Republic;* hpurs@advanced-eng.cz

[4] *Ing. Vojtěch Rulc; Advanced Engineering s.r.o.; Na Ostrohu 2405/16, 160 00 Prague, Czech Republic;* vrulc@advanced-eng.cz

The rain-flow matrices and the estimated damage of the measured node are evaluated as well. The measured data can be visualised or used for further improvement of the virtual model (Fig. 3).

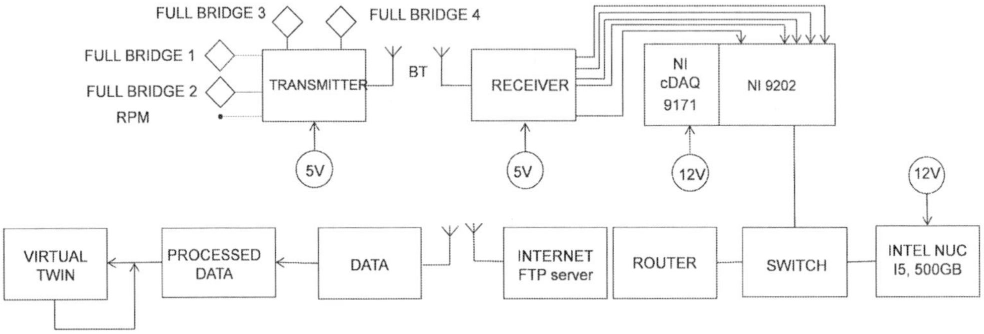

Fig. 1. Scheme of the on-line measuring system OLIN-4.

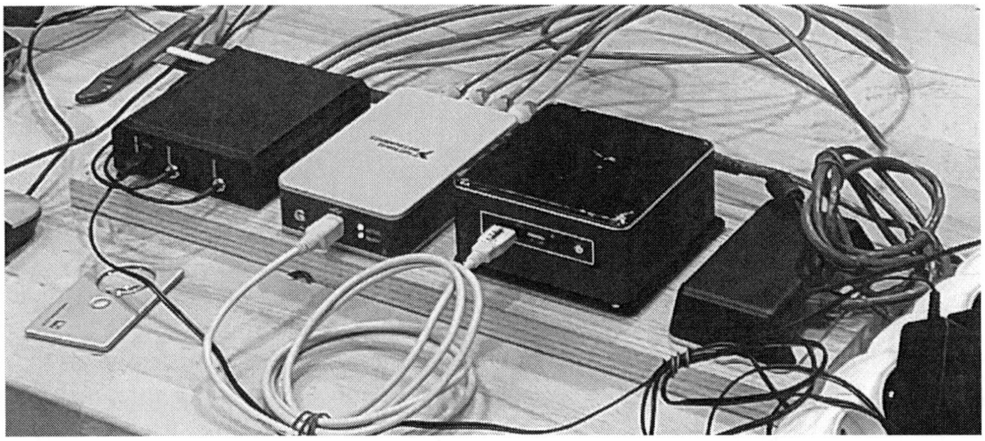

Fig. 2. Basic instruments of the on-line measuring system OLIN-4.

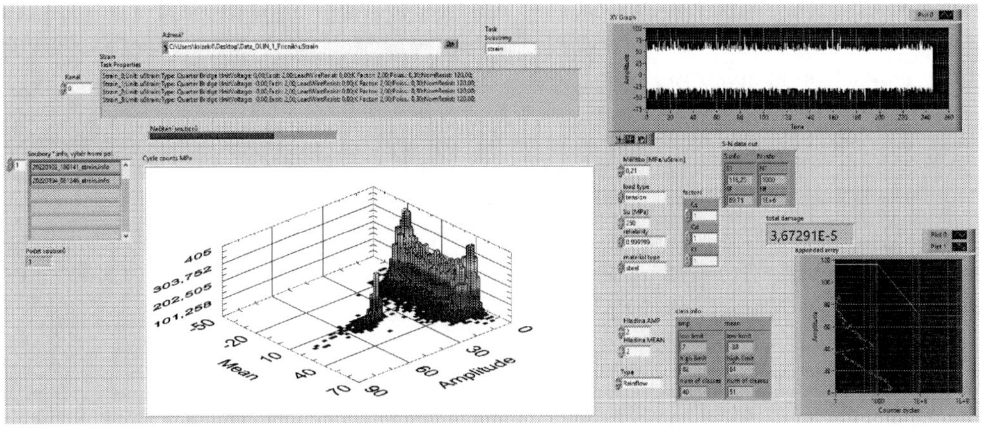

Fig. 3. Program interface for fatigue damage computation.

3. FEA Model

Altair tools from HyperMesh, OptiStruct and HyperView were used to build, calculate and evaluate the results of the virtual twin. The CAD model of investigated fun is given in Fig. 4. The positions for the application of strain gauges, which are the leading edges of the blades and the edges of the upper and lower edges of the blades, were also determined by these tools, see Fig. 5. One of goals of these investigations was also the comparison of two FEA accesses with the measured results.

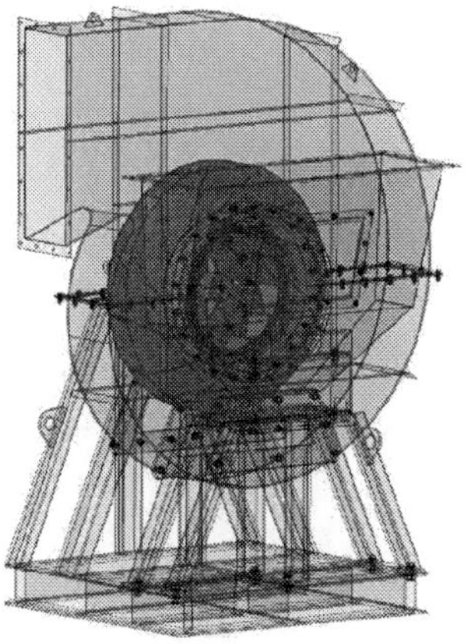

Fig. 4. CAD data of the investigated fan.

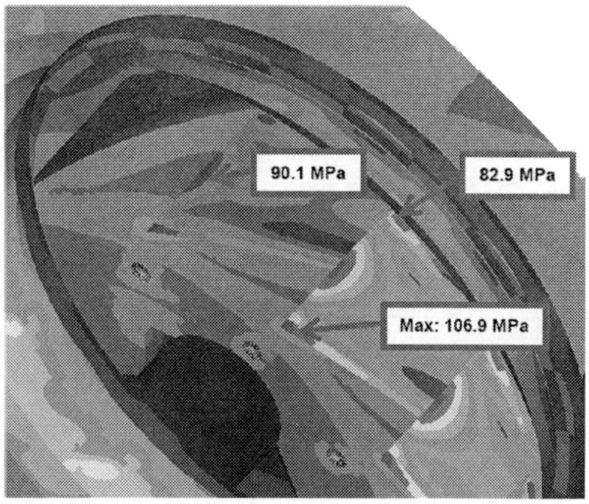

Fig. 5. Distribution of maximum HMH blade stresses.

4. Performed test

The validation of the system was performed on the fan test rig (Fig. 6). The transmitter including batteries was clamped to the plate of the impeller (Fig. 7). Three strain gauges were installed at the blade stuck just simulating the strain gauge rosette (SG1 to SG3). SG 4 was installed in the middle of the blade edge. The receiving antennas were located behind the closing steel lid. Following parameters were changed during the test: revolution of the impeller up to $n = 1650$ 1/min including start and stop at several revolution levels (parameter 1); change of the static pressure on the fan output at zero angle of control vanes up to $p = 5$ kPa (parameter 2); change of the tilt of control vanes from -30° to +75°(parameter 3) and the change of the tilt of control vanes at fan open output (parameter 4).

An example of measured stress change for steady revolution pressure is given in Fig. 8. Other investigated parameters had no substantial influence to the blade stress, see relations on Fig. 9, Fig. 10 and Fig. 11.

Fig. 6. Investigated industrial fan with open lid of the fan impeller.

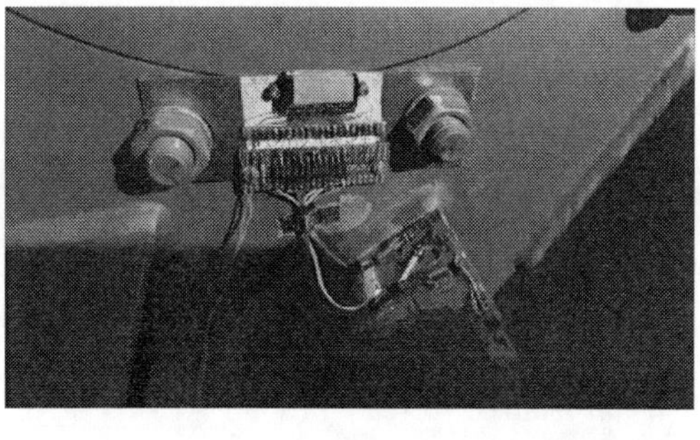

Fig. 7. Detail of the measuring amplifier and installed strain gauges on the blade.

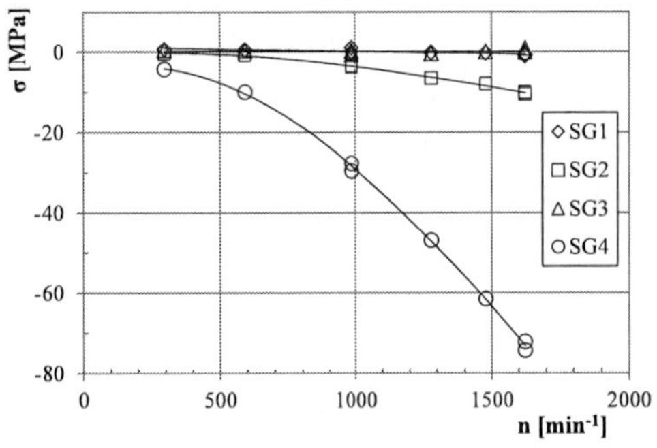

Fig. 8. Stress at four installed strain gauges as the function of fan revolutions.

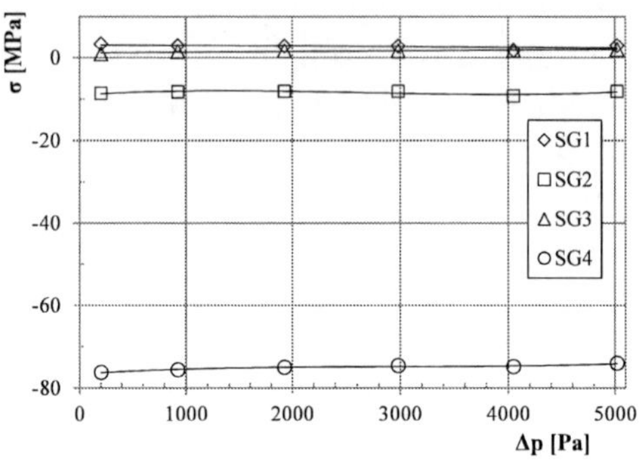

Fig. 9. Stress at four installed strain gauges as the function of the static output stress.

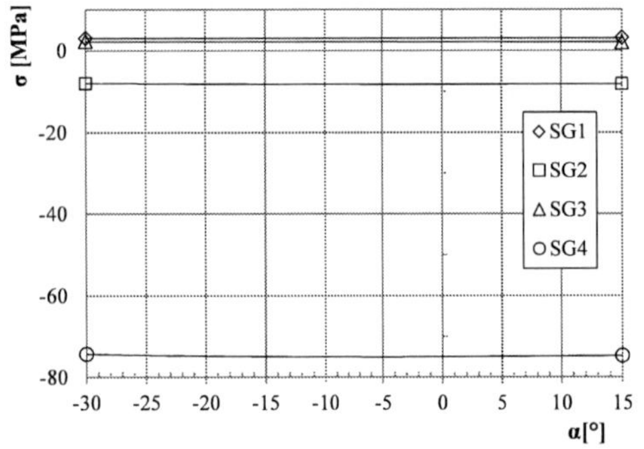

Fig. 10. Stress at four installed strain gauges as the function of the tilt of control vanes at the maximum static output pressurs.

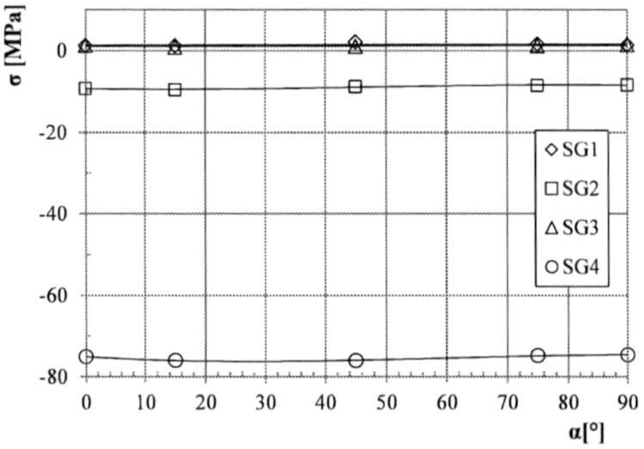

Fig. 11. Stress at four installed strain gauges as the function of the tilt of control vanes at open output.

5. Conclusion

It was found that the quasi-static loading of the blades due to centrifugal forces is significantly below the static endurance strength of the fan. The dynamic component under operating load does not reach significant values that could cause fatigue damage to the blade material, only repeated fan starts can contribute to the damage accumulation. The system is ready to use for measurement in real conditions.

Acknowledgements

The article was supported with Ministry of Industry and Trade under the program TRIO and project No. FV40260.

References

[1] Václavík J. and Musilová R., *OLIN 4 - Design and verification of an online system for monitoring stress on rotating parts*, Research report VZU-VZ-54/21/044, (VZÚ Plzeň, December 2021) (in Czech).

[2] Križan J. at all, *On-line measurement and analysis of operational loads of structures with adaptive virtual models*, Research report AE-VYZ-2021-FV40260-1, (Advanced Engineering s.r.o., Prague, December 2021) (In Czech).

Tensile and Compressive Properties of 3D-printed PETG Depending on Temperature

Tereza Vaňková[1], Tomáš Kroupa[2], Jan Krystek[3], Hana Zemčík[4] & Robert Zemčík[5]

Abstract: This paper focuses on the influence of temperature on the mechanical properties of 3D-printed PETG. Printed specimens were exposed to cyclic and monotonic tests in tension and compression. Young's modulus, yield stress, ultimate tensile stress and other properties were identified for five different temperatures. The obtained temperature dependences for different types of tests were subsequently compared.

Keywords: Experiment; PETG; Temperature; Tension; Compression; Cyclic; Plasticity

1. Introduction

The advancement of 3D-printing technology and its availability significantly influenced its use in production. Currently, the application of 3D-printing technology is immense, from small products with decorative purposes [1] to medical implants [4], or even whole buildings [3]. However, this wide applicability presents many different demands for the material properties of the final product.

This work deals with the influence of temperature on the mechanical properties of 3D-printed PETG (polyethylene terephthalate glycol-modified) during the tensile and compression tests. Several dependencies were obtained and subsequently evaluated on the basis of the analysis of experimental data. The identified parameters can be used for numerical simulation.

The influence of temperature on tensile properties of 3D-printed material was presented in [6]. In the cited article, the investigated material was PLA (polylactic acid) and the specimens with three different infill orientations were tested under static loading within the range between 20 °C and 60 °C. The decreasing character of Young's modulus and ultimate strength was observed for increasing temperature and the temperature influence was more significant than the influence of infill orientations. The tensile and compression mechanical and thermal properties of PETG and PLA were determined in [5]. The tested specimens with 20% infill density exhibited obvious tensile and compression asymmetry and the ultimate compressive stress exceeded the ultimate tensile stress.

A similar material as in the presented paper with the same infill density was of interest in [2], where the tensile monotonic and cyclic tests were carried out for four different infill orientations. The properties determined for infill orientation 45°/−45° (this orientation was selected at the presented paper) were mostly worse compared to the other orientations. Measured ultimate tensile strength ranged between 41.58 MPa and 48.04 MPa and Young's modulus varied from 1.48 GPa to 1.62 GPa. The authors designated this material as only slightly anisotropic.

[1] *Ing. Tereza Vaňková; Faculty of Applied Sciences, University of West Bohemia; Univerzitní 8, 301 00 Plzeň, Czech Republic;* vankovat@kme.zcu.cz
[2] *Ing. Tomáš Kroupa, Ph.D.; Faculty of Applied Sciences, University of West Bohemia;* kroupa@kme.zcu.cz
[3] *Ing. Jan Krystek, Ph.D.; Faculty of Applied Sciences, University of West Bohemia;* krystek@kme.zcu.cz
[4] *Ing. Hana Zemčík, Ph.D.; Faculty of Applied Sciences, University of West Bohemia;* hzemcik@kme.zcu.cz
[5] *doc. Ing. Robert Zemčík, Ph.D.; Faculty of Applied Sciences, University of West Bohemia;* zemcik@kme.zcu.cz

2. 3D printing specimens

The rectangular specimens with a length of 82 mm and width of 15 mm were printed horizontally with 100% density of infill and final thickness of 2 mm (see Fig. 1). The PETG filament and Original Prusa i3 MK3 printer were used. The main printing parameters are summarized in Table 1.

Table 1. Printing parameters

Parameter	Value
Layer height	0.2 mm
Line width	0.45 mm
Infill density	100 %
Infill pattern	linear
Raster angle	45°/-45°
Contour thickness	0.9 mm
Printing temperature	250 °C
Print speed	25 mm/s
Nozzle diameter	0.4 mm
Build plate temperature	90 °C

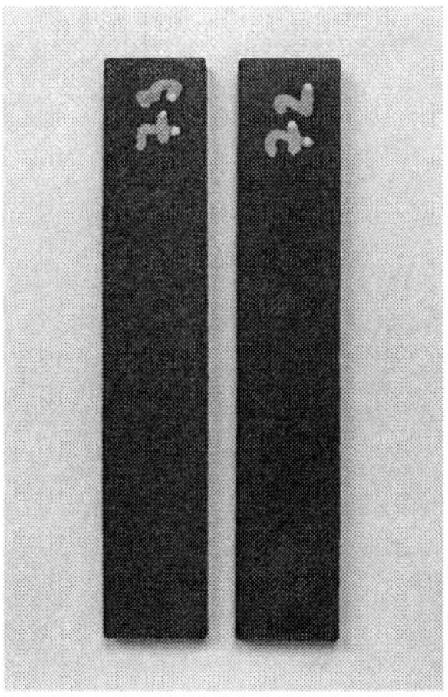

Fig. 1. 3D-printed specimens.

3. Experiments

The cyclic and monotonic tests were carried out on the specimens in both compression and tension. The universal testing machine Zwick/Roell Z050 was used for all experiments. Each type of experiment was performed at five different temperatures (-20 °C, 0 °C, 22 °C, 40 °C and 60 °C, in addition the monotonic tensile tests were also performed at 50 °C). The specimens were exposed to a quasi-static loading with loading velocity v = 1 mm/min and gage length 20 mm in tensile tests (resp. 25 mm in compression tests). The force-displacement dependencies were obtained from the experiments. A uniaxial (or biaxial in compression tests) extensometer was used for measuring the elongation.

A support jig according to standard ASTM D695 was used for compression tests. The load-unload cyclic tests were performed with increasing maximum load per cycle (resp. decreasing loading force in compression tests).

4. Calculated values

The processing of the measured data was performed using scripts written in Python 3.5. The stress-strain dependencies were calculated from the experimentally obtained value using

$$\varepsilon = \frac{\Delta l}{l_0} \tag{1}$$

and

$$\sigma = \frac{F}{A_0}, \tag{2}$$

where ε is strain, σ is stress, F is force, A_0 is initial cross section, Δl is elongation of measured length l_0. The maximum achieved stress $\sigma_{\max} = \max\{|\sigma|\}$ was determined from all experiments. The Young's modulus E was identified on the interval of strain $\varepsilon \in (0.001, 0.003\rangle$ from the monotonic tests (negative values for compression). The average values were determined from obtained data for each type of experiment and temperature.

The yield stress R_0 was calculated from the average curve of monotonic tests as the stress at which 0.05 % plastic strain occurs. Other behavior such as dependencies of tangents of load-unload cycles on total strain and dependencies of plastic strain on total strain were calculated from the load-unload cyclic tests (see Fig. 2 and Fig. 3). The total strain for tangents and plasticity dependencies was determined as the strain corresponding to the maximum (or minimum for compression) force of the cycle.

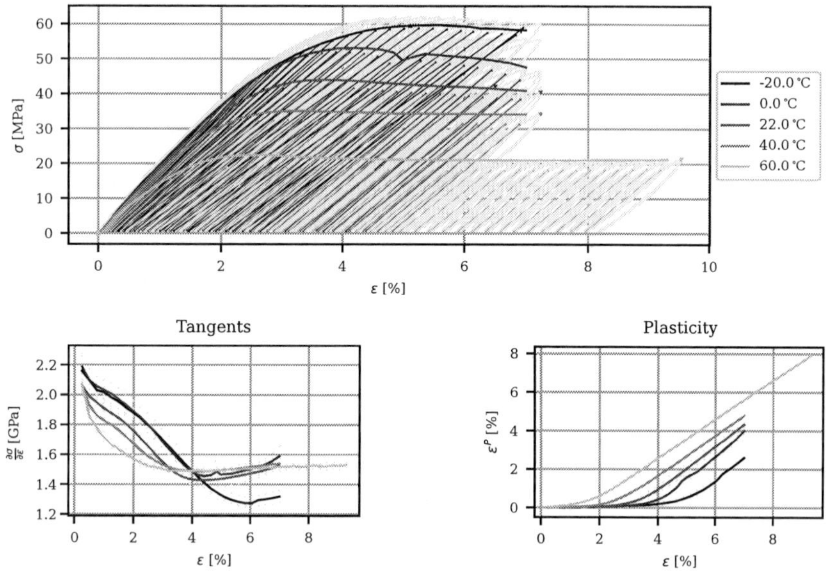

Fig. 2. Values determined from a cyclic tensile test (dependence of stress, load-unload tangents and plastic strain on total strain).

5. Results

In particular, the temperature dependencies of material parameters such as Young's modulus, maximum achieved stress and yield stress were determined and approximated by several different functions. The linear, quadratic and two exponential functions were selected. The difference between the experimentally obtained value and the approximated function was evaluated by the residual $\tilde{r}$ in the form

$$\tilde{r} = \frac{1}{n}\sum_{i=1}^{n}\left(\frac{A_{T_i}^{\text{exp}} - A^{\text{app}}(T_i)}{A_{\text{max}}}\right)^2, \tag{3}$$

where n is a number of temperatures from the experiment, $A_{T_i}^{\text{exp}}$ is the value from experiment at temperature T_i ($T_1 = -20$ °C, ..., $T_5 = 60$ °C), A^{app} is the approximated function and A_{max} is the maximum value from the experiment. The residuals were shown in Fig. 4 – Fig. 11. The deviation of value from the experiment was displayed for Young's modulus and maximum achieved stress (it can not be displayed for yield stress because the yield stress was determined from the average value).

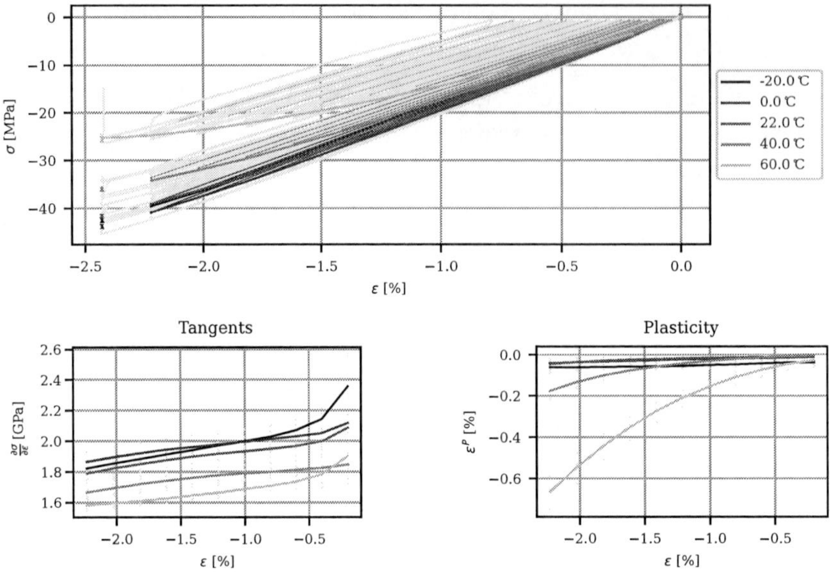

Fig. 3. Values determined from a cyclic compression test (dependence of stress, load-unload tangents and plastic strain on total strain).

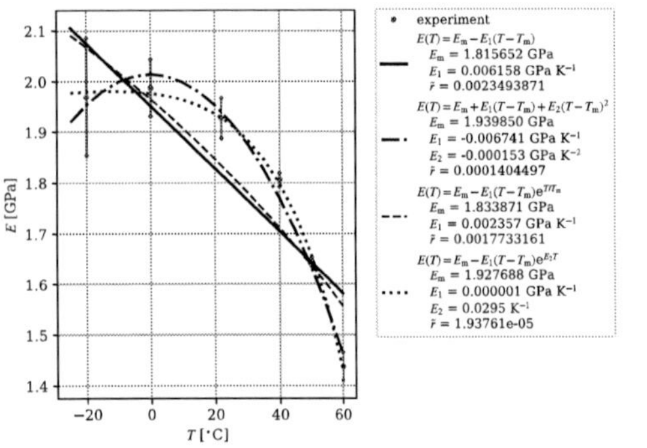

Fig. 4. Approximation of the dependence of tensile Young's modulus E on temperature.

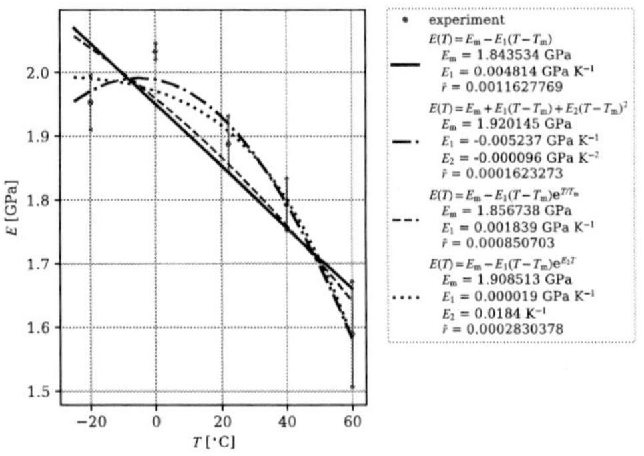

Fig. 5. Approximation of the dependence of compression Young's modulus E on temperature.

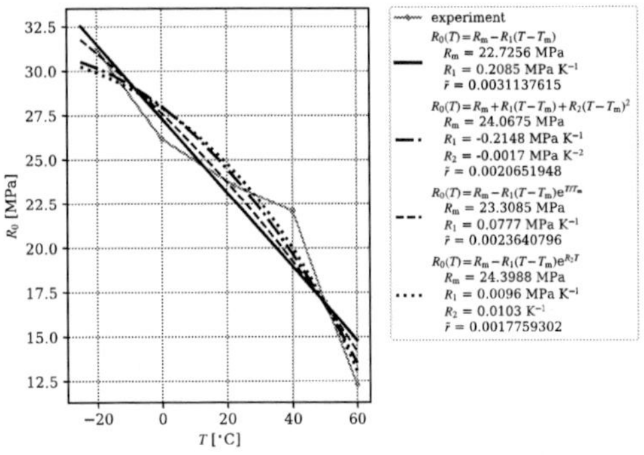

Fig. 6. Approximation of the dependence of tensile yield stress R_0 on temperature.

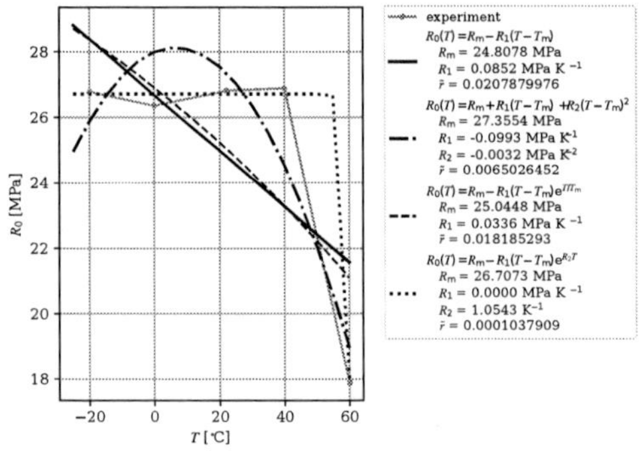

Fig. 7. Approximation of the dependence of compression yield stress R_0 on temperature.

In most cases, the temperature dependence of material parameters is best approximated by an exponential function with two parameters. In other cases (for compression values), the quadratic function seems better. The disadvantage of both of these functions is that they have two parameters. Therefore, for some applications, the better choice could be the exponential function with one parameter or even the linear function, which is the worst approximation of all the selected functions but it is the simplest.

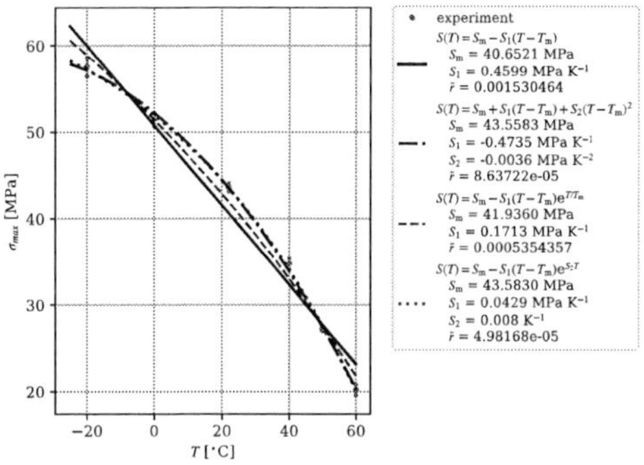

Fig. 8. Approximation of the dependence of tensile maximum achieved stress in monotonic test $\sigma_{\max}$ on temperature.

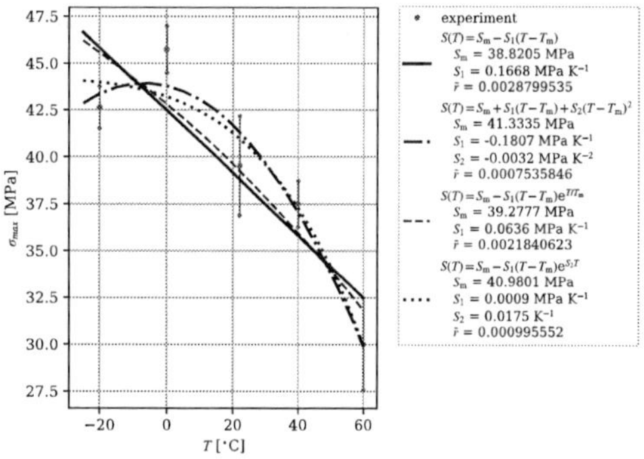

Fig. 9. Approximation of the dependence of compression maximum achieved stress in monotonic test $\sigma_{\max}$ on temperature.

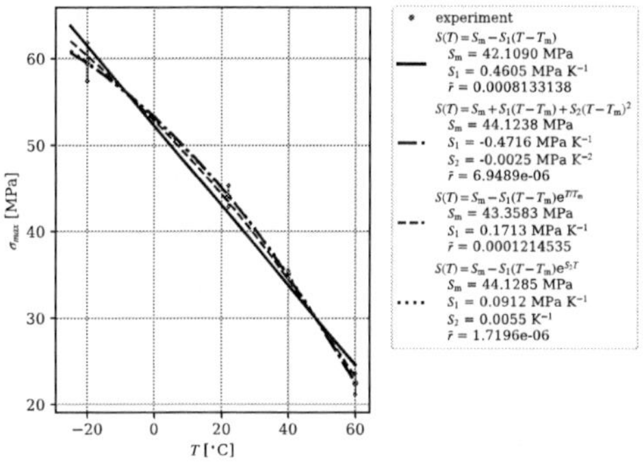

Fig. 10. Approximation of the dependence of tensile maximum achieved stress in cyclic test on $\sigma_{\max}$ temperature.

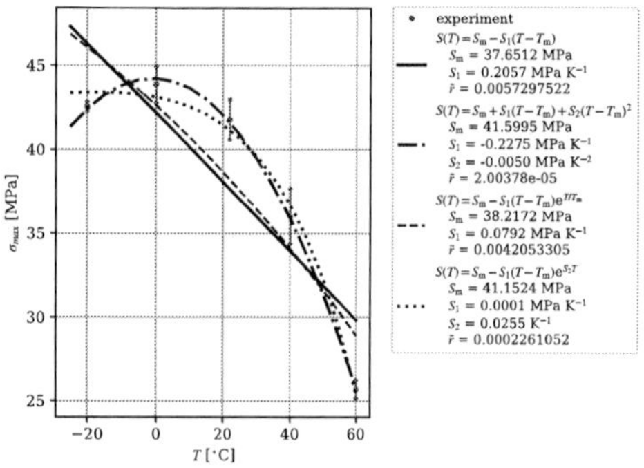

Fig. 11. Approximation of the dependence of compression maximum achieved stress in cyclic test $\sigma_{\max}$ on temperature.

The processed values of Young's modulus from tensile and compressive tests were similar but the maximum achieved stress varied. On the other hand, the maximum achieved stresses from cyclic and monotonic tests were similar which is obvious from Fig. 12. The tensile and compression yield stress were different, as is evident from a comparison of Fig. 6 and Fig. 7.

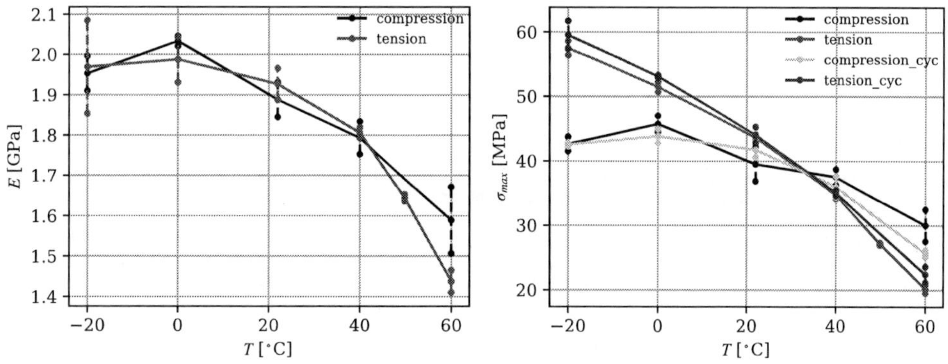

Fig. 12. Comparison of Young's modulus E and maximum achieved stress $\sigma_{\max}$ obtained from different tests.

6. Conclusions

The influence of temperature on the mechanical properties of 3D-printed PETG was investigated. Young's modulus, yield stress, maximum achieved stress, load-unload tangents and plastic strain were identified for five different temperatures from four types of experiments.

The processed values of maximum compressive stress were higher for some temperatures and lower for others than in tension. This may be due to the fact that the compression tests were terminated (for most specimens) by exceeding the maximum displacement measurable by the used extensometer. Another test should be performed to determine the ultimate compressive stress. The maximum achieved tensile stress values can be considered as the ultimate tensile stress.

The same reason could be the cause of a decrease in the maximum compressive achieved stress at negative temperatures (i.e. at -20 °C). However, there is no reason for the reduction of Young's modulus at negative temperatures. This was determined from tensile and compressive tests, although in tensile tests the decrease may be covert in a large variance of the values obtained at temperature of -20 °C. It would be desirable to carry out other measurements at a different temperature $T < 0$ °C to verify this character of temperature dependence.

The yield stress was specified as the stress at which 0.05% plastic deformation occurs. This offset was preferred to the commonly used value 0.2 because the yield stress at the usual offset was not found for compression tests at temperatures 0 and 22 °C (the yield stress was greater than the maximum achieved stress). This is in contrast to the fact that plastic deformation was determined in cyclic compression tests at the mentioned temperatures. However, the magnitude of the plastic deformation is not very large (as well as for the temperature -20 °C). Therefore, it is possible that the plasticity of this material loaded with compression at 0 and 22 °C (perhaps at -20 °C too) could be neglected.

Acknowledgements

This publication was supported by the project SGS-2022-008 of the Czech Ministry of Education, Youth and Sports.

References

[1] Anon., 40 Beautiful Home Decor 3D Printing Ideas. *All3DP: All about 3D printing*. Available from: https://all3dp.com/1/diy-room-decor-ideas-3d-print-home-decor-diy-ideas/ Accessed: 2022-04-05.

[2] Dolzyk G. and Jung S., "Tensile and Fatigue Analysis of 3D-Printed Polyethylene Terephthalate Glycol," *Journal of Failure Analysis and Prevention*, 19(2), pp. 511–518 (2019). ISSN 1864-1245. https://doi.org/10.1007/s11668-019-00631-z.

[3] Hager I., Golonka A. and Putanowicz R., "3D Printing of Buildings and Building Components as the Future of Sustainable Construction?," *Procedia Engineering*, Vol. 151, pp. 292–299 (2016). ISSN 1877-7058. https://doi.org/10.1016/j.proeng.2016.07.357.

[4] Haleem A. and Javaid M., "Polyether ether ketone (PEEK) and its 3D printed implants applications in medical field: An overview:," *Clinical Epidemiology and Global Health*, Vol. 7, no. 4, pp. 571–577 (2019). ISSN 2213-3984. https://doi.org/10.1016/j.cegh.2019.01.003.

[5] Hsueh M.-H., Lai C.-J., Wang S.-H., Zeng Y.-S., Hsieh C.-H., Pan C.-Y. and Huang, W.-C., "Effect of Printing Parameters on the Thermal and Mechanical Properties of 3D-Printed PLA and PETG, Using Fused Deposition Modeling," *Polymers*, Vol. 13 no. 11, p. 1758 (2021). ISSN 2073-4360.https://doi.org/10.3390/polym13111758.

[6] Grasso M., Azzouz L., Ruiz-Hincapie P., Zarrelli M., and Ren G., "Effect of temperature on the mechanical properties of 3D-printed PLA tensile specimens," *Rapid Prototyping Journal*, Vol. 24 no. 8, pp. 1337–1346 (2018). ISSN 1355-2546. https://doi.org/10.1108/RPJ-04-2017-0055

Finite Element Model of Laminate Made of Functionally-oriented Fabrics

Radek Zbončák[1]

Abstract: Functionally-oriented fabric (FOF) is a special reinforcement type. Fibres of basic and functional material are irregularly distributed in warp and weft. That locally improves mechanical performance and raises material usage efficiency. This type of fabric is suitable for a high volume manufacturing of large laminated parts. Basic material can be cheap fibres. Functional material can be purposefully weaved where it is needed, like around the future notches. That can reduce final waste and cost. Creating finite element model of laminate made of FOF is very complex as two materials with different mechanical properties are combined together in woven structure. The UD damage model, which is derived from Ladeveze's model, is chosen for describing the damage of laminate. Woven fabric is substituted by UD plies. Tensile test on specimen with circular notch is simulated with progressive failure analysis to map the evolution of damage.

Keywords: Functionally oriented woven fabric; UD ply damage model; Siemens NX

1. Introduction

In addition to conventional fabric, the term "hybrid" fabric is used in practice. It is a fabric with a regular structure that cannot be described as "functionally oriented". Manufacturers define a hybrid fabric as follows: "Hybrid fabrics, tapes and stockings consist of more than one type of fibre. Thanks to the combination of different types of fibres, the individual properties of the fibres can be used according to current needs, such as impact resistance of aramid fibres with the stiffness of carbon fibres or with the compressive strength of glass fibres." [1] Regular alternation of materials in the hybrid fabric leads to homogenization of the properties of the laminate layer. In contrast to these common types of fabrics, the new functionally oriented structure has irregularly distributed fibres of different materials in the warp and weft. The base material, in this case glass, predominates in the fabric. The functional material has a minority in the fabric, it is a material with better mechanical properties and in most cases it is in principle more expensive than the base material. The production of a functionally-oriented structure (FOS) allows lamination without cutting the continuous fibres in the layer and having better mechanical properties at the same time. That can be used in the layup to locally improve mechanical properties to increase strength or stiffness. Suitable applications are laminates with holes for fasteners, or with openings such as windows or mounting openings providing access to machine parts under the laminate part.

For the prototype production of functionally oriented structures, the VERA jet loom was rebuilt to a beam width of 1300 mm. The basic warp wound on a standard warp roll is supplemented by a system of warp threads brought to the weaving loom from two external laboratory bobbins. 200 spools of external bobbins are wound with carbon and glass fibres on a developed laboratory single-purpose machine. In this way, manual parameterization of the FOS is enabled. The pattern consists of 150 threads in the warp and 150 threads in the weft. The width of the carbon strip and the distance between the woven strips can be modified. The resulting weight of the woven pattern with dimensions of 250×250 mm ranges from 306 to 320 g/m^2 depending

[1] *Ing. Radek Zbončák; VÚTS, a. s.; Svárovská 619, 460 01 Liberec, Czech Republic;* radek.zboncak@vuts.cz

on the amount of carbon fibres.

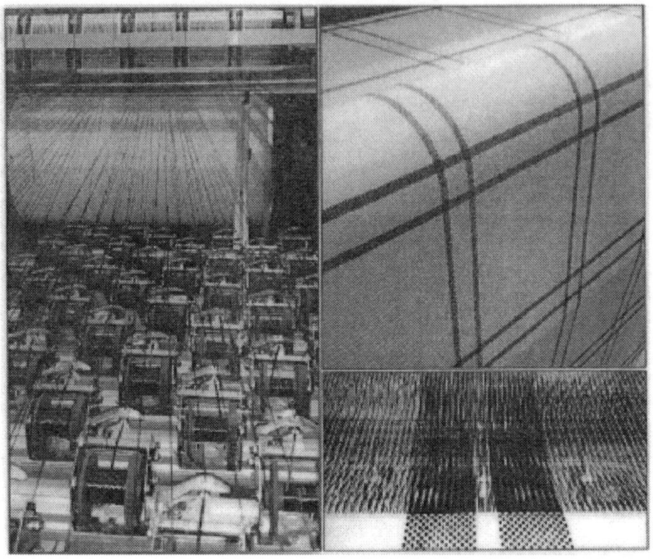

Fig. 1. Production of functionally-oriented fabric (FOF).

2. Experiment

Specimens for the experimental part of the project are made by infusion (VARTM). Specimens are machined and equipped with aluminum pads to prevent damage of the specimen in the jaws of testing machine. Each specimen is thus made in two versions – without a notch and with a notch. Notch means circular hole in the centre of the specimen. Subsequently, pattern of dots is applied on specimen's surface painted with white colour for better contrast, to measure in-plane deformation on the specimen surface. Non-standard size of the test specimen, the need to choose a suitable method of applying the pattern and other conditions are the reason for the creation of a separate methodological procedure, including the description of specimen preparation (internal company document MP2101) for this project. The load is realized by a uniaxial tensile test on the Instron 5982 device and the deformation is recorded using the Digital Image Correlation (DIC) method by the Monet3D system.

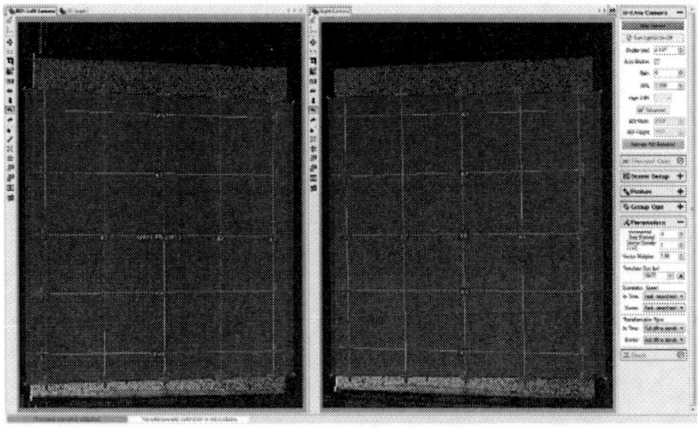

Fig. 2. Setting of virtual gauges in Mercury software for strain computation.

The figure shows how the virtual strain gauges (green grid lines) are placed in the measured area of the body. The relative deformation is evaluated from the 2nd and 4th strain gauges. The XX direction is from top to bottom and the YY direction is from left to right in Fig. 2.

The 100×200 mm test specimens are available in two layup variations: [0°/90°]s and [0°/45°]s. The width of the strip of functional material (carbon) is 10 mm and the distance between the strips is 10 mm, which corresponds to the diameter of the hole in the centre of the specimen.

Table 1. Specimens' sets.

Layup	[0°/90°]s	[0°/45°]s
E-glass	I-01	II-01
E-glass with notch	I-02	II-02
FOT	I-07	II-07
FOT with notch	I-14	II-14

The first two sets of each layup are made of the basic material (glass) only (Table 1). Each set has 10 test specimens, which are minimum for statistical evaluation of experimental data. Fig. 3 to Fig. 10 shows experimental results of tensile test.

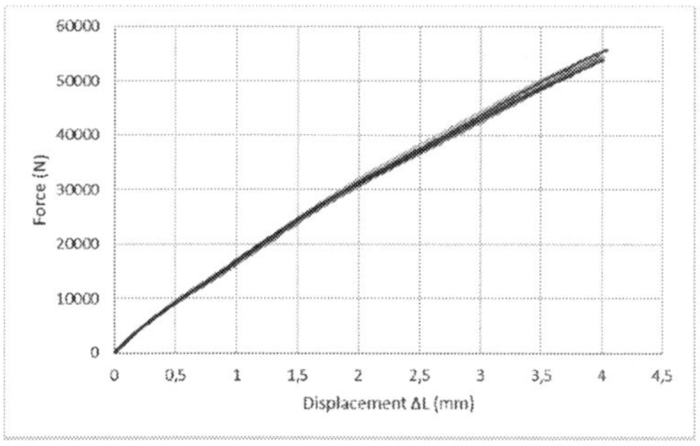

Fig. 3. Dependence of acting force on displacement for tensile test of set I-01.

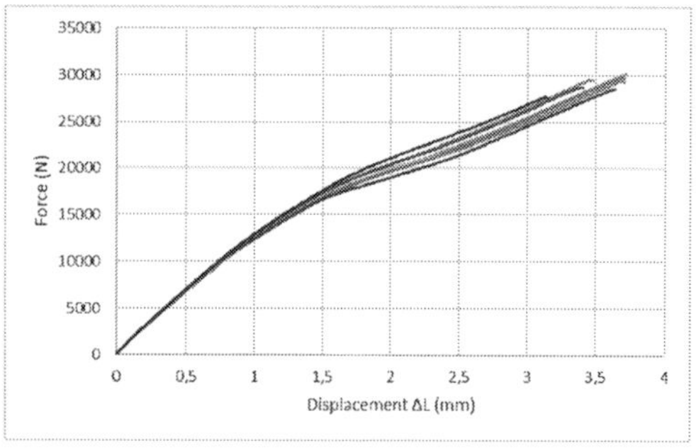

Fig. 4. Dependence of acting force on displacement for tensile test of set II-01.

There is 0°/90° layup in Fig. 3, where almost linear dependency occurs. While smaller value of acting force on 0°/45° layup at failure is perceptible in Fig. 4. That is predictable as there is just 50% of UDs in loading direction.

Fig. 5 and Fig. 6 show tensile curves of specimens with a notch. Nonlinearity of force-displacement dependency is again significant. Acting force of 0°/45° layup at the failure is about 30 % smaller than force of 0°/90°.

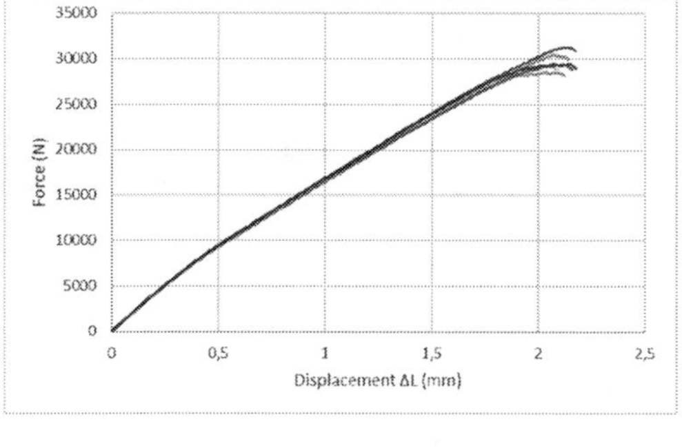

Fig. 5. Dependence of acting force on displacement for tensile test of set I-02.

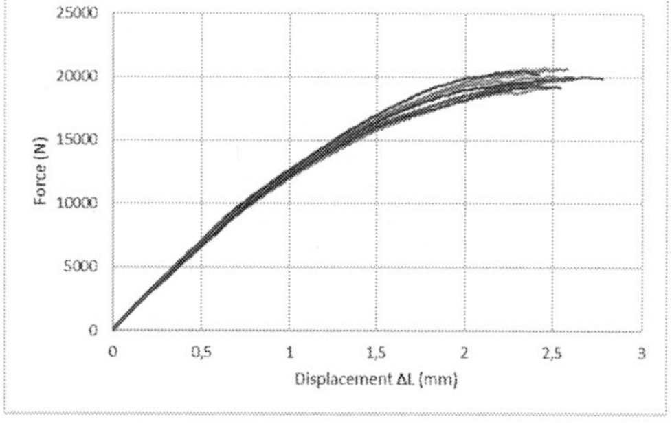

Fig. 6. Dependence of acting force on displacement for tensile test of set II-02.

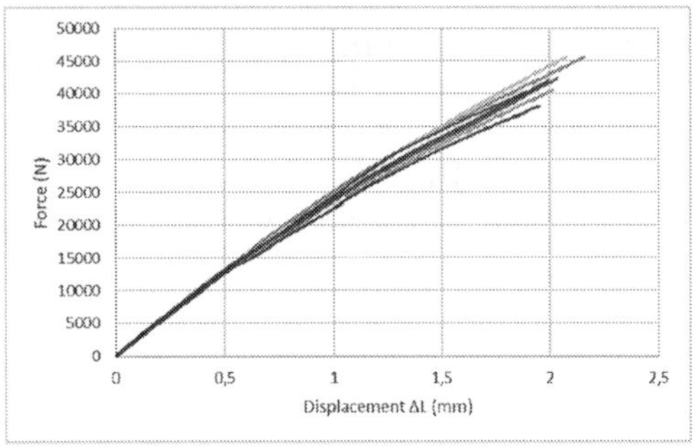

Fig. 7. Dependence of acting force on displacement for tensile test of set I-07.

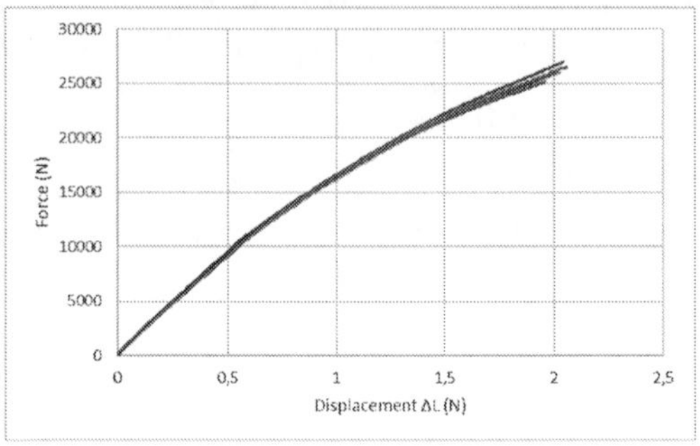

Fig. 8. Dependence of acting force on displacement for tensile test of set II-07.

Fig. 7 and Fig. 8 show tensile test performance of FOF specimens without a notch. Curves demonstrate higher stiffness than specimens made of glass fibres. The presence of carbon fibres negatively influences value of the force at the failure. Maximum value of acting force is about 5000 N smaller in both cases in comparison with pure glass specimens.

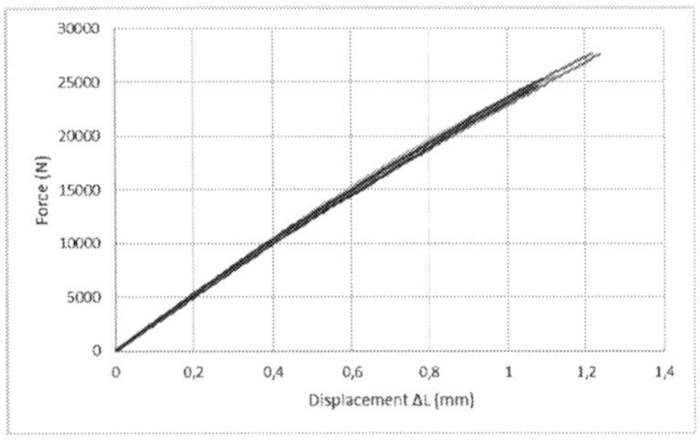

Fig. 9. Dependence of acting force on displacement for tensile test of set I-14.

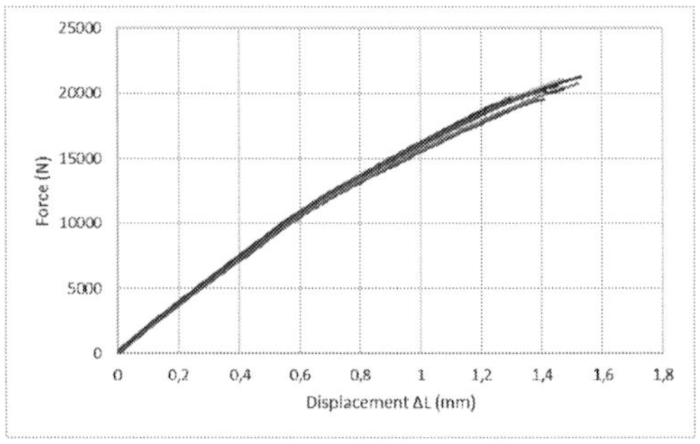

Fig. 10. Dependence of acting force on displacement for tensile test of set II-14.

Fig. 9 and Fig. 10 demonstrate carbon fibres benefit at the end of the curves. In comparison with pure glass specimen with notch, FOF specimens with notch have smaller displacement and reduced value of acting force at the break.

Table 2 and Table 3 summarize force and displacement values at failure for both layups. In case of 0°/90°, the presence of 10 mm carbon fibre strips does not increase the value of applied force at failure and decreases the elongation of the specimen as carbon makes the laminate stiffer.

Table 2. Force and displacement at failure for [0°/90°]s laminate.

[0°/90°]s	I-01	I-02	I-07	I-14
Failure force [N]	53 200	29 700	41 700	25 600
Failure displacement [mm]	3.87	2.07	2.00	1.13

Table 3. Force and displacement at failure for [0°/45°]s laminate.

[0°/45°]s	I-01	I-02	I-07	I-14
Failure force [N]	29 100	20 100	25 900	20 500
Failure displacement [mm]	3.54	2.50	2.00	1.14

For 0°/45°, the presence of carbon fibres is much beneficial. The difference between loading forces at failure for laminate with and without carbon strips is much smaller. The displacement has similar trend and values for 0°/90° and 0°/45° with carbon strips are almost the same.

3. Finite-element model

Specimens are made of E-glass fibres EC13 272 Z20 T63C and carbon fibres Tenax E HTA 40 E13 6K, which belongs to the fibres with standard modulus. The system is an epoxy resin LH 285 with hardener H 386.

The parametric finite-element laminate model of functionally-oriented fabric laminate is created in the Siemens NX program by the ply-based modelling method. The lamina defined as fabric is computed by software as three layers defined by unidirectional fibres. This approach is already taken into account during creating a laminate, where the fabric layer is defined as two outer layers UD (warp) and one intermediate layer UD rotated by 90° (weft). There are four different material areas with respect to the warp and weft in the FOF pattern: glass – glass, glass – carbon, carbon – glass and carbon – carbon. This can be well addressed by the above mentioned simplification (three UD layers). Outer layers are warp direction and inner layer is weft direction or the other way round. Important is to keep the format same within the entire model.

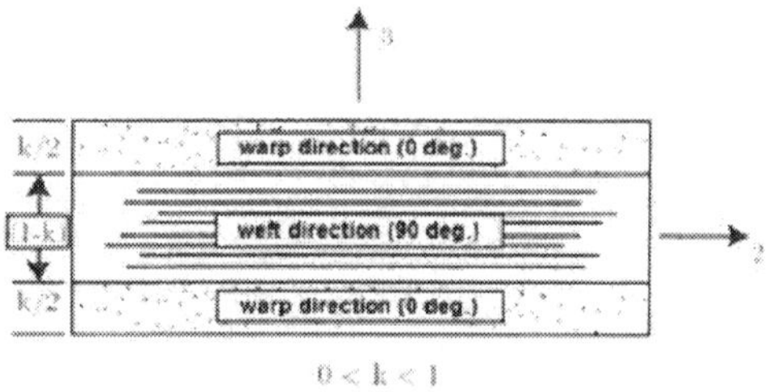

Fig. 11. The way the software transforms a woven fabric ply to an UD ply.

Progressive failure analysis uses model developed by Ladeveze and Le Dantec [1]. Laminate damage in this model is linked with the transverse direction 22 and the shear in the plane 12 of the laminate. Model is called UD ply damage model in Siemens NX Nastran and is implemented in nonlinear SOL 401. The model allows reducing the stiffness of the layer material in an iterative way during the simulation until the whole laminate fails.

Strain energy density at a point in a ply is

$$\begin{aligned} U_0 = & \frac{\sigma_{11}^2}{2(1-d_{11})E_1^0} - \frac{\nu_{12}^0}{E_1^0}\sigma_{11}\sigma_{22} - \frac{\nu_{13}^0}{E_1^0}\sigma_{11}\sigma_{33} - \frac{\nu_{23}^0}{E_2^0}\sigma_{22}\sigma_{33} + \\ & + \frac{\langle\sigma_{22}\rangle_+^2}{2(1-d_{22})E_2^0} + \frac{\langle\sigma_{22}\rangle_-^2}{2E_2^0} + \frac{\langle\sigma_{33}\rangle_+^2}{2(1-\lambda d_{22})E_3^0} + \frac{\langle\sigma_{33}\rangle_-^2}{2E_3^0} + \\ & + \frac{\tau_{12}^2}{2(1-d_{12})G_{12}^0} + \frac{\tau_{13}^2}{2(1-\lambda d_{12})G_{13}^0} + \frac{\tau_{23}^2}{2(1-\lambda d_{12})G_{23}^0}. \end{aligned} \tag{1}$$

Variables d_{ij} are damage variables which values are evaluated from experiment. Coefficient λ is an indicator of "out-of-plane" damage. If its value is 0, the damage is not linked to 3rd direction. If the value is 1, the influence of 3rd direction is included. Thermodynamic forces Y are obtained by taking derivative of the elastic strain E_0 with respect to damage variables.

$$Y_{11} = \frac{\partial U_0}{\partial d_{11}} = \frac{\sigma_{11}^2}{2(1-d_{11})^2 E_1^0} \tag{2}$$

$$Y_{22} = \frac{\partial U_0}{\partial d_{22}} = \frac{\langle \sigma_{22} \rangle_+^2}{2(1-d_{22})^2 E_1^0} + \lambda \frac{\langle \sigma_{33} \rangle_+^2}{2(1-d_{22})^2 E_3^0} + \lambda \frac{\tau_{23}^2}{2(1-d_{22})^2 G_{23}^0} \tag{3}$$

$$Y_{12} = \frac{\partial U_0}{\partial d_{12}} = \frac{\tau_{12}^2}{2(1-d_{12}) G_{12}^0} + \lambda \frac{\tau_{13}^2}{2(1-d_{12})^2 G_{13}^0} \tag{4}$$

If λ is equal to 0 only first term of the equations Y_{22} and Y_{12} remains. Superscript 0 means related to undamaged state. Evolution of "static" damage is a function of the thermodynamic forces.

$$d_{11}^s \begin{cases} 0 \text{ if } Y_{11} \leq Y_{11}^{\text{lim+}} \text{ and } Y_{11} \leq Y_{11}^{\text{lim-}} \\ 1 \text{ if } Y_{11} > Y_{11}^{\text{lim+}}, \sigma_{11} > 0 \text{ or } Y_{11} > Y_{11}^{\text{lim-}}, \sigma_{11} < 0 \end{cases} \tag{5}$$

$$d_{22}^s \begin{cases} \frac{\sqrt{\underline{Y}} - \sqrt{Y_{12}^0}}{\sqrt{Y_{12}^C} - \sqrt{Y_{12}^0}} \text{ if } d_{11} < 1, d_{22} < 1, Y_{12} < Y_{12}^F \\ 1 \text{ else} \end{cases} \tag{6}$$

$$d_{12}^s \begin{cases} b_3 d_{12} \text{ if } d_{12} < 1, d_{22} < 1, Y_{22} < Y_{22}^F \\ 1 \text{ else} \end{cases} \tag{7}$$

$Y_{11}^{\text{lim+}}$ and $Y_{11}^{\text{lim-}}$ are the breaking thresholds in traction and compression. Y_{12}^0 is a fibre/matrix debonding threshold. Y_{12}^C is critical thermodynamic force. Y_{12}^F and Y_{22}^F are thermodynamic forces in fragile behaviour and depend on Y_{12}^S and Y_{22}^S.

$$\underline{Y}(t) = \max_{\tau \leq t} (Y(\tau)) \tag{8}$$

$$Y(\tau) = b_2 Y_{22} + Y_{12} \tag{9}$$

b_2 and b_3 are coupling coefficients between damage variables.

$$Y_{12}^F = \sup \left\{ \frac{\tilde{h}}{h} Y_{12}^S, Y_{12}^S \right\} \tag{10}$$

$$Y_{22}^F = \sup \left\{ \frac{\tilde{h}}{h} Y_{22}^S, Y_{22}^S \right\} \tag{11}$$

The alternative model is the Enhanced UD ply damage model which splits damage variables into two parts: diffuse damage related parts and crack density dependents parts. Additionaly cohesive elements can be used in the model to take into account the nonlinearity in the 3rd direction.

CHEXA or CPENTA solid elements must be used for analysis of laminate. Material definition requires usage of a MATDMG card, which extends MAT11 data by other material parameters, which must be determined experimentally or numerically. Due to simplification of woven ply modelling, literature can be used to get proper data of UD fibre laminate [4]. However small deviations in results must be taken into account as the data are not exactly of measured specimens.

The result of the analysis are 4 allowable laminate failure states, which are designated as the "Status" variable in the results. The first condition is undamaged laminate (0). The second condition occurs if some layers of the laminate are damaged but not failed (1). In the third state, some layers are failed and others might be damaged (2). Finally, the fourth condition that can occur is failure of all layers of the laminate (3). Status is an alternative representation to failure index and shows the ability of laminate to withstand the load.

4. Results

There are simulated force-displacement relationships in the Fig. 12 and Fig. 13. Curves in the Fig. 12 correspond with experimental data of 0°/90° layup and represent the fact that using of carbon fibres increases stiffness of specimen with a notch. However the strength drops in this case. Fig. 13 shows 0°/45° layup and its performance with and without the carbon strips in notched specimen.

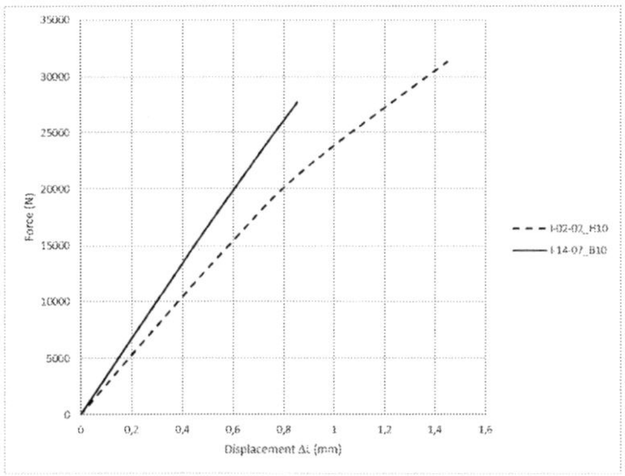

Fig. 12. Simulated Force-Displacement relationship of I-02 and I-14.

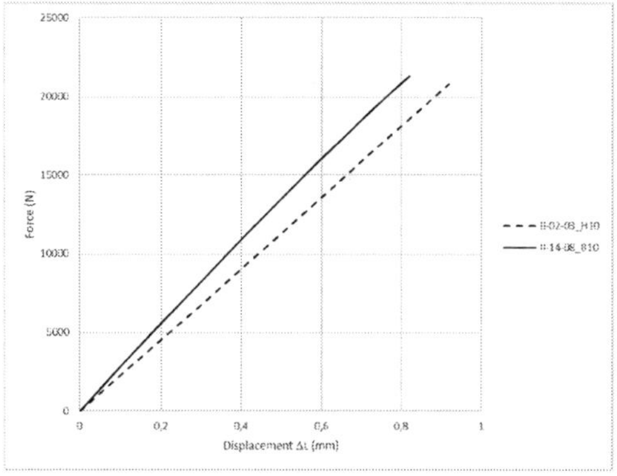

Fig. 13. Simulated Force-Displacement relationship of II-02 and II-14.

It must be stated, that general material damage data have been used in simulation and that is why absolute values of displacement in simulation does not correspond exactly with experimental data. To have comparable resultant data, it has been decided to specify threshold value of 15 kN for plain strain.

Results of UD ply damage model are presented on FOF laminates with a notch. The light blue colour in Fig. 14, represents damage status equal to 1, which means damage of some plies of 0°/90° layup. The location of start of the damage is close to notch edge. The way the damage develops is shown in Fig. 15. The red areas correspond to failing elements in the plies with fibres oriented perpendicularly to loading direction and light blue areas means, that some of the plies in the region are damaged. Red elements are the spot where the crack is going to start to grow from. Fig. 16 and Fig. 17 shows plain strain on specimen I-02 surface captured by DIC.

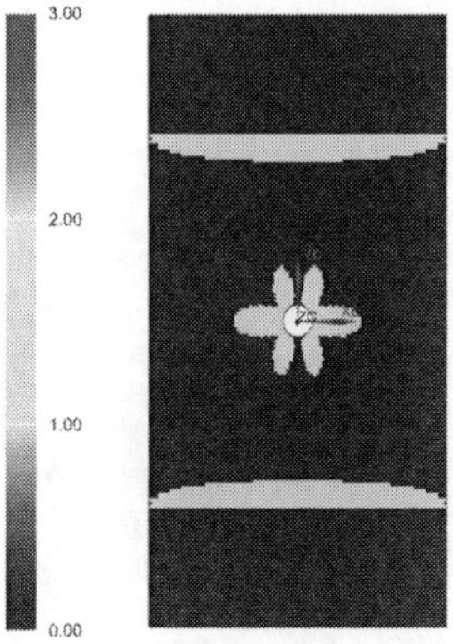

Fig. 14. Damage status of I-02 at 15 kN.

Fig. 15. Detail of failed elements in notch area of I-02 at 15 kN.

The issue is an applying pattern on the specimen surface close to the notch edge. DIC is sensitive to shadows which negatively influence contrast on the edge of the hole. The black hole area in combination with black dots in pattern make the edge of the notch almost not focusable.

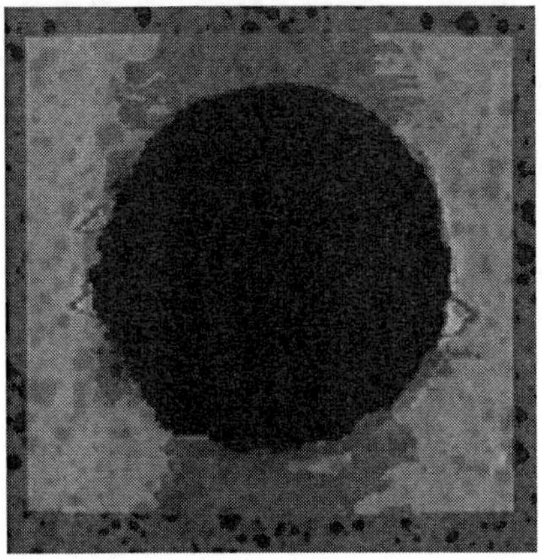

Fig. 16. Strain *XX* of I-02 at 15 kN.

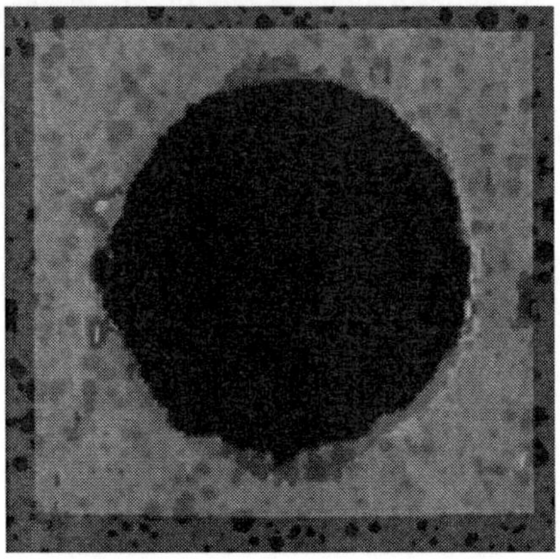

Fig. 17. Strain *YY* of I-02 at 15 kN.

Fig. 18 and Fig. 19 describes specimen I-14 in the same sense. What is important here is that presence of carbon strips reduces a number of damaged elements to none. In the detailed Fig. 17, the damaged area in transverse direction is obviously smaller.

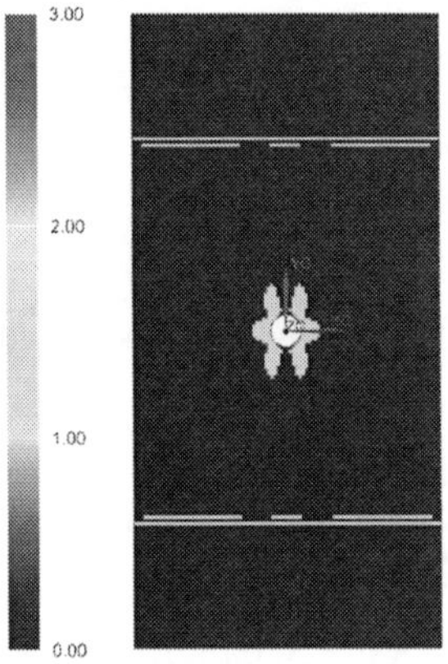

Fig. 18. Damage status of I-14 at 15 kN.

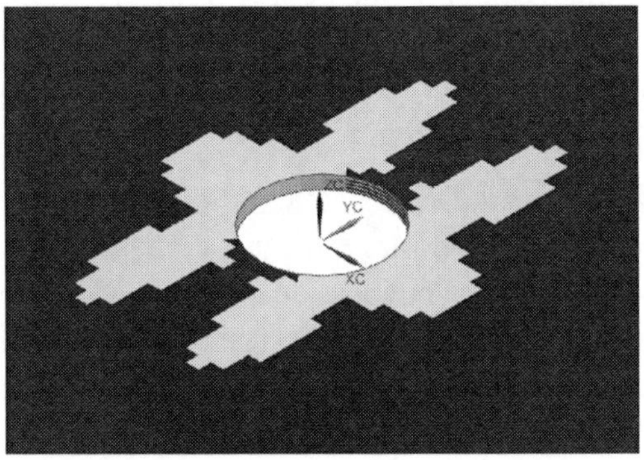

Fig. 19. Detail of failed elements in notch area of I-14 at 15 kN.

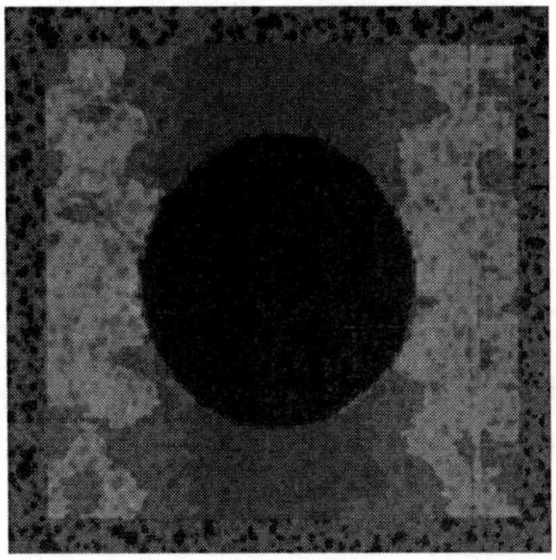

Fig. 20. Strain *XX* of I-14 at 15 kN.

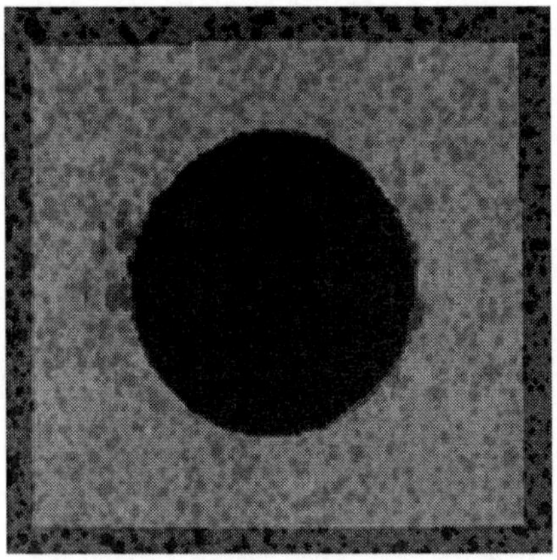

Fig. 21. Strain *YY* of I-14 at 15 kN.

The same behaviour can be observed for II-14 of 0°/45° layup in Fig. 22 and Fig. 23. Whole laminate is undamaged and first damaged elements occur just around the notch. After the rising the force, number of damaged elements increases and first failed elements occur around the notch in 90° layers where stress concentration is high (red).

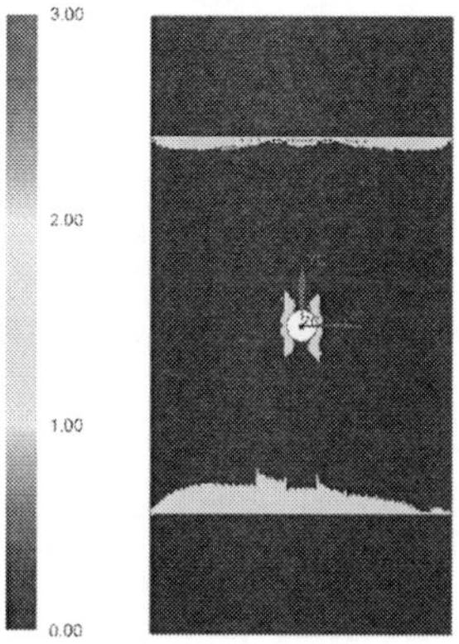

Fig. 22. Damage status of II-14 at 15 kN.

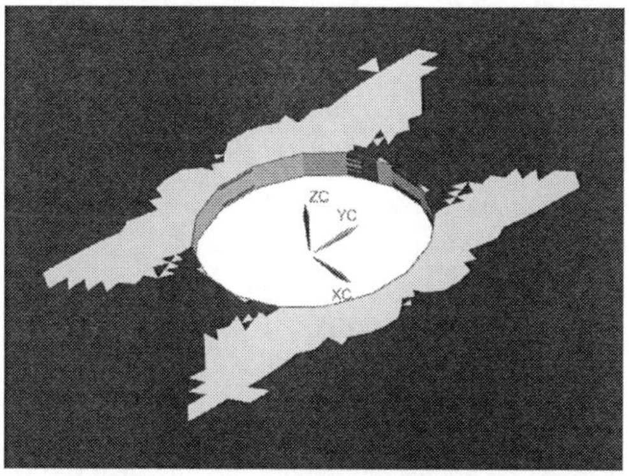

Fig. 23. Detail of failed elements in notch area of II-14 at 15 kN.

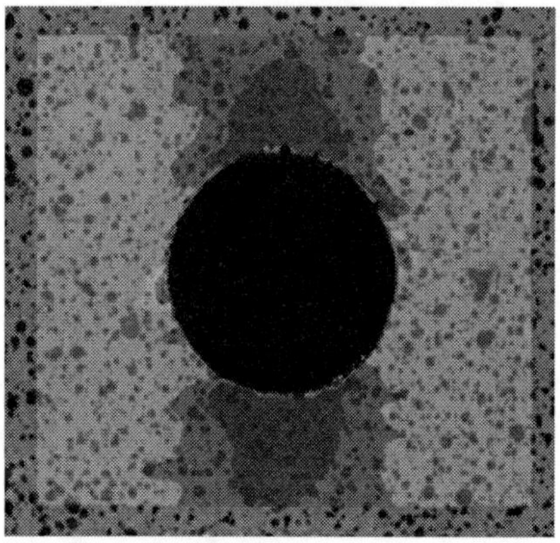

Fig. 24. Strain *XX* of II-14 at 15 kN.

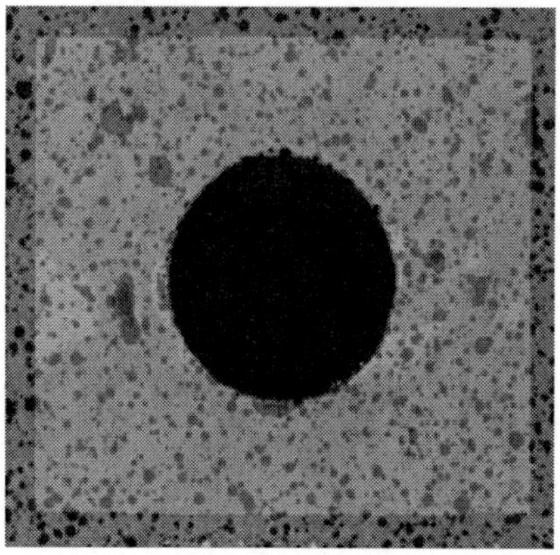

Fig. 25. Strain *YY* of II-14 at 15 kN.

5. Conclusion

This work is an introductory contribution of a long-term project focused on functionally-oriented fabrics and their use. The initial phase of the project is devoted to the production technology, which has its specifics mainly due to the fragility and conductivity of carbon capillaries. The next step is an extensive experimental part of the project, of which only the necessary part related to the creating of a finite-element model is presented here. For the purpose of creating the FE model, layups with fibre orientation of 0°/90° and 0°/45° and test specimens with and without a notch are selected. Selected laminate damage model is appropriate in this case. Damage develops in predicted regions, which correspond with experimental data. Attention must be focused on evaluation of material data, especially in laminates, where manufacturing technology and laminate quality strongly influence mechanical properties. To make the woven fabric model more precise, additional material damage properties' experiments need to be done to reduce the scatter of the experimental data.

Results of ongoing work show that functionally-oriented fabric has its added value. Mechanical properties around the notch are improved by weaving strips of functional material into the fabric. Another benefit is the fact that lamina consists of continuous fibres.

Detailed evaluations of all measurements will be published in subsequent articles as the project is still running. For example, other types of loading, not just one dimensional tensile test, will be tested including cyclic test.

Acknowledgements

This publication was supported by the Ministry of Industry and Trade (MPO) within the framework of institutional support for long-term strategic development of the research organization - provider MPO, recipient VÚTS, a. s.

References

[1] "http://www.grm-systems.cz," [Online]. Available: http://www.grm-systems.cz/hybridni-ostatni-vyztuze.

[2] Ladeveze P. and Le Dantec E., "Damage modelling of the elementary ply for laminated composites," *Composites Science and Technology*, n. 43., pp. 257-267, (1992), https://doi.org/10.1016/0266-3538(92)90097-M.

[3] Hochard C., Aubourg P.-A. and Charles J.-P., "Modelling of the mechanical behaviour of woven-fabric CFRP laminates up to failure," *Composites Science and Technology*, n. 61, pp. 221-230, (2001), https://doi.org/10.1016/S0266-3538(00)00199-8.

[4] Yi T., "The progressive failure analysis of uni-directional fibre reinforced composite laminates," *Journal of Mechanics*, n. 84, pp. 136-144, (2020), https://doi.org/10.1017/jmech.2019.55.

Simulation of Supercritical Bending of Cladded Pipes

Oskar Zemčík[1], Martin Slaný[2], Jan Zouhar[3] & Josef Sedlák[4]

Abstract: As a part of the project, experiments involving supercritical bending of cladded pipes for modern combustion boilers were performed. The supercritical bend in thick-walled pipes is given at a radius below $2.5d$ [1], when the cross-section of the bent pipe is already significantly deformed. Within the project, supercritical bending tests with radii $1d$ and $0.75d$ were performed. The simulation included the design of a tool model and bent cladded pipe, input parameters for tested materials including Inconel 625 and 16Mo3 steel and custom calculation using a finite element method in Ansys application. The resulting values were compared with subsequent experimental measurements.

Keywords: Experimental; Stress; Strain; Analysis; FEA; Bimetal tubes; Pipe bend; Steel 16Mo3; Inconel 625; Cladding; Rotary draw bending

1. Introduction

Modern energy operations require ever-increasing efficiency while adhering to strict environmental limits and ensuring longer equipment life. This trend is also reflected in the production and renovation of heat exchangers used in municipal waste incinerators. High temperature and chemically aggressive environment require an adequate protection of industrial boiler parts [2, 3]. Heat exchanger pipes are among the elements that are in a direct contact with that corrosive environment. As a result of the combustion of biomass with various industrial wastes in industrial boilers, aggressive chlorides or gas fluorides are formed, which leads to very intense high-temperature corrosion at temperatures from 700 to 1100 °C. [4–6] 16Mo3 steel (EN 10027-1) is widely used as a material for heat exchanger tubes and its life is approximately half a year in this environment [7, 8].

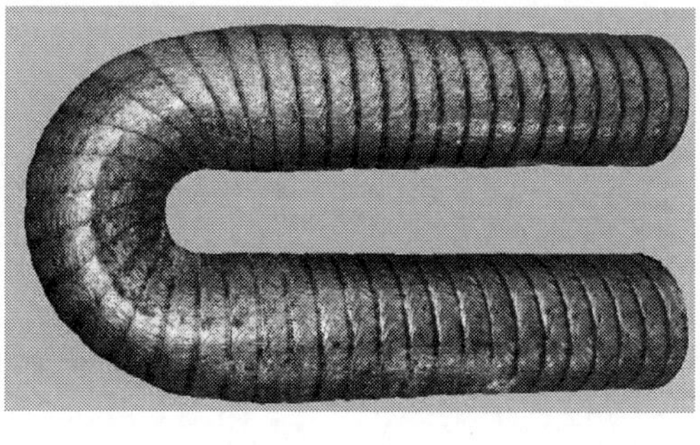

Fig. 1. A bent pipe elbow with the 2 mm Inconel 625 weld for welding to the heat exchanger.

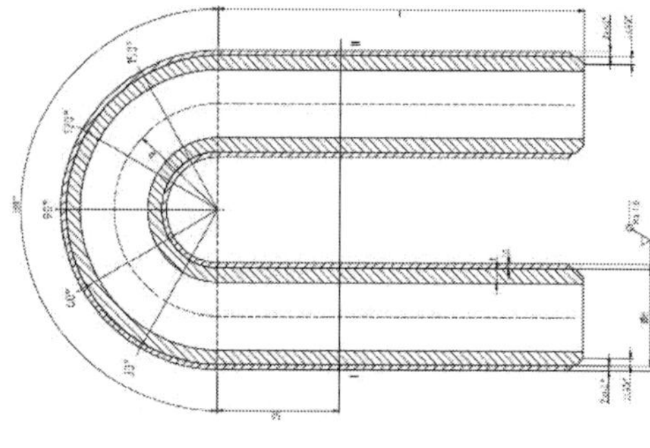

Fig. 2. Drawing of a 180-degree pipe elbow with the Inconel 625 weld.

[1] *Ing. Oskar Zemčík, Ph.D.; Institute of Manufacturing Technology, FME, Czech Brno University of Technology; Technická 2896/2, 61669 Brno, Czech Republic;* zemcik.o@fme.vutbr.cz

[2] *Ing. Martin Slaný, Ph.D.; Institute of Manufacturing Technology, FME, Czech Brno University of Technology; Technická 2896/2, 61669 Brno, Czech Republic;* slany.m@fme.vutbr.cz

[3] *Ing. Zouhar Jan, Ph.D.; Institute of Manufacturing Technology, FME, Czech Brno University of Technology; Technická 2896/2, 61669 Brno, Czech Republic;* zouhar@fme.vutbr.cz

[4] *Doc. Sedlák Josef, Ph.D.; Institute of Manufacturing Technology, FME, Czech Brno University of Technology; Technická 2896/2, 61669 Brno, Czech Republic;* sedlak@fme.vutbr.cz

To increase the boiler life 4 to 6 times, a layer of nickel superalloy with the thickness of 1–3 mm is usually welded to the base material of the pipe [7, 9]. Then, the pipes with this weld are shaped into the serpentine systems of the boiler. To ensure the optimal density of the elements of the serpentine system, it is often necessary to use a supercritical bend of these pipes with the radius up to $0.7D$, see Fig. 1 and Fig. 2. These bends of thick-walled pipes always lead to significant deformation of the pipe profile. [1] Therefore, the basic task is to ensure technological conditions that will enable this deformation to be kept below the allowable value of the ovality of the inner profile.

2. Bending simulation

A rotary draw bending method was chosen for the actual bending of welded pipes. In the introductory part, the bend was tested without the use of a filler or mandrel. The diagram of the bending tool can be seen in Fig. 3. It is possible to use both the variants – with and without the pipe filling. A similar principle can be found, for example, in Heftrich [10].

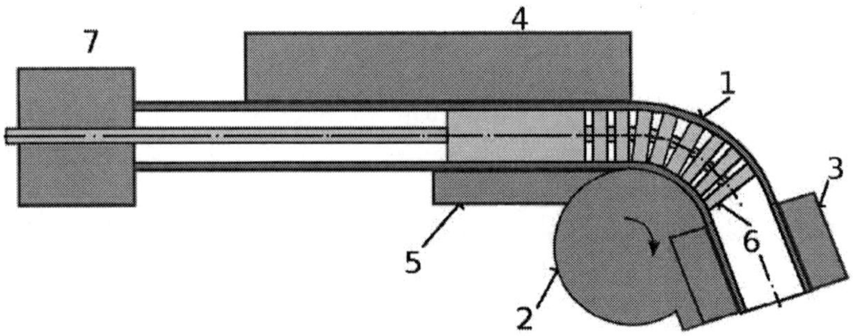

Fig. 3. Scheme of rotary draw bending, 1. tube, 2. bend die, 3. die clamp, 4. pressure die, 5. wiper, 6. mandrel, 7. collet.

2.1. Input parameters

The model was designed from 4 parts – a bend die, a pressure die, a die clamp and a cladded tube while the pipe and the weld were designed as separate bodies with a strong mutual bond. In addition to the basic mechanical quantities, tensile diagrams obtained from the material cut from the weld (according to the standard ISO 6892-1) [11] with a wire EDM machine, with the thickness of 1 mm were used for the input values of the FEA calculation (see Fig. 4 and Fig. 5). Data from the Yang study [12], which deals with a similar issue, were used for comparison. The values of Input parameters and yield strength are similar. Due to the use of the tensile stress diagram as input, i.e. the nonlinear stress-strain dependence, the calculation itself was nonlinear [13, 14]. The program Ansys v19.1 and v21.1 in the academic and commercial version was used for the simulation.

Table 1. Chemical composition of the tested metals.

Material	Fe	Ni	Cr	Mo	C	Mn	Si	Cu	Nb	Ti	Al
16Mo3	98	0.3	0.03	0.3	0.16	0.8	0.35	0.3	–	–	–
Inconel625	2.5	64.7	23.7	7.5	0.05	–	–	–	3.8	0.2	0.2

Chemical composition of the metal elements (see Table 1) was continuously verified using the X-MET 8000 X-ray spectrometer.

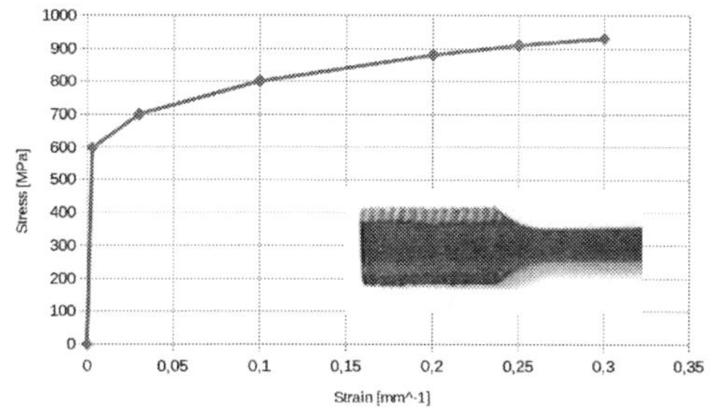

Fig. 4. A tensile diagram of the Inconel 625 material, a sample obtained from the weld fo Ansys.

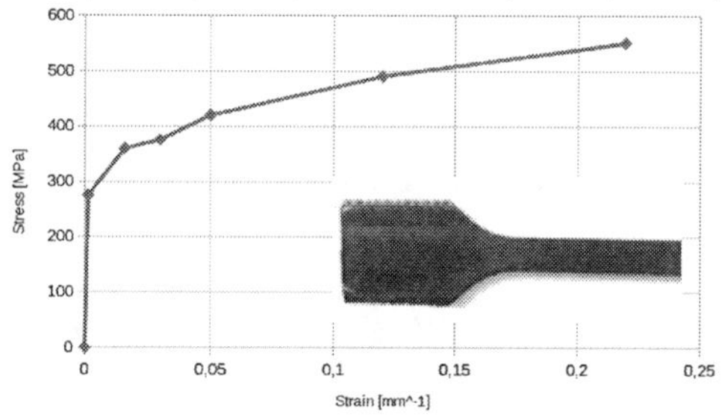

Fig. 5. A tensile diagram of the 16Mo3 steel, a sample obtained from the weld for Ansys.

The bending model assumes a mutual rotational movement of the bent die and pressure die of 182.4 degrees, where 2.4 degrees corresponded to the assumed value of suspension. The rotation was performed in 110 successive steps. Both the parts of a mold were chosen as absolutely rigid bodies, or as bodies with an order of magnitude larger value of Young modulus E. The basic parameters for the pipe and weld material are mentioned in Table 2. The working temperature was set at 22 °C, which corresponded to nonlinear dependences of the stress strain (Fig. 4 and Fig. 5).

Table 2. Material properties of pipe for Ansys.

Material	Inconel625	16Mo3
Young Module E [GPa]	207	211
Poison coefficient ν [–]	0.312	0.3
Yield Strength R_p [MPa]	598	275
Ultimate Tensile Strength R_m [MPa]	974	550

The meshing of the bodies contained 18,000 elements, of which the pipe and weld contained 8640 elements. At the point of contact, i.e. between the tool and weld, the meshing was refined with the coefficient 3. The contacts between the weld and tool were chosen as frictional with the friction coefficient 0.1 to 0.3. The details of the mesh are visible in Figures 6 (for the whole model) and 7 (for the detail in the cross section of the pipe). To compare the effect of a possible pipe filling, the inner pressure of 100 MPa was set. The unevenness on the welded surface of the pipe were neglected in the calculations.

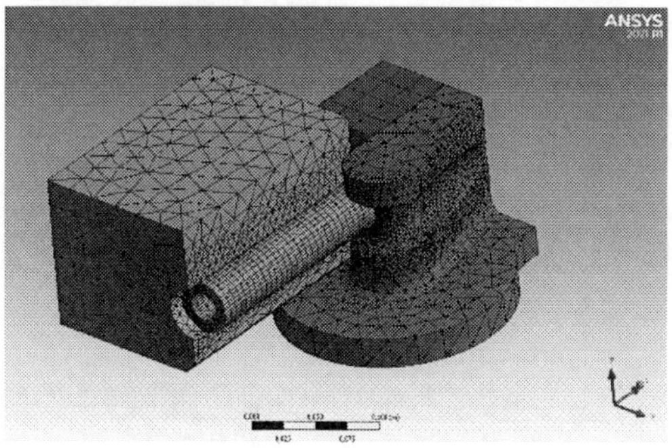

Fig. 6. A tool model and bent pipe model including meshing.

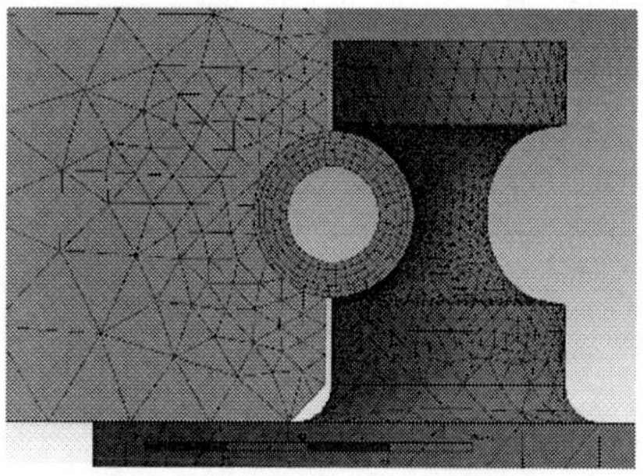

Fig. 7. A profile detail of the bent pipe, including meshing.

2.2. Calculation using FEA

The resulting shape of the pipe with the weld, bent into an arc of 180°, including the distribution of the e-strain, is shown in Fig. 8 and Fig. 9. Due to significant plastic deformation, the inner profile narrowed especially in the center of the arc, where inspection according to the standard VGB [15] is required. The von-Mises criterion was used to evaluate the amount of deformation, and the so-called equivalent strain according to the equation (1) [14] is shown in Fig. 8 and Fig. 9.

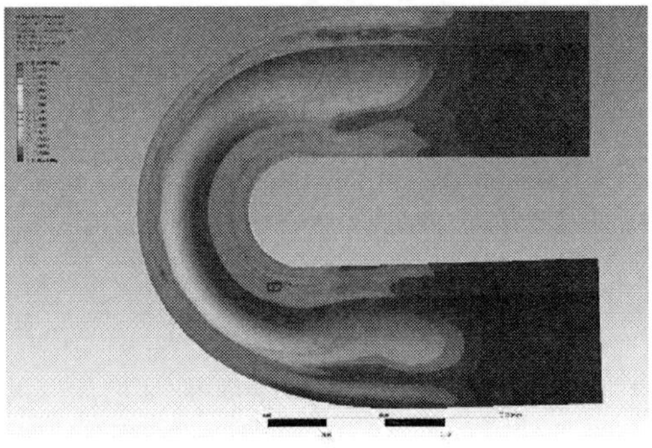

Fig. 8. Simulation of a 180° bent pipe in a longitudinal section – e-strain, $R = 1D$.

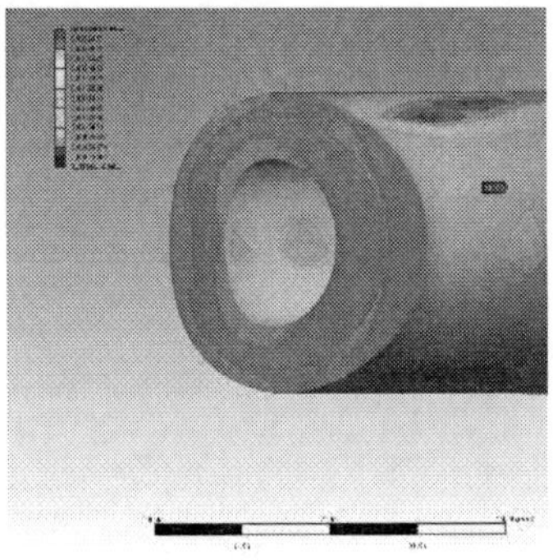

Fig. 9. Simulation of a 180° bent pipe with a cross-sectional weld – e-strain, $R = 1D$.

$$\varepsilon_{\mathrm{eqv}} = \left(\frac{1}{1+\nu}\right)\sqrt{\frac{(\varepsilon_1-\varepsilon_2)^2+(\varepsilon_2-\varepsilon_3)^2+(\varepsilon_3-\varepsilon_1)^2}{2}} \tag{1}$$

Table 3. Simulated inner tube size for different bend angle

Bending angle [°]	D_{min} [mm]	D_{max} [mm]
0	25.1	25.1
45	16.,9	24.93
90	12	25.23
135	10	24.95
180	9.79	24.92

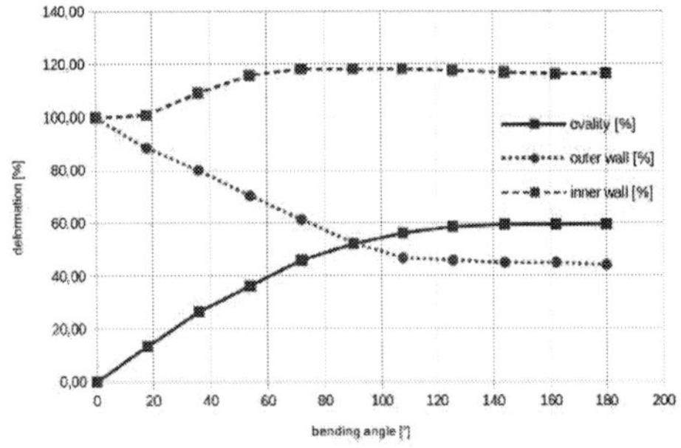

Fig. 10. Estimation of average wall deformation and ovality to the pipe bending angle.

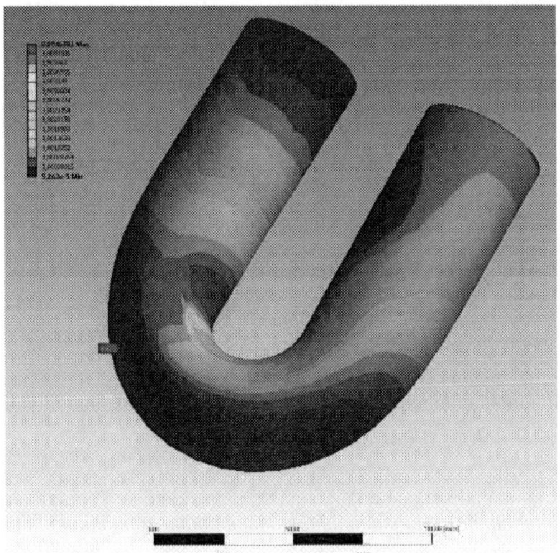

Fig. 11. Equivalent deformation distribution on a bent pipe for $R = 0.75d$.

The resulting value of the deformation of the inner profile with regard to its dimensions and the consequent change in the thickness of the outer and inner wall of the pipe depending on the bending angle are shown in Table 3 and the graph in Fig. 10.

The value of e-strain on the outer arc was at the critical deformation limit, which warned of the danger of violating the integrity of the bent arc. The distribution of e-strain at supercritical bend $R = 0.75d$ for a pipe with a 3 mm bend weld is evident in Fig. 11 with the marked place with the maximum value.

3. Experimental part

For this purpose, rotary draw bending tests were performed on samples of welded pipes. Semi-finished products with 3 mm and 2 mm welds of Inconel 625 were tested. The required shape of the pipe bend was according to Fig. 2.

The 16Mo3 steel was chosen for the material of the pipe. It is a heat-resistant non-alloy steel with increased heat resistance and resistance to high-temperature corrosion and creep up to 530 °C. It is well malleable in a hot and cold state with a ferritic-pearlitic structure and it is commonly used for heat exchanger pipes. The weld material Inconel 625 is a nickel-based superalloy that possesses high strength properties and resistance to elevated temperatures. It also demonstrates remarkable protection against corrosion and oxidation. It has the ability to withstand high stress and a wide range of temperatures.

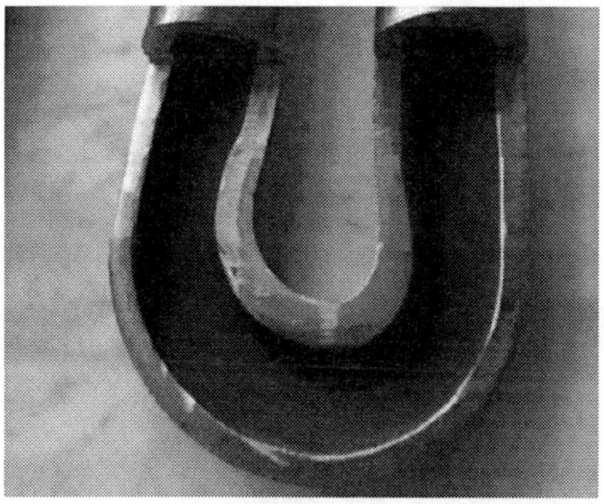

Fig. 12. The resulting shape of the bent pipe with the Inconel 625 weld in a longitudinal section for 3D scan.

Fig. 13. Fracture on the outer arc of the weld at the point of critical deformation.

The pipes were cold bent on a rotary draw bending machine - type CH 120 from the company Amob. The Cold Metal Transfer (CMT) welding technology was used for welding Inconel 625, which ensures a lower value of heat introduced into the weld and only about 0.2 kJ/mm. [16] For the purposes of the experiment, special bending tools were constructed, on which it was especially necessary to solve the extreme thinning of the cross-section in its center. The pipes with an inner diameter of 25.1 mm and an outer diameter of 38 mm with a 2 and 3 mm thick Inconel 625 weld were tested. Due to the previous simulations, it was assumed that the profile would deform extensively and therefore the pipes were tested both without filling and with a loose core infill and sealing.

After the bending, at first the condition of the bent pipe was verified. The passability of the pipe and compliance with the maximum ovality of the outer profile of 18% was tested with the help of a metal ball, according to the VGB standard [17], respectively using a 3D scan. For the inner diameter of the pipe 25.1 mm and $R = 0.7D$, the diameter of the test ball corresponds to 19 mm. This corresponds to the maximum ovality of 24%. Subsequently, the integrity of the outer surface of the pipe was verified visually and with the help of a penetration test according to EN ISO 3452-1 and EN ISO 23277 [18, 19], see Fig. 14.

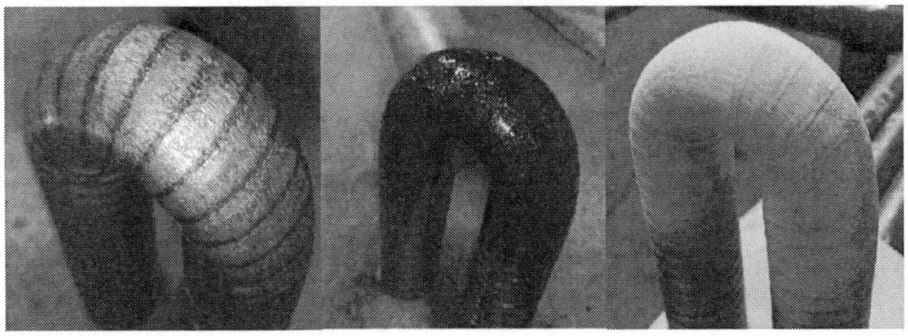

Fig. 14. The course of the penetration test of the pipe bend $R = 0.75d$.

Equation (2) was used to evaluate the internal and external ovality of the cross-sectional samples.

$$K_0 = \frac{D_{\max} - D_{\min}}{D_0} \cdot 100[\%] \tag{2}$$

In the case of the roundness deviation of the outer profile at the center of the arc, the relation (3) was used, where the control is prescribed by the standard ČSN EN 12952-5 [15].

$$u = 2\frac{D_{\max} - D_{\min}}{D_{\max} + D_{\min}} \cdot 100[\%] \tag{3}$$

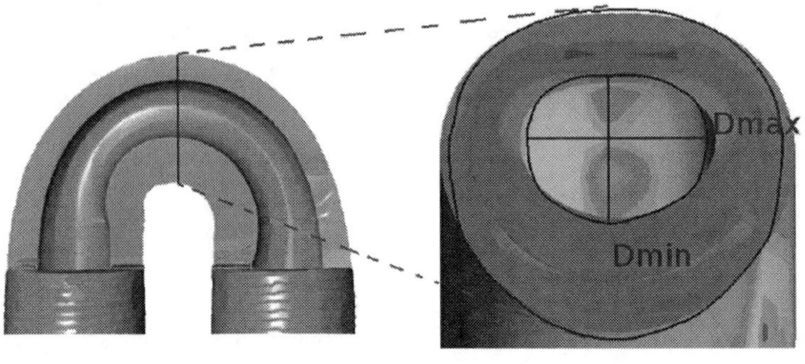

Fig. 15. Comparison of deformation of the tested bend for $R = 0.75d$, inner infill with simulation. A cut at the narrowest point at the center of the arc.

4. Results and discussion

A 3D scanner of the Atos Compact scan 2M, MV 250 type from the company GOM was used for the analysis of shape deviations. A similar comparison of CAD model differences from the actual pipe was also performed in the Li study [20]. The wall thickness of the pipes was also analyzed in the case of the explosive cladding method presented in the Mróz study [21]. Testing samples for tensile tests from cladded elbow samples were manufactured on a FANUC Robocut Alpha-0iD EDM wire cutter (WEDM) using a 0.25 mm radius Cut-G coated electrode from the company Penta Trading. The tensile test was performed using a universal Zwick Z 100 bursting machine from the company ZwickRoell Group. The resulting profile of cold-bent pipes with internal filling was compared with the theoretical profile and the profile obtained by finite element simulation. Deviations in the cross-sectional center of the arc are shown in Fig. 15. The deviation did not exceed 5% of the value in both the horizontal and vertical directions. In the case of the longitudinal section, which can be seen in Fig. 16, where the ideal bent pipe shape (black outline) is compared with three bent samples, and the variant estimated by finite element calculation, the deviation of the outer and inner arc of the bent pipe in the simulation was within individual test samples.

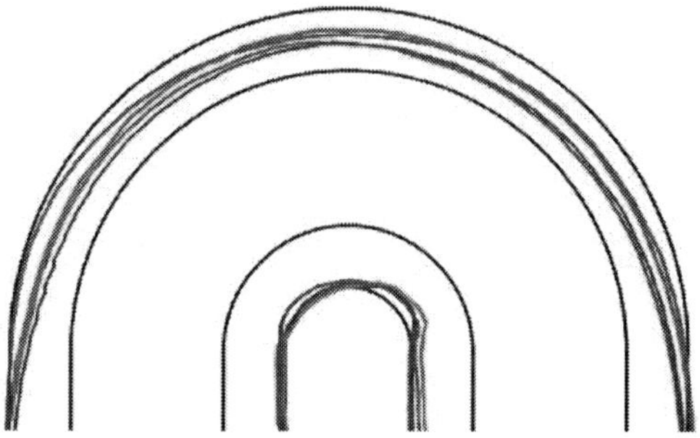

Fig. 16. Comparison of external and internal arc deviation for bend samples with internal infill for $R = 0.75d$ with simulation (green color) and ideal shape (black color).

5. Conclusion

In order to estimate the behaviour of bimetallic pipe bends with a supercritical bend of up to $0.7D$ with a welded layer of Inconel 625, many computational analyses and subsequent verification tests examining the bend of the pipe arc were performed. Based on these analyses, the following conclusions were reached:

- In the case of the supercritical bending, significant deformation of both the outer and inner profile of the pipe can be assumed,
- the thickness of the welded layer has a significant effect on the feasibility of the supercritical bending of the welded pipe,
- the analysis of the roundness deviations showed that the permissible deviations of both the outer and inner profile of the pipe were significantly exceeded using a rotary draw bending without mandrel or infill,
- in practical experiments, the risk of cracks on the outer arc of the weld and a significant reduction in the deformation of the profile was confirmed when using inner loose reinforcement and sealing,
- the condition of the weld and its thickness had a significant effect on the formation of cracks.

From the previous conclusions we can state that to ensure the successful production of a pipe bend from a bimetallic pipe with a bend of $0.7D$, without accompanying defects, it is necessary to ensure not only a new shape of the forming tool but also suitable technological conditions, especially the condition of the weld, its maximum thickness and a suitable pipe core infill.

Acknowledgements

This research was supported by the project No. TK04020189 within the public tender in the THETA program of the Czech Ministry of Industry and Trade.

References

[1] Samek R., Lidmila Z. and Šmehlíková E. *Speciální technologie tváření: část II.* 1. vyd. Brno: Akademické nakladatelství CERM, 2011. ISBN 978-80-214-4406-5.

[2] Wang Y., Yuan J., Zhang Y. and Guo C., "Technology status and development direction of bimetal pipe," *Welded Pipe and Tube*, Vol. 36, no. 2, pp. 05–09 (2013).

[3] Torbati A. M., Miranda R. M., Quintino L. and Williams S., "Welding bimetal pipes in duplex stainless steel," *Int. J. Adv. Manuf. Technol.*, Vol. 53, pp. 1039–1047 (2011). http://dx.doi.org/10.1007/s00170-010-2889-7.

[4] Uussitalo M.A, Vuoristo P.M.J. and Mäntylä T.A, "High temperature corrosion of coatings and boiler steels in reducing chlorine – containing atmosphere," *Surf. Coat., Technol.*, Vol. 161, pp. 275 – 285 (2002). https://doi.org/10.1016/S0257-8972(02)00472-3.

[5] Adamiec J. and Kierzak A., "Padding of the components of waste combustion boilers with the use nickel alloys," *Inżynieria Materiałowa*, Vol. 4, pp. 380 – 385 (2008).

[6] Chen W. H., "Waste burning and heat recovery characteristics of a mass burn incineration system," *J. Air Waste Manag. Assoc.*, Vol. 53, no. 2, pp. 136–142 (2003). https://doi.org/10.1080/10473289.2003.10466136.

[7] Golański G., Lachowicz M., Słania J., Jasak J., and Marszałek P. , "Research on 16Mo3 (16M) Steel Pipes Overlaid with Haynes Nicr o625 Alloy Using MIG (131) Method," *Archives of Metallurgy and Materials*, Vol. 60, no. 4, pp. 2521–2524 (2015). https://doi.org/10.1515/amm-2015-0408.

[8] ČSN EN 10028-2 (420938) *Ploché výrobky z ocelí pro tlakové nádoby a zařízení. Část 2: Nelegované a legované oceli pro vyšší teploty*, (2004)

[9] Rozmus-Górnikowska, M., Blicharski, M., Kusiński, J., Kusiński, L., and Marszycki, M., "Influence of boiler pipes cladding methods on their microstructure and properties, " *Archives of Metallurgy and Materials*, Vol. 58, no 4, pp. 1093–1096 (2013). https://doi.org/10.2478/amm-2013-0131.

[10] Heftrich, Ch., Steinheimer R. and Engel B., "Rotary-draw-bending using tools with reduced geometries," *Procedia Manufacturing*, Vol. 15, pp. 804-811 (2018). ISSN 2351-9789. https://doi.org/10.1016/j.promfg.2018.07.410.

[11] ISO 6892-1. *Metallic materials – Tensile testing – Part 1: Method of test at room temperature.* (2020).

[12] Yang Z., Kumar S. and Tronskar J.P., "ECA of Pipeline With Girth Weld Strength Mis-Matching Subjected to Large Strain," *International Conference on Offshore Mechanics and Arctic Engineering*, Vol. 6, pp. 75–82 (2009). http://dx.doi.org/10.1115/OMAE2009-79376.

[13] ANSYS Inc. (2005). *Advanced Nonlinearities - Geometric Instability.* Estados Unidos. ANSYS Inc. (2016).

[14] *ANSYS Mechanical 17.0 - Theory Reference.* Estados Unidos. ESSS. (2016). RAP_Distensionamento_02. Rio de Janeiro

[15] ČSN EN 12952-5. Vodotrubné kotle a pomocné zařízení: Část 5: Provedení a konstrukce částí kotle namáhaných tlakem. Praha: Úřad pro technickou normalizaci, metrologii a státní zkušebnictví, 2012.

[16] Talalaev R., Veinthal R., Laansoo A. and Šarkans M., "Cold metal transfer (CMT) welding of thin sheet metal products," *Estonian Journal of Engineering*, Vol 18, pp. 243–250 (2012). https://doi.org/10.3176/eng.2012.3.09.

[17] VGB-S-013-00-2014-12-EN. *Construction and installation supervision in the manufacture and assembly of water-tube boilers and associated systems in thermal power plants.* Essen: VGB PowerTech e.V., 2014.

[18] EN ISO 3452-1. *Non-destructive testing – Penetrant testing – Part 1: General principles.* (2015).

[19] ČSN EN ISO 23277 (051176) Nedestruktivní zkoušení svarů - Zkoušení kapilární metodou - Stupně přípustnosti, (2016)

[20] Li J., Zhou C. Y. and Zhu J. G., "Limit loads for 180° pipe bends under in-plane bending moment considering geometric nonlinearity," *International Journal of Pressure Vessels and Piping*, Vol. 183, p. 104100 (2020). https://doi.org/10.1016/j.ijpvp.2020.104100.

[21] Mróz S., Stradomski G., Dyja H. and Galka A., "Using the explosive cladding method for production of Mg-Al bimetallic bars," *Archives of Civil and Mechanical Engineering*, Vol. 15 no. 2, pp. 317–323 (2015). https://doi.org/10.1016/j.acme.2014.12.003.

Vibration Tests of Sandwich Panel with Inserts before and after Thermal-vacuum Conditioning

Robert Zemčík[1], Ladislav Zuzjak[2] & Petr Průcha[3]

Abstract: The analysis of spectral characteristics of a sandwich panel developed and manufactured by LA composite company was carried out. The testing was performed according to ECSS-E-ST-10-03C standard issued by the European Space Agency. The panel was attached to an electromagnetic shaker and excited either horizontally or vertically. The attachment to the shaker base or the slip table was either symmetrical or nonsymmetrical. Combinations of bolts and nuts were used as added masses. The tests were performed both before and after TVC.

Keywords: Experiment; Shaker; Random loading; Frequency sweep loading

1. Introduction

The presented work focuses on the analysis of spectral characteristics of a sandwich panel performed using sinusoidal and random vibrations. The panel was developed and manufactured by LA Composite company as a demonstrator component within the project FV 10400 – Development of processes applicable to development and production of components for the space industry. The testing was performed according to ECSS-E-ST-10-03C standard issued by the European Space Agency [1]. This standard aims at a consistent application of on ground testing requirements to allow proper qualification and acceptance of space products.

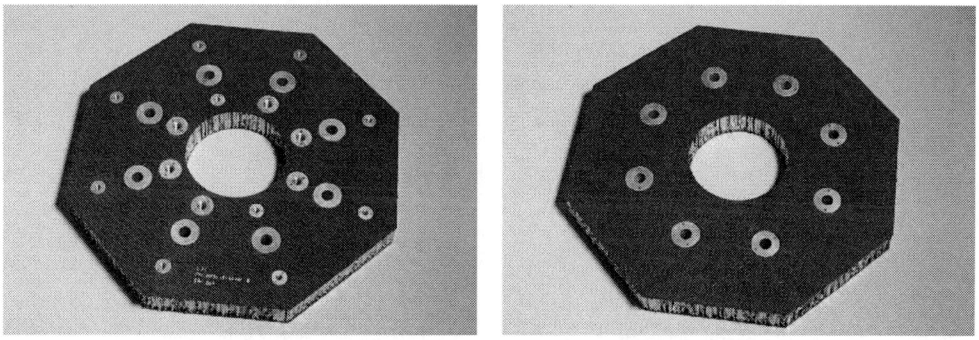

Fig. 1. Photographs of the sandwich panel (left – top view, right – bottom view).

2. Specimen

The sandwich panel contained threaded blind inserts (M5 and M8 Helicoil inserts) and thru clearance inserts [2, 3]. The lay-up of the two face-sheets was composed of foil adhesive with copper mesh and 8 plies of prepreg (EX-1515 AFC49 M55JB 6K 50B 80 g 33%) in symmetrical and quasi-isotropic sequence. The face-sheets were bonded to a perforated aluminum

[1] *doc. Ing. Robert Zemčík, Ph.D., Faculty of Applied Sciences, University of West Bohemia, Univerzitní 8, 30100 Pilsen, Czech Republic;* zemcik@kme.zcu.cz

[2] *Ing. Ladislav Zuzjak, Ph.D., Faculty of Electrical Engineering, University of West Bohemia, Univerzitní 8, 30100, Pilsen, Czech Republic;* zuzjak@fel.zcu.cz

[3] *Ing. Petr Průcha, Ph.D., LA composite, s.r.o., Beranových 65, 199 02 Prague 9, Czech Republic;* prucha@lacomposite.com

honeycomb core (PAMG-XR1-5056-3/16-0.001p) using foil adhesive on both sides. The inserts were potted using Stycast 1090 epoxy resin.

3. Apparatus and Procedure

The panel was excited using Dongling electrodynamic system ES-10-240 with GT-500 slip table and Vibration Research Medallion II control software. Brüel & Kjær 3150-A-060 unit with 2 accelerometers B&K 4524-B-004 (triaxial) and 1 accelerometer B&K 4507-B-004 (uniaxial) were used to generate and record signals with the sampling frequency of 32 kHz. The triaxial accelerometers were attached to the panel and the uniaxial accelerometer was attached to the shaker base (Fig. 2).

Fig. 2. Photographs of the accelerometers (left – shaker base, right – sandwich top surface).

The panel was attached either to the shaker base that moved vertically (for the Z-axis loading configurations A and B) or to the slip table (X-axis loading configurations C and D) that moved horizontally (Fig. 3).

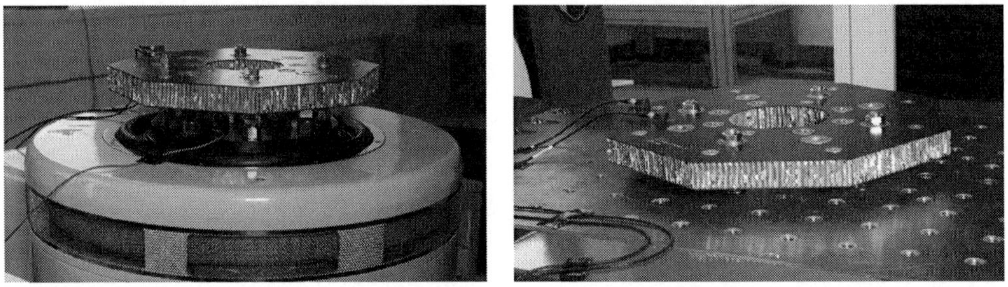

Fig. 3. Photographs of the sandwich panel attached to the shaker base (left) and slip table (right).

Four different configurations (2 load orientations, 2 panel fixings, added masses) designated as A, B, C, and D were analyzed (Fig. 4) as follows:

1. Panel attached using 4 M10 bolts at positions 10, 12, 14 and 16 (symmetrically). Loading along Z-axis.
2. Panel attached using 3 M10 bolts at positions 10, 14 and 16. Loading along Z-axis. Added mass: bolt M8/40 with 4 nuts at position 19 (mass 35.61 g) and bolt M5/20 with 2 nuts at position 04 (mass 6.04 g).
3. Panel attached using 4 M10 bolts at positions 09, 11, 13 and 15 (symmetrically). Loading along X-axis.
4. Panel attached using 3 M10 bolts at positions 09, 13 and 15. Loading along X-axis. Added

mass: bolt M8/40 with 4 nuts at position 18 (mass 35.61 g) and bolt M5/20 with 2 nuts at position 04 (mass 6.04 g).

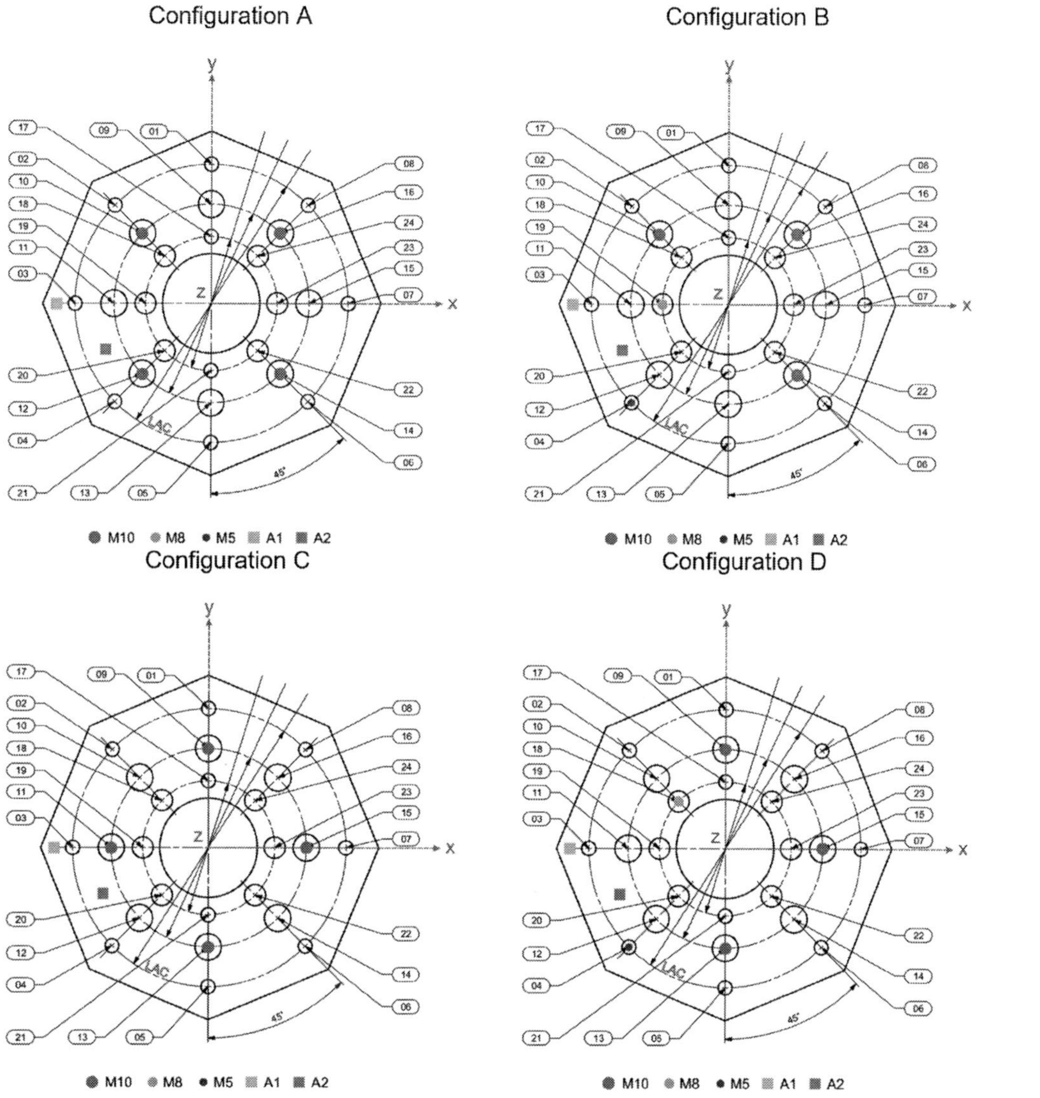

Fig. 4. Schemes of the four configurations used. M10 bolts provide attachments of the sandwich panel to the shaker base (configurations A and B) or slip table (configurations C and D), M5 and M8 bolts provide the added masses and A1 and A2 represent the triaxial accelerometers.

Each M10 bolt consisted of a threaded rod (100 mm) each with 3 nuts and 3 washers). The M10 bolts were tightened using 5 Nm torque, the M8 bolts using 4.5 Nm, and the M5 bolts using 3.0 Nm. Lubricant WD40 was used for the threads of the blind inserts. The X-axis is chosen as the vector connecting hole 03 and 07, the Y-axis connects hole 05 and 01, and the Z-axis is perpendicular to X and Y axes.

Each test comprised of 3 steps. The first step was a frequency sweep (2 Oct/min, 6–100 Hz with 2.50g and 100–140 Hz with 1.25g, 2 orthogonal axes), the second step was a random vibration (20–2000 Hz, 2 minutes, 2 orthogonal axes), and the third step was again the frequency sweep. The frequency sweep of low-level sinusoidal vibrations is used to characterise main resonant modes for preparing the higher level runs, and to show possible deficiencies in workmanship, as a consequence of high-level runs.

The random vibrations simulate the harsh mechanical conditions present during rocket launch. The success criteria for the resonance search shall be: less than 5% in frequency shift, for modes with an effective mass greater than 10%; less than 20% in amplitude shift, for modes with an effective mass greater than 10% [1]. The tests were performed at room temperature ranging between 24 °C and 25 °C. Detailed visual checks were carried out after each test. The panel was tested before and after the thermal-vacuum conditioning (TVC).

4. Results

The frequency sweep data were analysed up to 3200 Hz (see example for configuration C in Fig. 5). No natural frequencies were found within the loading spectrum. No significant change in the frequency spectra was observed for each configuration before and after the random vibrations. Only small changes in amplitudes were observed for frequency sweeps when comparing the tests performed before and after TVC. Example of random vibration results is shown in Fig. 6 for configuration C.

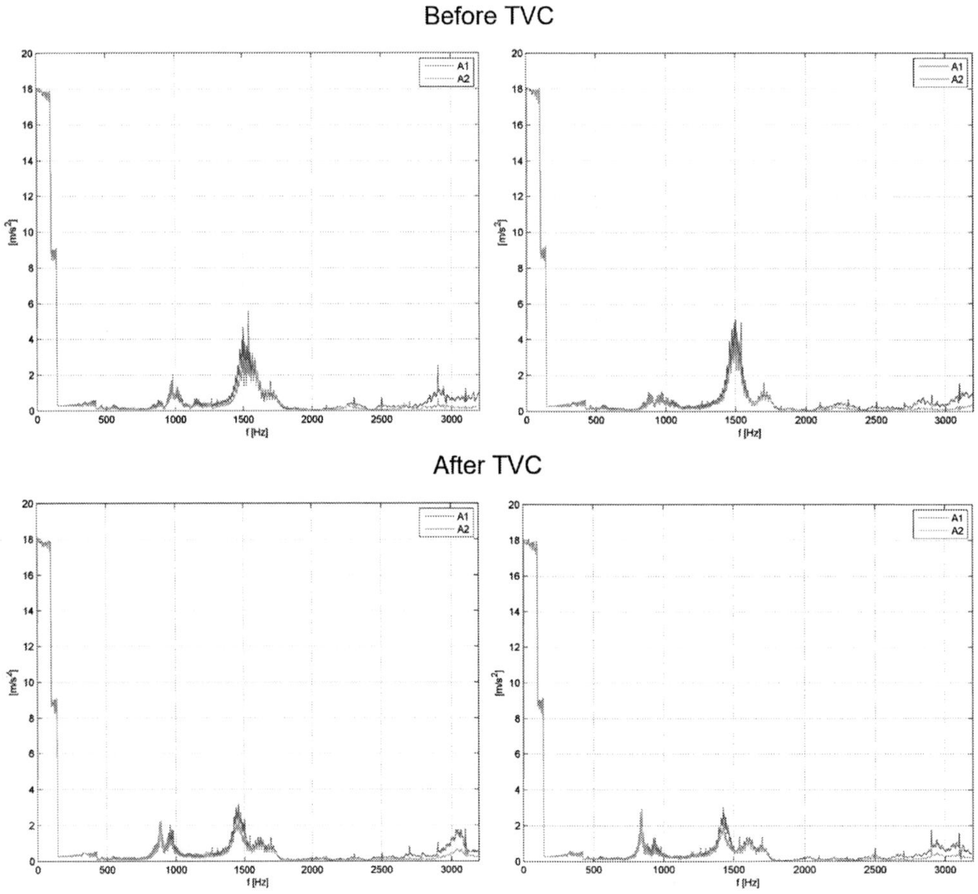

Fig. 5. Frequency sweep results for configuration C before (left) and after (right) random vibrations.

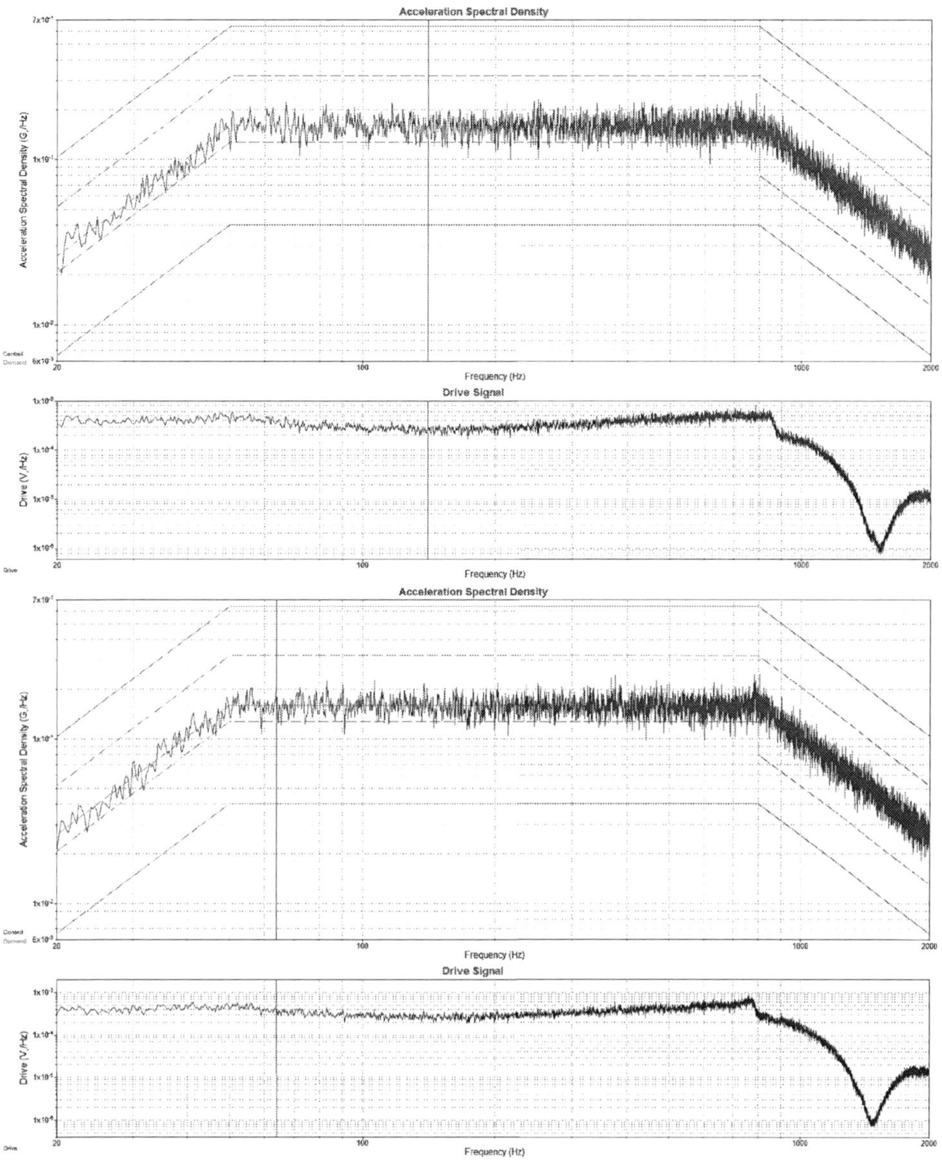

Fig. 6. Random vibration results for configuration C before (top) and after (bottom) TVC.

Acknowledgements

This work was supported by the project TRIO FV10400 "Development of processes applicable to development and production of components for the space industry" of the Czech Ministry of Industry and Trade.

References

[1] ECSS-E-ST-10-03C, Space engineering, Testing, ESA Requirements and Standards Division, (European Space Agency, 2012).

[2] Mandys T., Laš V., Zemčík R. and Krystek J., "Testing of honeycomb sandwich panel with potted insert," In: Experimental Stress Analysis – Conference Proceedings, pp. 273-278, Luhačovice, 2019.

[3] Zemčík R., Krystek J., Mandys T., Laš V. and Průcha P., "Testing of sandwich specimens for space applications," In: 36th Danubia-Adria Symposium on Advances in Experimental Mechanics, University of West Bohemia, Pilsen, pp. 145-146, 2019.

Long-term Monitoring of the Development of Temperatures in the Composition of the External Wall Realized by Dry Construction

Ondřej Zobal[1], Petr Bittnar[2], Zdeněk Bittnar[3],
Tomáš Koudelka[4], Miroslav Nyč[5] & Milan Pokrivčák[6]

Abstract: The experimental object was realized to verify the functionality and possible optimization of the composition of the facade. The long-term monitoring of thermal technical properties takes place in this experimental building. This paper deals with the development of temperatures in the section on the profile and in the section outside the profile, taking into account the influence of the season and alternating day and night. The results will be used, among other things, to verify the calculation tool.

Keywords: Long-term monitoring; Facade; Temperature; Calculation tool

1. Introduction

Dry construction began to gain ground in the Czech Republic at the turn of the 20th and 21st centuries, when this technology was significantly improved. The application of this technology is not only in the implementation of temporary constructions, but also in reconstructions and new constructions. Plasterboard is used for vertical and horizontal structures. The main advantages are fast construction, elimination of the wet process, good acoustic properties and fire resistance. The disadvantage is the higher sensitivity to humidity [1].

In practice, we encounter a difference between the expected state and the real behavior of the structures, which causes, for example, significant differences in the overall energy balance of the building. Status in common engineering practice is, that design according to empirical expressions or 1D heat transfer law. It is not considered influence of moisture. Another problematic aspect is, that steel profiles are not taken into account. Goals of this work are design of computational tool for common engineers, 2D analysis of couple heat and moisture transport on layered L-shape domain with influence of steel profiles.

[1] *Ing. Ondřej Zobal, Ph.D.; Czech Technical University in Prague, Faculty of Civil Engineering, Department of Mechanics; Thákurova 7, 166 29 Prague 6, Czech Republic; ondrej.zobal@fsv.cvut.cz*

[2] *Ing. Petr Bittnar, Ph.D.; Czech Technical University in Prague, Faculty of Civil Engineering, Department of Mechanics; Thákurova 7, 166 29 Prague 6, Czech Republic; petr.bittnar@fsv.cvut.cz*

[3] *prof. Ing. Zdeněk Bittnar, DrSc.; Czech Technical University in Prague, Faculty of Civil Engineering, Department of Mechanics; Thákurova 7, 166 29 Prague 6, Czech Republic; bittnar@fsv.cvut.cz*

[4] *Ing. Tomáš Koudelka, Ph.D.; Czech Technical University in Prague, Faculty of Civil Engineering, Department of Mechanics; Thákurova 7, 166 29 Prague 6, Czech Republic; tomas.koudelka@fsv.cvut.cz*

[5] *Ing. Miroslav Nyč; Knauf Praha spol. s r.o.; Mladoboleslavská 949, 197 00 Praha 9, Czech Republic; miroslav.nyc@knauf.com*

[6] *Ing. Milan Pokrivčák; Knauf Insulation spol. s r.o., Bucharova 2641/14, 158 00 Praha 5, Czech Republic; milan.pokrivcak@knaufinsulation.com*

2. Methodology

2.1. Methodology of the measurement

The experimental object consists of four measuring rooms and each room has its own facade, see Fig. 1 and Fig. 2. Facades are made of classic structural elements for dry construction – plasterboard, mineral insulation and steel thin-walled profiles [2–4]. The measuring sensors are located so that it is possible to analyze the course of temperatures and humidity in the entire cross-section of the facade. The data can be tracked online and stored on the cloud.

Other important parameters are obtained from the measurements of the meteorological station located directly in the UCEEB complex in Buståhrad. The current state of temperature and humidity, the intensity of sunlight or wind speed can also be monitored online. For the range of measurements and specifications of the weather station, see Snĕhota et al. [5].

Fig. 1. Experimental object with marked test facades and their orientation.

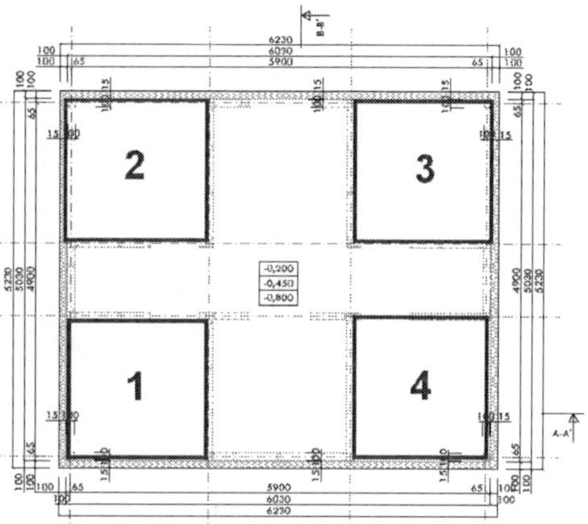

Fig. 2. Floor plan of the experimental object with marked measuring rooms.

2.2. Methodology of the simulation

It is well known that the moisture and heat transfers influence mutually significantly. In building physics, Kunzel-Kissel's model [6] represents a reasonable choice for the description of coupled heat and moisture transfer with respect to the number of required model parameters. The model states the most relevant effects in the water transfer represent water vapour diffusion in large pores and liquid transport in the small capillaries. The model is defined by the system of 2 partial differential equations that can be solved with the help of the finite element method (FEM). The weighted residual method is applied in this case. The layered composition of walls of the experimental object together with thin steel reinforcement profiles would lead to a very dense finite element mesh as well as increase the complexity of the mesh generator. The solution to that complexity can represent the concept of hanging nodes which allows for the placement of the substructure to the main mesh where the nodes of the substructure do not constitute additional degrees of freedom over ones of the main mesh. The measurements of temperature fields in the experimental object can provide boundary conditions required in the model together with data for the model verification and validation. The computational model has been implemented in the open-source finite element computer code SIFEL [7] and accomapnied by the generator FE mesh for layered wall graphical user interface allowing the simple problem setup.

3. Results

3.1. Long-term monitoring

The scheme of placement of measuring sensors (red color – temperature, blue color – combination of temperature and humidity) can be seen in Fig. 3. The sensors are located at two heights (1 and 3 m) and on the profile (T1; T3) and outside the profile (T4; T6).

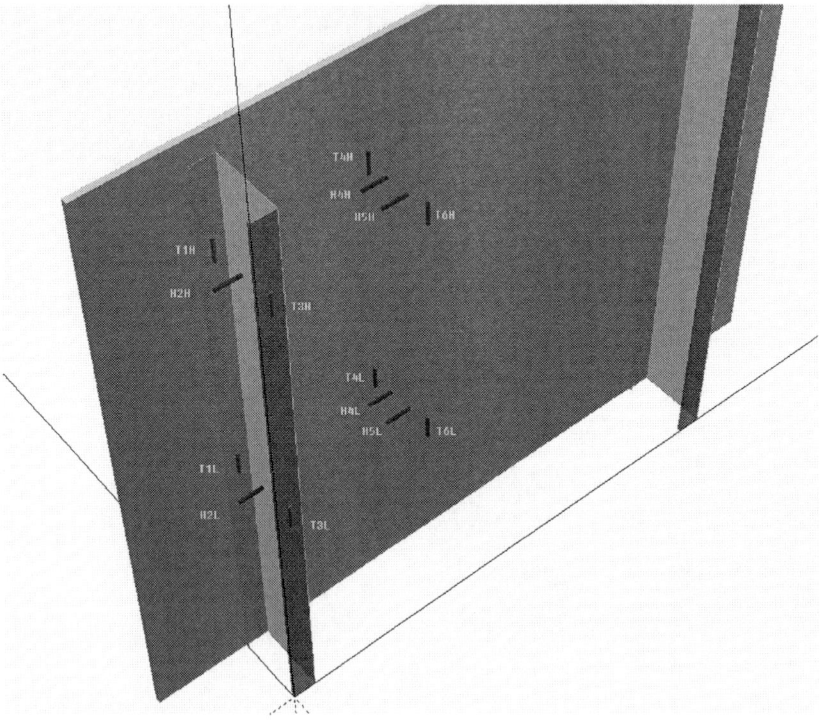

Fig. 3. The scheme of placement of measuring sensors (red color – temperature, blue color – combination of temperature and humidity).

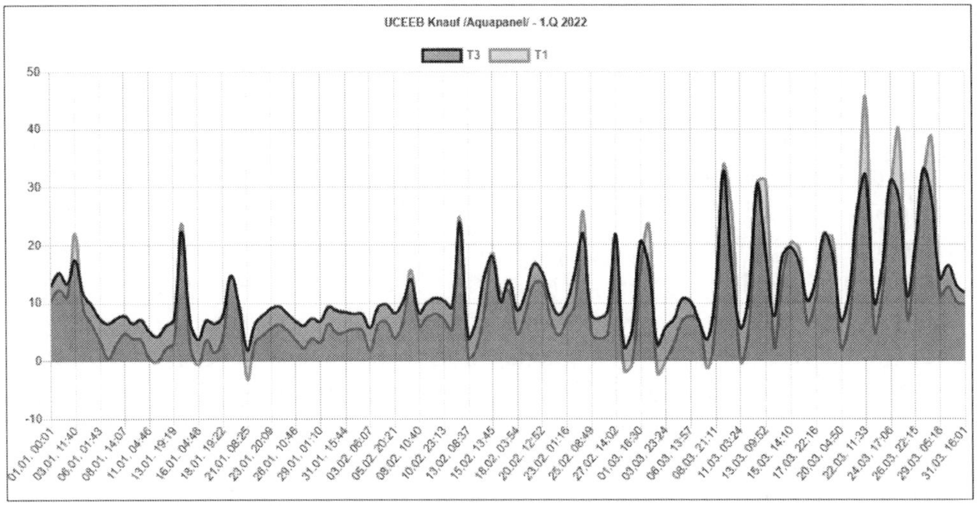

Fig. 4. The course of temperatures for the period of the 1st quarter of 2022 – section with the profile.

The course of temperatures for the period of the 1st quarter of 2022 (January to March 2022) can be seen in Fig. 4 and Fig. 5. The temperature is recorded every 10 minutes and the resulting temperatures are the average of the values from the two measured heights. It was found that the height does not affect the temperature, the temperature difference at the height of 1 and 3 m is maximal 0.1 °C. The development of temperatures in the cross section with the profile can be seen in Fig. 4 and we can observe that the temperature from the outside (T1) drops inwards only minimally (T3). This phenomenon can be dangerous with respect to the dew point value in the structure and can lead to moisture condensation and subsequent mold spread. This phenomenon does not occur in the area without the profile, see Fig. 5. The great influence of sunshine can be seen in both monitored areas, even in winter. The heat difference on the facade can reach more than 50 °C in a short period of time.

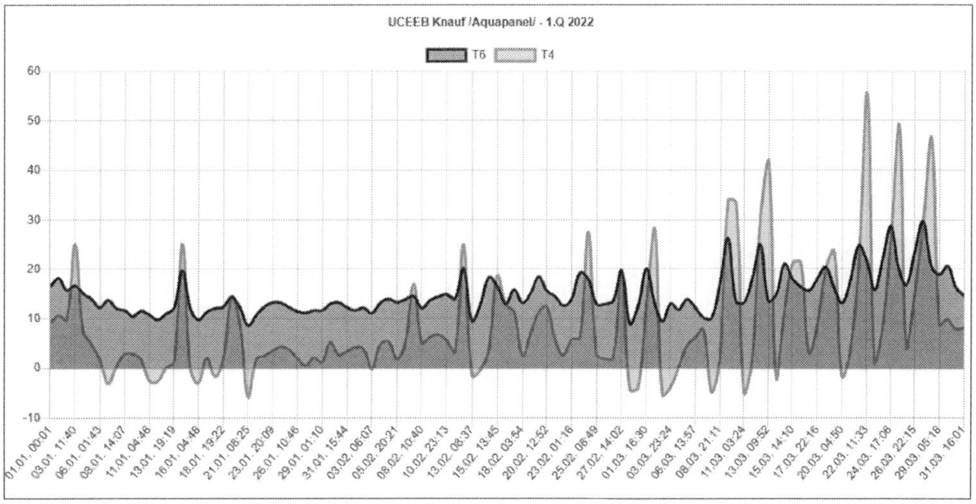

Fig. 5. The course of temperatures for the period of the 1st quarter of 2022 – section without the profile.

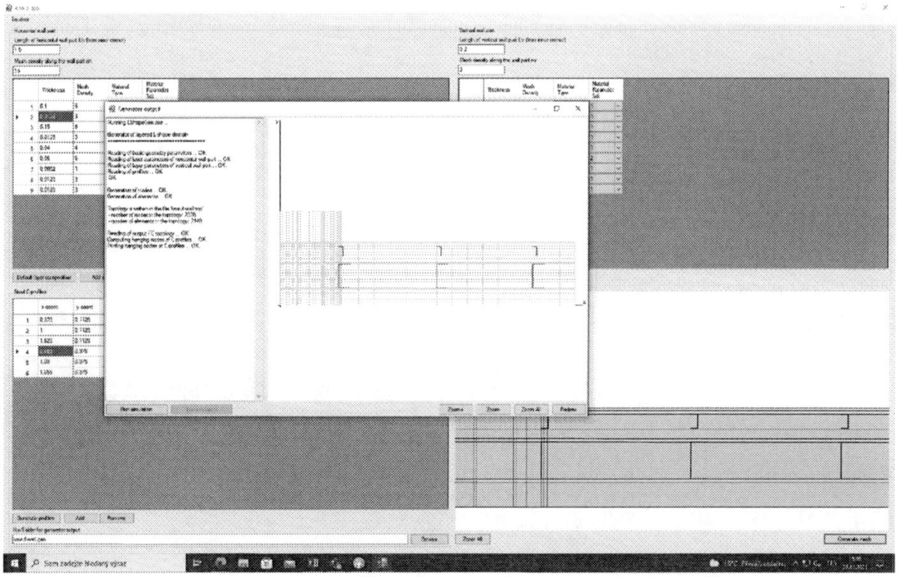

Fig. 6. The computing tool and its user interface.

3.2. *Calculation tool*

The chosen solution of the problem proves to be very effective. For a task with multiple profiles with offset location, it would be necessary for a simplified generator to generate a network with hundreds of thousands of nodes, which is difficult to use for deployment in common engineering practice. The computing tool and its user interface can be seen in Fig. 6. The resulting temperature profile obtained from the calculation tool can be seen in Fig. 7.

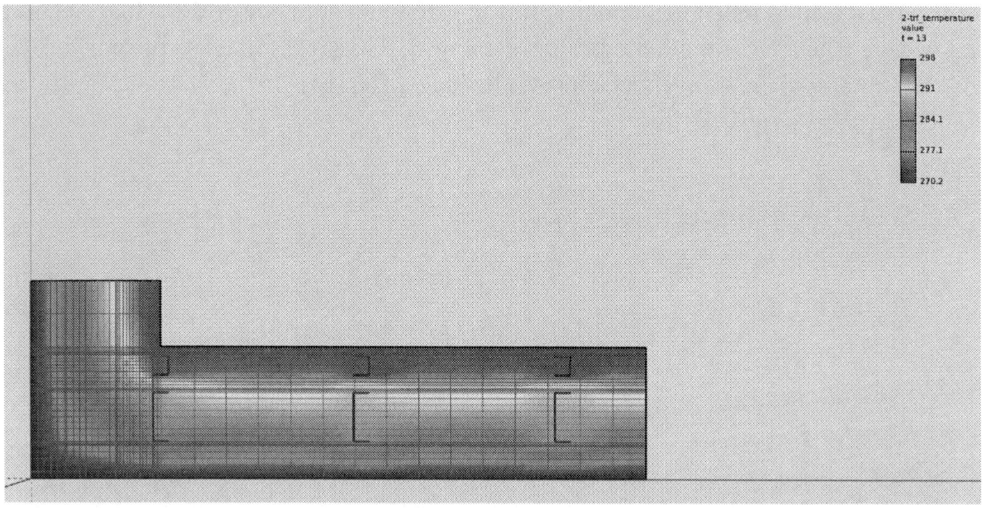

Fig. 7. The resulting temperature profile obtained from the calculation tool.

4. Conclusion

The experimental object was built and is used to collect real data. The measurement is compared with the obtained data in a laboratory environment. The calculation model is created thanks to the obtained parameters. The design and use of dry construction facades will be able to be optimized.

Acknowledgements

This paper has been supported by the Technology Agency of the Czech Republic TA ČR, project No. TN01000056 "Centre for Advanced Materials and Efficient Buildings".

References

[1] Nyč M., *Plasterboard*, (Grada, 2005). ISBN 80-247-0986-4.

[2] *Technical sheet: K716 - GKFI hard gypsum boards for special use*, Knauf Praha spol. s r.o., Prague, 2019.

[3] *Technical sheet: Mineral Plus EXT 035*, Knauf Insulation spol. s r.o., Prague, 2019.

[4] *Technical sheet: STRUCTURES BORABEL SYSTEM*, Borabela, Prague, 2019.

[5] Sněhota M., Jelínková V., Šácha J., Bareš V. and Vogel T., "Meteorologická měření v okolí budovy UCEEB," in Sympozium Energeticky efektivní budovy 2015. Praha: Společnost pro techniku prostředí, 2015, pp. 271–276. ISBN 978-80-02-02615-0.

[6] Künzel H. M., *Simultaneous heat and moisture transport in building components*, Tech. Rep. (Fraunhofer Institute of Building Physics, Fraunhofer IRB Verlag Stutgart, 1995). ISBN 3-8167-4103-7.

[7] Open source finite element (FE) computer code SIFEL available on: mech.fsv.cvut.cz/ sifel/.

EAN 2022 60th annual conference on experimental stress analysis

Determination of the Johnson-Cook Material Model of EN AW 6061-T6511 with evaluation for grooving machining

Jan Zouhar[1], Andrej Gabriš[2], Miroslav Jopek[3] & Kamil Podaný[4]

Abstract: This paper deals with modelling of a grooving machining process that includes chip formation and tool loading. The finite element method (FEM) is used for a simulation based on the Lagrangian formulation, including ANSYS/LS-DYNA software. The constitutive Johnson-Cook material model regarding strain and strain rate to predict material plasticity and chip separation is used. The model was determined based on the Taylor test, where a cylindrical specimen is shot against a solid substrate. Based on the final reshaping of the test specimen and the resulting stresses on the impact plate and comparing it with the numerical simulation, the individual parameters of the Johnson-Cook constitutive equation can be determined. Loading forces, chip morphology, stress and strain have been obtained. The Results are compared with experimental results and other references.

Keywords: Orthogonal cutting; Taylor anvil test; Johnson-Cook; Explicit Lagrangian; Machining; Simulation

1. Introduction

Grooving is one of the important operations in machining, widely used in many areas of industry, like the car and aerospace industry. Progressive development and nowadays cost requirements are highly interested in new technologies like HSM, HFM, etc. New cutting tools are needed for these progressive technologies, so that development of their design with understanding of machining process is important. Laboratory optimization of cutting conditions cost a lot of time and money usually. To cut down these costs and times implementation of FEA to the virtual development cycle could be useful technique.

The Finite Element Method (FEM) was developed in the early 70's. In technologies, it was first pioneered on simulations of the orthogonal machining process. Those research works used self-developed finite element code usually. From 1990, a massive use of commercial software's start, which are capable of modelling the machining process in details, such as the NIKE2, ABAQUS/Standard, MARC, DEFORM 2D, FORGE 2D, ALGOR, FLUENT, ABAQUS/Explicit, LS DYNA. History and major research works presents [1–4].

Most of those works is focused on the orthogonal cutting process. However, a few research works about milling process are done with new software's and powerful hardware equipment's lately. Many authors are focused on 2D simulation [5], HSM flat milling [6], micro milling [7], etc. Advanced 3D simulations are presented in [8–10]. Maurell [11] used FE simulation of cutting for determination of material parameters. New approach and reviews are presented in [12–15].

[1] *Ing. Jan Zouhar, Ph.D.; Institute of manufacturing, Fac. of Mech. Engineering, Brno University of Technology; Technická 2, 616 69 Brno, Czech Republic; zouhar@fme.vutbr.cz*

[2] *Bc. Andrej Gabriš; Institute of manufacturing, Fac. of Mech. Engineering, Brno University of Technology; Technická 2, 616 69 Brno, Czech Republic; 200019@vutbr.cz*

[3] *Ing. Miroslav Jopek, Ph.D.; Institute of manufacturing, Fac. of Mech. Engineering, Brno University of Technology; Technická 2, 616 69 Brno, Czech Republic; jopek@fme.vutbr.cz*

[4] *Ing. Kamil Podaný, Ph.D.; Institute of manufacturing, Fac. of Mech. Engineering, Brno University of Technology; Technická 2, 616 69 Brno, Czech Republic; podany@fme.vutbr.cz*

2. Methods

FEM simulation of the machining is a typical dynamic non/linear problem for application of the explicit method, because a calculation of an integration step using implicit integration methods is too slow for fast dynamic processes. For this reason, such processes (like explosions, penetration simulations, crash tests etc.) are calculated using explicit integration methods typically [16]. A difference between the implicit and explicit methods is in the way of calculating of each time step from equation (1). Implicit method uses e.g. Newmarks-method and the explicit method uses a so-called central difference scheme.

$$M \cdot \ddot{U} + C \cdot \dot{U} + K \cdot U = F(t) \tag{1}$$

In the Eq. (1), M is the mass matrix, C is the damping matrix, K is the stiffness matrix, U is the displacement matrix and F is the force matrix.

Most important advantages of explicit FEM are:

- a simple code;
- a building of stiffness matrix K is not required;
- the non-linearities included into the vector of internal forces (simple setup of a contact);
- the matrix inversions are not necessary;
- less memory required.

The most used explicit FEM programs today are: LS-DYNA, ABAQUS Explicit, PAM-CRASH, Deform 3D and MSC-Dytran.

2.1. FEA model formulation

There are four formulations in FEM continuum modelling, which we can be used for simulation: Lagrangian, Eulerian, Arbitrary Lagrangian Eulerian (ALE) and Smooth Particle Hydrodynamics (SPH) [14].

Lagrangian formulation: In the Lagrangian formulation the FE mesh is attached to the workpiece and deformed together. This case is useful for relatively low distortions and possibly large deformation. The history of state of the material in each element is known completely. Compared to the Eulerian method the Lagrange method tends to be faster in calculations as no transport of material through the mesh needs to be calculated. For the metal cutting simulations, the Lagrangian formulation is preferable due to the more convenient modelling of the evolution of the chip from the incipient stage to a steady form [3, 6]. The geometry of the material boundaries (or chip shape) does not have to be predetermined, but is developed during the course of the analysis entirely as a function of the physical deformation process, machining parameters, and material properties. Disadvantage of Lagrangian mesh is increasing time step or a stability loss when the distortion of elements is too high.

Eulerian formulation: In the Eulerian formulation the material flows through the mesh. The unknown material variables are calculated at pre-set locations as the material flows through the mesh. Eulerian mesh is free from distortion problems and remeshing algorithms are not necessary. Other advantage is that a mix of material through cell is allowed. A large deformation is possible in the simulation. A disadvantage of the Eulerian mesh can be seen in higher time consumption as a fine mesh is needed. For the machining simulation chip shape must be known before the simulation. Other separation criteria are not necessary. This formulation is often used for hydrodynamic and aerodynamic problems [18].

Arbitrary Lagrangian-Eulerian (ALE) formulation: The ALE formulation is an extension of the Lagrangian formulation that, with additional computational steps, moves the grid and remaps the solution onto a new grid. One advantage of this technique is the freedom in dynamically defining the mesh configuration that allows a combination of the best features of Lagrange and Eulerian. For metal cutting simulations, the Eulerian approach is convenient for modelling of the area around the tool tip, while the Lagrangian approach can be used for modelling of the unconstrained material flow at the free boundaries [7, 19].

Smoothed Particle Hydrodynamics (SPH) formulation: The SPH technique uses no grid. In other words, it is a pure Lagrangian formulation. The absence of a mesh and the calculation of interactions main particles based on their separation alone means that large deformation can be computed. This method is primarily used in fluid mechanics [20].

2.2. Material model and Methods

Nevertheless, the explicit algorithm has to be specified in solving the problem of material separation - the condition when damage/fracture of each element occurs. Commercially used programs have several models of damage described in [6,21]. For simulations of machining Johnson-Cook criterion [22] appears to fit as best and the criterion of equivalent strain similarly.

Johnson-Cook material model was developed in the 80's for study of impacts, penetrations and explosives. It is favoured in studies of problems of fast deformations, large strain, e. g. simulations of machining as well [22].

The Johnson-Cook formulation – Eq. (2) – involves the yield stress at non-zero strain rate, equivalent plastic strain $\bar{\epsilon}^p$, strain hardening index n, equivalent plastic strain rate $\dot{\bar{\epsilon}}^p$, initial dimensionless plastic strain rate $\dot{\epsilon}^0$, the melting temperature of the workpiece T_{melt}, workpiece transitive temperature T_{room}, and strain rate sensitive exponent m as shown at Eq. (2). A, B and C are constants.

$$\sigma_y = [A + B(\bar{\epsilon}^p)^n] \left[1 + c \ln\left(\frac{\dot{\bar{\epsilon}}^p}{\dot{\epsilon}^0}\right)\right] \left[1 - \left(\frac{T - T_{room}}{T_{melt} - T_{room}}\right)^m\right] \tag{2}$$

A study of Umbrello [23] deals with an influence of the constants used in the J-C model. However, his model has some disadvantages, as a failure to describe the material softening [24].

A dilatation of the element is based on the value of equivalent plastic strain at element integration points [17]. Failure is reached when damage parameter D exceeds one element. The damage parameter follows a cumulative damage law and is given by Eq. (3)

$$D = \sum \frac{\Delta\bar{\epsilon}^p}{\epsilon^f} \tag{3}$$

Where $\Delta\bar{\epsilon}^p$is the increment of the equivalent plastic strain during an integration cycle and ϵ^f the equivalent strain to fracture under current conditions of strain rate, temperature, pressure and equivalent stress, see Eq. (4) [22]:

$$\varepsilon^f = \left[D_1 + D_2 \exp\left(D_2 \frac{p}{q}\right)\right] \left[1 + D_4 \ln\left(\frac{\dot{\bar{\varepsilon}}^p}{\dot{\varepsilon}^0}\right)\right] \left[1 + D_5 \left(\frac{T - T_{room}}{T_{melt} - T_{room}}\right)\right] \tag{4}$$

where p is the pressure stress, q is the Von Mises stress and other parameters were defined above.

2.3. Johnson-Cook parameters experimental determination

The J-C parameters experimental determination is divided into three interconnected stages. In the first of them, a compression test is performed under quasi-static conditions. In the second part, a TAT experiment is performed to determine the correct strain rate sensitivity coefficients for loading under dynamic conditions. Numerical simulations are then used to verify the correctness of these values using finite elements method. Finally, the third part of the manuscript is devoted to the practical validation of the proposed Johnson-Cook (J-C) material model.

Taylor Test, Theoretical Background. The compressive Taylor anvil test (TAT), first presented in 1948 [25], is advantageously used to investigate behaviours of materials characterised with high imposed strains at high strain rates, i.e., strain rates about 103 s^{-1} [26]. TAT is designed to simulate real industrial conditions of loading and can favourably be used to test bulk metallic specimens of simple geometries. During TAT, a cylindrical testing specimen of density ρ, initial length L_0, and initial diameter d_0 is shot using expanding air with the initial speed v_0, typically varying between 30 and 240 m/s, on a rigid impact plate (see Fig. 1) or a schematic depiction of a TAT testing specimen). The impact plate can also be a high-strength rod or dynamometer, on which the elastic deformations and impact forces develop and the result of impact of the testing specimen with the rigid surface are detected. Immediately after the collision of the testing specimen with the rigid surface, the acting compressive force creates elastic and plastic waves, which begin to spread across the specimen. The elastic wave, spreading within the material at the speed of sound, bounces back from the free end of the specimen and returns as a tensile wave. The elastic wave can perform multiple bouncing from a surface, or interface. In a certain location within the specimen, both the waves interact. The typical shape of a TAT tested specimen is depicted in (Fig. 1), where L_f is the final length of the specimen, d_f is the deformed diameter of the impacted forehead of the specimen, and X is the length of its undeformed cylindrical par.

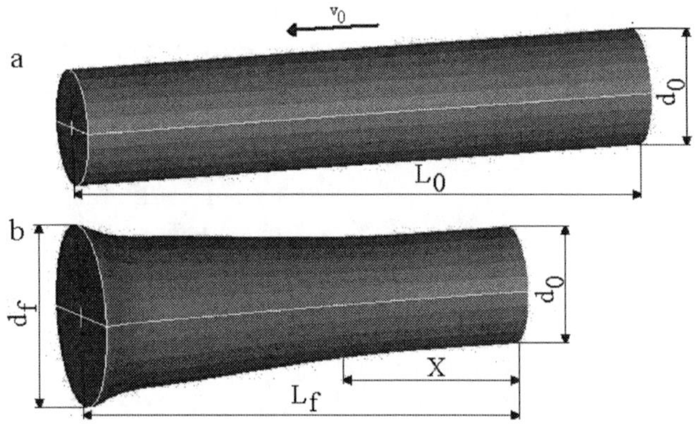

Fig. 1. Depiction of TAT testing specimen: before and after TAT.

TAT Specimen Analyses. The subsequent analyses focused on observing the geometry of the tested specimen were at first performed experimentally, then using image analysis methods, and finally using numerical simulations. Mutual correlation of all the results subsequently enabled verification of the computed values of the J-C constitutive equation parameters. Fig. 2 shows the deformed geometry of the tested specimen. Fig. 3 shows the final curve characterising the dimensions of the tested specimen after TAT.

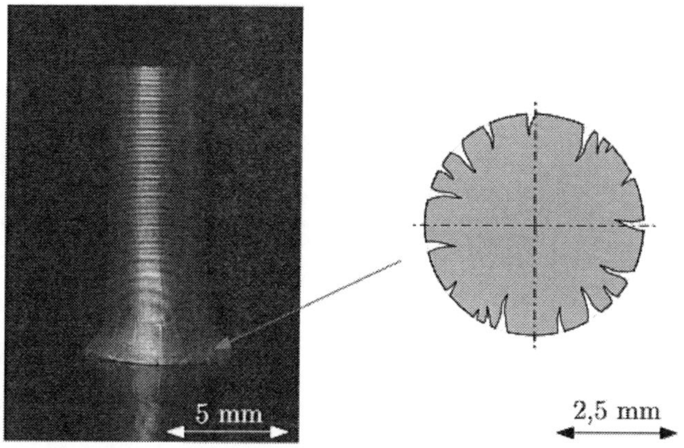

Fig. 2. Tested specimen after TAT.

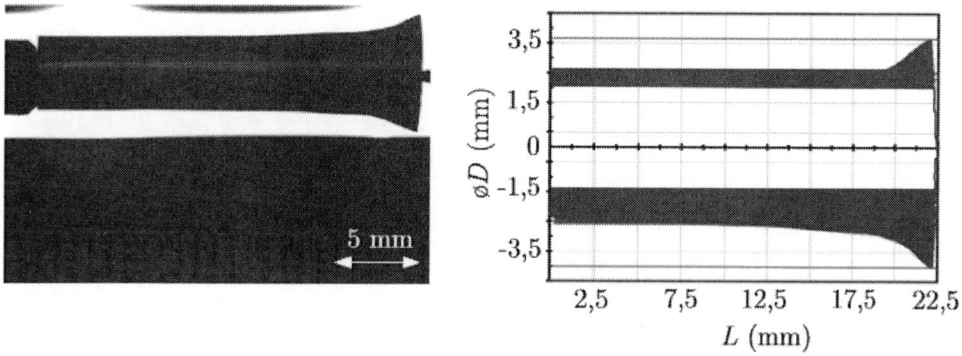

Fig. 3. Acquired image of the surface of the tested specimen.

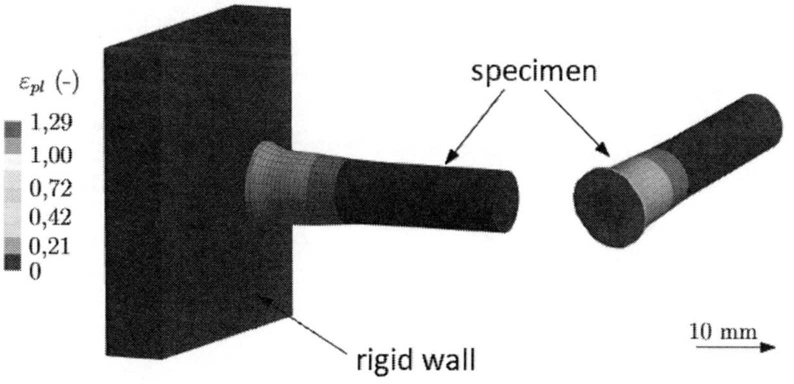

Fig. 4. Numerically predicted distributions of effective imposed strain.

TAT Finite Element Analyses. Having experimentally investigated the geometry, and dynamic behaviour of the EN AW-6061 T6511 testing specimens, numerical simulations were performed and optimised to determine the ideal parameters of the J–C constitutive equation. At first, the experimentally acquired geometry of the tested specimen was correlated with the geometry of the deformed specimen from the numerically simulated TAT – see Fig. 4. The simulation and experiment show a good correlation of both the sets of results as their maximum deviation was between 1% to 3%.

A review of the available material parameters and experimentally given is in Table 1. It is evident that the materials achieve very similar values for measurements by different authors. The only deviation is the ultimate strength, which is higher for the experimental material than for the others due to the slightly different material composition. The failure constants D_1–D_5 were determined experimentally by [30], see Table 2.

Table 1. Comparison of J-C parameters.

Author	A [MPa]	B [MPa]	n	C	m
Experimental	371.2	162.4	0.493	0.004	1.2
Lesuer et al [27]	324	114	0.42	0.002	1.34
Manes et al [28]	270	138.2	0.1792	0.1301	0.8
Chen [29]	270	154.3	0.239	0.029	1.42

Table 2. J-C deformation parameters.

Author	D_1	D_2	D_3	D_4	D_5
Akram [30]	-0.77	1.45	-0.47	0.00	1.60

2.4. Models of chip formation

There are several possibilities how this state of material can be implemented into the FEM algorithm. Three most often used methods are presented below.

Adaptive meshing: During crack propagation a new mesh is created around the moving crack tip. All parameters have to be obtained for the new mesh and it leads to a long computation time [19].

Element deletion: After achieving the criterion of fracture the element is removed from the calculation. Disadvantage of this method is disappearing of mass during solution. With large losses of material obtained results are non-credible. In this work, J-C law and the criterion of equivalent strain [9] as the selected fracture criteria were used.

Damage of contact link at the elements: In this approach area of chip formation has to known before the calculations. A contact link and rest of the workpiece with condition of failure has to be created for the area. This condition is based on normal and shear forces applied on the contact elements or using normal and shear stresses between elements [7]. This approach is published in [5].

2.5. Evaluation of the computational model

Evaluation of computational model was done by means of the orthogonal cutting experiment. Orthogonal cutting can be seen like a parting of a big diameter workpiece (with a minor error). The force measurement was carried with a piezo-electrical dynamometer Kistler 9257B, equipped with the charge amplifiers Kistler 9011A, fully controlled by a PC. (Fig. 5, Fig. 6).

Fig. 5. Orthogonal cutting experiment.

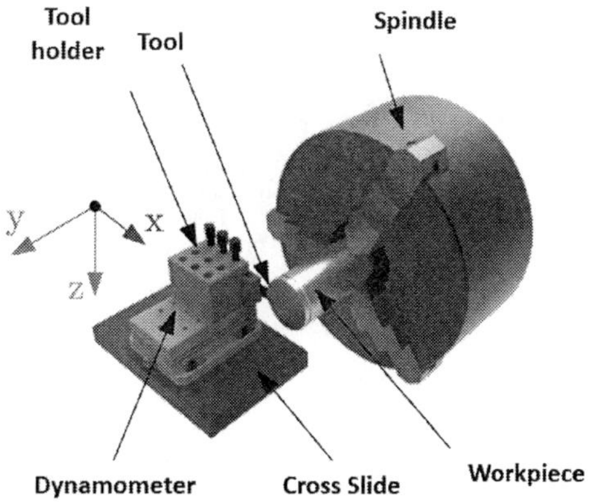

Fig. 6. Setup of experiment.

First evaluation model was done by steel material EN C45 (ČSN 12050). The measurement and observation were focused on cutting forces and shape of chips. Cutting speed 180 m/min and depth of cut 0.2 mm were used. Workpiece dimensions for FEM simulation were: length 10 mm, height 2.2 mm and width 0.2 mm. Width of the real workpiece was ten times bigger. Calculations were done supposing that the cutting force in orthogonal cutting is linearly proportional to the width of the cut. Cutting tool was modelled with a different rake angle in the scale of -5°, 0°, 5°, 10°, see Fig. 7–Fig. 14 [31].

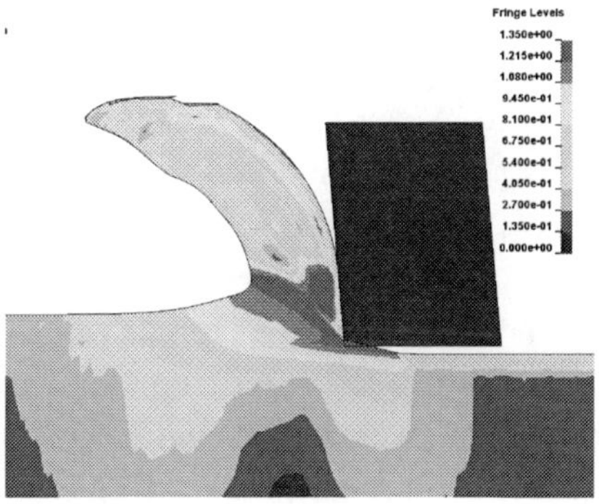

Fig. 7. Simulation of rake angle -5°.

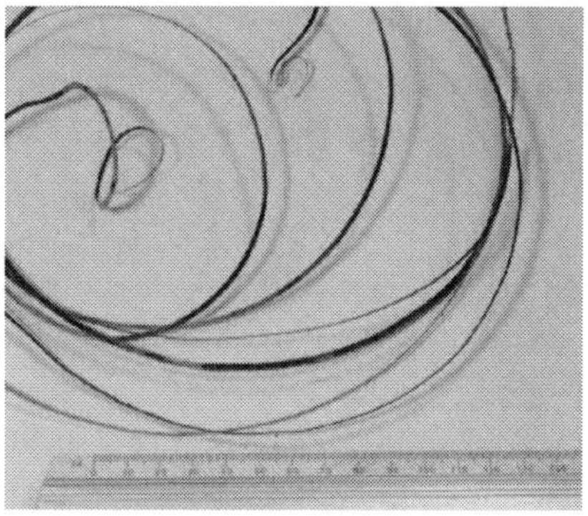

Fig. 8. Chip shape of rake angle -5°.

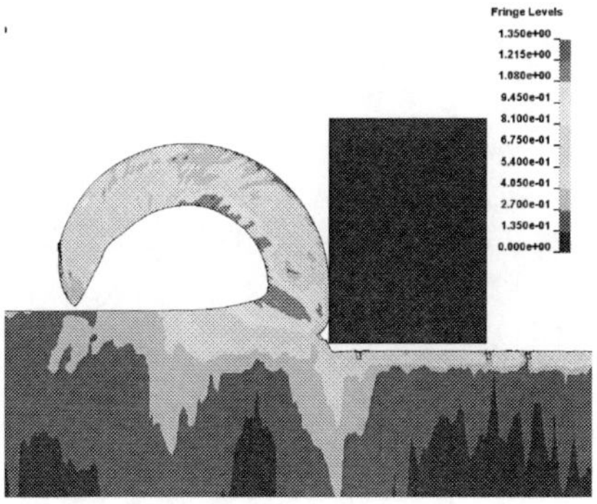

Fig. 9. Simulation of rake angle 0°.

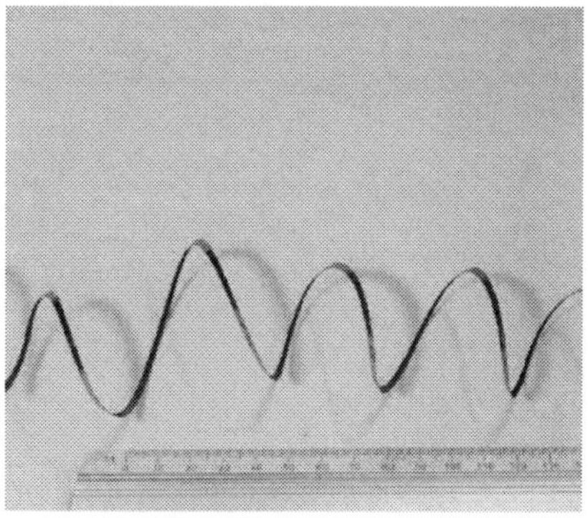

Fig. 10. Chip shape of rake angle 0°.

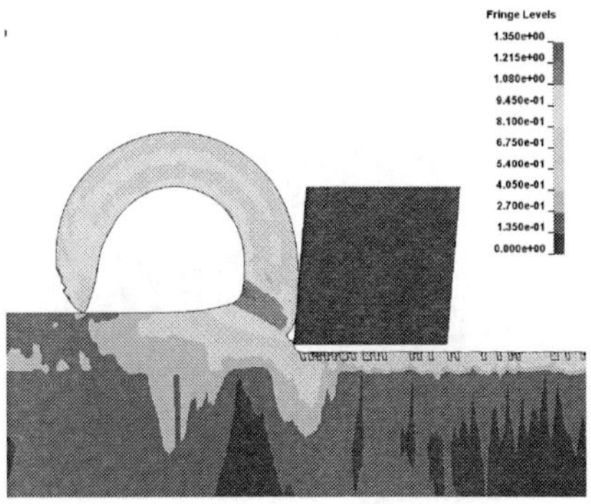

Fig. 11. Simulation of rake angle 5°.

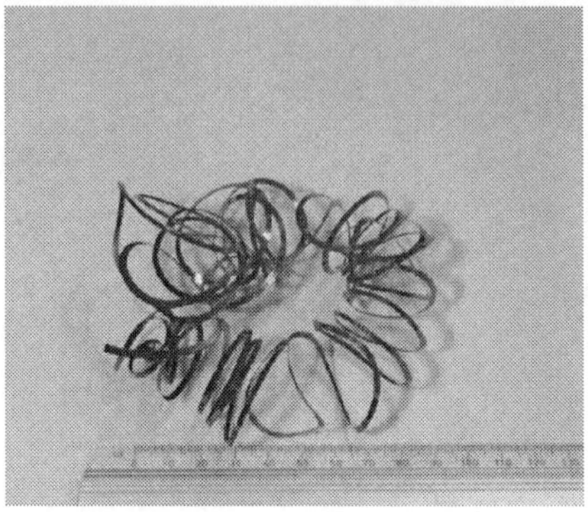

Fig. 12. Chip shape of rake angle 5°.

Fig. 13. Simulation of rake angle 10°.

Fig. 14. Chip shape of rake angle 10°.

The resulting values of the cutting forces are listed in Table 3. It is evident that deviations are in range of 6–10%. These exceptions may be made due to inaccuracies of the machine-tool-workpiece system and inaccuracies of simulation methods. The SPH method for simulation of this case was used also. Fig. 15 shows time series of the cutting forces. A general course of the curves is the same, but the values differ a bit due to some inaccuracies also. The experimental cutting force had a smoother start at machining then the simulated force, where some peaks revealed. At the area of steady state cutting all forces has the same course. The SPH method showed very good results. The differences between the experiments and SPH were 3% in average, but computational time of the simple task was much higher (about 117 h compared to 31 h) [31].

Table 3. Forces given by different rake angles.

Rake Angle	-5°	0°	5°	10°
Experimental F [N]	826	743	676	590
Simulated F [N]	884	801	743	614
Relative error [%]	6.6	7.2	9.1	11.7

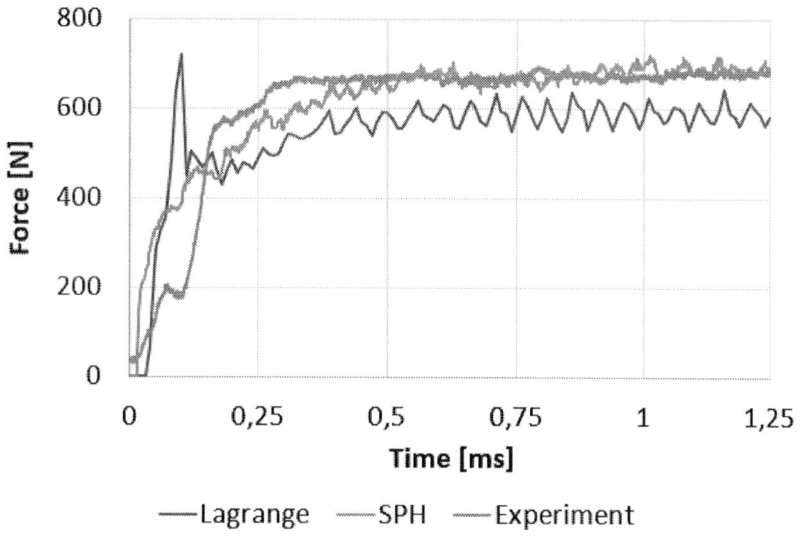

Fig. 15. Force simulated by Lagrangian, SPH and comparison with experimental results.

The results on steel confirmed the applicability of the lagrange method for simulation, although SPH achieves better results. Therefore, in terms of computational times, the Lagrange method was used for the simulation of EN AW 6061 material.

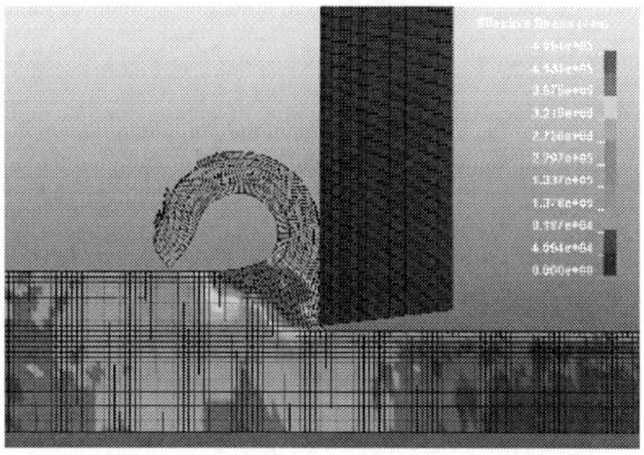

Fig. 16. Equivalent Von-Misses stress plot during chip formation – detail shear zone.

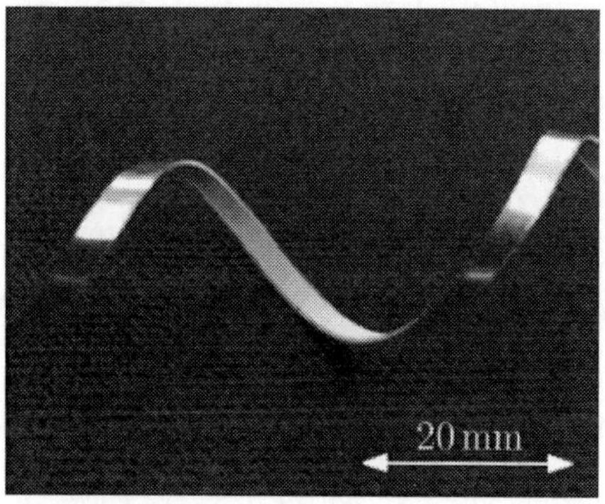

Fig. 17. The real shape of chip.

The shapes of the simulated real and chips for cutting tool with tool rake angle 0° is shown on Fig. 16 and Fig. 17. The equivalent von Mises stress is on the Fig. 16. Distribution of the stresses matched well with the theory of orthogonal cutting [32]. Primary shear zone is visible clearly and the secondary shear zone is involved due to a friction can be seen also. Third deformation zone (at the flank contact of the tool) is very small, what is caused with the sharp geometry of the tool tip and the method of element deletion. The shape of the chip corresponds to the simulation, it is continuous with built-up spots that are particularly noticeable on the inside of the chip and the tool. The chip width is smaller than the real one, it is caused by deleted elements in the calculation due to the deformation parameters given by the J-C model. Cutting forces simulated. The cutting force calculated by simulation was 98±10 N, here is a big difference compared to the real measured force of 180±6 N, this is almost double. The real force was checked by analytical calculation [32] there is a deviation of up to 10%. The differences between simulation and reality are largely due to the use of the D_1–D_5 deformation coefficients [26] for a slightly different material with a lower yield strength. Due to this, a large number of features are deleted in the simulation, especially at the tool face where the main cutting force is applied. The difference could also be due to the effect of build-up on the tool, which is mainly due to the tribological material-tool pair and other circumstances such as process fluid, vibration, etc.

3. Conclusions

The Johnson/Cook material model was used and evaluated for orthogonal machining with a good cogency – depend on fracture criteria. The SPH formulation appeared to be more appropriate and confident to the real machining than the Lagrangian formulation, but with very long computational time demands. The 2D simulation of grooving showed similar stress distribution and chip formation as the orthogonal cutting – continuous pattern. A force analysis grooving of material EN AW 6061 of and experimental indicated similar time progression and differences up to 50% differences of the values, with use of Lagrangian method. While the simulation on steel material showed a compliance of up to 10%. To further refinement of the calculation, it is necessary to determine the deformation parameters D_1–D_5 for the given material and to use a calculation method where no elements are deleted. The deletion of an element fundamentally affects the calculation.

Acknowledgements

This research work was supported by the grant "Modern technologies for processing advanced materials used for interdisciplinary applications", FSI-S-22-7957.

References

[1] Mackerle J., "Finite-element analysis and simulation of machining: a bibliography (1976–1996)," *Journal of Materials Processing Technology*, Vol. 86, pp. 17–44 (1999). https://doi.org/10.1016/S0924-0136(98)00227-1.

[2] Mackerle J., "Finite element analysis and simulation of machining: an addendum A bibliography (1996–2002)," *International Journal of Machine Tools & Manufacture*, Vol 43, pp. 103–114 (2003). https://doi.org/10.1016/S0890-6955(02)00162-1.

[3] Sadeghifar M, Sedaghati R, Jomaa W, and Songmene V., "A comprehensive review of finite element modeling of orthogonal machining process: chip formation and surface integrity predictions," *The International Journal of Advanced Manufacturing Technology*, Vol. 96, no. 9-12, pp. 3747–3791 (2018). ISSN 0268-3768. https://doi.org/10.1007/s00170-018-1759-6.

[4] Shi, B and Attia H., "Current status and future direction in the numerical modeling and simulation of machining processes: a critical literature review," *Machining Science and Technology*, Vol. 14 no. 2, pp. 149–188 (2010). http://dx.doi.org/10.1080/10910344.2010.503455.

[5] Watanabe, K. and Umezu Y., "Cutting simulation using LS-DYNA3D," in Third International LS-DYNA3D Conference, Kyoto, Japan, pp. 1–17 (1995).

[6] Umer, U., Lijing X. and Wang X., "Finite element chip forrmation analysis for high speed milling operations," *Transactions of the CSME/ dela SCGM*, Vol. 32, pp. 513–522 (2008). http://dx.doi.org/10.1139/tcsme-2008-0034.

[7] Asad, M., Hignette C., Mabrouki T. and Rigal J.F., "Numerical and experimental approach for an aluminium alloy micro-milling," in *Intercut 2008 – 2.nd International conference*, France, Cluny, pp. 1–8 (2008).

[8] Soo, S.L., Aspinwall D.K. and Dewes R.C., "3D FE modeling of the cutting of Inconel 718," *Journal of Materials Processing Technology*, Vol. 150, pp. 116–123 (2004). https://doi.org/10.1016/j.jmatprotec.2004.01.046.

[9] Pantalé O., Bacaria J.-L., Dalverny O., Rakotomalala R. and Caperaa S.," 2D and 3D numerical models of metal cutting with damage effects," *Computer methods in applied mechanics and engineering*, Vol. 193, pp. 4383–4399 (2004). https://doi.org/10.1016/j.cma.2003.12.062.

[10] Jalili Saffar R., Razfar R.M., Zarei O. and Ghassemieh E., "Simulation of three-dimension cutting force and tool deflection in the end milling operation based on finite element metod," *Simulation Modelling Practice and Theory*, No. 16, pp. 1677–1688 (2008). https://doi.org/10.1016/j.simpat.2008.08.010.

[11] Maurell, A. et al., "Experiments and FEM Simulations of Milling Performed to Identify Material Parameters, " *Int J Mater Form*, Vol. 1, pp. 1435 –1438 (2008). http://dx.doi.org/10.1007/s12289-008-0106-0.

[12] Zhang L., Wu Z., Wu C. and Wu Q., "On the numerical modelling of composite machining", *Composites Part B: Engineering*, Vol. 241, (2022). https://doi.org/10.1016/j.compositesb.2022.110023.

[13] Singha D., Mandal A. and Kar S., "Simulation, analysis and comparison of cutting parameter during turning of Ti-6Al-4V alloy with assorted industrial coated carbide", *Materials Today: Proceedings*, Vol. 62, pp. 6136–6142 (2022). https://doi.org/10.1016/j.matpr.2022.05.025.

[14] Paturi U., Methuku S., Siripragada S., Sangishetty Y. and Gunda R., "Finite element simulations of machinability parameters in turning of Inconel 718,"*Materials Today: Proceedings*, Vol. 38, pp. 2658–2663 (2021). https://doi.org/10.1016/j.matpr.2020.08.275

[15] Soliman H., Shash A., El Hossainy T.M. Abd-Rabou M., "Investigation of process parameters in orthogonal cutting using finite element approaches," *Heliyon*, Vol. 6, no. 11 (2020). https://doi.org/10.1016/j.heliyon.2020.e05498.

[16] Petruška J. *FEM in Engineering Computations*, Learning Texts, Brno: Brno University of Technology, Institute of solid mechanics, mechatronics and Biomechanics (2003).

[17] Hallquist J. O., "LS-DYNA theoretical manual". *Available from* Livermore, California, USA: Livermore Software technology Corporation, (2006).

[18] Kim J.D., Marinov V.R. and Kim D.S., "Built-up edge analysis of orthogonal cutting by the visco-plastic finite-element method," *J. Mater. Process. Technol.*, Vol. 71, pp. 367–372 (1997). https://doi.org/10.1016/S0924-0136(97)00099-X.

[19] Özel T. and Zeren E., "Finite Element Modeling of Stresses Induced by High Speed Machining with Round Edge Cutting Tools," in *Proc. of IMECE 05*, Paper No.81046, Orlando, Florida, USA(2005).

[20] Limido J., Espinosa C., Salaün M. and Lacome J.L., "SPH method applied to high speed cutting modelling," *International Journal of Mechanical Science*, Vol. 49, pp. 898–908 (2007). https://doi.org/10.1016/j.ijmecsci.2006.11.005.

[21] Liang, R. and Khan A.S., "A critical review of experimental results and constitutive models for BCC and FCC metals over a wide range of strain rates and temperatures," *International Journal of Plasticity*, Vol. 15, pp. 963–980 (1999). https://doi.org/10.1016/S0749-6419(99)00021-2.

[22] Johnson, G.R. and Cook W.H., "A constitutive model and data for metals subjected to large strains, high strain rates and high temperatures," in *7th International Symposium on Ballistics*, Hague, Netherlands, pp. 541–547 (1983).

[23] Umbrello D., M'Saoubi R. and Outeiro J.C., "The influence of Johnson-Cook material constants on finite element simulation of machining of AISI 316L steel," *Int. J. Mach. Tools Manuf.*,Vol. 47, pp. 462–470 (2007). https://doi.org/10.1016/j.ijmachtools.2006.06.006.

[24] Calamaz M., Coupard D. and Girot F., "A new material model for 2D numerical simulation of serrated chip formation when machining titanium alloy Ti-6Al-4V," *Int. J. Mach. Tools Manuf.*, Vol. 48, pp. 275–288 (2008). https://doi.org/10.1016/j.ijmachtools.2007.10.014.

[25] Chapman D.J., Radford D.D. and Walley, S.M. "A history of the Taylor test and its present use in the study of lightweight materials," in *Proceedings of the 3rd Design and Use of Light-Weight Materials Conference*, University of Aveiro, Aveiro, Portugal, 8–10 October 2005, pp. 12–24.

[26] Taylor G.I., "The use of flat-ended projectiles for determining dynamic yield stress I. Theoretical considerations," *Proc. R. Soc. London Ser. A Math. Phys. Sci.*, Vol. 194, pp. 289–299 (1948). https://doi.org/10.1098/rspa.1948.0081.

[27] Lesuer D., Kay G. and LeBlanc M., "Modeling Large-Strain, High-Rate Deformation in Metals," in *Third Biennial Tri-Laboratory Engineering Conference on Modeling and Simulation, Pleasanton*, 1999.

[28] Manes A., Peroni L., Scapin M. and Giglio M., "Analysis of strain rate behavior of an Al 6061 T6 alloy", *Procedia Engineering*, Vol. 10, pp. 3477–3482 (2011). https://doi.org/10.1016/j.proeng.2011.04.573.

[29] Chen Q., Alizadeh A., Xie W., Wang X., Champagne V., Gouldstone A., Lee J. and Müftü S., "High-Strain-Rate Material Behavior and Adiabatic Material Instability in Impact of Micron-Scale Al-6061 Particles," *J Therm Spray Tech*, Vol. 27, no. 4, pp. 641–653 (2018). https://doi.org/10.1007/s11666-018-0712-4

[30] Akram S., Jaffery S., Khan M., Fahad M., Mubashar A. and Ali L., "Numerical and experimental investigation of Johnson–Cook material models for aluminum (Al 6061-T6) alloy using orthogonal machining approach," *Advances in Mechanical Engineering*, Vol. 10, no. 9, (2018). https://doi.org/10.1177/1687814018797794

[31] Zouhar J., *On the development of powerfull milling tools with CAD/CAM and mechanism of chip production use*, Ph.D. thesis, (University of Technology, Institute of manufacturing technology, Brno (2009).

[32] Shaw M.C., *Metal Cutting Principles*, (Oxford University Press, 2nd ed., pp. 651 2005).

Index of Authors